Manufacturing Processes for Design Professionals

Rob Thompson

Manufacturing Processes for Design Professionals

Manufacturing Processes for Design Professionals
© 2007 Rob Thompson

First published in 2007 in hardcover
in the United States of America by
Thames & Hudson Inc., 500 Fifth Avenue,
New York, New York 10110

www.thamesandhudsonusa.com

Reprinted 2018

Library of Congress Control Number 2007921449

ISBN 978-0-500-51375-0

Designed by Christopher Perkins

Printed and bound in China by Midas Printing
International Ltd.

Contents

Part One

Forming Technology

Part Two

Cutting Technology

Part Three

Joining Technology

Part Four

Finishing Technology

Part Five

Materials

How to use this book

Manufacturing Processes for Design Professionals **explores established, emerging and cutting-edge production techniques that have, or will have, an important impact on the design industry. There is a danger today of designers becoming detached from manufacturing as a result of CAD, globalization and design education. This book aims to restore the balance with a hands-on and inspiring approach to design and production. It is a comprehensive, accessible and practical resource that focuses on providing relevant information to aid fast and efficient decision-making in design projects.**

STRUCTURE

This book is organized into 2 main sections: Processes and Materials. These can be used separately or in combination. Each section contains design guidance supplied by manufacturers to ease the transition between design and production and provides information that will inspire decision-making, encourage experimentation and support design ideas.

HOW TO USE PROCESSES

The Processes section is organized into 4 parts, each focusing on a specific type of technology, and each process is explained with photographs, diagrams and analytical and descriptive text. The 4 parts (colour coded for ease of reference) are: Forming Technology (blue), Cutting Technology (red), Joining Technology (orange) and Finishing Technology (yellow). Each featured manufacturing process is fully illustrated and provides a comprehensive understanding of

the process through 3 key elements. The text gives an analysis of the typical applications, competing or related processes, quality and cost, design opportunities and considerations, and environmental impacts of a process. There is also a full technical description of the process and how the machinery involved works, with diagrams, and a case study showing products or components being made by a leading manufacturer using the featured process.

On the opening spread of each process you will find a data panel, which provides a bullet-pointed summary of factors such as the typical applications, quality and cost, as well as function diagrams (see opposite) which, when highlighted, indicate the particular functions and design outcomes of each process. These function diagrams quickly enable the reader to compare a wide range of similar processes to see which is the most effective in producing a given item or component.

HOW TO USE THE CASE STUDIES

The Processes section features real-life case studies from factories around the world. The processes are explained with photographs and analytical and descriptive text. All types of production are included, from one-off to batch and mass. For cross-comparison the case studies can be set against each other on many levels, including functions, cost, typical applications, suitability, quality, competing processes and speed. This information is accessible, logical and at the forefront of each process.

HOW TO USE MATERIALS

Each manufacturing process can be used to shape, fabricate and finish a number of different materials. The main objective of the material profiles is to support the processes, expand opportunities for designers and provide relevant information for potential applications. The layout of the Processes and Materials sections is designed to encourage cross-

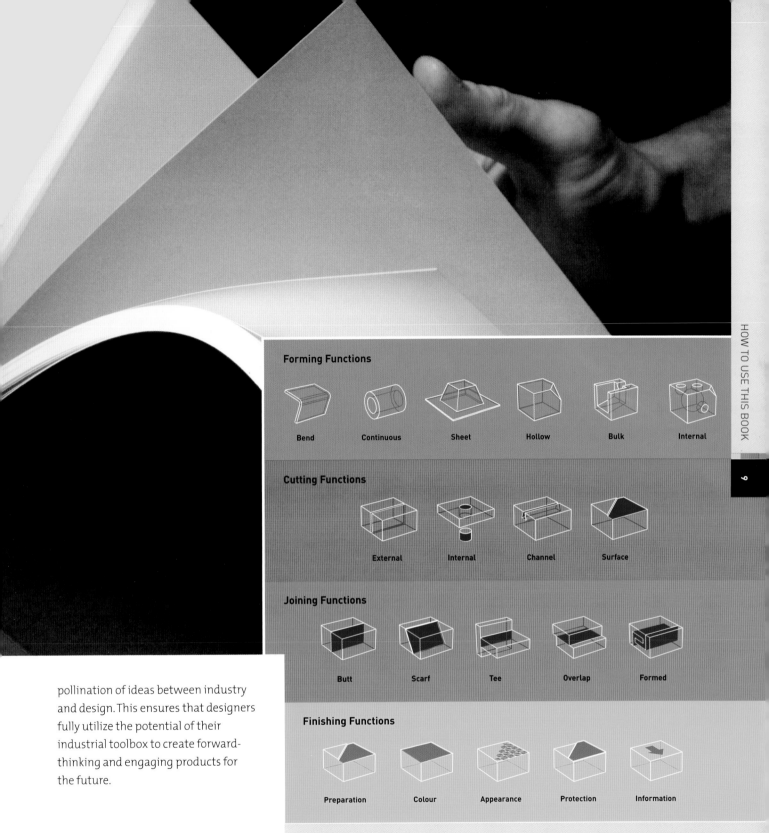

Forming Functions

Bend · Continuous · Sheet · Hollow · Bulk · Internal

Cutting Functions

External · Internal · Channel · Surface

Joining Functions

Butt · Scarf · Tee · Overlap · Formed

Finishing Functions

Preparation · Colour · Appearance · Protection · Information

pollination of ideas between industry and design. This ensures that designers fully utilize the potential of their industrial toolbox to create forward-thinking and engaging products for the future.

HOW TO USE THE FUNCTION ICONS
These icons represent the function that each process performs. The functions are different for forming, cutting, joining and finishing processes. Likewise, different materials are more suitable for certain functions than others. These icons guide designers in the early stage of product development by highlighting the relevant processes and materials for their project.

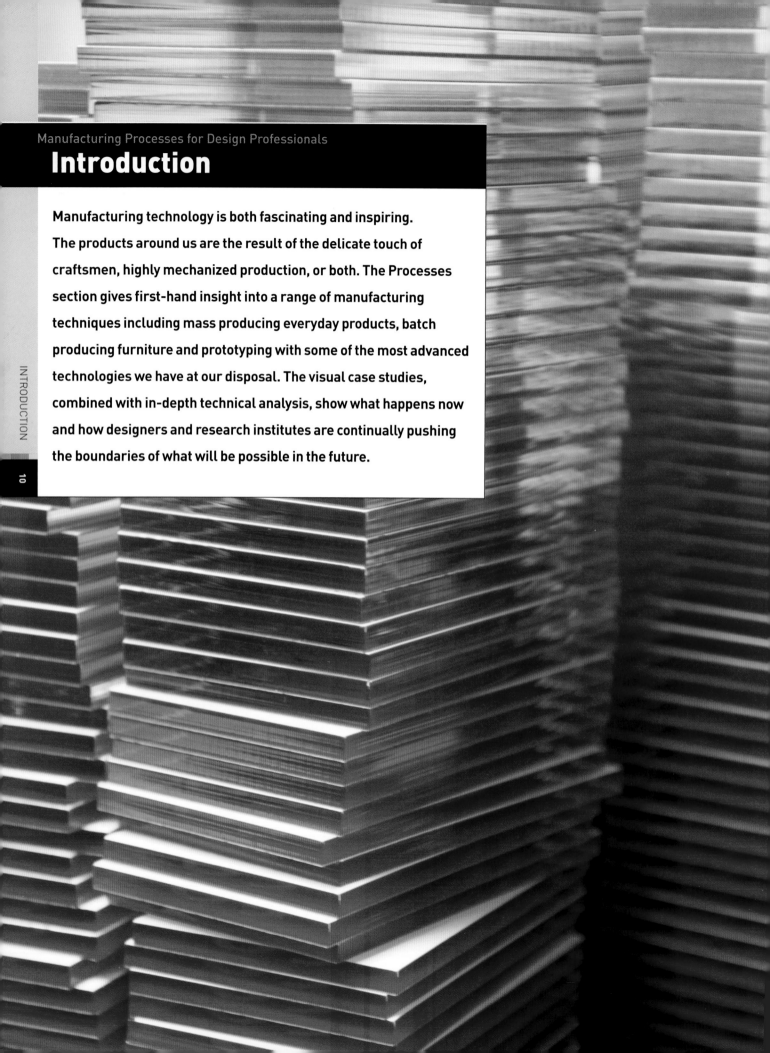

Introduction

Manufacturing technology is both fascinating and inspiring. The products around us are the result of the delicate touch of craftsmen, highly mechanized production, or both. The Processes section gives first-hand insight into a range of manufacturing techniques including mass producing everyday products, batch producing furniture and prototyping with some of the most advanced technologies we have at our disposal. The visual case studies, combined with in-depth technical analysis, show what happens now and how designers and research institutes are continually pushing the boundaries of what will be possible in the future.

Manufacturing is continually in a state of transition. The level of technology is different in various industries, so whilst some manufacturers are leading the way, such as in the production of carbon fibre composites (page 214) and rapid prototyped plastics and metals (page 232), others are maintaining highly skilled traditional crafts. The combination of craft and industrial techniques in processes such as panel beating metal (page 72), jiggering and jolleying ceramics (page 176) and steam bending wood (page 198) produces articles that unite the user and maker with a sense of pride and ownership.

The examples in this book demonstrate the inner workings of a large range of manufacturing processes. In some cases the tasks are carried out by hand to demonstrate the techniques more clearly. The continued importance of an operator is evident in many processes. Even mass-production techniques, such as die cutting and assembling packaging (page 266), rely on an operator to set up and fine tune the production line. But, where possible, manual labour is being replaced by computer-guided robotic systems. The aim is to reduce imperfections caused by human error and minimize labour costs. Even so, many metal, glass, wood and ceramic processes are based on manufacturing principles that have changed very little over the years.

DEVELOPMENTS IN FORMING

Plastic products have come a long way since they were first formed by compression molding (page 44) in the 1920s. Injection molding (page

Bellini Chair

Designer/client:	Mario Bellini/Heller Inc.
Date:	1998
Material:	Polypropylene and glass fibre
Manufacture:	Injection molded

Panasonic P901iS smartphone

Designer:	Panasonic, Japan
Date:	2005
Manufacture:	Injection molded plastic covers using Yoshida Technoworks in-mold decoration technology

50) is now one of the most important processes for designers, and probably the most widely used. It is utilized to shape thermoplastics and thermosetting plastics, waxes for investment casting (page 130) and even metals (page 136). It is continually developing and in recent years has been revolutionized by in-mold decoration (page 50) and gas assist technologies (page 50). In-mold decoration is the application of graphics during the molding process, eliminating finishing operations such as printing. With this technology it is possible to apply graphics on 1 side, both sides or onto multishot injection molded parts

(page 50). It is also possible to integrate fabric, metal foils and leather (see image, above right) into plastic moldings. Gas assist injection molding produces hollow, rigid and lightweight plastic parts (see image, above left). The introduction of gas reduces material consumption and the amount of pressure required in the molding cycle. Surface finish is improved because the gas applies internal pressure while the mold is closed.

Another area of important progress within injection molding is multishot. This is the process of injecting more than 1 plastic into the same die cavity to produce parts made up of materials with

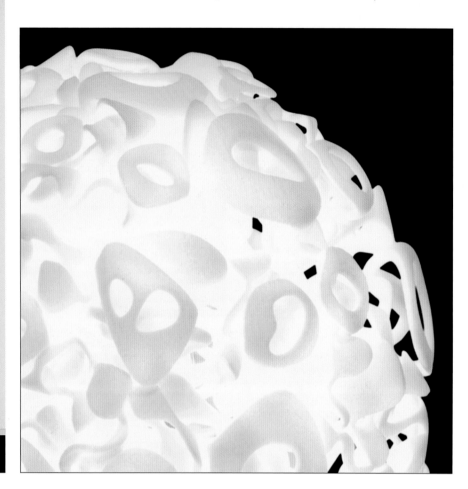

Entropia

Designer/client:	Lionel Dean, FutureFactories/ Kundalini
Date:	2006
Material:	Polyamide (PA) nylon
Manufacture:	Selective laser sintering (SLS)

widely used to describe the matt texture produced by EDM. Nowadays, mobile phones and similar products are more often produced with in-mold decoration, which requires a gloss finish.

Whereas mass production is limited by high tooling costs and thus high volume production of identical parts, each product directly manufactured from CAD data is limited only by the imagination and capabilities of the designer. Rapid prototyping (page 232) is one of these processes: it is not yet suitable for mass production, but it is capable of producing similar volume parts whose shape is different each time without significant cost implications. Combined with the possibility of making shapes not possible with any other process, these techniques are giving rise to a new design language. For example, one-off and low volume products have emerged in recent years that are designed by the customers and merely facilitated by the designer. Also, the US military are using rapid prototyping to make spare parts for their equipment, as opposed to waiting for delivery.

Recently metal powders have been added to the list of materials that can be shaped by rapid prototyping. At present, rapid prototyping metal is best suited to the production of parts no larger than about 0.01 m³ (0.35 ft³). Even so, the opportunities of this process are vast, because larger parts can be made by investment casting (page 130) a rapid prototyped wax or plastic pattern.

Thermoforming (page 30) is a process generally associated with plastic packaging. However, Superform Aluminium in the USA and UK have

different colours, hardnesses, textures or transparency. Over-molding is a similar process; the difference is that over-molding is not carried out in the same tool. Using this technique, materials other than plastic can be integrated into injection molding.

Developments in plastic molding are also affected by improvements in metalworking technologies. The surface finish on mobile phones and other small consumer electronic equipment became almost standardized due to the development of electrical discharge machining (EDM) for plastic mold making. This process makes it possible to machine concave profiles (molds) to the same high degree of precision as relief profiles. High voltage sparks between a copper electrode (tool) and metal workpiece vaporize surface material. The rate of spark erosion determines the surface finish and so it is used to simultaneously cut and finish metal parts. Hence the term 'sparked' finish was

Roses on the Vine

Designer/client:	Studio Job/Swarovski Crystal Palace Project
Date:	2005
Material:	Aluminium base with red and peridot coloured Swarovski® crystal
Manufacture:	Base laser cut and gold anodized

Laser Vent polo shirt

Designer: Vexed Generation
Date: Spring/Summer 2004
Material: Quick dry polyester microfibre
Manufacture: Laser cutting and stitching

Biomega MN01 bike

Designer: Marc Newson
Date: 2000
Material: Aluminium alloy frame
Manufacture: Superforming and welding

Outrageous painted guitar

Paint by:	Cambridgeshire Coatings Ltd/US US Chemicals and Plastics
Date:	2003
Material:	Illusion Outrageous paint
Manufacture:	Spray painting

Camouflage printed rifle stock

Designer:	Hydrographics
Material:	Plastic stock
Manufacture:	Hydro transfer printing
Notes:	The transfer films can be decorated with artwork, photographs or patterns.

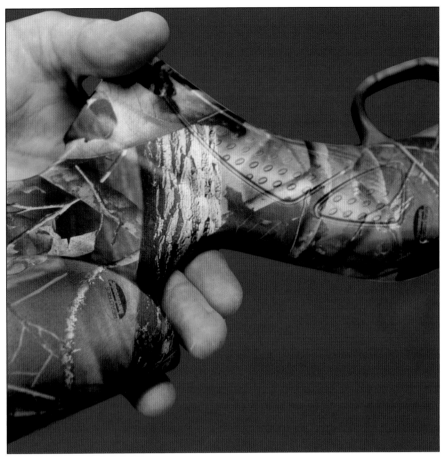

developed a range of processes, known as superforming (page 92), that are capable of shaping aluminium alloys (page 457) and magnesium alloys (page 458) using a similar technique. At around 450°C (840°F) certain grades of these metals become superplastic and so can be stretched to many times their length without breaking. Since the development of commercially viable metals and processes in the mid 1970s, superforming has had a major impact in the automotive, aerospace and rail industries. Recently, designers such as Marc Newson have begun to explore the possibilities of using this technology to produce consumer products such as bicycles (see image, opposite).

DEVELOPMENTS IN CUTTING

Like rapid prototyping, laser cutting (page 248) works directly from CAD data. This means that data can be translated very readily from a designer's computer onto the surface of a wide range of materials. Indeed, laser cutting is used a great deal by architects for model making within short timeframes.

Roses on the Vine by Studio Job (see image, page 13), is an example of how laser cutting can be used to produce intricate, complex and otherwise impractical shapes with very high precision. This cutting process is not limited to rigid materials. In 2004 Vexed Generation launched Laser Vent. The aim of the designers was to create ultra lightweight clothing for cyclists, but with the aesthetic of low-key leisurewear and so suitable for the office. They achieved this by laser cutting synthetic fibre. The cut edge is sealed by the heat of the laser, eliminating conventional hems and reducing material usage. Another benefit of specifying laser cutting was that vents could be integrated into areas that needed better ventilation or greater freedom of movement.

DEVELOPMENTS IN JOINING

Powerbeam technologies (page 288), which include laser beam and electron beam, are making an impact in joining as well as cutting applications. Electron beam welding is capable of producing coalesced joints in steels up to 150 mm (5.9 in) thick and aluminium up to

450 mm (17.7 in) thick. Laser welding is not usually applied to thick materials, but a recent development known as Clearweld® makes it possible to laser weld clear plastics and textiles (page 288). This technique has the potential to transform applications that are currently limited to coloured materials.

Joinery (page 324) and timber frame construction (page 344) have changed very little over the years. Developments have mostly been concentrated in materials such as new types of engineering timbers (page 465) and biocomposites. However, in 2005 TWI assessed the possibility of joining wood with techniques similar to friction welding metals (page 294) and plastics (page 298). Beech and oak were successfully joined by linear friction welding (see image, page 295). Friction welding revolutionized metalwork by eliminating the need for mechanical fasteners, and in the future the same could happen in woodwork.

DEVELOPMENTS IN FINISHING

There have been many developments in finishing, but spray painting remains one of the most widely used processes, from one-off to mass production. Over the years a range of sprayed finishes have evolved including high gloss, soft touch, thermochromatic, pearlescent and iridescent. The cost of paint varies dramatically depending on the type and can be very high for specialist paints such as the Outrageous range (see image, page 15 above left). A film of aluminium is incorporated into spray painted finishes by vacuum metalizing (page 372) to give the appearance of chrome, silver or anodizing, or for functional purposes, such as heat reflection.

Hydro transfer printing (page 408) has recently transformed spray painting. With this process it is possible to wrap printed graphics around 3D shapes. This means anything that can be digitally printed can be applied to almost any surface. Applications include car interiors, mobile phones, packaging and camouflaging gun stocks (see image, page 15 above right).

SELECTING A PROCESS

Process and material selection is integral to the design of a product. Economically, it is about striking a balance between the investment costs (research, development and tooling) and running costs (labour and materials). The role of the designer is to ensure that the available technology will deliver the expected level of quality.

High investment costs are usually only justifiable for high volume products, whereas low volume products are limited by high labour and material costs. Therefore, the tipping point comes when expected volumes outweigh the initial costs. This can happen before a product has reached the market, or after years of manufacturing at relatively low volumes.

The cost of materials tends to have a greater impact in high volume processes. This is because labour costs are generally reduced through automation. Therefore, fluctuations in material value caused by rising fuel prices and increased demand affect the cost of mass produced items.

Occasionally the cost of production is irrelevant, such as composite laminating carbon fibre racing cars (page 214). Until recently, this has been too expensive for application in commodity products, but due to recent developments in the production techniques it is now becoming more common in sports equipment and automotive parts. This is also partly due to the obvious benefits of improved strength to weight.

The design features that can be achieved with high volume processes cannot always be produced with lower volumes ones. For example, blow molding plastic (page 22) and machine glassblowing (page 152) are limited to continuous production due to the nature of the process. Therefore, there is very little room for experimentation. In contrast, the qualities of injection molding can be reproduced with vacuum casting (page 40) and reaction injection molding (page 64), for instance. This means that low volumes can be produced with much lower initial costs and therefore a higher level of experimentation is usually possible.

Process selection will affect the quality of the finish part and therefore the perceived value. This is especially important when 2 processes can produce the same geometry of part. For example, sand casting (page 120) and investment casting (page 130) can both be used to manufacture 3D bulk shapes in steel. However, due to the lower levels of turbulence, investment casting will produce parts with less porosity. It is for this reason that aerospace and automotive parts are investment cast.

DESIGN SOFTWARE

Predicting and testing the quality of the finished part has become more reliable with developments in computer simulation software, such as finite element analysis (FEA). There are many different programmes, which are becoming more widely used in the process of design for manufacture (pages 50–63, 64–7 and 130–5), and it is no longer limited to high volume production. The forming of many products is simulated using FEA software

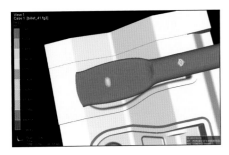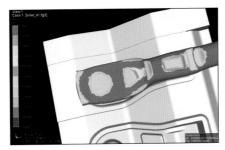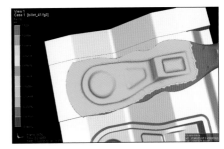

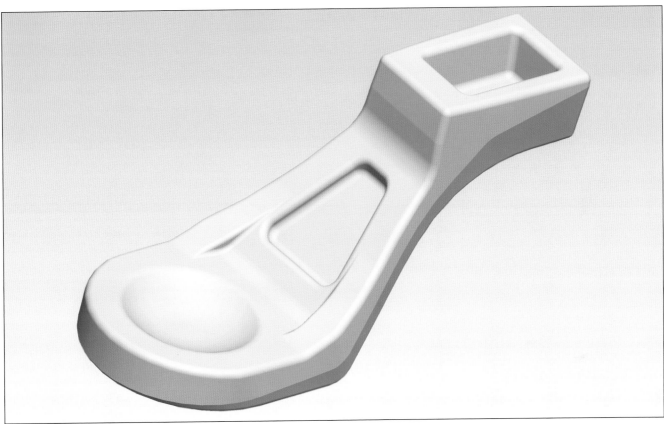

to maximize the efficiency of the operation (see images, above); previously, tools were engineered and then tested and adjusted accordingly. FEA software is not used in all forming applications, but it does have many advantages. Most importantly, it reduces tooling costs, because parts can be molded 'right first time'.

As well as mold flow simulation, FEA is used to predict accurately how the part will perform in application (pages 124–9, 214 and 226–7). This does not de-skill the process of engineering; it is another tool that can help to minimize material consumption and double-check calculations made by designers and engineers.

FEA simulation of an aluminium forging

Software:	Professional Engineer (Pro E) and Forge3
Simulation by:	Bruce Burden, W.H. Tildesley Ltd
Material:	Aluminium alloy
Manufacture:	Forging

→ # Binding this book

This book demonstrates the techniques used to manufacture many of the products that surround us in our day-to-day lives. Like most commercial books, it is manufactured using a process similar to traditional bookbinding. And whether hardcover or paperback, it is made of folded sections stitched together in a process known as section-sewn binding (**image 1**). An alternative method is perfect binding, which is the process of adhesive bonding individual pages into a scored paper cover. It is less durable and so is limited to thinner books, but the advantage is that the pages can be opened out flatter than section-sewn bindings.

This sequence of images demonstrates section-sewn case binding on a smaller book than this, but the principles are the same. In this case, the sections are made up of 16 pages (**image 2**), which consist of a large signature (printed sheet) folded 4 times and cut to size on a guillotine. Therefore, the extent of books bound in this way is divisible by the number of pages in a section, which is typically 4, 8 or 16. Perfect binding is not limited by the same factors and so can be any number of pages.

Linen tape is adhesive bonded onto the sewn edges of the sections and the sewn assembly is cut to size in a guillotine (**image 3**). This produces a neat edge (**image 4**). Hard covers are made up of heavy-duty grey board 2.5 mm (0.1 in.) thick concealed in paper, cloth or leather. By contrast, paperback covers are heavy-duty paper, which is typically around 0.25 mm (0.01 in.) thick. Any printed decoration, such as hot foiling (pages 412–5) is applied prior to assembly. The cover is adhesive bonded to the first and last page, the assembly is clamped in a press and the adhesive cures (**image 5**).

1

2

3

5

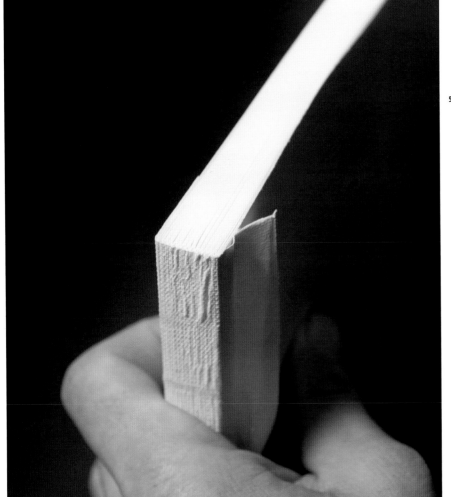

4

Featured Company

R S Bookbinders
www.rsbookbinders.co.uk

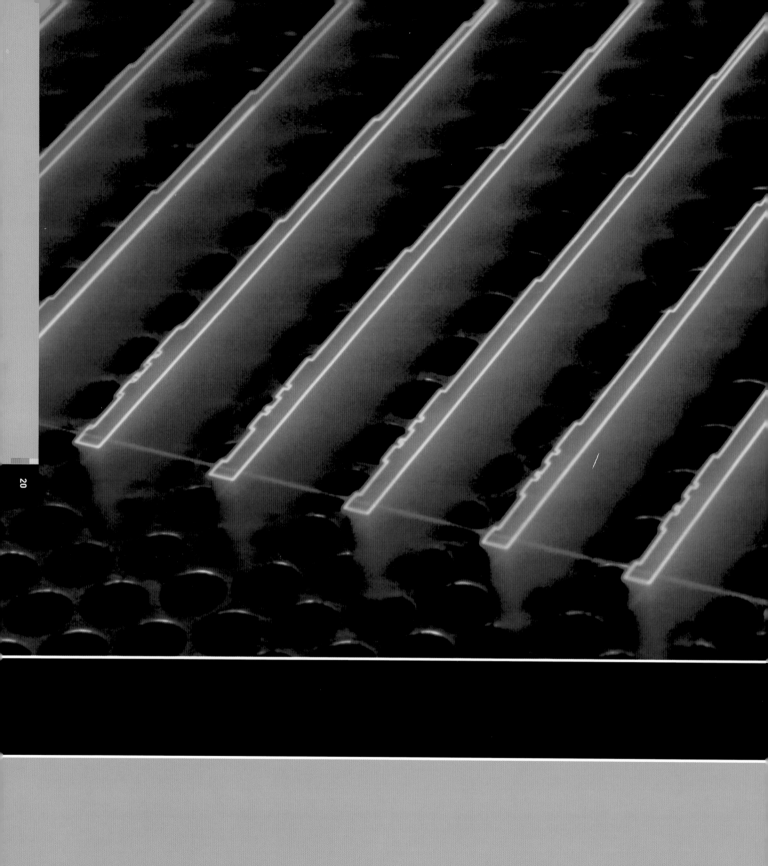

Part

1

Forming Technology

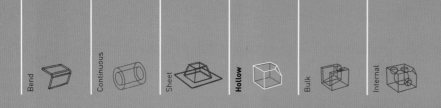
Forming Technology

Blow Molding

This group of processes is typically used to mass produce hollow packaging containers. They are a very rapid production method for large volumes of thin walled parts.

Costs	Typical Applications	Suitability
• Moderate tooling costs • Low unit costs	• Chemical packaging • Consumer packaging • Medical packaging	• Suitable only for high volume production runs

Quality	Related Processes	Speed
• High quality, uniform thin walled parts • High quality surface finish that can be gloss, textured or matt	• Injection molding • Rotation molding • Thermoforming	• Very rapid cycle time (typically 1–2 minutes)

INTRODUCTION

Blow molding is carried out in 3 different ways: extrusion blow molding (EBM), injection blow molding (IBM) and injection stretch blow molding (ISBM). Each of the processes has its particular design opportunities and is suitable for different industries.

EBM is favourable for many applications because it has low tooling and running costs. It is a versatile process that can be used to produce a wide variety of shapes in an extensive choice of materials. Containers can be molded with integral handles and multiple layered walls.

IBM is a precise process that is suitable for more demanding applications such as medical containers and cosmetic packaging. It is used to produce containers with very accurate neck finishes as well as wide mouths.

ISBM is typically used to produce high quality glass clear PET containers such as water bottles. The injection cycle ensures

very accurate neck finishes and the stretch cycle gives superior mechanical properties. ISBM is particularly suitable for beverage, agrochemical and personal care applications.

TYPICAL APPLICATIONS

EBM is used mainly in the medical, chemical, veterinary and consumer industries to produce intravenous containers, medicine bottles and vials, and consumer packaging.

IBM is utilized especially for consumer packaging and medical packaging (medicine bottles, tablet and diagnostic bottles and vials).

ISBM is predominant in the personal care, agrochemicals, general chemicals, food and beverages and pharmaceutical industries to produce carbonated and soft drink bottles, cooking oil containers, agrochemical containers, health and oral hygiene products, bathroom and toiletry products, and a number of other food application containers.

RELATED PROCESSES

Thermoforming (page 30), rotation molding (page 36) and injection molding (page 50) can all be used to form the same geometry parts. Even so, blow molding is the process of choice for large volumes of hollow thin walled packaging.

QUALITY

The surface finish is very high for all of these processes. The IBM and ISBM technologies have the additional advantage of precise control over neck details, wall thickness and weight.

DESIGN OPPORTUNITIES

All of the blow molding processes can be used to produce thin walled and strong containers. The neck does not have to be vertical or tubular. Features such as handles, screw necks and surface texture can be integrated into all 3 processes.

The principal reason to select IBM is that there is more control over wall thickness and neck details. This means

Extrusion Blow Molding Process

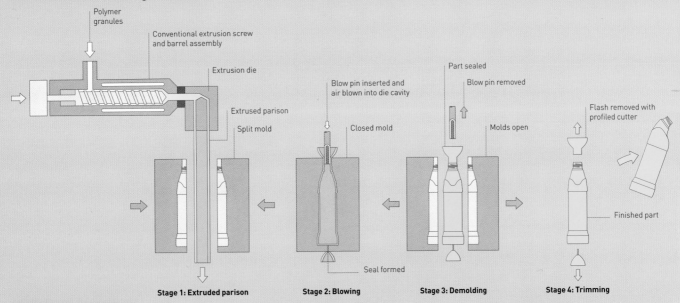

Stage 1: Extruded parison **Stage 2: Blowing** **Stage 3: Demolding** **Stage 4: Trimming**

TECHNICAL DESCRIPTION

In stage 1 of the EBM process, a conventional extrusion assembly feeds plasticized polymer into the die. The polymer is forced over the mandrel and emerges as a circular tube, known as an extruded parison. The extrusion process is continuous. In stage 2, once the parison has reached a sufficient length the 2 sides of the mold close around it. A seal is formed along the bottom edge. The parison is cut at the top by a knife and moved sideways to the second station, where air is blown in through a blow pin, forcing the parison to take the shape of the mold. The hot polymer solidifies as it makes contact with the cold tool. In stage 3, when the part is sufficiently cool the mold opens and the part is ejected. In stage 4, the container is deflashed using a trimmer.

---

that a wider range of anti-tamper and other caps can be introduced.

The main advantages of EBM are that a wide choice of materials can be used in this process, and complex and intricate shapes manufactured.

ISBM can be used to produce clear containers with very high clarity. Stretching the pre-form during blowing greatly increases the mechanical strength of the container by aligning the polymer chains longitudinally. These containers also have good gas and solvent barrier properties and so can be used to package aggressive foods, concentrates and chemicals.

DESIGN CONSIDERATIONS

A major difference between these blow molding techniques is the capacity that each can accommodate. IBM is generally limited to the production of containers between 3 ml and 1 litre (0.005–1.760 pints) and ISBM can produce containers between 50 ml and 5 litres (0.088–8.799 pints). EBM can create the largest variety of containers ranging between 3 ml and 220 litres (0.005–387 pints). Blow molding is a complex process with which to work. Expert advice from engineers and toolmakers is required to guide the design process through to completion. There are many considerations that need to be taken into account when designing for blow molding, including the user (ergonomics), product (light sensitivity of contents and viscosity), filling (neck, contents and filling line), packaging (shelf height) and presentation (labelling using sleeves or print, for example).

COMPATIBLE MATERIALS

All thermoplastics can be shaped using blow molding, but certain materials are more suited to each of the technologies. Typical materials used in the EBM process include polypropylene (PP), polyethylene (PE), polyethylene terephthalate (PET) and polyvinyl chloride (PVC), while the IBM process is suitable for PP and HDPE among other materials. Typical materials for the ISBM process include PE and PET.

COSTS

Tooling costs are moderate. EBM is the least expensive, the tooling for IBM is typically twice as much and ISBM is the most expensive.

Cycle time is very rapid. A single mold may contain 10 or more cavities and eject a batch of parts every 1–2 minutes.

Labour costs are low, as production is automated. Set-up and changeover can be expensive, however, so machines are often dedicated to a single product.

ENVIRONMENTAL IMPACTS

All thermoplastic scrap can be directly recycled. Process scrap is recycled in-house. Post-consumer waste can also be recycled and turned into new products. Recycled PET is used in the production of certain items of clothing, for example. Blow molding plastics is more energy efficient than glassblowing.

→ Extrusion blow molding a cleaning agent container

The PE polymer granules are stored in a communal hopper and coloured individually for each machine (**image 1**). In this case a small percentage of blue granules are added just prior to extrusion. The extrusion process is continuous and produces an even wall thickness parison (**image 2**). The 2 halves of the mold close around the parison to form a seal and the parison is cut to length (**image 3**). A blow rod is then inserted into the neck, and air is blown into the mold at 8 bar

(116 psi) forcing the parison to take the shape of the mold (**image 4**). The molds separate to reveal the blown part with the blow rods still inserted (**image 5**). The rods retract and the part is deflashed with a profiled trimmer (**image 6**).

Each batch is conveyed from the blow molding machine to labelling and capping via pressure testing (**image 7**). The EBM bottles pass through the filling line (**image 8**). The caps are screwed on automatically (**image 9**)

and the labels adhesive bonded to the bottle (**image 10**). The finished product is packaged and shipped.

1

2

3

4

5

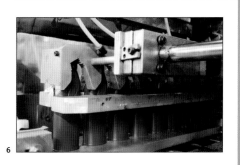

6

7

8

9

10

Injection Blow Molding Process

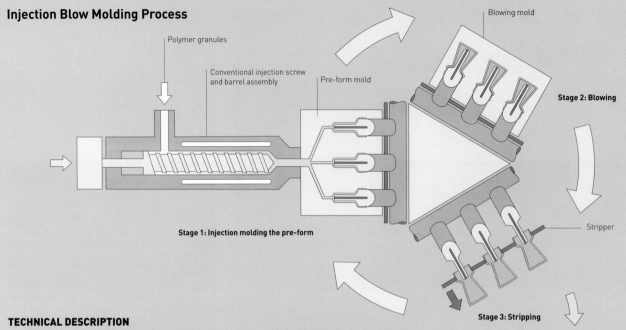

Polymer granules

Conventional injection screw
and barrel assembly

Pre-form mold

Blowing mold

Stage 2: Blowing

Stage 1: Injection molding the pre-form

Stripper

Stage 3: Stripping

Stage 4: Final product

TECHNICAL DESCRIPTION

The IBM process is based on a rotary table that transfers the parts onto each stage in the process. In stage 1, a pre-form is injection molded over a core rod with finished neck details. The pre-form and core rod are transferred through 120° to the blowing station. In stage 2, air is blown into the pre-form forcing the parison to take the shape of the mold. In stage 3, after sufficient cooling, the part is rotated through 120° and stripped from the core rod complete. No trimming or deflashing is needed.

1

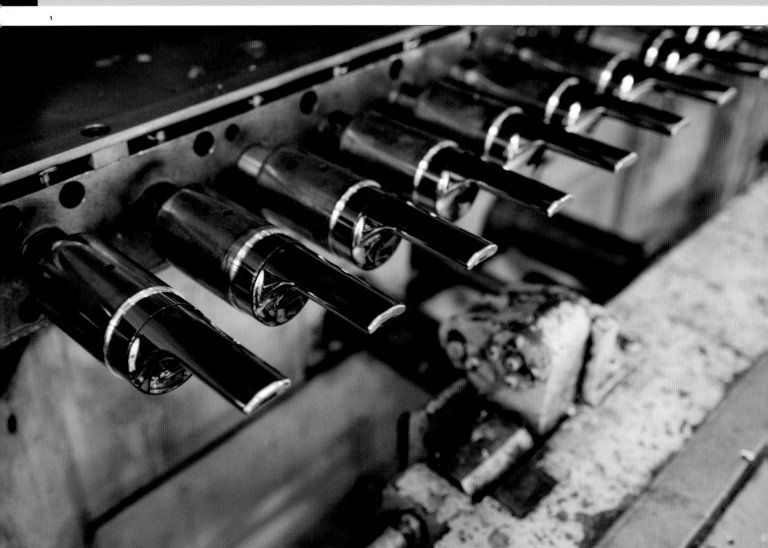

→ Injection blow molding a roll-on deodorant bottle

The polished core rods are prepared so the pre-forms can be injection molded onto them (**image 1**). Each core rod is inserted into a split mold and hot molten white PP is molded around it. The neck is fully formed (**image 2**). The parts are rotated through 120° and are inserted into the blowing mold. Air is blown in through the core rod and the plastic is forced to take the shape of the mold cavity. The polymer solidifies when it makes contact with the relatively cooler walls of the mold (**image 3**). The parts are stripped from the core rods (**image 4**), counted by a laser sensor (**image 5**) and pressure tested (**image 6**). The parts (**image 7**) are then fed into a filling and capping system similar to the EBM process.

2

3

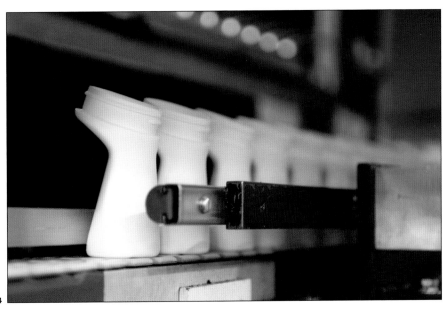

4

5

6

7

Featured Manufacturer

Polimoon
www.polimoon.com

1

2

3

4

TECHNICAL DESCRIPTION

In stage 1, the ISBM process uses the same technique as IBM, in that the pre-form is injection molded over a core rod. In stage 2, however, in ISBM the core rod is removed and replaced by a stretch rod. The pre-form is inserted into the blow mold, which is clamped shut. In stage 3, air is blown in through the stretch rod, which simultaneously orientates the pre-form longitudinally. In stage 4, the mold opens and the parts are stripped from the stretch rod.

Injection Stretch Blow Molding Process

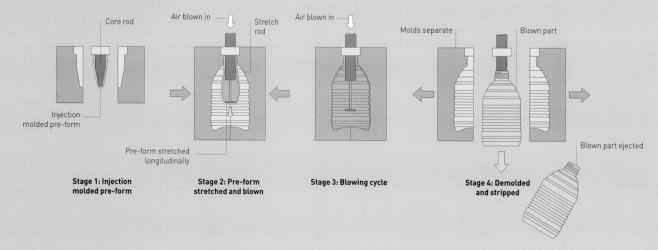

Core rod

Air blown in | Stretch rod

Air blown in

Molds separate

Blown part

Injection molded pre-form

Pre-form stretched longitudinally

Blown part ejected

Stage 1: Injection molded pre-form

Stage 2: Pre-form stretched and blown

Stage 3: Blowing cycle

Stage 4: Demolded and stripped

→ Injection stretch blow molding a chemical container

The pre-form is injection molded over a core rod, which is removed prior to blowing. The injection molded part is thin walled, as demonstrated by the operator (**image 1**). There is not usually operator contact with the parts during production. The pre-form is transferred to the blow molds (**image 2**). The molds close around the pre-form and it is simultaneously stretched longitudinally and blown to form the container (**image 3**). The blown product is ejected and does not require any trimming. It is demolded (**image 4**), pressure tested (**image 5**) and a handle is fitted around the neck of the injection molded container, using an annular snap fit (**image 6**). The container is then sent for capping (**image 7**).

5

6

7

Featured Manufacturer

Polimoon
www.polimoon.com

Forming Technology
Thermoforming

In this group of processes, thermoplastic sheet materials are formed with the use of heat and pressure. Low pressures are inexpensive and versatile, whereas higher pressures can produce surface finishes and details similar to injection molding.

Costs	Typical Applications	Suitability
• Low to moderate tooling costs • Low to moderate unit costs, roughly 3 times material cost	• Baths and shower trays • Packaging • Transportation and aerospace interiors	• Roll fed: batch to mass production • Sheet fed: one-off to batch production
Quality	**Related Processes**	**Speed**
• Depends on material, pressure and technique	• Composite laminating • Injection molding • Rotation and blow molding	• Roll fed cycle time: 10 seconds to 1 minute • Sheet fed cycle time: 1–8 minutes

INTRODUCTION

There are 2 distinct categories of thermoforming: sheet fed and roll fed. Sheet fed thermoforming is for heavy duty applications, such as pallets, baths, shower trays and luggage. The sheet material is typically cut to size and loaded by hand. Roll fed processes, on the other hand, are supplied by a roll of sheet material from a reel. It is sometimes referred to as the 'in-line' process because it thermoforms, trims and stacks in a continuous operation.

Thermoforming includes vacuum forming, pressure forming, plug-assisted forming and twin sheet thermoforming.

Vacuum forming is the simplest and least expensive of these sheet forming processes. A sheet of hot plastic is blown into a bubble and then sucked onto the surface of the tool. It is a single-sided tool and so only one side of the plastic will be affected by its surface. In pressure forming the hot softened sheet is forced into the mold with pressure. Higher pressure means that more complex and intricate details can be molded, including surface textures. For relatively low volumes this process is capable of producing parts similar to injection molding (page 50).

The above 2 processes are suited to forming shallow sheet geometries. For deep profiles the process is plug assisted. The role of the plug is to push the softened material into the recess, stretching it evenly.

Twin sheet thermoforming combines the qualities of these processes with the production of hollow parts. In essence, 2 sheets are thermoformed simultaneously and bonded together

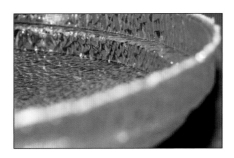 

Above left
Textured light diffuser manufactured by vacuum forming.

Above
Foam filled Terracover® ice pallet made by twin sheet thermoforming.

while they are still hot. This complicated process is more expensive than conventional thermoforming.

TYPICAL APPLICATIONS

Thermoforming is used extensively to produce everything from disposable food packaging to heavy-duty returnable transit packaging. Some typical examples include clear plastic point-of-sale packaging, clamshell packaging, cosmetic trays, drinking cups and briefcases.

This process is also used to produce lighting diffusers; bath and shower trays; garden pots; signage; vending machines; small and large tanks; motorcycle fairings; interiors for cars, aeroplanes and trains; housing for consumer electronics; and protective head gear.

Blister packaging (bubble wrap) is made by thermoforming. It is produced continuously on a roller, which is the vacuum former, and is laminated to seal air into the individual blisters and provide protective cushioning for packaged goods.

Vacuum forming tooling can be very inexpensive and so is suitable for prototyping and low volume production.

RELATED PROCESSES

Low pressure thermoforming techniques are versatile and inexpensive. This is because the plastic is formed as a softened sheet, as opposed to a mass of molten material. This sets thermoforming apart from many other plastic forming processes. However, the use of sheet increases the material cost slightly. To minimize this some factories extrude their own materials.

Pressure forming can produce surface finishes similar to injection molding. The tooling is more expensive, but for some applications it will be less than for injection molding. This is because a single-sided tool is used, as opposed to matched tooling which is required for injection molding.

Twin sheet thermoforming is used to produce 3D hollow geometries. Similar parts can be produced by blow molding (page 22) and rotation molding (page 36), yet the benefit of thermoforming is that it is ideal for large and flat panels. Also, the 2 sides are not limited to the same colour or even type of material.

Material developments mean that twin sheet thermoformed products will sometimes have suitable characteristics for parts that were previously made by composite laminating (page 206).

QUALITY

Heating and forming a sheet of thermoplastic stretches it. A properly designed mold will pull it in a uniform manner. Otherwise, the properties of the material will remain largely unchanged. Thus a combination of the mold finish, molding pressure and material will determine surface finish.

The side of thermoformed plastic sheet that comes into contact with the tool will have an inferior finish to pressure formed parts. However, the reverse side will be smooth and unmarked. Therefore, parts are generally designed so that the side that came into contact with the tool is concealed in application. Tools can be outward (male) or inward (female) curving.

Pressure forming produces a fine surface finish with excellent reproduction of detail.

DESIGN OPPORTUNITIES

Thermoforming is typically carried out on a single mold. In vacuum forming the mold can be made from metal, wood or resin. Wood and resin are ideal for prototyping and low volume production. A resin tool will last between 10 and 500 cycles, depending on complexity of shape. For more products, aluminium molds are cast or machined.

Similar to other molding operations inserts can be used to form re-entrant angles. They are typically inserted and removed manually. Routing or cutting into the part after molding can sometimes produce the same effect. Otherwise, parts can be molded separately and welded together.

Many thermoplastic materials are suitable. Therefore, there are numerous decorative and functional opportunities associated with each of them. Also, materials have been developed for thermoforming. These include preprinted sheets (such as carbon fibre-effect), which are a coextrusion of acrylonitrile butadiene styrene (ABS) and poly methyl methacrylate (PMMA) with a printed film sandwiched in between. These materials are usually affected by a minimum order in the region of 3 tonnes.

TECHNICAL DESCRIPTION

Vacuum forming is a straightforward process and provides the foundation for the other thermoforming techniques. A sheet of material is heated to its softening point. This is different for each material. For example, the softening point of polystyrene (PS) is 127–182°C (261–360°F) and PP is 143–165°C (289–329°F). Certain materials, such as HIPS, have a larger operating window (that is, the temperature range in which they are formable), which makes them much easier to thermoform.

The softened plastic sheet is blown into a bubble, which stretches it in a uniform manner. The airflow is then reversed and the tool is pushed up into the sheet. The material is drawn onto the surface of the mold by vacuum at about 0.96 bar (14 psi). To assist the flow of air, channels are drilled into the tool. They are located in recesses and across the surface of the mold to extract the air as efficiently as possible.

Pressure forming is the reverse of vacuum forming: the sheet is formed onto the surface of the mold under approximately 6.9 bar (100 psi) of air pressure. This means that a greater level of detail can be achieved. Surface details on the mold will be reproduced with much more accuracy than vacuum forming. Surface finish can be more accurately controlled and is therefore functional. However, like vacuum forming only one side of the sheet will be functional.

Thermoforming Processes

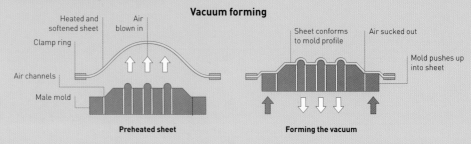

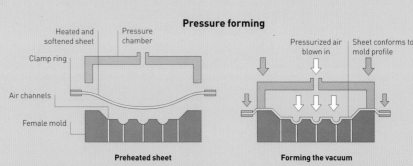

Plug-assisted forming is used to bring the benefits of male mold forming to female parts because blowing a softened sheet into a bubble stretches it evenly while draping the sheet into a female mold produces more localized stretching. The plug stretches the sheet prior to forming and so ensures that there is adequate wall thickness for deep profiles. Otherwise the material will tear. When the air is drawn out, the sheet conforms to the mold profile and the hydraulic plug retracts.

In twin sheet thermoforming, 2 sheets are thermoformed and clamped together.

This forms an enclosed and thin walled product. Both sides of the product are functional, unlike the single sheet processes.

The machines are rotary. The clamps transfer the sheet into the heating chamber, where it is raised to softening temperature; it is then thermoformed and clamped; and finally rotated to the unloading station. The 2 sheets are thermoformed individually, one above the other. Once fully formed they are clamped together. Residual heat from the thermoforming enables the bond to form with prolonged contact. The bond has similar strength to the parent material.

Multilayered materials are coextruded to provide benefits: for example, providing a barrier to moisture or bacteria; different colours; and sandwiching recycled material between films of virgin (aesthetic) material to reduce energy.

Textured sheets can be thermoformed (see image, page 31, above left). Typically sheets are textured on one side and so the smooth side is formed against the mold. There is a range of standard textures, which are cast or extruded by the material manufacturer. Examples include frosted, haircell, lens, embossed and relief.

Twin sheet thermoforming produces 3D hollow parts. This has added benefits including lightness and rigidity, insulation and 2 separate materials (upper and lower). Single sheet thermoforming is typically limited to only a single side of functionality. Twin sheet, on the other hand, provides functional surfaces on both sides.

Foam can be molded into twin sheet, or be injected post-forming for added rigidity and strength (see image, page 31, above right).

DESIGN CONSIDERATIONS

Thermoforming is ideal for shallow thin walled parts. It is not usually practical for the depth to exceed the diameter. Materials can be stretched more evenly over the surface of a deep profile using plug-assisted forming.

Air channels on the surface of the mold will leave slight pimples. These can be eliminated on aesthetic surfaces with the use of microporous aluminium molds (see image, top). This material has a much shorter molding lifespan because the pores clog up eventually. However, they are very useful for design details (see image, middle), which would otherwise require hundreds of air channels (holes).

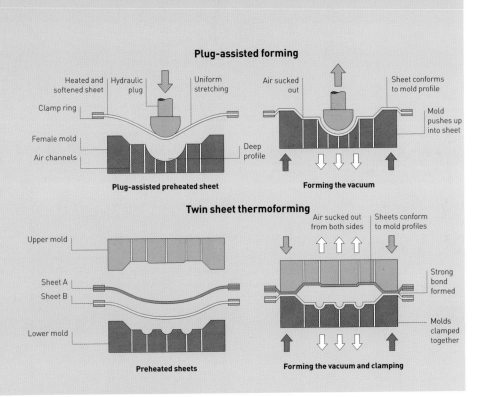

Plug-assisted forming

Heated and softened sheet | Hydraulic plug | Uniform stretching

Clamp ring

Female mold

Air channels

Deep profile

Plug-assisted preheated sheet

Air sucked out | Sheet conforms to mold profile

Mold pushes up into sheet

Forming the vacuum

Twin sheet thermoforming

Upper mold

Sheet A
Sheet B

Lower mold

Preheated sheets

Air sucked out from both sides | Sheets conform to mold profiles

Strong bond formed

Molds clamped together

Forming the vacuum and clamping

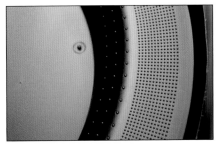

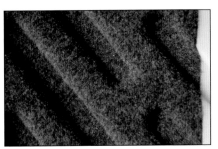

Top
Microporous aluminium moldmaking material, shaped by CNC machining.

Above
Carpet covered thermoforming produced in a single operation.

Middle
Textured design feature vacuum formed on mold made of microporous aluminium.

Textures are sometimes necessary on the surface of the mold to assist the flow of air and avoid air pockets forming. These are known as 'open textures'.

Sheet fed thermoforming is used to process sheet materials ranging from 1 mm to 12 mm (0.04–0.47 in.). Roll fed is supplied by a reel and so is limited to 0.1 mm to 2.5 mm (0.004–0.1 in.). The size of part is typically restricted to 1.5 x 3.5 m (5 x 11.5 ft), although some machines are capable of dealing with sheets of 2.5 x 4 m (8 x 13 ft).

Draft angles are essential when thermoforming over protrusions, or male molds. This is because the heated plastic sheet is expanded; as it cools it will shrink by up to 2%. Different materials have different levels of shrinkage: for example, ABS will shrink 0.6% and HDPE 2%. A draft angle of 2° is usually recommended.

The material stretches as it is thermoformed, and this will be more prominent in deeper and undulating profiles. Therefore, care must be taken to avoid sharp corners and points where three corners meet. These will cause excessive thinning and thus weak points.

COMPATIBLE MATERIALS

Although almost all thermoplastic materials can be thermoformed, the most common are ABS, polyethylene terephthalate (PET) (including PETG, which is PET modified with glycol), polypropylene (PP), polycarbonate (PC), high impact polystyrene (HIPS) and high density polyethylene (HDPE). The glycol in PETG reduces brittleness and premature ageing. PETG is clear (almost like PC) and is therefore often the material of choice for lighting diffusers and medical packaging, for example.

COSTS

Tooling costs are typically low to moderate, depending on the size, complexity and quantity of parts. The most expensive tools are machined aluminium. Tooling for pressure forming is 30–50% more expensive than vacuum forming, but is still considerably cheaper than tooling for injection molding.

The speed of thermoforming depends on the selected process and material thickness. Sheet fed processes typically produce 1–8 parts per minute. Roll fed machines are generally faster and multiple cavity tools can make hundreds of parts per minute. Roll fed machines are automated, while sheet fed machines are generally loaded by hand. This increases labour costs.

ENVIRONMENTAL IMPACTS

This process is only used to form thermoplastic materials, so the majority of scrap can be recycled. Kaysersberg Plastics, who provided the extruding the sheet material case study (page 34), produce only 1.4% scrap material; the rest is recycled.

Extruding the sheet material

Kaysersberg Plastics extrude sheet material for thermoforming. They can very accurately control the quality of the material and make adjustments almost instantly as a result of their findings in the thermoforming process.

As with injection molding, the thermoplastic is melted and mixed by an Archimedean screw. It is squeezed out through a die that is 3,084 mm (121 in.) wide, after which it is forced between highly polished steel rollers to a set thickness (**image 1**). The extrusion process is continuous and in operation the sheet is pulled through the rollers. However, when the process begins, there is nothing to pull through. To overcome this problem, straps are fixed to the extrusion flow front, and these are pulled through the rollers (**image 2**). The extrusion die is made up of many independently controllable segments, which are adjusted to make a uniform wall thickness (**image 3**). The continuous sheet of HDPE (**image 4**) is slit along its length with knives (**image 5**) and then cut to length. The finished sheets are stacked in preparation for thermoforming.

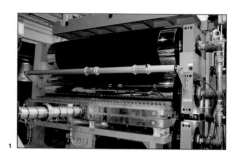

1

2

3

4

5

Twin sheet thermoforming the Terracover® ice pallet

Twin sheet thermoforming demonstrates the capabilities of thermoforming with the added benefit of bonding together to form hollow parts. The sheet material, which is 4 mm (0.16 in.) thick HDPE (**image 1**), was extruded and cut to size as outlined above.

The thermoforming machine consists of 4 stations: loading and unloading; primary heating; secondary heating; and thermoforming. Two sheets are thermoformed and then clamped together; in this case study they will be referred to as 'sheet A' and 'sheet B'. First of all, sheet A is manually loaded onto the rotary thermoforming machine and clamped tightly around its perimeter (**image 2**). It is then loaded into the primary heating chamber, where the temperature is set at 160°C (320°F).

Meanwhile, sheet B is loaded. As sheet A moves into the secondary heating chamber (**image 3**), sheet B is rotated into the primary heating chamber, where it is heated to softening point (**image 4**) and suspended between the thermoforming tools. Because this is twin sheet thermoforming there are 2 tools in this station, as opposed to only 1 for single sheet parts. Sheet A is vacuum formed over the lower tool, and sheet B is formed over the upper tool. As the vacuum draws the material onto the surface of the tool the profile becomes more pronounced (**image 5**). Just as they are both formed the molds come together and clamp the sheets together (**image 6**). The residual heat bonds the 2 materials together.

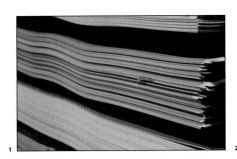

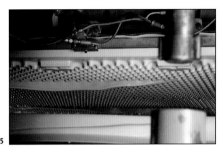

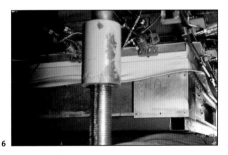

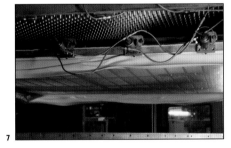

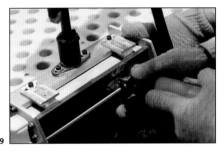

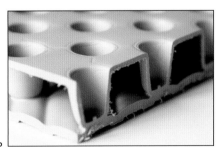

After 4½ minutes the tools separate (**image 7**). Meanwhile, 2 more sheets have been loaded and preheated, and replace the formed part as it is rotated to the unloading station. In this way production is as continuous as possible. The formed part is trimmed by hand (**image 8**). Re-entrants are cut using a handheld router guided by a jig (**image 9**), while complex trimming operations are carried out by CNC machining.

Sheets A and B (**image 10**) are welded together around the perimeter and across the surface with a series of spikes (approximately 2,400), so an homogenous bond is formed between the 2 sheets. This is essential for the strength of the Terracover® ice pallet, which is engineered to accommodate 5 tonnes of load over an area of only 150 x 150 mm (5.91 x 5.91 in.). It is used to cover ice rinks temporarily for events such as concerts. Filling the cavity with foam (see image, page 31, above, right) further increases rigidity, strength and thermal insulation.

Featured Manufacturer

Kayserberg Plastics
www.kayplast.com

Bend
 Continuous
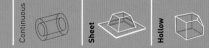 **Sheet**
 Hollow
 Bulk
 Internal

Forming Technology

Rotation Molding

Rotation molding produces hollow forms with a constant wall thickness. Polymer powder is tumbled around inside the mold to produce virtually stress free parts. Recent developments include in-mold graphics and multi-layered wall sections.

Costs	Typical Applications	Suitability
• Medium cost tooling • Low to medium unit cost (3 to 4 times material cost)	• Automotive • Furniture • Toys	• Low to medium volumes up to 10,000 units

Quality	Related Processes	Speed
• Surface finish is good • Low pressure during molding produces low stress concentrations	• Blow molding • Thermoforming	• Long cycle time (between 30 and 60 minutes)

INTRODUCTION

Rotation molding is a versatile process that can be used to create hollow and sheet geometries. It is cost effective to produce large and small products in low to medium volume production runs.

Tooling is comparatively inexpensive because it does not have to be matched to internal cores or engineered to withstand high pressure. Even so, this process can be used to produce parts with tight tolerances on closures and fixtures. Similar to injection molding (page 50), in-mold graphics can be applied to reduce finishing operations.

Rotation Molding Process

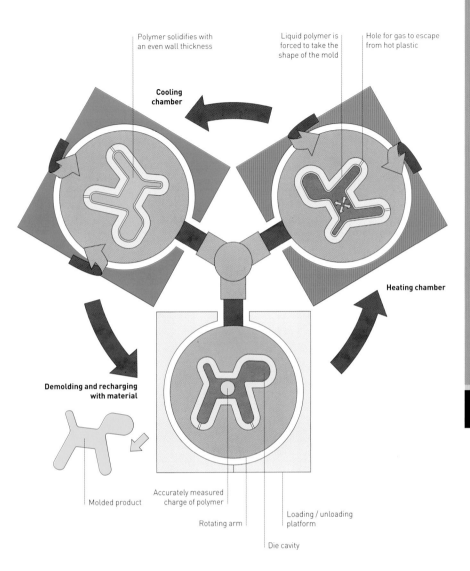

Polymer solidifies with an even wall thickness

Liquid polymer is forced to take the shape of the mold

Hole for gas to escape from hot plastic

Cooling chamber

Heating chamber

Demolding and recharging with material

Molded product

Accurately measured charge of polymer

Rotating arm

Loading / unloading platform

Die cavity

TYPICAL APPLICATIONS

Rotation molding is used to produce a variety of products, such as boat hulls, canoes and kayaks, furniture, containers and tanks, road signs and bollards, planters, pet houses and toys.

RELATED PROCESSES

Thermoforming (page 30) and blow molding (page 22) can also produce hollow and sheet profiles. Hollow parts produced by twin sheet thermoforming will have a seam where the 2 sheets of material come together. Blow molding is typically used for high volume production of relatively smaller parts, such as thin walled packaging.

QUALITY

Surface finish is good even though no pressure is applied. The molded product is almost stress free and has an even wall thickness. The plastic will shrink by 3% during the process, which may cause warpage in parts with large flat areas.

DESIGN OPPORTUNITIES

This process is relatively inexpensive for the production of low to medium volume runs and is suitable for small products and large products up to 10 m³ (353 ft³). Some products can be molded in pairs and then separated post-molding to create sheet geometries.

Tooling can be taken directly from a full size prototype in wood, aluminium or resin, which aids the transition between design and production. There is no core in a rotation molding tool, so changes are relatively easy and inexpensive.

Several different types of powder are used to mold shapes that have different

TECHNICAL DESCRIPTION

The rotation molding process starts with the assembly of the metal molds on the rotating arm. A predetermined measure of polymer powder is dispensed evenly into each mold. It is closed, clamped and rotated into the heating chamber. There it is heated up to around 250°C (482°F) for 25 minutes and is constantly rotated around its horizontal and vertical axes.

As the walls of the mold heat up the powder melts and gradually builds up an even coating on the inside surface. The rotating arm passes the molds onto the cooling chamber, where fresh air and moisture is pumped in to cool them for 25 minutes. They continue to rotate at about

20 revolutions per minute throughout the entire process to ensure even material distribution and wall thickness.

Once the parts have cooled sufficiently they are removed from the molds so that the process can start again. The heating and cooling times, and speed of rotation, are controlled very carefully throughout the process.

levels of complexity. Micro-pellets and fine powders are more suitable for tight radii and fine surface details; however, fine powders show up surface defects more readily. Foam-filled, hollow or other multi-layered walls can be produced with composite polymers, which are triggered at varying temperatures. Wall thicknesses in solid wall sections are typically no more than 6 mm (0.24 in.). Maximum thickness is determined by the temperature of the mold and thermal conductivity of the polymer.

Inserts and preformed sections of different colour, threads, in-mold graphics and surface details can be integrated into the molding process. Overmolding of 1 material onto another premold reduces assembly costs and produces seamless finishes. Additives can be used to make the materials UV and weather resistant, flame retardant, static-free or food safe.

DESIGN CONSIDERATIONS

Low pressure produces low molecular mass materials that have low mechanical strength. However, this can be overcome by integrating ribs into the design. Abrupt changes in wall section are not possible, and sharp angles and tight corners should be avoided. Small radii can be achieved across bends in 1 direction, but are not suitable for corners. Low pressure also means that high gloss finishes are not practical.

Product length is restricted to 4 times the diameter, to prevent uneven material distribution in the molding cycle.

COMPATIBLE MATERIALS

Polyethylene (PE) is the most commonly rotation molded material. Many other thermoplastics can be used, such as polyamide (PA), polypropylene (PP), polyvinyl chloride (PVC) and ethylene vinyl acetate (EVA).

COSTS

Tooling costs are relatively inexpensive because tools do not have to be

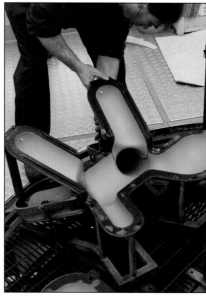

1

2

3

4

5

6

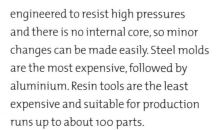

engineered to resist high pressures and there is no internal core, so minor changes can be made easily. Steel molds are the most expensive, followed by aluminium. Resin tools are the least expensive and suitable for production runs up to about 100 parts.

Cycle time is usually between 30 and 90 minutes depending on wall thickness and choice of material. Several tools are mounted on the rotating arms at the same time, which reduces cycle time.

Rotation molding is labour intensive. Fully automated molding is available for small parts and high volumes to reduce the costs.

ENVIRONMENTAL IMPACTS

There is very little scrap material in this process because predetermined measures of powder are used. The molds stay closed and clamped throughout the entire molding and cooling process. Any thermoplastic scrap can be recycled.

→ Rotation molding the Grande Puppy

Each mold is used to produce batches in a single colour to avoid contamination. The Grande Puppy is being molded in green PE. The powder is weighed (**image 1**) and the mold assembled on the rotating arm (**image 2**). Valves are inserted into the holes in the puppy's feet, which control the flow of gas in and out of the mold as the polymer heats up. A generous 3.6 kg (7.94 lb) of powder is used in the Grande Puppy to ensure a wall thickness of 6 mm (0.236 in.), which makes it safe as a seat and children's toy. The mold is made up of 4 parts, which have to be carefully cleaned between each molding cycle. The mold is charged with the predetermined measure of powder (**image 3**) and clamped shut (**image 4**). The molds, on either side of the rotating arm (**image 5**), are passed onto the heating chamber (**image 6**) for 25 minutes. Once the powder has had sufficient time to melt and adhere to the walls of the mold, the rotating arm moves through 120° to the cooling chamber for a further 25 minutes. After cooling, the molds are separated (**image 7**) and the final products exposed (**image 8**). The puppy is demolded carefully to avoid any surface damage while the polymer is still warm (**image 9**) and placed on a conveyor belt on its way to being deflashed and packaged (**image 10**).

7

8

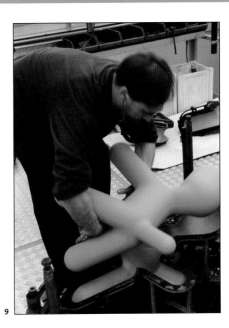

9

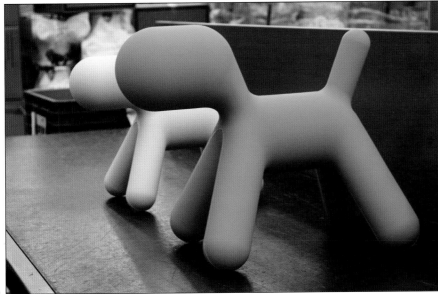

10

Featured Company

Magis
www.magisdesign.com

Bend
 Continuous
Sheet
 Hollow
 Bulk
 Internal

Forming Technology

Vacuum Casting

Used for prototypes, one-offs and low volumes, vacuum casting can replicate almost all the properties of injection molding. It is primarily used to mold 2-part polyurethane (PUR), which is available in a vast range of grades, colours and hardnesses.

Costs	Typical Applications	Suitability
• Low tooling costs • Moderate unit costs	• Automotive • Consumer electronics • Sports equipment	• Prototypes, one-offs and low volume production

Quality	Related Processes	Speed
• Very high surface finish and reproduction of detail	• Injection molding • Reaction injection molding	• Variable cycle time (typically between 45 minutes and 4 hours, but up to several days depending on size of part)

INTRODUCTION

This is a method used to mold thermosetting polyurethane (PUR) for prototyping and one-off production. There are many different types of PUR, which make this process suitable for the majority of plastic prototyping. It is used for a diversity of prototyping applications including mobile phone covers, automotive components, sports equipment and medical instruments.

Flexible silicone molds are produced directly from the master (pattern). Vacuum casting is then used to reproduce the product with properties

Vacuum Casting Process

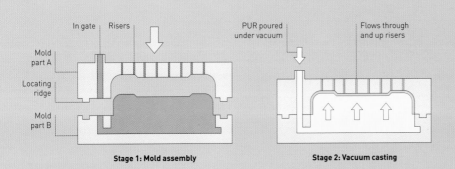

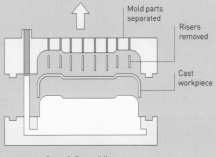

Stage 1: Mold assembly   Stage 2: Vacuum casting   Stage 3: De-molding

TECHNICAL DESCRIPTION

In stage 1, the two halves of the mold are brought together. Molds are typically made in two halves: parts A and B. Inserts, cores and additional mold parts can also be incorporated for complex parts.

The two separate parts of PUR are stored at 40°C (104°F). Mixing them together causes a catalytic and exothermic reaction to take place, and the PUR warms up to between 65°C and 70°C (149–158°F).

In stage 2, the liquid PUR is drawn into the mold by a vacuum. This ensures that the material has no porosity, will flow through the mold cavity and will not be restricted by air pressure. When it is poured in, it flows down through the gate and into the runner system. The PUR is held high above the mold, so as it is poured into the mold and runs underneath gravity forces it to rise upwards. The runner is designed to supply

a curtain of liquid PUR. It flows through the mold and up the risers in part A. There are lots of risers to ensure that the mold is evenly and completely filled.

After a few minutes the mold cavity is filled and the vacuum equalized. The mold is left closed for the complete curing time, which is typically 45 minutes to 4 hours. In stage 3, the part is ejected and the flash, risers and excess material are removed.

very similar to the mass-produced part. The full colour range is available, and it can be made soft and flexible like thermoplastic elastomer (TPE), or rigid like acrylonitrile butadiene styrene (ABS).

PUR is a 2-part thermosetting plastic. When the 2 parts are mixed in the right proportions, the polymerization process takes place. It is an exothermic reaction. The type of material is adjusted to suit the volume of the casting because very large parts have to be cured very slowly, otherwise they will overheat and distort. In contrast, small parts can be cured very rapidly without any problems.

TYPICAL APPLICATIONS

Applications for vacuum casting are widespread and used in the automotive, consumer electronic, consumer product, toy and sport equipment industries. As well as prototyping, it is used when low volumes rule out expensive tooling for injection molding (page 50). Applications in the automotive

industry include the production of inlet manifolds, water tanks, air filter housings, radiator parts, lamp housings, clips, gears and live hinges.

Consumer electronic applications include keyboards and housings for mobile phones, televisions, cameras, MP3 players, sound systems and computers.

RELATED PROCESSES

Vacuum casting is used to simulate the material and processing capabilities of injection molding. Vacuum casting has some advantages over injection molding because the tooling is flexible and there is less pressure involved in the process, but it is not suitable for the same production volumes.

Reaction injection moulding (RIM) (page 64) is also used for prototyping, one-offs and low volume production. Like vacuum casting, it is used to mold foam, solid and elastomeric PUR materials. Due to the nature of the process, it is best suited to parts with simpler geometries

and smoother features. This is due to the speed that the PUR cures in the RIM process. Even so, RIM is used for low volume production of car bumpers and dashboards, for example. Until there are sufficient volumes to justify the tooling for injection molding, this process is a very suitable alternative.

QUALITY

The tolerance is typically within 0.4% of the dimensions of the pattern. The surface finish will be an exact replica of the pattern. It is not possible to alter the surface finish of the silicone mold.

The material can be adjusted to suit the application. PUR is available in water clear grades and the full colour range. It can be very soft and flexible (shore A range 25–90) or rigid (shore D range).

DESIGN OPPORTUNITIES

This is a versatile process that has many advantages for designers. The tooling is flexible silicone. This means that slight re-entrant angles and undercuts can be ejected by flexing the tool rather than increasing the number of parts of the tool. It is also possible to incorporate cores and inserts to make larger re-entrants and internal features.

The PUR prototyping material range is designed to mimic injection molded materials such as polypropylene (PP), ABS and polyamide (PA) nylon. Snap fits, live hinges and other details that are otherwise limited to injection molding can be made. Additives can be used to improve flame retardant qualities, heat resistance and UV stability.

The silicone tooling can be taken from almost any non-porous pattern material. Reproducing the master in silicone greatly reduces cost and cycle time.

Because this is a low pressure process, it is possible to make step changes in the wall thickness.

The properties of different plastics can be incorporated as over-moldings. This is similar to multi-shot injection molding, except that the part is removed from the mold and placed in another for the second material.

DESIGN CONSIDERATIONS

Parts can be any size from a few grams up to several hundred kilos, but size will affect the cycle time. Larger parts

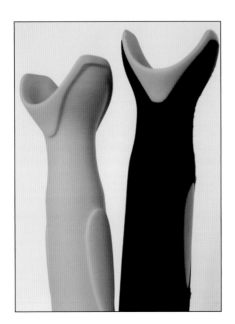

Left
Here a PUR vacuum casting of rigid yellow ABS mimic is over-molded with flexible black TPE mimic.

have to be cured slowly to avoid the exothermic reaction causing shrinkage and distortion. Parts are typically less than 2.5 kg (5.51 lb), which can be cured in 1 hour or less.

Wall thicknesses down to 0.5 mm (0.02 in.) can be produced in small areas. However, it is recommended that wall thickness be greater than 1 mm (0.04 in.) and ideally no less than 3 mm (0.118 in.).

Very sharp details, such as knife edges, are difficult to reproduce in any molding application and so should be avoided.

COMPATIBLE MATERIALS

There are several hundred different grades of PUR. This process is also used to cast polyamide (PA) and wax patterns for investment casting (page 130).

COSTS

Tooling costs are generally low, but this is largely dependent on the size and complexity of the pattern. A mold will usually take between 0.5–1 day to make in silicone and need to be replaced after 20–30 cycles.

Cycle time is good, but depends on the size of the part and type of material.

Labour costs are moderate, depending on the level of finishing required. Mold making, vacuum casting and finishing are all carried out manually.

ENVIRONMENTAL IMPACTS

Accurately measuring the material reduces scrap, which cannot be recycled because it is a thermosetting plastic.

The process is carried out in a vacuum chamber, so any fumes and gases can be extracted and filtered.

→ Vacuum casting a computer screen housing

This housing for a computer screen used in hospitals is a low volume production part.

First the silicone mold is made. The pattern is suspended in a box by the gate, runners and risers. The box is then filled with silicone rubber, which completely encapsulates it. When fully cured, the mold halves are separated by cutting into the silicone to meet the pattern (**image 1**). It is a skilful process, and the geometry of the part will determine the mold-making technique.

The pattern is removed from the mold (**image 2**). Fine details, such as logos and surface texture, are reproduced exactly. The split line was continued across the open central area by a thin film, which was attached to the product prior to casting in silicone. This technique produces a split mold, which has incorporated gate, runners and risers (**image 3**).

The silicone is flexible and so will typically only last for 20–30 cycles before it has to be remade. It is taped together to stop any movement during casting (**image 4**). Two pipes are attached, which will deliver the liquid PUR.

The casting process takes place under vacuum. The PUR pours into the mold and is unrestricted by air pressure (**image 5**). As the mold fills up the PUR emerges from the risers. The mold is left in an oven at 40°C (104°F) until the PUR has fully cured. This particular grade takes 45 minutes.

The ejection process begins by blowing pressurized air into the risers (**image 6**). This helps to release the casting from the mold. The 2 halves of the mold are separated to reveal the casting (**image 7**). It is removed with the risers intact (**image 8**). These are detached by

hand, and the 2 halves of the computer screen housing assembled (**image 9**). The finished product (**image 10**) is very similar to a mass-produced injection molded part, with ribs, perforations, logos and matt surface finish.

4

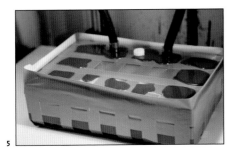

5

Featured Manufacturer

CMA Moldform
www.cmamoldform.co.uk

6

7

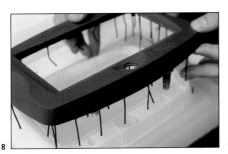

8

9

10

Bend | Continuous | **Sheet** | Hollow | **Bulk** | Internal

Forming Technology

Compression Molding

In this rapid process, rubber and plastic are shaped by compressing them into a preheated die cavity. Compression molding is generally used for thermosetting materials.

PLASTICS AND RUBBER

44

Costs	Typical Applications	Suitability
• Moderate tooling costs • Low unit costs (3–4 times material cost)	• Automotive under-the-bonnet • Electrical housing and kitchen equipment • Seals, gaskets and keypads	• Medium to high volume production

Quality	Related Processes	Speed
• High strength parts with high quality surface finish	• DMC and SMC molding • Injection molding • Vacuum casting	• Plastic: Rapid (2 minute cycle time) • Rubber: Slower (10 minute cycle time)

INTRODUCTION

This process is used to mold rubber and plastic into sheet and bulk geometries. It is suitable for molding both thermosoftening and thermosetting materials. It is also used to produce parts in the dough and sheet molding compound (DMC and SMC), in DMC and SMC molding (page 222).

Compression molding thermosetting plastics has played an important role in the transition between metal and plastic parts in engineering. Plastics have been used to replace metals ever since phenolic resin (Bakelite) was

Compression Molding Rubber Process

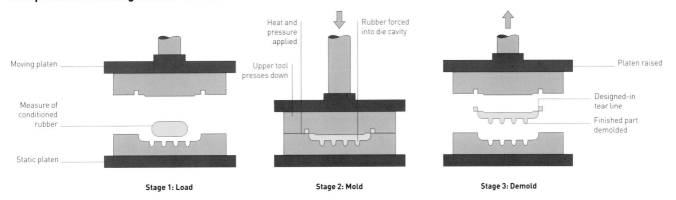

Stage 1: Load

Stage 2: Mold

Stage 3: Demold

first compression molded in the 1920s. Bakelite marks a significant point in the history of plastic manufacturing because it was the first synthetically produced polymer that could be molded. Since then huge developments in plastic technology have meant that injection molding (page 50) is now used to produce parts that were traditionally compression molded. Even so, compression molding is used to form certain rubber and thermosetting plastic parts that are not suitable for injection molding. DMC and SMC molding technologies followed in the 1960s and provided industry with thermosetting plastic parts which compete with metal die castings in terms of strength, durability and resilience.

The process of compression molding is simple: a measure of material is placed in between preheated matched tools, which come together and force the material into the die cavity.

TYPICAL APPLICATIONS

Compression molding is reserved for specific materials with particularly demanding applications such as heat and electrical insulation. Some typical applications include electrical housing, kitchen equipment, ashtrays, handles and light fittings. Demand has recently increased for under-the-bonnet applications in battery-powered cars because thermosetting materials can provide the electrical insulation and stability that are required for these parts.

TECHNICAL DESCRIPTION

The sequence for compression molding rubber is identical to compression molding plastic (page 48) except that the cycle time is slightly longer. To show the different compression molding techniques, a sheet geometry has here been used instead of a side action tool.

In stage 1, the rubber is conditioned to remove any crystallinity that might have built up since its production. Then a measure of conditioned rubber is placed in the lower mold. In stage 2, the 2 halves of the mold are brought together and pressure is applied gradually to encourage the material to flow. After 10 minutes the rubber is fully cured and its molecular structure is formed. In stage 3, the molds separate and the parts are peeled from the die cavity. The tear lines, which are integrated into the design to reduce secondary operations, ensure that the flash separates in a consistent manner when it is removed and so leaves a tidy edge detail.

Thermosetting rubber can be formed by compression molding, injection molding and vacuum casting (page 40). It is used to produce an array of products including flexible keyboards, keypads, seals and gaskets. Logos and other decorative moldings for running shoes, shoe soles and other sports equipment are also made in this way. Electronics housing can be molded in a single piece of rubber, which provides protection against damage as well as water. This is especially useful for handheld navigation devices and other portable electronic equipment.

RELATED PROCESSES

Compression molding and injection molding are closely related. Both require matched tooling (although tooling for compression is cheaper) and the molding process is done under pressure with heat. The difference is that injection molding is predominantly used for thermoplastics and compression molding for thermosetting plastics. However, engineering thermoplastics are suitable for demanding applications and can also be injection molded; thermoplastic elastomers (TPEs) can be injection molded to give the same look and feel as rubber.

Compression molding long strand fibre reinforced plastics (FRP) is known as DMC and SMC molding.

Vacuum casting is typically used for prototypes, one-offs and lower volumes than compression molding. It is used to form polyurethane resin (PUR), which is available is a range of densities and hardnesses.

QUALITY

This is a high quality process. Many of the characteristics can be attributed to the materials, such as heat resistant and electrically insulating phenolics or flexible and resilient silicones. Thermosetting plastics are more crystalline and so are more resilient to heat, acids and other chemicals.

Surface finish and reproduction of detail is very good. By compressing rather than injecting material in the die cavity, the parts have reduced stress and are less prone to distortion.

DESIGN OPPORTUNITIES

The main design opportunities are associated with the material properties. Thermosetting materials have many advantageous qualities when compared with thermoplastics. They can be filled with glass fibre (see also DMC and SMC molding), talc, cotton fibre or wood dust to increase their strength, durability, resistance to cracking, dielectric resilience and insulating properties.

Rubber compression molding is used to produce parts with various levels of flexibility. Live hinges and tear lines can be integrated into the design to eliminate secondary operations. Another major advantage of working with rubber is that draft angles can be eliminated and slight re-entrant angles are possible because the material is flexible and so can be stretched over a mold core. A further advantage is that it is possible to integrate a range of colours in rubber compression molding. They are either introduced as a pre-form, such as buttons or logos, or are molded in the same sequence. Preformed rubber inserts provide a cleaner joint between colours. However, the colour joint is quite often covered up, for instance, by a control panel in a keypad application.

Thermosetting resins, on the other hand, are much less colourful, especially phenolics. Phenolic resin is naturally dark brown and so only dark colours can be

achieved, such as the traditional dark brown Bakelite products of the 1920s. If necessary, vibrant colours are applied in the finishing operations.

Another major advantage of compression molding is its relatively inexpensive tooling, especially for rubber molding. Also metal inserts and electrical components can be molded into both rubber and plastic parts.

DESIGN CONSIDERATIONS

As with injection molding, there are many design considerations that need to be taken into account when working with compression molding.

Draft angles can be reduced to less than 0.5° if the tool and ejector system are designed carefully.

The size of the part can be anything from 0.1 kg up to 8 kg (0.22–17.64 lb) (on a 400 tonne/441 US ton press). The overall dimensions are limited by the pressure that can be applied across the surface area, which is affected by part geometry and design. Another major factor that affects part size is the method by which gases are vented from the thermosetting material as it cures and heats up. This plays an important role in tool design, which aims to get rid of gases through the use of vents and by incorporating clever rib design in the tool.

Wall thickness for parts can range from less than 1 mm (0.04 in.) (0.3 mm/ 0.012 in. in rubber) up to 50 mm (1.97 in.) or more. Step changes between different wall thicknesses are not a problem; the transition can be immediate. Wall thickness in plastic parts is limited by the nature of the thermosetting reaction because it is exothermic. Thick wall

sections are prone to blistering and other defects as a direct result of the catalytic reaction. It is therefore generally better to reduce wall thickness and minimize material consumption. For this reason, bulky parts are hollowed out or inserts are added. However, some applications require thick wall sections, such as parts that have to withstand high levels of dielectric vibration.

COMPATIBLE MATERIALS

Compatible thermosetting materials include phenolic, polyester, urea, melamine and rubber. Even though it is possible to compression mold thermoplastics it is not recommended.

There are many rubbers that can be molded in this way. The most common are silicones because they are readily available in small and large batches and take colour very well.

COSTS

Tooling costs are moderate and much less expensive than for injection molding, especially for certain rubber sheet geometries, which can be made using simple and inexpensive tooling that is manually operated.

For plastics, cycle time is very rapid and usually about 2 minutes per part. By contrast, rubbers take considerably

→ Compression molding silicone keypads

In this case transparent silicone keypads are being molded in a multiple cavity tool. This sort of tooling can only be used for sheet geometries; it would virtually impossible to force the material into a tall cavity. However, the tooling for rubber molding can be as complex as that for plastic molding, which is demonstrated in the case study on page 48.

Silicone rubber is extruded, conditioned and preformed into pellets in preparation for molding (**image 1**). Excess material is used to ensure even and thorough material distribution. The pellets are inserted into each mold cavity (**image 2**), generally using human hands – automation being necessary only for large volumes.

The 2 halves of the tool are brought together and the complete tool is placed under a press (**image 3**). The tool takes about 10 minutes to reach 180°C (356°F). Crosslinking takes place as a result of time and pressure. The time it takes depends on the thickness of material and cure system used. The mold is removed from the press and opened (**image 4**). The parts are removed by blowing compressed air between the flash and surface of the mold. Once they have been demolded, the parts are packed for shipping (**image 5**). They will be 'torn' from the flash when they are finally assembled. The thinner sections of material that surround each button determine the resistance of the button when it is pressed.

3

4

longer and often need to be left in the heated press for 10 minutes or more.

Labour costs can be quite high.

ENVIRONMENTAL IMPACTS

The main environmental impacts arise as a result of the materials used. Thermosetting plastics require higher molding temperatures, typically between 170°C and 180°C (338–356°F). It is not possible to recycle thermosets directly, due to their molecular structure, which is cross-linked. This means that any scrap produced, such as flash and offcuts, has to be disposed of.

5

Featured Manufacturer
RubberTech2000
www.rubbertech2000.co.uk

Compression Molding Plastic Process

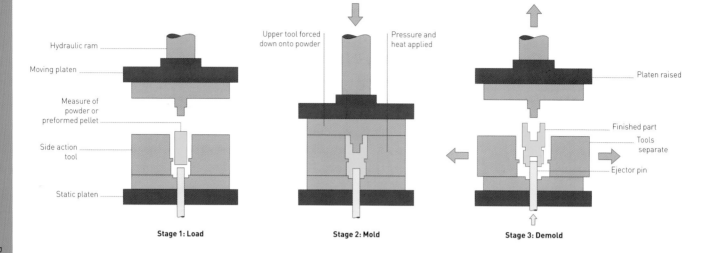

Stage 1: Load

- Hydraulic ram
- Moving platen
- Measure of powder or preformed pellet
- Side action tool
- Static platen

Stage 2: Mold

- Upper tool forced down onto powder
- Pressure and heat applied

Stage 3: Demold

- Platen raised
- Finished part
- Tools separate
- Ejector pin

TECHNICAL DESCRIPTION

In stage 1, a measure of powder, or a preformed pellet, is loaded into the die cavity in the lower tool. Prior to loading, the preformed pellets will have been conditioned in a heating chamber at approximately 100˚C (212˚F), in order to improve production rates and molding quality. In stage 2, the upper tool is gradually forced into the die cavity in a steady process, which ensures even distribution of material throughout the die cavity. The material plasticizes at approximately 115˚C (239˚F) and is cured when it reaches 150˚C (302˚F). This takes about 2 minutes. In stage 3, the parts of the mold separate in sequence. If necessary, the part is forced from the lower tool by the ejector pin.

Compression molding plastic is a simple operation, yet it is suitable for the production of complex parts. It operates at high pressure, ranging from 40 to 400 tonnes (40–441 US tons), although 150 tonnes (165 US tons) is generally the limit. The size and shape of the part will affect the amount of pressure required. Greater pressures will ensure better surface finish and reproduction of detail.

1

2

3

4

5

6

7

Featured Manufacturer

Cromwell Plastics
www.cromwell-plastics.co.uk

Case Study

→ Compression molding thermosetting plastic

The tooling for compression molding can be very complex, especially for large volumes and automated production. For this lamp housing for outdoor lighting, a multiple impression tool has been used, which produces 6 parts in every cycle. The product is being manufactured in phenolic resin because the material has to be able to withstand weathering and have exceptional electrical insulation qualities.

The lower half of the compression tool is made up of 3 parts: 1 static and 2 side action

(**image 1**). Phenolic powder is prepared by compressing it into pellets, which are heated up to around 100°C (212°F) in preparation for molding (**image 2**). The 3 parts of the lower tool are brought together to form the die cavity, into which the phenolic pellets are dropped (**image 3**). The upper mold is forced into the die cavity and after 2 minutes the mold separates to reveal the formed and fully cured resin. The parts are then stripped from the upper tool (**image 4**). Flash can be seen around the split lines

and has to be removed manually or in a vibration chamber.

For this product, small electrical contacts made from brass (**image 5**) are then inserted into the molding and maintained in position by friction (**image 6**), unlike inserts in the molding, which are not removable from the part. The final lamp housing parts demonstrate the high level of finish that can be achieved with this process (**image 7**).

Bend

Constant

Sheet

Hollow

Bulk

Internal

Forming Technology

Injection Molding

This is one of the leading processes used for manufacturing plastic products, and is ideal for high volume production of identical products. Variations on conventional injection molding include gas assisted, multishot and in-mold decoration.

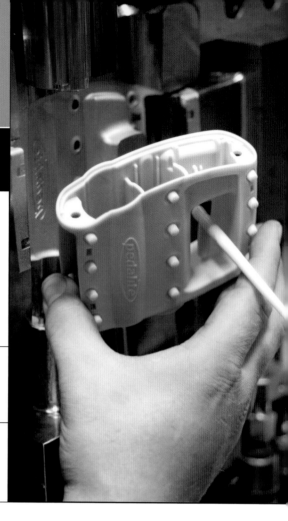

Costs	Typical Applications	Suitability
• Very high tooling costs but depends on complexity and number of cavities • Very low unit costs	• Automotive • Consumer electronics and appliances • Industrial and household products	• High volume mass production

Quality	Related Processes	Speed
• Very high surface finish • Highly repeatable process	• Reaction injection molding • Thermoforming • Vacuum casting	• Injection cycle time is generally between 30 and 60 seconds

INTRODUCTION

Injection molding is a widely used and well-developed process that is excellent for rapid production of identical parts with tight tolerances. It is used to create a huge diversity of our day-to-day plastic products. Accurately engineered tools and high injection pressures are essential for achieving excellent surface finish and reproduction of detail. Consequently, this process is suitable only for high volume production runs.

There are many different variations on injection molding technology. Some of the most popular include gas-assisted injection molding (page 58), multishot injection molding (page 60) and in-mold decoration (page 62).

TYPICAL APPLICATIONS

Injection-molded parts can be found in every market sector, in particular in automotive, industrial and household products. They include shopping baskets, stationery, garden furniture, keypads, the housing of consumer electronics, plastic cookware handles and buttons.

RELATED PROCESSES

The relative suitability of related processes depends on factors such as part size and configuration, materials being used, functional and aesthetic requirements, and budget.

Although injection molding is very often the most desirable process due to its repeatability and speed, thermoforming (page 30) is a suitable alternative for certain sheet geometries, and extrusion is more cost-effective for the production of continuous profiles.

Parts that will ultimately be made by injection molding can be prototyped and produced in low volumes by vacuum casting (page 40) and reaction injection molding (page 64). Both of these processes are used to form polyurethane resin (PUR). This is a thermosetting plastic that is available in a wide range of grades, colours and hardnesses. It can be solid or foamed. Reaction injection molding is used for a diverse range of products, including foam moldings for upholstering furniture and car seats, and low volume production of car bumpers and dashboards.

QUALITY

The high pressures used during injection molding ensure good surface finish, fine reproduction of detail and, most importantly, excellent repeatability.

The downside of the high pressure is that the resolidified polymer has a tendency to shrink and warp. These defects can be designed out using rib details and careful flow analysis.

Surface defects can include sink marks, weld lines and streaks of pigment. Sink marks occur on the surface opposite a rib detail, and weld lines appear where the material is forced to flow around obstacles, such as holes and recesses.

Injection Molding Process

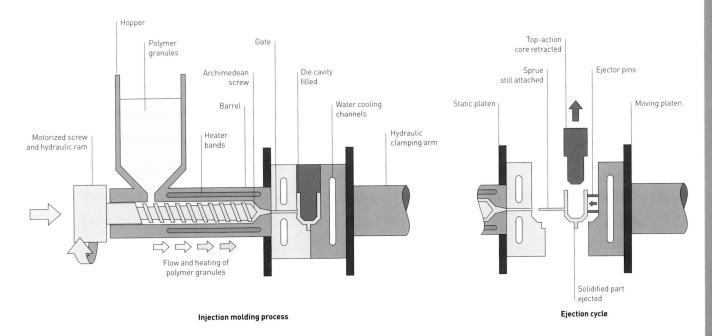

Injection molding process

Ejection cycle

DESIGN OPPORTUNITIES

So much is practically possible with injection molding that restrictions generally come down to economics. The process is least expensive when using a simple split mold. Most expensive are very complex shapes, which are achievable in a range of sizes, from large car bumpers to the tiniest widgets. Retractable cores controlled by cams or hydraulics can make undercuts from the sides, top or bottom of the tool simultaneously and will not affect the cost significantly, depending on the complexity of the action.

In-mold and insert film decoration are often integrated into the molding cycle, so eliminating finishing processes such as printing. There is also a range of pigments available to produce metallic, pearlescent, thermochromatic and photochromatic effects, as well as vibrant fluorescent and regular colour ranges. Inserts and snap-fits can be molded into the product to assist assembly.

Multishot injection molding can combine up to 6 materials in one product. The combination possibilities include density, rigidity, colour, texture and varying levels of transparency.

TECHNICAL DESCRIPTION

Polymer granules are dried to exactly the right water content and fed into the hopper. Any pigments are added at this stage at between 0.5% and 5% dilution.

The material is fed into the barrel, where it is simultaneously heated, mixed and moved towards the mold by the rotating action of the Archimedean screw. The melted polymer is held in the barrel momentarily as the pressure builds up ready for injection into the mold cavity.

The correct pressure is achieved and the melted plastic is injected into the die cavity. Cycle time is determined by the size of the part and how long the polymer takes to resolidify, and is usually between 30 and 60 seconds.

Clamping pressure is maintained after injection to minimize warpage and shrinkage once the part is ejected.

To eject the part, the tools move apart, the cores retract and force is applied by the ejector pins to separate the part from the surface of the tool. The part is dispensed onto a conveyor belt or holding container, sometimes by a robotic arm.

Tools and cores are generally machined from either aluminium or tool steel. The tools are very complex parts of the injection molding process. They are made up of water cooling channels (for temperature control), an injection point (gate), runner systems (connecting parts) and electronic measuring equipment which continuously monitors temperature. Good heat dispersal within the tool is essential to ensure the steady flow of melted polymer through the die cavity. To this end, some cores are machined from copper, which has much better conductive qualities than aluminium or steel.

The least expensive injection molding tooling consists of 2 halves, known as the male tool and female tool. But engineers and toolmakers are constantly pushing the boundaries of the process with more complex tooling, retractable cores, multiple gates and multishot injection of contrasting materials.

DESIGN CONSIDERATIONS

Designing for injection molding is a complex and demanding task that involves designers, polymer specialists, engineers, toolmakers and molders. Full collaboration by these experts will help realize the many benefits of this process.

Injection molding operates at high temperatures and injects plasticized material into the die cavity at high pressure. This means that problems can occur as a result of shrinkage and stress build-up. Shrinkage can result in warpage, distortion, cracking and sink marks. Stress can build up in areas with sharp corners and draft angles that are too small. Draft angles should be at least 0.5° to avoid stressing the part during ejection from the tool.

The injected plastic will follow the path of least resistance as it enters the die cavity and so the material must be fed into the thickest wall section and finish in the areas with the thinnest wall sections. For best results wall thickness should be uniform, or at least within 10%. Uneven wall sections will produce different rates of cooling, which cause the part to warp. The factors that determine optimal wall thickness include cost, functional requirements and molding consideration.

Ribs serve a dual function in part design: firstly, they increase the strength of the part, while decreasing wall thickness; and secondly, they aid the flow of material during molding. Ribs should not exceed 5 times the height of the wall thickness. Therefore, it is often recommended to use lots of shallow ribs, as opposed to fewer deep ribs.

All protruding features are treated as ribs and must be 'tied-in' (connected) to the walls of the part to reduce air traps and possible stress concentration points. Holes and recesses are often integrated in part design in order to avoid costly secondary operations.

Injection molded parts are often finished with a fine texture, which disguises surface imperfections. Large

gloss areas are more expensive to produce than matt or textured ones.

COMPATIBLE MATERIALS

Almost all thermoplastic materials can be injection molded. It is also possible to mold certain thermosetting plastics and metal powders in a polymer matrix.

COSTS

Tooling costs are very high and depend on the number of cavities and cores and the complexity of design.

Injection molding can produce small parts very rapidly, especially because multicavity tools can be used to increase production rates dramatically. Cycle time is between 30 and 60 seconds, even for multiple cavity tools. Larger parts have longer cycle times, especially because the polymer will take longer to resolidify and so will need to be held in the tool while it cools.

Labour costs are relatively low. However, manual operations, such

as mold preparation and demolding, increase the costs significantly.

ENVIRONMENTAL IMPACTS

Thermoplastic scrap can be directly recycled in this process. Some applications, such as medical and food packaging, require a high level of virgin material, whereas garden furniture may require only 50% virgin material for adequate structural integrity, hygiene and colouring capability.

Injection molded plastic is commonly associated with mass produced short term products, such as disposables. However, it is possible to design products so that they can be disassembled easily, which is advantageous for both maintenance and recycling. If different types of materials are used, then snap fits and other mechanical fasteners make it more convenient to disassemble and dispose of the parts with minimal environmental impact.

→ Manufacturing and assembling a Pedalite

This bicycle pedal light is powered by an energy storage capacitor and microgenerator rather than any form of chemical battery. It was designed by Product Partners, in conjunction with the client (Pedalite Limited), toolmaker/molder (ENL Limited), gearbox manufacturer (Davall Gears) and polymer distributor (Distrupol Limited).

A close working partnership with ENL ensured Product Partners' ideas were successfully translated through the design, development and toolmaking process,

ensuring a 'right first time' product. The technically challenging combined overmold assembly, for instance, was approved at the first-off tool trials.

To ensure the gearbox concept was feasible Davall Gears were recruited to provide expertise in gear ratios, gear detail design, materials specification and manufacture. Polyamide (PA) nylon was chosen for its exceptional wear characteristics and self-lubrication properties.

The cutaway drawing (below) shows the anatomy of the injection-molded parts as well as the internal mechanisms and parts. Injection-molded casings very often have to accommodate a fixed-space package. In this case, specific spindle bearings have been used to satisfy legislations and the gear system is designed for optimum energy generation.

MOLDFLOW ANALYSIS

Polymer distributor Distrupol was consulted about materials selection and moldflow analysis. Feedback on the developing design led to the modification of the components in CAD to reduce potential problems of sinkage, flow marks, weld lines and so forth.

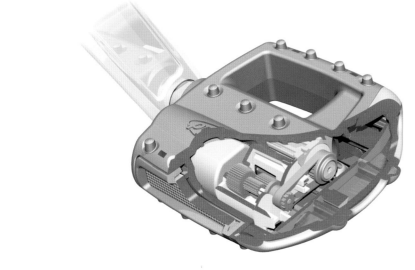

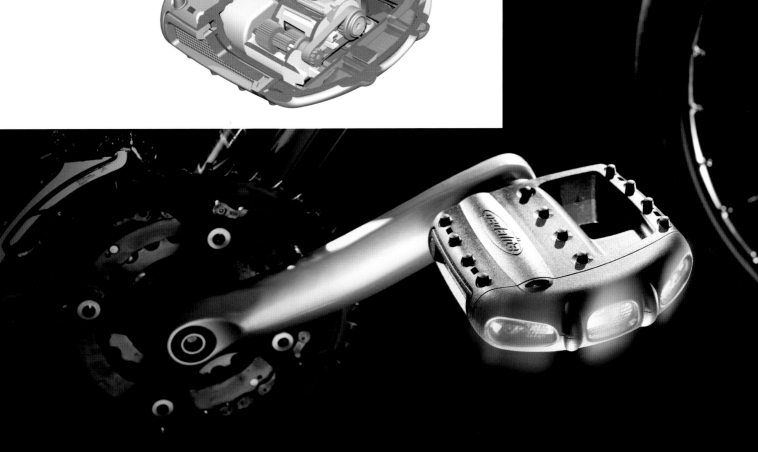

MOLDING THE PEDALITE

The raw material is a glass-filled nylon that is white in its raw state. If a colour is required, then pigment is added. In this case a small quantity of Clariant masterbatch yellow was used (**image 1**). The end result is a surprisingly vivid yellow colouring.

In normal operations, the injection molding process takes place behind a screen within the machine (**image 2**), but for the purposes of this case study the dies are shown in close-up with the screen open.

The polymer is melted and mixed before injection into the die cavity. Once the die cavity has been filled, packed and clamped, and the polymer has resolidified, the male and female halves of the mold move apart. The product is held in the moving tool by the upper and lower retractable cores and the 2 side-action cores (**image 3**). The injection point is indicated by the sprue, which has remained intact, to be removed either by hand or robotically. The 4 cores are retracted in sequence, to reveal the true complexity of this molding (**image 4**). Finally, the part is ejected from the mold by a series of ejector pins (**image 5**).

1

2

3

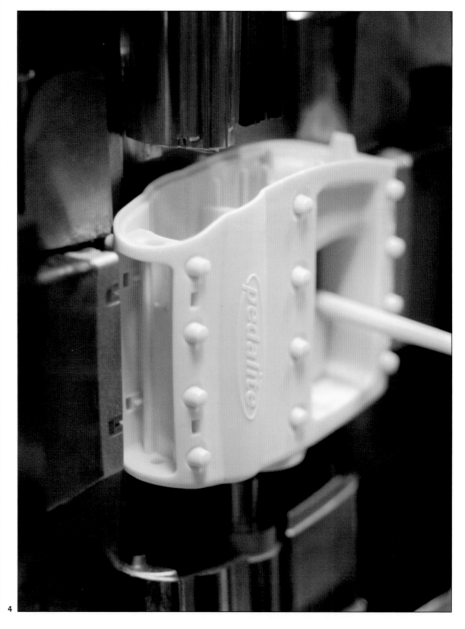

4

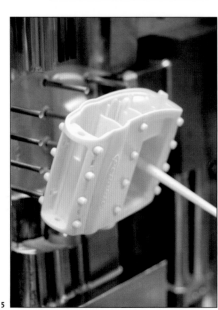

5

ASSEMBLING THE PEDALITE

There are many parts that make up the Pedalite (see image, page 54). All of the plastic parts are injection molded. The bearing locator is a friction fit, which requires more precise tolerance than can be achieved with injection molding. Therefore, it is drilled post-forming (**image 6**) and the bearing locator and bearing inserted. The overmolded end cap is fixed to the pedal housing with screw fixings (**image 7**). The reflectors snap fit into place (**image 8**), ensuring that all the components are held together securely. The snap fits can be released so that the Pedalite can be dismantled for maintenance and recycling (**image 9**). The finished product is installed by conventional means onto a bicycle (**image 10**).

Cycle pedal design is subject to considerable safety and technical restrictions. Pedalite eliminates the expense and inconvenience of battery replacement, as well as the negative environmental impact of battery disposal.

The 24/7 light output of Pedalite does not replace existing cycle safety lighting, but supplements it and also provides a unique light 'signature' (lights moving up and down) that helps motorists judge their distance from the cyclist.

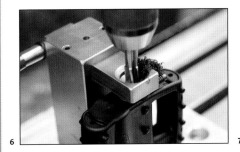

6

7

8

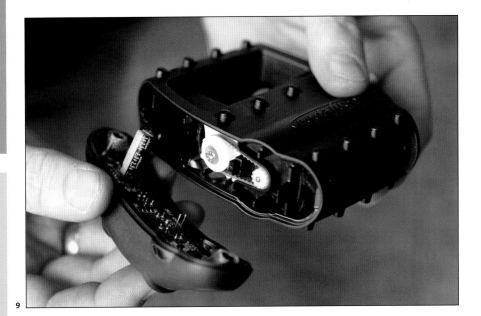

9

10

Featured Manufacturer

ENL
www.enl.co.uk

→ Moldflow analysis

Prior to manufacturing, Moldflow software is used to analyse and maximize the efficiency of a design. The software is suitable for all types of plastic injection molding and metal die casting. It brings together part design, material selection, mold design and processing conditions to determine the manufacturability of the part. This reduces the costs and time delays associated with otherwise unforeseen manufacturing problems. It also maximizes the efficiency of production and can reduce material consumption with significant savings.

A 3D model of the required part is generated in a suitable computer aided design (CAD) or computer aided engineering (CAE) software package.

Moldflow is a predictive analysis tool used to simulate the 3D model in production to analyse filling, packing and cooling.

The examples below illustrate analysis of flow, warp, fibre orientation, cooling and stress.

MPI/FLOW

MPI/Flow simulates the filling and packing phases in the molding process, helping to predict the behaviour of the material as it flows through the die cavity. This is used to optimize the location of the gate, balance runner systems and predict potential problems. Different versions are used to simulate different plastic and metal molding techniques.

The MPI/Flow is here demonstrated on 3 products. Two stages of an Abtec part are simulated (**images 1** and **2**) to demonstrate confidence of fill, which is colour coded. MPI/Flow was used to simulate various gate positions and runner system configurations.

By changing the location of the gate on the automotive hubcap for PolyOne (**image 3**) it was possible to reduce stress and make sure there were no air traps in critical areas. The colour scale indicates bulk stress.

The colour scale on the automotive interior product manufactured by Resinex and Gaertner & Lang (**image 4**) indicates fill time in seconds. Appearance is very important, so the flow analysis software was used to eliminate weld lines and colour variation in critical areas. This was achieved by changing the location of the gate and temperature of the runner system.

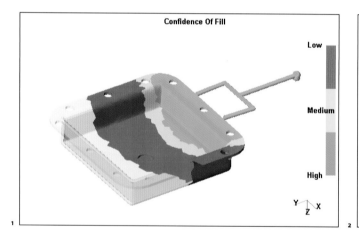

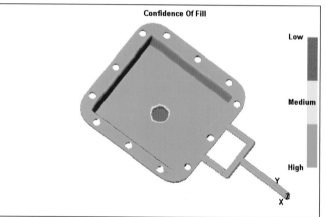

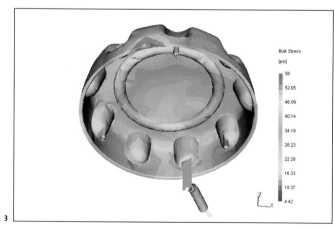

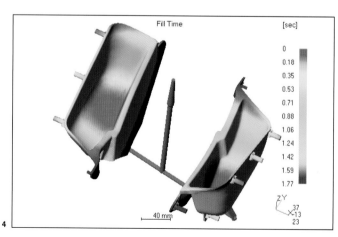

MPI/WARP

This analysis tool is used to predict shrinkage and warping, which are the result of stresses built up during the molding process. The information is used to specify material selection and processing parameters to minimize potential problems.

Any more than 5 mm (0.2 in.) warp was unacceptable on this Efen electronic switchboard cabin (**image 5**). The analysis found that by reducing wall thickness warp could be reduced by 90%.

On the CAD model for a Jokon automotive lamp assembly (**image 6**), it was essential that the part did not warp so that it would maintain a water-tight seal in application. Warpage was reduced by 50% by optimizing wall thickness (**images 7** and **8**).

MPI/COOL

MPI/Cool is used to analyse the design of mold cooling circuits. Uniform cooling is important to make sure that the part does not warp and to minimize cycle times.

A filter housing manufactured by Hozelock shows the configuration of the mold cooling circuits (**images 9** to **11**). Changing the layout of the circuit reduced cycle time by 2 seconds and so saved more than 4% of the production cost; reducing the wall thickness reduced cycle time by 7.3 seconds and shot weight by 19.6%, saving 24% of the production costs. Combining the 2 produced 26.1% overall savings.

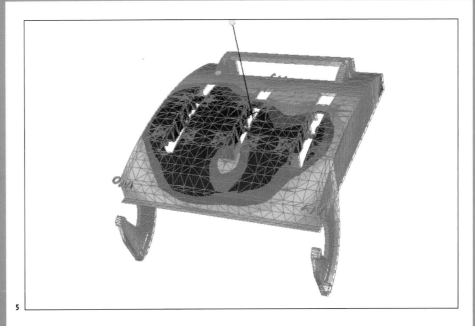

5

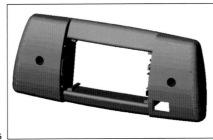

6

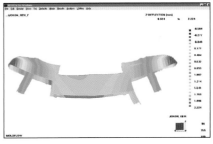

7

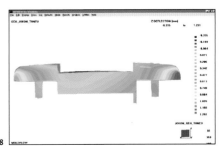

8

9

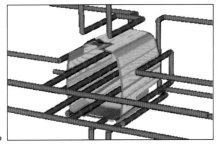

10

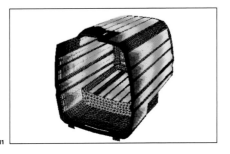

11

Featured Manufacturer

Moldflow
www.moldflow.com

TECHNICAL DESCRIPTION

Gas-assisted injection molding techniques were first used in mass production in 1985. Since then the technology has steadily improved and is now into the third generation of development. Initially it was developed to overcome the problem of sink marks caused by shrinking. A small amount of gas was blown in during the injection cycle to apply internal pressure as the polymer cooled before the tool opened. With very precise computer control, it is now possible to gas fill long and complex moldings. Each cycle will be slightly different because the computer makes adjustments for slight changes in material properties and flow.

The process uses modified injection molding equipment. In stage 1, plastic is injected into the mold cavity but does not completely fill it. In stage 2, gas is injected, which forms a bubble in the molten plastic and forces it into the extremities of the mold. The plastic and gas injection cycles overlap. This produces a more even wall thickness because as more plastic is injected the air pressure pushes it through the mold like a viscous bubble. The gas bubble maintains equal pressure even over long and narrow profiles. Wall thickness can be 3 mm (0.118 in.) or more.

In stage 3, as the plastic cools and solidifies, the gas pressure is maintained. This minimizes shrinkage. Less pressure is applied to the plastic because the gas assists its flow around the die cavity.

Gas-Assisted Injection Molding Process

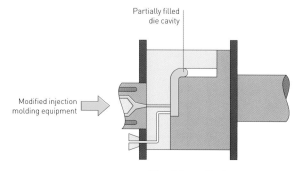

Partially filled
die cavity

Modified injection
molding equipment

Stage 1: Conventional
injection molding

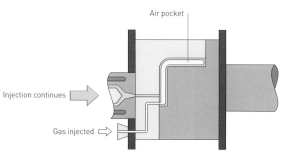

Air pocket

Injection continues

Gas injected

Stage 2: Gas injected

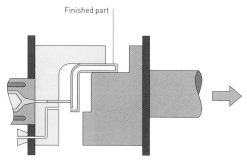

Finished part

Stage 3: Finished product

1

Gas injection molding the Magis Air Chair

The Air Chair was designed by Jasper Morrison and production began in 2000 (**image 1**). The gas injection molding sequence takes approximately 3 minutes (**images 2–5**).

The sample cut from the leg of the Air Table shows 2 technologies (**image 6**). The first is gas injection, which creates the hollow profile. The second technology is the thin skin around the outside of the material: there is a clear division between the outer unfilled PP and the glass filled structural internal PP.

Two materials are used because the outer skin is aesthetic and therefore should not be filled. However, unfilled PP is not strong enough to make the entire structure.

The 2 layers of material in this sample are produced in a similar way to gas injection. The outer skin is injected first. The glass filled PP is injected behind it in a technique known as 'packing out'. The second material acts like a bubble of air and pushes the first material further into the die cavity without

breaching it. Finally, the gas is injected to produce a hollow section and make it rigid but lightweight.

The gas injection molding technique produces a plastic chair with a very good surface finish. It weighs only 4.5 kg (9.92 lb) and is capable of withstanding heavy use.

2

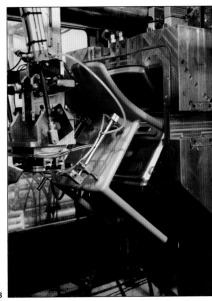

3

4

5

6

Featured Company

Magis
www.magisdesign.com

Multishot Injection Molding Process

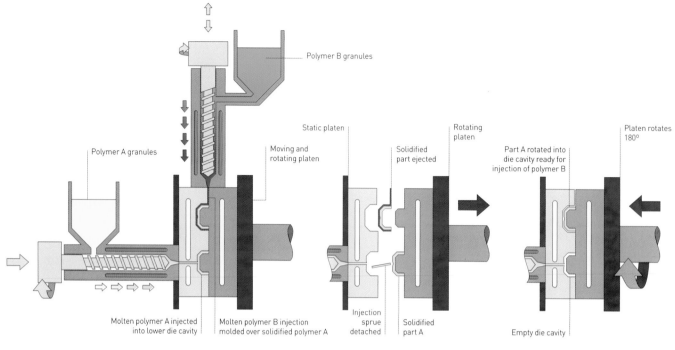

Polymer B granules

Static platen

Rotating platen

Platen rotates 180°

Polymer A granules

Moving and rotating platen

Solidified part ejected

Part A rotated into die cavity ready for injection of polymer B

Molten polymer A injected into lower die cavity

Molten polymer B injection molded over solidified polymer A

Injection sprue detached

Solidified part A

Empty die cavity

Stage 1: Injection

Stage 2: Ejection

Stage 3: Rotation

TECHNICAL DESCRIPTION

Injection molding 2 or more plastics together is known as multishot or overmolding. The difference is that multishot is carried out in the same tool. Overmolding is a term used to describe injection molding over any preformed material, including another thermoplastic, or metal insert, for example.

The process of multishot injection molding uses conventional injection molding machines. It is possible to multishot up to 6 different materials simultaneously, each one into a different die cavity in the same tool.

The tool is made up of 2 halves: one is mounted onto a static platen, the other onto a rotating platen. Like conventional injection, this process can have complex cores, inserts and other features.

In stage 1, polymers A and B are injected at the same time into different die cavities: polymer A is injected into the lower die cavity; meanwhile, polymer B is injected over a previously molded polymer A in the upper cavity. The molten polymers form a strong bond because they are fused together under pressure.

In stage 2, the molds separate and the sprue is removed from molded polymer A. Meanwhile the finished molding is ejected from the upper die cavity. The rotating platen spins to align molded polymer A with the upper die cavity. In stage 3, the mold closes again and the sequence of operations is repeated.

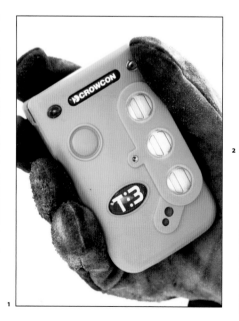

Multishot injection molding a handheld gas detector

This product is molded by Hymid for Crowcon. It is a handheld gas detection unit (**image 1**). Multishot injection molding has very important benefits that are essential for the effectiveness of this device, which is potentially lifesaving. The part is made up of a water clear polycarbonate (PC) body and thermoplastic electrometric (TPE) covering. The features of these materials are combined by multishot injection molding.

This is a tricky combination because the materials operate at different temperatures. Therefore, the runner system for the PC is heated with oil, whereas the TPE runner is cooled with water.

Of the 2 die cavities (**image 2**), the closest has just been injected with water clear PC. This gives the product rigidity, toughness and impact resistance. The farthest die cavity is the PC with a TPE

molded over it. The TPE provides integral features with hermetic seals, such as flexible buttons and a seal between the 2 halves.

The mold half with the moldings rotates through 180° (**images 3** and **4**). In doing so, it brings the solidified PC into alignment with the second injection cavity. Then the finished molding is ejected (**image 5**) ready for the next injection cycle to commence.

The knurled metal inserts (**image 6**) are incorporated in the PC by overmolding. These

are inserted into the mold by hand prior to each injection cycle.

The finished moldings are stacked (**image 7**) ready for assembly.. The integral and flexible button detail is shown in the final product (**image 8**).

3

4

5

6

Featured Manufacturer

Hymid Multi-Shot
www.hymid.co.uk

7

8

TECHNICAL DESCRIPTION

The in-mold decoration process is used to apply print to plastic products during injection molding, thus eliminating secondary operations such as printing and spraying. However, the cycle time of injection molding is increased slightly. The process is used in the production of nearly every modern mobile phone, camera and other small injection molded product.

In stage 1, a printed PC film is loaded into the die cavity prior to injection molding. The print side is placed inwards, so that when it is injection molded the print will be protected behind a thin film of PC.

In stage 2, when the hot plastic is injected in the die cavity, it bonds with the PC film. (This is similar to multishot injection molding.) In stage 3, the film becomes integral with the injection molded plastic and has a seamless finish, with a printed surface.

If the surface of the mold is not flat or slightly curved then the film is thermoformed to fit exactly. When the hot plastic is injected it is forced against the mold face at pressures between 30 and 17,000 N/cm² (20–11,720 psi). The pressure is determined by the type of material and surface finish.

Another technique, known as insert film molding, differs because the film is supplied as a continuous sheet. It is sucked into the die cavity by a strong vacuum (similar to the thermoforming process), the mold closes and the injection process follows.

Featured Company

Luceplan
www.Luceplan.com

In-Mold Decoration Process

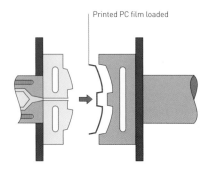

Printed PC film loaded

Stage 1: Film inserted

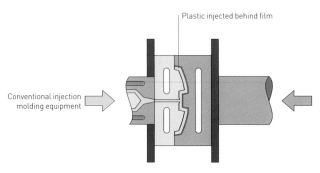

Plastic injected behind film

Conventional injection molding equipment

Stage 2: Conventional injection molding

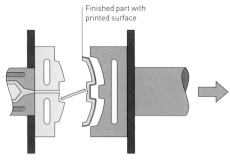

Finished part with printed surface

Stage 3: Finished product

1

2

3

→ In-mold decoration Luceplan Lightdisc

This case study demonstrates the production of the Luceplan Lightdisc (**image 1**). It was designed by Alberto Meda and Paolo Rizzatto in 2002. By incorporating in-mold decoration, the diffuser acts as shade, too. Graphics and instructions are included on the in-mold film and this eliminates all secondary printing.

The process is similar to conventional injection molding, except that a printed film is placed into the die cavity.

The film is prepared and placed into the mold by hand (**images 2** and **3**). The opposite side of the mold is textured to provide the diffusing effect in the light (**image 4**). The mold is clamped shut under 600 tonnes (661 US tons) of hydraulic pressure and the injection molding takes place (**image 5**). The part is demolded by hand (**image 6**), but this can also be carried out using a robot.

This finished molding is inspected prior to assembly (**image 7**). Screw bosses are incorporated into the injection molding, so the assembly procedure is relatively straightforward. The electrics are inserted and fixed in place (**image 8**). Then the 2 halves of the Lightdisc are screwed together (**image 9**). The fixings are covered with a snap fit enclosure (**image 10**), which means the whole product can be taken apart for maintenance and recycling.

4

5

6

7

8

9

10

Bend   Continuous   **Sheet**   Hollow   **Bulk**   Internal

Forming Technology

Reaction Injection Molding

Reaction injection molding (RIM) includes cold cure foam molding. Both processes are used to shape thermosetting foam by injecting thermosetting polyurethane resin (PUR) into a mold, where it reacts to form a foamed or solid part.

PLASTICS AND RUBBER

64

Costs	Typical Applications	Suitability
• Low to moderate tooling costs, depending on the size and complexity of molding	• Automotive • Furniture • Sporting goods and toys	• One-off to high volume production
Quality	**Related Processes**	**Speed**
• High quality moldings with good reproduction of detail	• CNC machining • Injection molding • Vacuum casting	• Rapid cycle time (5–15 minutes), depending on complexity of molding

INTRODUCTION

These are low pressure, cold cure processes. Cold cure foam molding is generally attributed to molding PUR foam for upholstery and sports equipment, whereas RIM covers molding all types of PUR including foam. In both processes, the density and structure of PUR are chosen to suit the application.

As well as low, medium and high volume production, this process is commonly used to prototype parts that will be injection molded (page 50) because the tooling is much less expensive than for injection molding, while repeatability and accuracy are high.

TYPICAL APPLICATIONS

Popular uses include domestic and commercial furniture such as chairs, car, train and aeroplane seats, armrests and cushions. These processes are also suitable for cushioning functions in footwear, such as in the soles, and for safety and tactility in toys.

RIM is chosen a great deal in the automotive industry for products such as bumpers, under-bonnet applications and car interiors. It is utilized in the medical and aerospace industries for niche products and low volume production runs, too.

RELATED PROCESSES

Bulk foam geometries are also formed by CNC machining (page 186) or foam fabrication. This technique is often utilized to cover wooden structures in upholstery (page 342). Foam molding is becoming more widely used, for example, in the production of furniture and car seats – new formulations of PUR foam producing fewer isocyanates, so they are therefore less toxic.

Vacuum casting (page 40) is used to form similar geometries in PUR, but is typically applied to smaller and more complex shapes. Vacuum casting and RIM are chosen to prototype and manufacture low volumes. The

properties of the part are similar to injection molding.

QUALITY

The quality of the surface finish is determined by the surface of the mold. The tooling can be produced in glass reinforced plastic (GRP), etched or alloyed steel. Even though this is a low pressure process, the liquid PUR reproduces fine surface textures and details very well.

DESIGN OPPORTUNITIES

This is an extremely versatile process as the mechanical properties of the cured PUR can be designed to suit the application. The flexibility of foam can range from semi-rigid to very rigid, and the density can be adjusted from 40 kg/m^3 to 400 kg/m^3 (2.5–25 lb/ft^3). The outer skin and inner foam can have contrasting properties to form parts with a rigid skin and lightweight foam core, for example.

RIM is similar to injection molding: parts can be textured and the surface

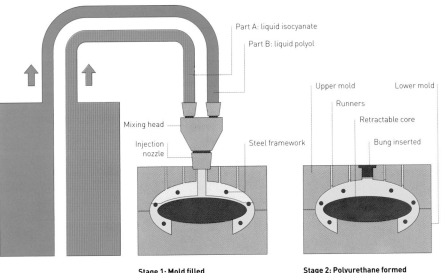

Cold Cure Foam Molding Process

Part A: liquid isocyanate
Part B: liquid polyol

Mixing head
Injection nozzle
Steel framework

Stage 1: Mold filled

Upper mold
Lower mold
Runners
Retractable core
Bung inserted

Stage 2: Polyurethane formed

Upper mold raised
Core retracted
Finished part demolded

Stage 3: Demolded and trimmed

TECHNICAL DESCRIPTION

In stage 1, the molds are cleaned and a release agent is applied. Inserts and frames are then put into place and the mold is clamped shut. The 2 ingredients that react to form PUR are stored in separate containers. The polyol and isocyanate are fed into the mixing head, where they are combined at high pressure. The predetermined quantities of liquid chemicals are dispensed into the mold at a low pressure. As they are mixed they begin to go through a chemical exothermic reaction to create PUR.

During stage 2 the polymer begins to expand to fill the mold. The only pressure on the mold is from the expanding liquid, so molds have to be designed and filled to ensure even spread of the polymer while it is still in its liquid state. As the polymer expands the runners allow trapped air to escape. A bung is inserted in the gate to maintain internal pressure in the mold.

In stage 3, the product is demolded after 5–15 minutes. Shot and cycle times vary according to the size and complexity of the part. The upper and lower sections of the mold are separated and the cores are removed. The mold is then cleaned and prepared for the next cycle.

Above
A predetermined measure of polyol and isocyanate is dispensed into a plastic bag to demonstrate the reaction process.

Above
The 2 liquids react and expand to form lightweight and flexible foam. The reaction is 1-way, so once the material has been formed it cannot be modified except by CNC machining.

1

2

3

4

5

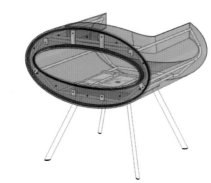

Above
The Eye chair is molded
in cold cure MDI
formulation of around
55 kg/m³ (3.4 lb/ft³).

This CAD visual shows
the supporting metal
framework, which is
overmolded with foam.
Plastic panels

are included in the
molding for upholstery.

printed with in-mold decoration.
Preformed materials are used to decorate
the surface of the part, and colour is
specified by Pantone reference.

RIM has many advantages over other
plastic molding processes. For example,
both thick and thin wall sections from
5 mm (0.2 in.) upwards can be molded
into the same part. Also, inserts such as
plywood laminates, plastic moldings,
threaded bushes and metal structures
can be molded into the part (see
image, left). Fibre reinforcement can be
incorporated into the plastic to improve
strength and rigidity of the product. This
is known as SRIM (structural injection

Case Study

→ Cold cure foam molding the Eye chair

The Eye chair has a parallel internal core, complex internal steel structure and plastic back plate for upholstery, so has a challenging geometry for foam molding.

The first step in the molding cycle is the mold preparation. A release agent is sprayed onto the internal surfaces of the mold (**image 1**). The steelwork is then loaded into the mold over the internal core and the whole assembly is closed and clamped shut (**image 2**). A predetermined quantity of polyol and isocyanate are mixed and injected into the mold through the gate at the top (**image 3**). After 12 minutes the chemical reaction is complete and the part can be demolded. The 2 halves of the mold are separated to reveal the foam product (**image 4**). The chair is removed from the internal core and checked for any defects (**image 5**). Excess flash is then removed with a rotating trimmer (**image 6**). This foam seat is part of the Eye chair (**image 7**), which is upholstered (page 342) by Boss Design.

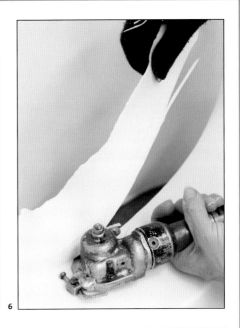

6

molding) or RRIM (reinforced reaction injection molding). Finally, fibre mats and other textiles can be molded in, for improved resistance to tearing and stretching.

The molds can be made in various materials, depending on the quantities and required surface finish. GRP is often used to produce molds for prototyping; it is relatively inexpensive and reduces lead times. For production runs above 1,000 units, aluminium or steel tools are used.

DESIGN CONSIDERATIONS

The size of part that can be molded ranges from minute to very large (up to 3 m/10 ft long). The polymer is very liquid in its non-catalysed state, so will easily flow around large and complex molds.

When foam molding over steel framework, the foam must be at least 10–15 mm (0.4–0.59 in.) larger, to ensure the structure is not visible on the surface of the foam.

COMPATIBLE MATERIALS

PUR is the most suitable material because it is available in a range of densities, colours and hardnesses. It can be very soft and flexible (shore A range 25–90) or rigid (shore D range). The cell structure of foam materials is either open or closed. Open-cell foams tend to be softer and are upholstered or covered. Closed-cell foams are self-skinning and used in applications such as armrests.

COSTS

Tooling costs are low to moderate. They are considerably less than when tooling for injection molding due to reduced pressure and temperature. GRP tooling is cheaper than aluminium and steel.

Cycle time is quite rapid (5–15 minutes). A typical mold will produce 50 components per day.

Labour costs are low to moderate and automated processes reduce labour costs significantly. Prototyping and low volume production require more labour input.

ENVIRONMENTAL IMPACTS

A predetermined quantity of PUR is mixed and injected in each cycle to ensure minimal waste. Flash has to be trimmed from the molded part. Reconstituted foam blocks are often incorporated into the molded part to reduce virgin material consumption.

The isocyanates that are off gassed during the reaction are harmful and known to cause asthma. The polyol/MDI system produces fewer isocyanates than the TDI method.

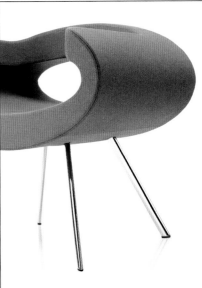

7

Featured Manufacturer

Interfoam
www.interfoam.co.uk

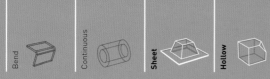
Forming Technology
Dip Molding

This low-cost method of producing thermoplastic products is used to produce hollow and sheet geometries in flexible and semi-rigid materials. As a coating method, this process can build up a thick, bright, insulating and protective layer on metal parts.

Costs	Typical Applications	Suitability
• Very low tooling costs • Low to moderate unit costs	• Caps and sleeves • Electrical insulation covers • Tool handgrips	• One-off to high volume production

Quality	Related Processes	Speed
• Gloss or matt finish • No flash or split lines	• Injection molding • Powder coating • Thermoforming	• Rapid cycle time (typically, manual 5–6 minutes, automated 1–2 minutes)

INTRODUCTION

The processes of dip molding and dip coating have been used for commercial applications since the 1940s. Many electrical, automotive and hand tool applications surround us in our daily lives. Even so, it remains an unfamiliar process to many.

Dip molding is versatile and low cost. Flexible materials can be formed into hollow profiles and severe re-entrant angles, and all materials can be formed into sheet profiles. It can easily be converted into a coating process by exchanging the release agent for a primer. Polyvinyl chloride (PVC) coatings are common for electrical applications, metal tools and handles.

It uses only a single male tool, which keeps costs to a minimum. As in thermoforming (page 30), the side that does not come into contact with the tool will be smooth and free of split lines, flash or marks.

TYPICAL APPLICATIONS

Dip moldings are used in a wide range of industries, including automotive, mining, marine, medical, aerospace, and promotion and marketing. Roughly 60% of dip molding is for electrical insulation covers, due to PVC's high electrical insulation properties.

Dip coatings can be found on tools and handle grips, playground equipment, outdoor furniture, wire racks (in fridges and dishwashers for example) and as electrical housing.

RELATED PROCESSES

Dip molding produces parts with similar geometries to thermoforming and

Dip Molding Process

Tool rack

Pre-heated
metal tool

PVC gels
over tool

Liquid PVC
plastisol

Tank

injection molding (page 50). The main difference between thermoforming and dip molding is that dip molding is used to produce flexible parts. Dip molding is less expensive for low volumes. A further advantage is that flexible materials can be molded into hollow profiles and parts with re-entrant angles.

Plastic coating methods include dip coating and powder coating (page 360). Dip coating is used in applications that demand comfort and flexibility, such as handle grips.

QUALITY

Dip molding and coating produce parts with a smooth and seamless finish. There is a single male tool and so there are no split lines, flash or other related imperfections. The outside surface (that does not come into contact with the tool) tends to be glossy, but can also be matt or foam-like.

The side that comes into contact with the tool is precise; embossed details and textures will be reproduced on the tool-side of the molding exactly. It is possible to revert the part after molding, so that textures are on the outside.

Because the material is liquid until it gels onto the tool, it can run and sag like paint. Therefore, the wall thickness at the bottom of the part is likely to be thicker than the top. To minimize this tools are inverted after dipping, which also reduces the formation of a drip at the base of the molding.

The speed of the dipping operation and temperature of the tool will affect the quality of the molded part. 'Creep' lines are caused by high tool temperature and slow dipping speed;

TECHNICAL DESCRIPTION

The dip molding process consists of pre-heating, dipping and baking. Automated and continuous production is rapid; these 3 stages overlap as more than 1 batch is processed simultaneously.

First of all the tool is cleaned and pre-heated. The oven is set to between 300°C and 400°C (572–752°F). The length of time pre-heating takes depends on the tool mass, but is typically between 5 and 20 minutes. The tools are commonly made of cast or machined aluminium, although steel and brass are also used.

The hot metal tool is coated with a dilute silicone solution for dip molding and with a primer for dip coating. The tool, which is now between 80°C and 110°C (176–230°F), is positioned above the tank of liquid plastisol PVC.

The tank rises up to submerge the tool to the fill line. On contact the plastisol gels to form polymerized PVC on the tool surface. The wall thickness rapidly builds up, reaching 2.5 mm (0.1 in.) within 60 seconds. The PVC polymerizes at 60°C (140°F), so as the tool cools and the wall

thickness builds up, polymerization slows down. Once the PVC has polymerized, it cannot be returned to a liquid and so cannot be recycled directly.

Dwell time in the liquid polymer is usually between 20 and 60 seconds. To increase wall thickness the tool is heated up for longer, or steel is used to maintain a higher temperature for longer.

The gelled part is removed from the plastisol and placed in an oven to solidify fully. Sometimes parts are inverted so that drips run back in, but this is not always necessary. The curing oven is set between 120°C and 240°C (248–464°F), depending on tool mass and wall thickness. The material remains hot and pliable, which makes it easier to remove from the tool, especially if it has severe re-entrant angles. It is prepared for removal by cooling in a bath of water, which brings it down to handling temperature. The parts are then removed with compressed air. The PVC fully age hardens in about 24 hours.

→ Dip molding a flexible bellow

The pre-heated aluminium tools are dipped in dilute silicone solution (**image 1**). The silicone promotes the flow of liquid over the surface of the tool and then acts as a release agent once it has solidified. The water steams from the surface of the tool to leave a very thin film of concentrated silicone.

The tools are mounted above the tank of room temperature liquid plastisol PVC (**image 2**). They are submerged steadily, ensuring that the viscous liquid does not fold over on itself and trap air (**image 3**). After 45 seconds the tank is lowered to reveal the dip molded parts (**image 4**), which are quickly inverted (**image 5**) and placed in an oven to fully cure.

They are removed from the oven and lowered into a bath of cold water (**image 6**). The water lowers the temperature of the tool enough for the operator to handle the parts.

Compressed air is blown in, which frees the plastic part from the metal and allows it to be removed (**image 7**).

The inside of the final part is matt (replicating the tool) and the outside is smooth and glossy (**image 8**).

on the other hand, air bubbles can form if the dipping speed is too fast. Prototyping is essential to calculate the optimum dip speed and tool temperature.

DESIGN OPPORTUNITIES

A major advantage is that tooling for this process is very low cost. There is no pressure applied to the tool and wear is minimal with flexible materials, so the same tooling can be used for prototyping and actual production. Multiple tools can be mounted together for simultaneous dipping to reduce cycle time considerably. Typically, the tooling is machined or cast aluminium, and it is a male tool, which is simpler to machine than a cavity.

Wall thickness is determined by 2 factors: the temperature of the tool and the dip time (dwell). Within reason, high temperatures and long dips produce thick wall sections. Wall thickness is generally between 1 mm and 5 mm (0.04–0.2 in.).

It is possible to dip twice to build up 2 layers of material. The advantages include dual colour and different shore hardness. As well as the obvious aesthetic advantages, twin dip can provide functional benefits, such as electrical insulation that wears to highlight material thinning.

There are many vivid colours available and parts can be produced with gloss, matt or foam-like finishes. PVC is also available in clear, metallic, fluorescent and translucent grades.

DESIGN CONSIDERATIONS

This process is only suitable for molding over a single tool, therefore the accuracy of the outside details is difficult to maintain. The PVC will smooth over features such as sharp corners. This will produce variable wall thicknesses, which can be a problem with protruding details because sufficient material may not build over them .

Tool design is affected by the nature of the liquid material. Plastisol PVC is viscous and gels on contact with hot metal. Therefore, flat surfaces, undercuts and holes become air traps if they are not designed carefully. Air traps will stop the material coming into contact with the hot metal tool, resulting in depressions and even holes where the material has not gelled sufficiently. In contrast, air traps might be deliberately designed into a tool to produce holes that would otherwise need to be punched as a secondary operation.

To avoid air traps, a draft angle of between 5° and 15° is recommended on faces that are parallel to the surface of

1

2

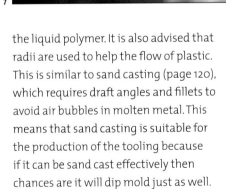

ENVIRONMENTAL IMPACTS

PVC is the most suitable material for the dip process and so makes up the majority of dip molded and coated products. The environmental credibility of PVC has been under investigation in recent years due to dioxins, harmful organic compounds that are given off during both the production and the incineration of the material.

Gelled material cannot be reused in dip molding because the reaction process is 1-way. It can be ground up and used in other applications.

the liquid polymer. It is also advised that radii are used to help the flow of plastic. This is similar to sand casting (page 120), which requires draft angles and fillets to avoid air bubbles in molten metal. This means that sand casting is suitable for the production of the tooling because if it can be sand cast effectively then chances are it will dip mold just as well.

COMPATIBLE MATERIALS

PVC is the most common material used for dip molding and coating. Other materials, including nylon, silicone, latex and urethane, are also used, but only for specialist applications.

There are many additives, which are used to improve the material's flame retardant qualities, chemical resistance, UV stability and temperature resistance, and to reduce its toxicity for food-approved grades. The level of plasticizer affects the hardness of PVC. It is available from shore hardness A 30 to 100; whereby 30 is very soft and 100 is semi-rigid.

COSTS

Tooling costs are minimal. Cycle time is rapid and multiple tools reduce cycle time dramatically.

Labour costs are moderate.

Featured Manufacturer

Cove Industries
www.cove-industries.co.uk

Bend　Continuous　Sheet　Hollow　Bulk　Internal

METAL

72

Forming Technology

Panel Beating

Smooth curves and undulating shapes in sheet metal can be produced with this sheet forming process. Combined with metal welding technologies, panel beating by a skilled operator is capable of producing almost any shape.

Costs	Typical Applications	Suitability
• Low to moderate tooling costs • Moderate to high unit costs	• Aerospace • Automotive • Furniture	• One-off to low volume production

Quality	Related Processes	Speed
• High quality handmade	• Deep drawing • Stamping • Superforming	• Long cycle time, dependent on the size and complexity of part

INTRODUCTION

Panel beating is controlled stretching and compressing of sheet metal. Many techniques are used, including press braking (page 148), dishing, crimping, wheel forming (English wheeling) and jig chasing (hammerforming). These processes in conjunction with arc welding (page 288) produce almost any profile in sheet metal. Panel beating is used in the automotive, aerospace and furniture industries for prototyping, preproduction and low volume production runs. It is used to produce the entire chassis and bodywork of cars.

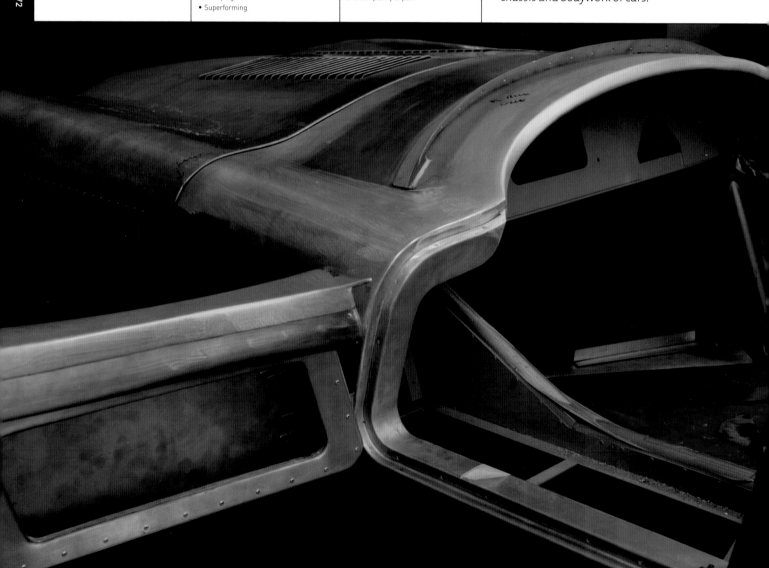

Manual panel beating is a highly skilled process. Coventry Prototype Panels operate a 5-year apprenticeship, which is required to learn all of the necessary skills. Wheeling and jig chasing are combined to form sheet materials into both smooth and sharp multi-directional curves, embosses, beads and flanges.

TYPICAL APPLICATIONS

Panel beating is used in prototyping, production and repair work for the automotive, furniture and aerospace industries. Examples of cars that are manufactured in this way include Spyker, Rolls Royce, Bentley, Austin Martin and Jaguar. Designers Ron Arad and Ross Lovegrove harness the opportunities of these techniques to produce seamless metal furniture, interiors and sculpture.

RELATED PROCESSES

Stamping (page 82), deep drawing (page 88) and superforming (page 92) are used to produce similar geometries. The difference is that panel beating is labour intensive and thus has higher unit costs. Stamping and deep drawing require matched tooling, which means very high investment costs but dramatically reduced unit costs and improved cycle time. Therefore, panel beating is usually reserved for production volumes of under 10 parts per year. Any more than this and it becomes more economical to invest in matched tooling or superforming.

QUALITY

Shaped metal profiles use the ductility and strength of metals to produce lightweight and high strength parts.

Above
Dishing into a sand bag is now largely confined to prototyping.

Right
This Austin Healey 3000 was given a completely new body by Coventry Prototype Panels.

Surfaces are planished and polished (page 388) and a skilled operator can achieve a superior 'A-class' finish. These techniques are used to finish stainless steel brightwork for Bentley production cars because the requirements of the surface finish are so high.

DESIGN OPPORTUNITIES

The most important benefit for designers is that almost any shape can be produced in metal by panel beating. Large and small radius curves are produced with similar ease by a skilled operator. Sheets can be embossed, beaded or flanged to improve their rigidity without increasing their weight.

Parts are not limited to the size of the sheet metal because multiple forms can be seamlessly welded together. In fact, most shapes are produced from multiple panels because they would be too impractical to make from a single piece of material.

A range of sheet materials can be formed, including stainless steel, aluminium and magnesium. Even though magnesium is more expensive than aluminium, it has superior strength to weight and is approximately one-third lighter.

For low to medium volume production, panel beating is used in combination with superforming. These processes complement each other because superforming produces sheet profiles with a high surface finish in a single operation. Panel beating is used to produce details such as fins, air intakes and beading that have re-entrant angles and are not suitable for superforming in a single operation.

DESIGN CONSIDERATIONS

A significant consideration is cost. A great deal of skill is required to produce accurate profiles with a high surface finish. This means high labour costs.

Panel Beating Process

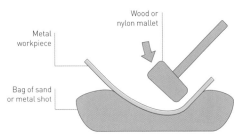

Dishing into a sandbag

Labels: Metal workpiece; Wood or nylon mallet; Bag of sand or metal shot

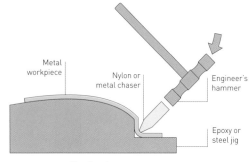

Jig chasing

Labels: Metal workpiece; Nylon or metal chaser; Engineer's hammer; Epoxy or steel jig

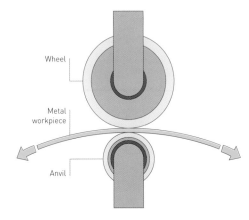

Wheel forming

Labels: Wheel; Metal workpiece; Anvil

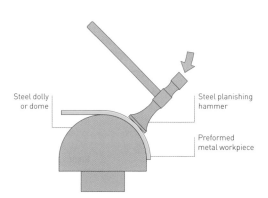

Planishing

Labels: Steel dolly or dome; Steel planishing hammer; Preformed metal workpiece

TECHNICAL DESCRIPTION

Panel beating is made up of different operations, such as dishing, jig chasing and wheel forming. Planishing is used to produce a smooth finish on panel beaten sheet metal.

Bags of sand or metal shot are still used for certain applications. They are useful in the forming of deep profiles such as motorcycle mudguards. Dishing is rapid, but it is the least accurate and controllable of the panel beating techniques. A wooden, leather or plastic mallet (generally shaped like a teardrop) is used to beat the metal into shape. With each blow the sand or metal shot displaces and conforms to the shape of the profile of the hammer, which allows the metal to be formed. It requires a great deal of skill to stretch and compress the metal accurately into the bag. Hammering into a shaped dolly or over a stake is more accurate but less versatile.

Jig chasing (also known as hammerforming) is the process of stretching and compressing (gathering) sheet metal to conform to the shape of a CNC machined tool. The tool is either 'soft' and made of

epoxy, or 'hard' and made of steel. Epoxy tools will typically only produce up to 10 parts or so before the edge details are too worn. Large circumference curves are typically formed by another method prior to jig chasing. For example, a sheet may be wheel formed to the general shape of the tool and then jig chased to form accurate sheet metal shapes. An engineer's hammer is used to beat a plastic or metal chaser against the surface of the metal. Plastic chasers are used for soft profiles and dish shapes, while metal chasers are used for tight bends, sharp angles and flat surfaces.

Wheel forming is also known as English Wheeling. It was developed and mastered by panel beaters in early automotive production in England. The metal workpiece is passed back and forth between a wheel and an anvil. The wheel is flat faced and the anvil has a profile (crown). The role of the anvil is to stretch the sheet progressively with each overlapping pass. Low-crown anvils produce a large radius curve and high-crown anvils produce a tighter bend.

Planishing is a finishing operation and is essentially smoothing over the surface with repeated and overlapping hammer blows. A flat-faced planishing hammer or slap hammer are used to hammer the surface gently against a dolly or dome. The process stretches the metal slightly but is not considered a forming operation. After a succession of taps with the hammer, the operator abrades the surface with a metal file to highlight remaining undulations. This process is repeated until the desired surface finish is required. After this the metal work is sanded and polished.

It also means that there are very few facilities that can carry out such work.

Soft tooling (epoxy) is only suitable for production of up to 10 parts. After this hard tooling in steel is required because the soft tooling will continuously have to be replaced. Material thickness is limited to 0.8 mm to 6 mm (0.024–0.236 in.) for aluminium and 0.8 mm to 3 mm (0.024–0.118 in.) for steel.

COMPATIBLE MATERIALS
Most ferrous and non-ferrous metals can be shaped in this way. Aluminium, magnesium and all types of steel are the most commonly used sheet materials.

COSTS
Tooling costs are low to moderate depending on the size and complexity. For one-offs and low volumes they are 5-axis CNC machined (page 186) from blocks of epoxy and can be used for up to 10 parts. For higher volumes tools are machined from steel, which is considerably more expensive but still a great deal less expensive than tooling for stamping or superforming.

Cycle time is long but depends on the size and complexity of the part. It is possible to construct the chassis and bodywork of a car from 3D CAD drawings in approximately 6 weeks.

This is a labour intensive process and the level of skill required is very high. Therefore, labour costs are high, but, combined with low tooling, costs are considerably cheaper than start-up costs for the related processes.

ENVIRONMENTAL IMPACTS
Panel beating is an efficient use of materials and energy. There is no scrap produced in the forming operations, although there may be scrap produced in the preparation (of the blank, for instance) and subsequent finishing operations.

Case Study

→ Panel beating the Spyker C8 Spyder

This case study illustrates the production of the Spyker C8 Spyder in aluminium (**image 1**). In 1914 Spyker cars merged with the Dutch Aircraft Factory to combine their skills in automotives, aircraft and aerodynamics. Since then they have been producing lightweight and high performance cars. They are manufactured in low volumes and so can be modified to bespoke customer requirements. It is even possible for customers to watch cars being built by the skilled operators via a dedicated webcam in the factory.

The chassis and bodywork are handcrafted aluminium. Sheet aluminium is cut to size (**image 2**). A crimping machine is used to compress (gather) the metal together in a selected area for increased curvature on the wheel former (**image 3**). It works like 2 pairs of pliers, gripping and forcing the sheet together in tandem.

1

2

3

The wheel former is made up of a flat-faced wheel and a low- or high-crowned anvil (**image 4**). The sheet is passed back and forth between the rolls in overlapping strokes (**image 5**). Each pass stretches the metal slightly and so forms a 2-directional bow in the sheet.

When the correct curvature is approximately achieved in the sheet, it is transferred onto the jig chaser. It is clamped onto the surface of the epoxy tool and gradually stretched and compressed to conform to the shape (**image 6**). The sheet is worked gently with a variety of chasers until it matches the shape of the epoxy tool precisely. Polyamide (PA) nylon chasers are used to stretch selected areas of the sheet into embossed details (**image 7**). Aluminium chasers are used to form flat areas (**image 8**).

The sheet is removed from the tool and holes are cut and filed (**image 9**). After this it is mounted onto the tool once again and reworked with an aluminium chaser (**image 10**).

The panels are shaped individually and then brought together on a jig. They are TIG welded (page 290) to form strong and seamless joints (**image 11**). Each weld is carefully levelled by grinding, planishing, filing and polishing. A slap hammer is used to planish the curved metal onto a dolly, which is held on the other side of the metal by the operator (**image 12**). It takes a great deal of skill and patience to achieve the desired finish. The finished panels are polished to a very smooth finish and spray painted. In the meantime, the brightwork is polished on conventional polishing wheels (**image 13**).

Assembly of the car engine, suspension and interior (**images 14** and **15**) is carried out by Karmann in Germany.

4

5

6

7

8

9

10

11

12

13

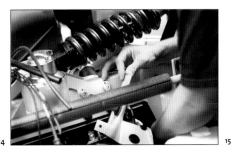

14

15

Featured Manufacturer

Coventry Prototype Panels
www.covproto.com

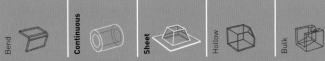

Forming Technology
Metal Spinning

Spinning is the process of forming rotationally symmetrical sheet metal profiles. It is carried out on a single-sided tool, as progressive tooling or – in a process known as spinning 'on-air' – without tooling at all.

Costs	Typical Applications	Suitability
• Low tooling costs • Moderate unit costs	• Automotive and aerospace • Jewelry • Lighting and furniture	• One-offs and low to medium volume production

Quality	Related Processes	Speed
• Variable: surface finish is largely dependent on the skill of the operator and the speed of the process	• Deep drawing • Metal stamping	• Moderate to rapid cycle time, depending on part size and complexity, and the type and thickness of material

METAL

78

INTRODUCTION

Although metal spinning is an industrial process, it has retained elements of craft. The combination of these qualities makes this a very satisfying process with which to work.

Metal spinning can be utilized to form sheet profiles including cylinders, cones and hemispheres. It is frequently combined with punching, fabrication or pressing to provide a wider range of design opportunities such as flanged edges, asymmetric profiles and perforated shapes.

Spinning Process

Lathe

Aluminium tool (mandrel)

Split tool for re-entrant

Metal blank

Stage 1: Load

Rolling wheel or tool

Work in progress

Final shape

Stage 2: Spin

End trimmed

Finished part removed

Tool comes apart

End trimmed

Stage 3: Finish

TECHNICAL DESCRIPTION

In stage 1 of the spinning process, a circular metal blank is loaded onto the tool (mandrel). In stage 2, the metal blank and mandrel are spun on the lathe and a rolling wheel forces the metal sheet onto the surface of the mandrel. This stage of the process is similar to throwing clay on a potter's wheel. The metal is gradually shaped and thinned as it is pressed onto the mandrel. In this case the finished part cannot be removed from the mandrel until it is trimmed. In stage 3, the mandrel is split so that a re-entrant angle can be shaped in the neck of the part. The top and bottom are trimmed and the part is demolded. The whole process takes less than a minute.

DESIGN OPPORTUNITIES

Metal spinning is an adaptable process. Tooling for manually operated low production runs is relatively inexpensive. This means that designers can realize their ideas with this process early on and make structural and aesthetic adjustments to the 3D form. Ribs, undulations and surface texture can be integrated into a single spinning operation. Steps or ribs can reduce wall thickness and so produce thinner and more cost-effective parts.

Textures can be integrated onto only 1 side of the part because a low profile texture on the mold will not be visible on the opposite surface. Metal spinning can also form meshes and perforated sheet materials.

The tool (mandrel) can be male or female depending on the part geometry. In both cases only 1 tool is required, unlike matched tooling used in deep drawing and stamping. Therefore, changes are relatively inexpensive and the tooling costs are considerably reduced.

DESIGN CONSIDERATIONS

Metal spinning is limited to rotationally symmetrical parts. The ideal shape for this process is a hemisphere, where the diameter is greater than or equal to twice the depth. Parts that have a greater

TYPICAL APPLICATIONS

This process is utilized in the furniture, lighting, kitchenware, automotive, aerospace and jewelry industries. Some typical products made in this way include anglepoise lampshades, lamp stands, flanged caps and covers, clock façades, bowls and dishes.

RELATED PROCESSES

Metal stamping (page 82) and deep drawing (page 88) are capable of producing similar profiles in sheet metal. Metal spinning is often combined with pressing and metal fabrication to achieve more complex geometries, a wider range of shapes and technical features. Simple profiles can be extruded or fabricated and time scales, quantities and budget determine the appropriate choice.

QUALITY

A very high finish is achievable by skilled operators with manual and automated processes. The quality of the inside surface finish is determined by the mold and the outside surface finish is shaped by the tool. Manual and automated techniques are often combined for optimum quality. There is no tooling for spinning on-air and the inside surface finish is therefore not affected.

Case Study

→ **Spinning the Grito lampshade**

A combination of automated and manual techniques is used to shape the Grito lampshade. The metal blank is cut and loaded onto the mandrel (**image 1**) and the CNC metal rolling wheel guides the blank over the surface of the mandrel in stages (**image 2**). With each pass the metal sheet is manipulated and controlled in much the same way as clay throwing (page 172). The metal can be drawn out and thinned, or simply pressed onto the surface of the spinning mandrel. It takes only 30 seconds to complete the first stage of the spinning process (**image 3**).

The parts are made in batches, which are stacked up in preparation for the next stage of spinning (**image 4**). The pre-spun parts are now loaded onto a manually operated spinning lathe and worked by hand (**image 5**). This requires a profiled tool (**image 6**), the shape of the tool being

determined by the design of the part. In this case manual metal spinning is required to achieve a re-entrant angle on the neck of the lampshade. This is made possible by splitting the mandrel inside the part so that each piece can be removed independently.

The shape is worked by the craftsman, who gradually improves the surface finish with a polished profiled metal tool (**image 7**). Once spinning is complete, the surface finish is improved further by polishing with an abrasive pad (**image 8**). The top and bottom are trimmed and the 2 parts of the mandrel removed. Post-spinning operations include punching in 2 stages (**image 9**) and painting. The parts are masked (**image 10**) and then sprayed (**image 11**). The finished part (**image 12**) is slipped over the cord and bulb and is supported by the undercut in the neck.

1

2

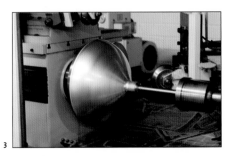

3

depth than diameter are achievable but will greatly increase costs. Parallel sides and re-entrant angles are also possible. However, this can entail multiple interlocking tools and will increase cost.

Work hardening in steels is another consideration for a designer. This factor can prove advantageous: for example, work hardening can increase the durability of a part. However, parts that need to be worked on after spinning have to be stress-relieved by heat treatment processes, which is a disadvantage.

Small to very large parts can be metal spun up to 2.5 m (8 ft) in diameter and tolerances down to 1.5 mm (0.059 in.).

COMPATIBLE MATERIALS
Mild steel, stainless steel, brass, copper, aluminium and titanium can all be formed by metal spinning.

COSTS
Tooling costs can be very low, especially for prototyping and low production

runs. Materials include wood, plastic, aluminium and steel. For large production runs the metal tooling costs are considerably cheaper than for metal stamping and deep drawing because metal spinning uses a single-sided tool. Spinning on-air needs no tooling, but the labour costs tend to be higher.

Cycle time is determined by the size and complexity of the part and the choice of material. Forming aluminium is far quicker than steel because of its ductility and malleability. Steel may also require heat treatment, which further increases cycle time.

Labour costs can be moderate to high for manual operations, especially if a very high quality, A-class aesthetic finish is required. Automation reduces labour costs considerably.

ENVIRONMENTAL IMPACTS
The metal blank is often supplied as a square sheet, so there is waste produced at the beginning of each cycle when the

4

sheet is cut into a circle. However, waste metal is readily recycled. In operation, a small amount of metal is trimmed from the part, in order to finish the edges.

Energy requirements for this process are quite low, especially if it is manually operated. The energy used is equivalent to turning a lathe.

5

6

7

8

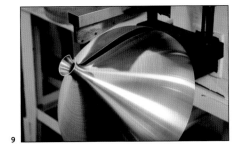

9

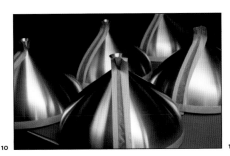

10

11

12

Featured Company
Mathmos
www.mathmos.co.uk

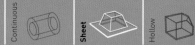

Forming Technology
Metal Stamping

This cold metal pressing technique is used to form shallow sheet and bend profiles in sheet metal. It is rapid and precise, and is used to produce a wide range of everyday products from car bodies to metal trays.

Costs	Typical Applications	Suitability
• High tooling costs • Low to moderate unit costs	• Automotive • Consumer products • Furniture	• High volume production

Quality	Related Processes	Speed
• High quality and precise bends as a result of matched tooling	• Deep drawing • Metal spinning • Press braking	• Rapid cycle time (under 1 second to 1 minute)

INTRODUCTION

Metal stamping is used to describe forming shallow metal profiles between matched steel tools. Tooling costs are high and so this process is only suitable for high volume production.

It is a rapid process used to form precise shapes without significantly reducing the material's thickness. It is known as deep drawing (page 88) when the relationship between depth and diameter mean that controlled drawing is required, which slightly reduces the material's thickness during forming.

Mass produced parts that require multiple forming and cutting operations are produced with progressive dies, a process made up of a series of dies working very rapidly and in tandem. While a part is being formed the second operation is being carried out simultaneously on the part that was formed previously and so on. Parts may require 5 operations or more, which is reflected in the number of workstations.

TYPICAL APPLICATIONS

Stamping is widely used. Most mass production metalwork in the automotive industry is stamped or pressed. Examples include bodywork, door linings and trim.

Metal camera bodies, mobile phones, television housing, appliances and MP3 players are formed by stamping. Kitchen and office equipment, tools and cutlery are made in this way. Not only are the external features stamped, but also much of the internal metalwork and structure is shaped in the same way.

RELATED PROCESSES

Low volume parts are manufactured by panel beating (page 72) metal spinning (page 78) or press braking (page 148). These can produce similar geometries to metal stamping, but they are labour intensive and require skilled operators.

Even though stamping and deep drawing are similar, there are distinct differences between them. Parts with a depth greater than half their diameter have to be drawn out and their wall thickness reduced. This has to be carried out gradually and more slowly to avoid overstretching and tearing the material.

Superforming (page 92) can produce large and deep parts in a single operation. However, it is limited to aluminium, magnesium and titanium because they are the only materials with suitable superplastic properties.

QUALITY

Shaped metal profiles combine the ductility and strength of metals in parts with improved rigidity and lightness.

If appearance is not critical, parts are simply deburred after forming. Polishing (page 388) is used to improved surface finish. Parts can also be powder coated (page 356), spray painted (page 350) or electroplated (page 364).

DESIGN OPPORTUNITIES

These are rapid and precise methods for forming shallow metal profiles from sheet materials. Round, square and polygonal shapes can be made with similar ease.

Ribs can be integrated into parts to reduce the required wall thickness. This has a knock on effect in reducing weight and cost.

Shapes with compound curves and complex undulating profiles can be formed between matched tools. The only other practical way of forming these profiles is by panel beating, which is labour intensive and a highly skilled process. Soft tooling can be used to progress from panel beating to metal stamping. This is where 1 side of the tooling is made of semi-rigid rubber, which applies enough pressure to form the metal blank over the punch.

DESIGN CONSIDERATIONS

Stamping is carried out on a vertical axis. Therefore, re-entrant angles have to be formed with a secondary pressing operation. Secondary operations include further press forming, cutting, deburring and edge rolling.

The first stamping operation can only reduce the diameter of the blank by up to 30%. Subsequent operations can reduce the diameter by 20%. This indicates the number or operations (workstations) required in the production of a part.

As in deep drawing, the limits of this process are often determined by the capabilities of the machine; bed size determines the size of blank and stroke determines the length of draw achievable. Speed is determined by stroke height and the complexity of the part.

The thickness of stainless steel that can be stamped is generally between

Metal Stamping Process

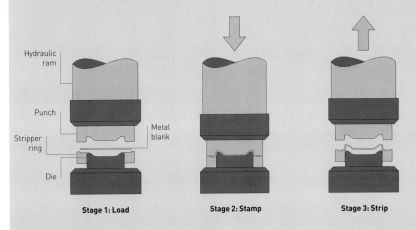

Stage 1: Load   Stage 2: Stamp   Stage 3: Strip

TECHNICAL DESCRIPTION

Metal stamping is carried out on a punch press. The power is transferred to the punch by either hydraulic rams or mechanical means such as a flypress. Hydraulic rams are generally preferred because they supply even pressure throughout the stamping cycle. Even so, flypresses still have their place in the metalworking industry.

The punch and die (matched tooling) are dedicated and generally carry out a single operation such as forming or punching. In operation, the metal blank is loaded onto the stripper. The punch then clamps and forms the part in a single stroke.

After forming, the stripper rises up to eject the part, which is removed. Sometimes the part is formed in a continuous strip, out of which it then has to be punched. This is common in progressive die forming.

In progressive die forming the stamped metal workpiece would now be transferred to the next workstation. This is either carried out manually, or by a transfer yoke. Most systems are automated for very high-speed production. The next operation may be further pressing, punching, beading or another secondary operation.

0.4 mm and 2 mm (0.02–0.08 in.). It is possible to stamp thicker sheet, up to 6 mm (0.236 in.), but this will affect the shape that can be formed.

COMPATIBLE MATERIALS

Most sheet metals can be formed in this way including carbon steel, stainless steel, aluminium, magnesium, titanium, copper, brass and zinc.

COSTS

Tooling costs are high because tools have to be machined precisely from high-strength tool steel. Semi-rigid rubber tooling is less expensive, but still requires

a single-sided steel tool and is only capable of low volume production.

Cycle time is rapid and ranges from 1 to over 100 parts per minute. Changeover and setup can be time consuming.

Labour costs are generally low because this process is mechanized and often fully automated. Finishing products by polishing will increase labour costs considerably.

ENVIRONMENTAL IMPACTS

All scrap can be recycled. Metal stamping produces long lasting and durable items.

Case Study

→ # Blank preparation

There are 2 main processes used to cut out the blanks for metal pressing. The first is punching. The tooling is expensive and so it is limited to parts that are produced in high volumes and are relatively simple shapes (**images 1** and **2**). This example is the Cactus! bowl, made by stamping and punching. It was designed by Marta Sansoni and production began in 2002.

The alternative is to laser cut the parts. This can produce very complex blanks and thus reduce secondary operations (**images 3** and **4**). The Mediterraneo bowl illustrated here was designed by Emma Silvestris in 2005.

1

2

3

4

Featured Manufacturer
Alessi
www.alessi.com

Case Study

→ # Metal stamping

Stamping is used across many industries and for a wide range of applications. Alessi is an example of a manufacturer who uses metal pressing (stamping and deep drawing) to produce parts of the highest standard.

The example that is used to demonstrate the stamping process is a serving plate, which is formed in stainless steel sheet. It was designed by Jasper Morrison in 2000 (**image 1**).

The metal blanks are laser cut in batches. They are coated with a thin film of oil just prior to stamping (**image 2**). This is essential to ensure that the metal will slide between the die and punch during operation.

The blank is loaded onto the tool and stamped (**images 3–6**). It is a rapid process and each part takes only a few seconds to complete. The part is removed from the stripper (**image 7**). It is transferred onto a spinning clamp, a rolling cutter is introduced from the side and the excess is trimmed (**image 8**). The formed parts are stacked up in preparation for secondary operations (**image 9**). In this case, a small bead of wire is rolled into the perimeter of the tray to give it rigidity. This helps to reduce the gauge of metal used.

1

2

3

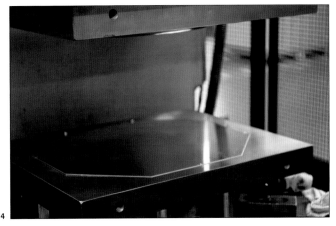

4

5

6

7

8

9

Featured Manufacturer

Alessi
www.alessi.com

TECHNICAL DESCRIPTION

Secondary pressing covers a range of processes, including further stamping, bending, rolling, beading or embossing. Side action tools are used to create re-entrant angles that are not possible in a single operation.

Exchanging the punch for a roller die and spinning the part makes it possible to apply a bead, rolled edge or other detail that runs continuously around the product.

Secondary Pressing Processes

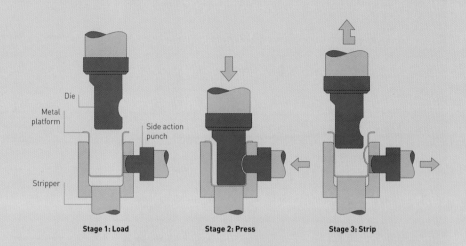

Die

Metal platform

Side action punch

Stripper

Stage 1: Load　　　**Stage 2: Press**　　　**Stage 3: Strip**

1

2

3

→ Secondary pressing operations

These images demonstrate some typical secondary pressing and deburring operations.

The first example is a product from the Kalistò family, designed by Clare Brass in 1992 (**image 1**). First of all the metal is deep drawn. Then it is placed onto separate tooling that applies sideways pressure (**image 2**). The parts are spun against the side of the tool, forming the side profile (**image 3**).

The second example is the Tralcio Muto tray (**image 4**), which was designed for Alessi by Marta Sansoni in 2000. It is punched on page 263. If this product was manufactured in very large volumes, these processes would be combined into adjacent workstations as progressive die forming.

The stamped plate is placed onto a spinning clamp (**image 5**). There is a sequence of 3 operations, which include cutting (see image 8, page 85), deburring (**image 6**) and then edge rolling (**image 7**).

The rolled edge increases the stiffness of the part and means thinner gauge steel can be used (**image 8**). It also improves the feel of the edge of the plate; without a roll it would feel very thin and would damage easily.

4

5

6

7

8

Featured Manufacturer

Alessi
www.alessi.com

Bend

Continuous

Sheet

Hollow

Bulk

Internal

Forming Technology

Deep Drawing

In this cold metal forming process, the part is made by a punch that forces a sheet metal blank into a closely matched die to produce sheet geometries. Very deep parts can be formed using progressive dies.

Costs	Typical Applications	Suitability
• High to very high tooling cost • Moderate unit cost	• Automotive and aerospace • Food and beverage packaging • Furniture and lighting	• Medium to high volume production

Quality	Related Processes	Speed
• Good surface finish	• Metal spinning • Metal stamping • Superforming	• Rapid cycle time (a few seconds to several minutes), depending on the number of operations

INTRODUCTION

Cold metal pressing is known as 'deep drawing' when the depth of the draw is greater then the diameter (sometimes when the depth is only 0.5 times greater than the diameter). This process can be used to produce seamless sheet geometries without the need for any further forming or joining operations.

There is a limit to how much sheet metal can be deformed in 1 operation, and the type of material and sheet thickness determine the level of deformation. A variety of techniques are therefore used to produce different

geometries. Simple cup-like geometries can be produced in a single operation, while very deep parts and complex geometries are made using progressive dies or the reverse drawing technique. Reverse drawing presses the sheet material twice in a single operation, inverting the shape after the first draw. Operating in this way accelerates cycle time and reduces the number of progressive tools required.

TYPICAL APPLICATIONS

The most common products manufactured by deep drawing include beverage cans and kitchen sinks. However, these techniques are also used to produce a variety of items in the automotive, aerospace, packaging, furniture and lighting industries.

RELATED PROCESSES

Shallow profiles are formed by metal stamping (page 82). Metal spinning (page 78), sheet ring rolling (see tube and section bending, page 98) and superforming (page 92) can be used to make similar geometries in sheet metal. Deep drawing and metal spinning produce seamless parts that typically do need to be welded post-forming.

QUALITY

Surface finish is generally very good, but depends on the quality of the punch and die. Wrinkling and surface issues usually occur around the edge, which is trimmed post-forming.

DESIGN OPPORTUNITIES

Various sheet geometries can be produced with the deep drawing, including cylindrical, box-shaped and irregular profiles, which can be formed with straight, tapered or curved sides.

Undercuts can be achieved with progressive dies or perpendicular action in the drawing press. However, this will greatly increase the tooling costs.

DESIGN CONSIDERATIONS

Depending on the type of material and thickness, parts ranging from less than 5 mm to 500 mm (0.2–19.69 in.) in diameter can be formed by deep drawing. The length of draw can be up to 5 times the diameter of the part. Longer profiles require thicker materials because material thickness is reduced in long draws.

The limits of deep drawing are often determined by the capabilities of the machine such as bed size (controls the size of blank), stroke (determines the length of draw achievable) and speed

(which is restricted by stroke height and complexity of part).

COMPATIBLE MATERIALS

Deep drawing relies on a combination of a metal's malleability and resistance to thinning. The most suitable materials are steels, zinc, copper and aluminium alloys.

TECHNICAL DESCRIPTION
The deep drawing process is carried out in different ways – the method of process being determined by the complexity of the shape, depth of draw, material and thickness. In stage 1, a sheet metal blank is loaded into the hydraulic press and clamped into the blank holder. In stage 2, as the blank holder progresses downwards the material flows over the sides of the lower die to form a symmetrical cup shape. In stage 3, the punch forces the material through the lower die in the opposite direction. The metal flows over the edge of the lower die to take the shape of the punch. In stage 4, the part is ejected.

The tonnage of the press is determined by the tooling. Anything up to 1,000 tonnes may be applied to shape a long or large profile.

Deep Drawing Process

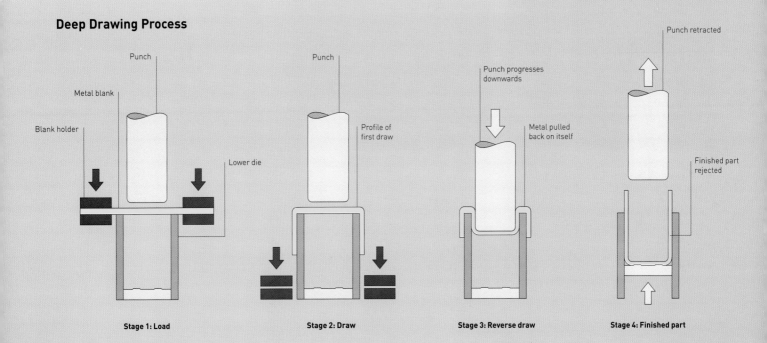

Stage 1: Load

Stage 2: Draw

Stage 3: Reverse draw

Stage 4: Finished part

Metals with high resistance to thinning are less likely to tear, wrinkle or fracture during processing, so thinner sheet material can be used to start with.

COSTS

Tooling costs are very expensive because the punch and die have to be engineered to precise tolerances. Progressive tooling, required to produce complex or especially deep parts, increases costs considerably for this process.

Cycle time is quite rapid but depends on the number of stages in the pressing cycle, while labour costs are moderate due to the level of automation.

ENVIRONMENTAL IMPACTS

Scrap is produced when the sheet material is cut to size and the finished part trimmed. Fortunately, all scrap material can be recycled into new sheet metals or other metal products.

→ Deep drawing the Cribbio

Blanks are cut to suit each deep drawing application. In this case, the Cribbio is circular, so a circular blank is cut from a sheet of 0.8 mm (0.031 in.) carbon steel (**image 1**). The final part has a reduced wall thickness of 0.7 mm (0.028 in.) as a result of thinning during drawing. A fine layer of oil is then applied to both sides of the blank, for lubrication (**image 2**).

The blank is loaded into the blank holder (**image 3**) on a 500 tonne press, which progresses downwards, forcing the sheet metal to flow over the lower die (**image 4**). The punch simultaneously forces the material inside the lower die (turning it inside out). The first stage of this part's forming is complete and it is removed (**image 5**).

The drawn part is loaded onto the second of the progressive dies (**image 6**). A punch forces the material into the lower die, turning it inside out once again. This process of reverse deep drawing means that fewer tools are required to achieve the same length of draw. The drawn part is then removed (**image 7**). At this point the metal blank has been forced through 2 progressive deep drawing cycles, which both applied reverse draw. Even though the Cribbio is a complex part to deep draw, production remains as high as 50 parts per hour. The top edge is then trimmed to remove any wrinkles and tearing that may have occurred to the perimeter of the metal blank during clamping and drawing (**image 8**) and so produce a clean edge detail. The parts are transferred onto a punch that perforates the surface (**image 9**).

Side actions are extremely expensive to incorporate into the deep drawing cycle, so are often carried out post-forming. After perforation a ring of pressed metal is crimped over the top edge to create a safe and ergonomic trim (**images 10** and **11**). The steel Cribbio is finished with a hardwearing epoxy coating (**image 12**).

9

10

11

12

Featured Manufacturer

Rexite
www.rexite.it

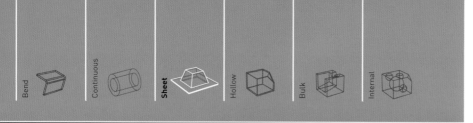
Forming Technology

Superforming

This recently developed hot forming process is used to produce sheet metal parts following similar principles to thermoforming: a metal blank is heated to softening point and formed onto a single-sided tool using air pressure.

METAL

92

Costs	Typical Applications	Suitability
• Low to moderate tooling costs • Moderate to high unit costs	• Aerospace • Automotive • Furniture	• Low to medium volume production

Quality	Related Processes	Speed
• Very good surface finish	• Deep drawing • Metal stamping • Thermoforming	• Rapid cycle time (5–20 minutes) • Trimming and assembly operations increase overall processing time

INTRODUCTION

Superform aluminium developed the technique of superforming aluminium alloys and more recently magnesium alloys. The process was developed to reduce the weight of metal components by minimizing fabrication operations and required wall thickness.

Superforming is a hot metal forming process that uses similar principles to thermoforming plastics (page 30). A sheet of aluminium is heated to 450–500°C (840–932°F) and then forced onto a single surface male or female tool using air pressure. There are 4 main types of superforming: cavity, bubble, backpressure and diaphragm. Each of these techniques has been developed to fulfil specific application requirements.

Cavity forming is good for large and complex parts such as automotive body panels and is excellent for shaping 5083 aluminium alloy.

Bubble forming is suitable for deep complex components, especially where wall thickness needs to remain relatively constant. This process can be used to manufacture geometries that are impossible to achieve using any other forming process.

The backpressure forming process was developed to produce structural aircraft components in 7475 alloys. Although similar to cavity forming, the process differs by using air pressure from both sides of the sheet. It is gradually pulled onto the surface of the tool using slight pressure differential. This maintains the integrity of the sheet and means that 'difficult' alloys can be formed.

Diaphragm forming is used to shape complex sheet geometries in non-superplastic alloys such as 2014, 2024, 2219 and 6061, making the process ideal for producing structural components.

Finite element analysis (FEA) flow simulation software helps to reduce the time needed to get from CAD design to superformed product.

TYPICAL APPLICATIONS
The use of these processes to create complex sheet geometries from a single piece of material has been rapidly growing in many applications, including aerospace, automotive, buildings, trains, electronics, furniture and sculpture.

RELATED PROCESSES
Metal stamping (page 82) and deep drawing (page 88) are used to make similar sheet metal geometries, while thermoforming (page 30) and composite laminating (page 206) produce similar geometries in thermoplastics and glass reinforced plastic (GRP) composites.

QUALITY
The surface finish on the tool and the accuracy of any post-forming operations affect the quality of a superformed part. Like thermoforming, the side of the sheet that does not come into contact with the mold will have the highest quality finish.

Typically, the aluminium alloys exhibit good corrosion resistance, mechanical strength and surface finish. The alloys that are suitable for superforming have differing characteristics, which make them useful for a variety of applications.

DESIGN OPPORTUNITIES
Like other aluminium parts, superformed components can undergo a range of post-forming operations to achieve the final desired part or assembly.

The diaphragm forming process can be used to form parts that are generally classified as 'non-superplastic'. This allows alloys used for aircraft structures, such as 2014, 2024, 2219 and 6061, to be superformed. In many cases diaphragm forming is the only practical way of

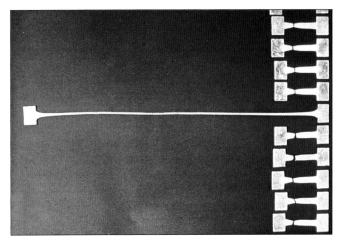

successfully shaping such materials, and aerospace designers are beginning to see the benefits that the superforming of such alloys brings to the design and manufacture process. For example, they eliminate costly fabrication work by forming large parts from a single sheet, thereby enhancing structural integrity while reducing costs and improving repeatability.

Another advantage of the superforming process is the range of aluminium alloys that can be formed, providing solutions to virtually any engineering problem.

The alloy 5083 contains aluminium, magnesium and manganese. It is used for applications requiring a weldable moderate strength alloy having good corrosion resistance. As such it is an excellent all-round alloy and ideal for many applications. Superformed components using 5083 alloy sheet are supplied in the O temper and offer many advantages over parts fabricated from 1200 or 3000 series and other 5000 series alloys. Simple or complex 3D sheet geometries can be manufactured as a single piece forming with high quality surface finish, making 5083 particularly suitable for automotive, rail, architectural and marine applications. This alloy allows a good compromise between formability and corrosion resistance, combined with moderate strength. Typical applications of 5083 include transportation and construction.

Superforming of 5083 is performed using the cavity or bubble forming technique.

A heat treatable alloy processed to give excellent superplastic forming properties – with component strains in excess of 200% – is 2004, which allows complex detail to be achieved. Typical uses include electronic enclosures, aerospace components and smaller complex form components. Unprotected 2004 has similar corrosion resistance to other copper containing aluminium alloys. In most service situations it is necessary to have some form of surface protection. Cladding with pure aluminium is commonly used to enhance the corrosion resistance of these alloys with Clad 2004 being designed to give enhanced corrosion resistance.

The alloy 7475 contains aluminium, zinc, magnesium and copper. It is suitable for applications requiring the high strength of 7075 and increased fracture toughness. The sheet's strength is approximately the same as that of 7075 combined with toughness similar to 2024-T3 at room temperature. Its high strength to weight ratio allows its extensive use within the aerospace industry for structural components. Resistance to stress corrosion cracking and exfoliation are similar to that of 7075. The T76 type temper provides for improved exfoliation resistance over T6 type temper, with some decrease in strength. Stress corrosion cracking in 7475 T76 is not anticipated if the total

Superforming Process

Cavity forming

Stage 1: Preheated sheet loaded

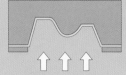

Stage 2: Pressure applied

Bubble forming

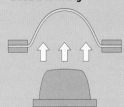

Stage 1: Preheated sheet loaded and blown

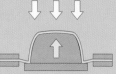

Stage 2: Vacuum applied

Backpressure forming

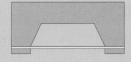

Stage 1: Preheated sheet loaded

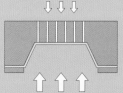

Stage 2: Pressure applied

Diaphragm forming

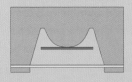

Stage 1: Preheated sheet loaded

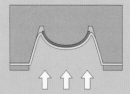

Stage 2: Pressure applied

TECHNICAL DESCRIPTION

In all 4 superforming processes a sheet of metal is loaded into the machine, clamped in place and heated to between 450°C and 500°C (840–932°F). The temperature is determined by the type and thickness of sheet material.

In cavity forming, the hot metal sheet is forced onto the inside surface of the tool by air pressure at 1–30 bar (14.5–435 psi). The hot metal is superplastic and so forms easily over complex and intricate shapes. This process is generally used to manufacture large shallow parts.

Bubble forming is similar to thermoforming. In stage 1, the hot metal sheet is blown into a bubble and the tool rises into the mold chamber. In stage 2, the pressure is then reversed and the bubble of metal is forced onto the outside surface of the tool. This process is ideal for deep and complex parts. The wall thickness is uniform because the bubble process stretches the material evenly prior to forming.

Backpressure forming is very similar to cavity forming. The difference is that in backpressure forming air pressure is also applied to the reverse side of the hot metal sheet as it is being superformed. In this way the forming process is more controlled and reduces the stress on the hot metal sheet.

Diaphragm forming was developed to superform so-called 'non-superplastic' alloys. It can achieve this because the metallic sheet diaphragm supports the hot metal sheet and aids the flow of material into complex 3D profiles, by

sustained tensile stress is less than 25% of the minimum specified yield strength.

DESIGN CONSIDERATIONS

The maximum size of part that can be formed is different for each superforming technique. In each case, 1 of the limiting dimensions can often be exceeded depending on part geometry and alloy selection.

Cavity forming can be used to produce parts up to 3,000 x 2,000 x 600 mm (118 x 79 x 24 in.) and 10 mm (0.4 in.) thick, while bubble forming can be used to produce parts up to 950 x 650 x 300 mm (37 x 26 x 12 in.) and 6 mm (0.236 in.) thick. Backpressure forming has a maximum plan area of roughly 4,500 mm² (6.97 in.²). and diaphragm forming can be used to produce parts up to 2,800 x 1,600 x 600 mm (110 x 63 x 24 in.).

Different alloys have different mechanical and physical properties, and this must be taken into consideration during the design process. It may well be possible to form a complex shape in 1 alloy, but that alloy may not have the properties required for in-service use.

Primary structure applications for the aerospace industry require high strength alloys accompanied with good service properties such as fatigue toughness and stress corrosion resistance. These requirements are adequately met with the alloy 7475, which has been successfully used to form air intake lip skins and access door assemblies, for example. For less demanding applications the heat-treated version of 2004 has found many applications for secondary structures such as aerodynamic fairings and stiffeners.

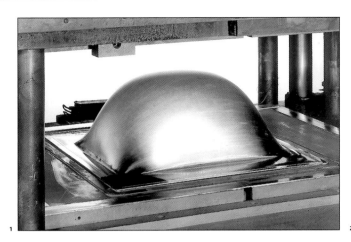

allowing unrestrained movement of the component sheet.

THE BUBBLE FORMING SEQUENCE
The bubble forming sequence shows how aluminium is superformed. The hot metal sheet is blown into a bubble in the molding chamber (image 1). The tool rises up into the metal bubble. **Only a virtual image of this stage of the process is shown in image 2 because the metal sheet is not usually translucent. As the tool rises up the hot metal is forced onto its surface by air pressure (image 3). The tool continues to rise and the metal is forced onto it until the forming process is complete and the tool** **can be retracted (image 4). The part is now formed and ready for trimming and any other post-forming operations.**

In the formed condition the alloy has suitable mechanical properties for internal fittings such as kicking panels and light fittings.

COMPATIBLE MATERIALS
Superplastic metals that can be shaped in this manner include aluminium, magnesium and titanium alloys. The most commonly formed aluminium sheet materials include 5083, 2004 and 7475.

COSTS
Although tooling costs can be quite low compared to matched die tooling, they do depend on the size and complexity of the part. Cycle time is rapid, typically 5–20 minutes.

Labour costs are moderate. Each part is trimmed and cleaned post-forming.

ENVIRONMENTAL IMPACTS
Scrap and offcuts are recycled to produce new sheets of aluminium and other aluminium products.

1

2

3

4

5

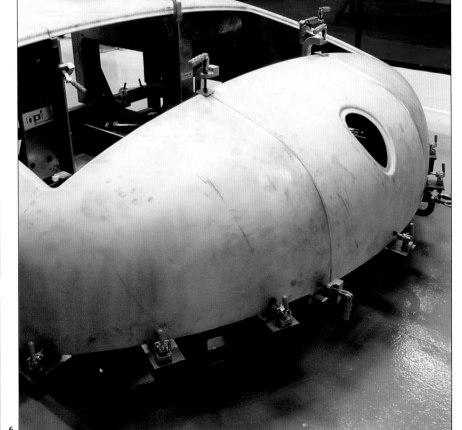

6

Featured Manufacturer

Superform Aluminium
www.superform-aluminium.com

Superforming the Siemens Desiro train façade

The aluminium sheet is loaded into the superforming mold (**image 1**). The clamps are brought down onto the perimeter of the sheet (**image 2**) and the temperature increased to 450°C (840°F). The superforming cycle takes approximately 50 minutes, after which the part is unloaded (**image 3**). The demolded part is loaded onto a support structure and CNC trimmed and machined (**image 4**). The scale and complexity of the part can vary greatly before it is assembled (**image 5**).

The train front is made in 2 halves so it has to be tungsten inert gas (TIG) welded to form the complete front unit. The 2 halves are brought together in a specially designed jig (**image 6**) and welded (**image 7**). CAD renderings are available of the Siemens Desiro train façade, which the panels are destined to become (**image 8**).

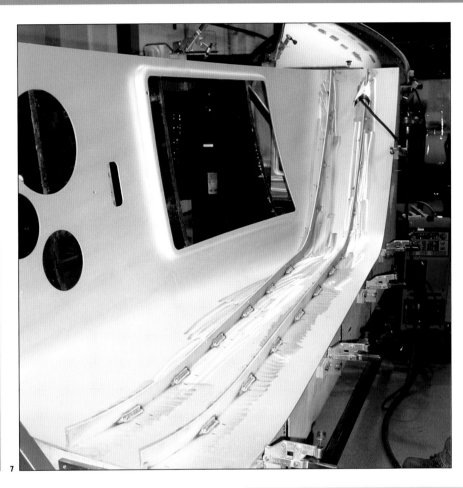

7

8

Forming Technology

Tube and Section Bending

Used mainly in the furniture, automotive and construction industries, this process is used to form continuous and fluid metal structures. Tight bends can be formed with a mandrel over a rotating die, or long and undulating curves between rollers.

Costs	Typical Applications	Suitability
• No cost for standard tooling • Moderate to high cost for specialized tooling • Low to moderate unit costs	• Construction • Furniture • Transport and automotive	• One-off to high volume production

Quality	Related Processes	Speed
• High	• Arc welding • Press braking • Swaging	• Rapid cycle time • Machine setup time can be long

INTRODUCTION

Architects and designers have long utilized the functional and aesthetic properties of bent metal, especially tubular steel. Bending methods are generally inexpensive and make use of the ductility and strength of metal. Bending minimizes cutting and joining in certain applications, which reduces waste and cost.

There are 2 main types of tube and section bending, which are mandrel bending and ring rolling (see images, opposite, above).

Mandrel bending is specifically designed for making small radii in metal tube. It is named after the mandrel that is inserted into the metal tube to prevent it from collapsing during bending. Ring rolling is used to form continuous and generally larger bends in both tube and section (extruded profile or bar). It is also commonly known as section bending.

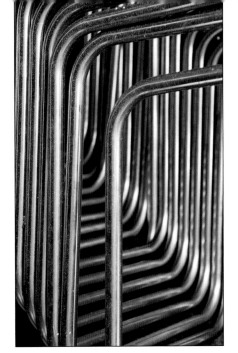

Steam bent (page 198) furniture was the precursor to tubular steel and uses similar bending techniques. Tubular steel was developed in Germany for the automotive and aerospace industries, and Thonet saw its potential. In the 1920s they began developing tubular steel furniture with members of the Bauhaus, including Mart Stam, Marcel Breuer and Mies van der Rohe. Many of the designs from this time are still in production.

TYPICAL APPLICATIONS

Metalwork is either bent as a continuous length, or fabricated using cast or pre-bent 'elbows' to join straight sections. Many products are formed by mandrel bending, including domestic and office furniture, security fencing and automotive applications (such as exhaust pipes).

Ring rolling can be used to bend a range of profiles (tube, extrusions and bar) into large-radius bends. Many products are formed using this technique such as structural beams (for bridges and buildings), architectural trims (on curved façades) and street furniture. In fact, most bent metalwork in construction will be processed using ring rolling. Rolling sheet materials is known as plate rolling.

It is even possible to produce 3D section bends such as the tracks on a rollercoaster, which can be achieved with careful manipulation on a ring roller.

RELATED PROCESSES

Tubular metalwork consists of mainly bending, press braking (page 148) and arc welding (page 282). Press braking is capable of producing tapered hollow profiles with multiple bends on a suitably shaped punch. An example of this is hexagonal lampposts that are tapered along their length.

Ring rolling can produce similar profiles to roll forging (page 114). The difference is that roll forged components

Above, left
The legs of tables and chairs are commonly produced by mandrel bending.

Above
Ring rolling can shape tube, extruded profile and metal sheets. The arc can be adjusted along its length to give non-circular bends.

Mandrel Bending Process

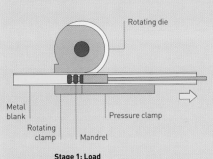

Rotating die

Metal blank

Rotating clamp

Mandrel

Pressure clamp

Stage 1: Load

Stage 2: 90° bend

TECHNICAL DESCRIPTION

The metal blank (tube) is loaded over the mandrel and clamped onto the die. Non-mandrel forming is only possible for certain parts with thicker walls.

The blank is drawn onto the rotating die as it spins, and the mandrel stops the walls collapsing at the point of bending. The pressure clamp travels with the tube to maintain an accurate and wrinkle free bend. An additional clamp is sometimes required to prevent wrinkling on the inside of the bend, especially for very thin walled sections.

The size of the rotating die determines the radius of the bend. The distance travelled determines the angle of bend.

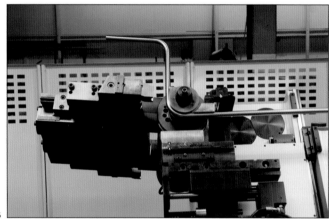

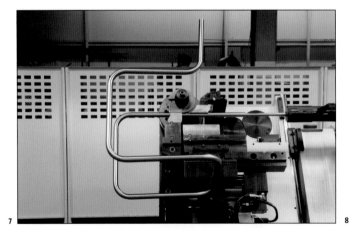

→ Mandrel bending the S43 chair

Mart Stam designed the S43 chair (**image 1**), which was introduced by Thonet in 1931. In 1926 it began as a chair made from straight sections of gas pipe connected with cast 'elbows'. The design was later refined and produced by Thonet from a single, continuous bent steel tube. It has since been made using the same technique; the only difference is that new technology is making the process faster and more accurate. The majority of mandrel bending is still performed on semi-automated machines because set up time is quicker and tooling is less expensive. However, it is much more cost effective in the long run to produce large volumes on fully automated machines.

The steel tube, or blank (**image 2**), is carefully cleaned and polished (**image 3**) prior to forming, because it is much more easily done at this stage. The blank is loaded by hand over the mandrel and into the pressure clamps (**image 4**).

The CNC machine aligns the tube and the bending sequence begins. The first bend is made at the start point (**image 5**). From there, the blank is extended and twisted to the second bend point and so on (**images 6, 7** and **8**). The operation is precisely controlled and there is no scrap; the bending process uses the entire length of tube. When the process is complete the bent structure is removed by hand and each piece is checked for accuracy (**image 9**) and then hung up (**image 10**).

The bent metal forms are metal plated prior to assembly (**image 11**). The wooden seat and back are laminated, CNC machined and lacquered in preparation (**images 12** and **13**). The final parts are brought together and assembled with rivets (**image 14**). The careful design, refinement and production of this chair make the process look easy. The final product is simple, lightweight and uses the minimum material (**image 15**).

12

13

14

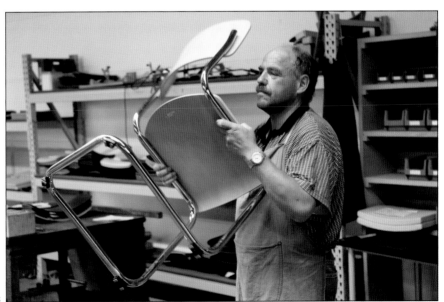
15

Ring Rolling Process

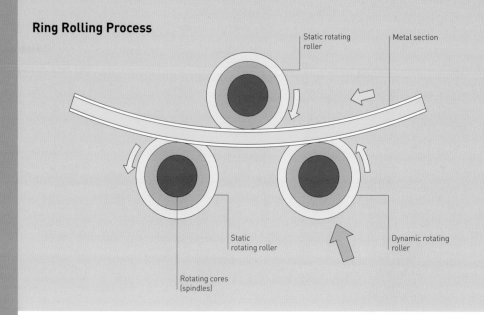

Static rotating roller

Metal section

Static rotating roller

Dynamic rotating roller

Rotating cores (spindles)

TECHNICAL DESCRIPTION
Ring rolling is much simpler than mandrel bending in operation. The tube, profile, or sheet is passed between 3 rollers. One of the rollers, in this case the bottom right, is moved inwards to make the bend tighter. The radius of the bend is decreased gradually over several cycles to avoid cracking or wrinkling the metal. The blank is rolled back and forth until the curve meets the required shape.

are solid and ring rolled parts are hollow or sheet. The thickest solid section that can be ring rolled is 80 mm (3.15 in.), whereas roll forging can produce solid parts of up to 150 tonnes (165 US tons). Roll forging is sometimes referred to as 'ring rolling', which can be confusing.

QUALITY

Applying a bend to a sheet of material increases its strength; these processes combine the ductility and strength of metals to produce parts with improved rigidity and lightness.

Left
Continuous circular metal profiles, such as these heating elements, are produced by ring rolling and then arc welding the seam.

The aesthetic quality of manual operations, such as mandrel bends, is largely dependent on the skill of the operator and their experience with the particular machine set up. The dies are often not exact, so an operator must know how to compensate for optimum bend angles. Automated and CNC operations are more precise.

DESIGN OPPORTUNITIES

Mandrel bending is more versatile than ring rolling for certain applications; a variety of bend radii can be applied to a single length of metal across any axis. However, it is limited to relatively small diameter tube and can only form bends up to 200 mm (7.87 in.) radius. It is suitable for one-offs, batch and mass production.

The advantage of ring rolling is that it is capable of producing a range of bends in almost any metal profile (although it must be remembered that specialized

tooling can be very expensive, especially for large section bending). It is possible to form either a specific section or the complete length; metal rings can be formed along their entire length and welded (see image, left). The radius does not have to be constant; this process can be used to produce arcs that do not fit into a circle. This technique is used predominantly in the construction industry for structural beams.

The 2 main functions of ring rolling are bending lengths or forming rings. Lengths of tube up to 1 m (3.3 ft) in diameter can be bent. Alternatively, metal sheet 4.5 m (14.8 ft) wide can be rolled to form a tube, or cone, in materials up to 80 mm thick (plate rolling). Lengths of metal, including profile, tube and sheet, can be rolled into rings, the maximum radius of which is determined by the capabilities of the manufacturer.

DESIGN CONSIDERATIONS

All of the bending processes stretch material on the outside of the bend and compress material on the inside of the bend. However, metal is more willing to stretch than compress. Therefore, a bent piece of metal will be slightly longer, especially the outside dimension. This means that the length of a piece of material on a drawing rarely corresponds

→ Ring rolling

In this process metal tube is cut to length (**image 1**) and fed into the rollers. The rollers are adjusted to gradually bend the pipe into the desired radius (**image 2**). As the rollers move closer together the radius of the bend will be decreased. These are parts of a larger structure, which do not require a tight radius (**image 3**). In this case the process is being used to bend tubular steel; a range of profiles and flat sections can be bent, but require rollers with fitting profiles.

to the length a piece of material once it has been bent.

In mandrel bending, the maximum size of tube is typically 80 mm (3.15 in.), but depends on the tooling available. Wall thickness ranges from 0.5 mm to 2 mm (0.002–0.079 in.).

Minimum bend radius is typically around 50 mm (2 in.). However, bends have to fit into the equipment; in other words, several bends in close proximity may not be feasible.

Mandrel bending is capable of producing tight bends because the internal mandrel prevents wrinkling and failure. Ring rolling does not use a mandrel and so is generally limited to bend radii above 200 mm (7.87 in.).

COMPATIBLE MATERIALS
Almost all metals can be formed in this way including steel, aluminium, copper and titanium. Ductile metals will bend more easily.

COSTS
Standard tooling is used to produce a wide range of bent geometries. Specialized tooling will increase the unit price considerably, but will depend on the size and complexity of the bend.

Cycle time is rapid in most operations.

Labour costs are high for manual operations, because a high level of skill and experience is required to produce accurate bends.

ENVIRONMENTAL IMPACTS
Bending, as opposed to cutting and welding, is generally less wasteful and a more efficient use of energy. There is no scrap produced in the bending operation, although there may be scrap produced in the preparation (of a blank for instance) and subsequent finishing operations.

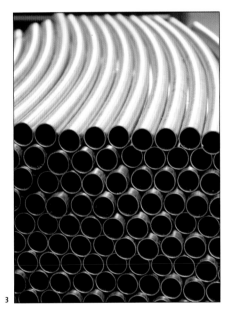

Featured Manufacturer
Pipecraft www.pipecraft.co.uk

Bend
Continuous
Sheet
Hollow
Bulk
Internal

Forming Technology

Swaging

There are 2 swaging techniques, hammering and pressing. These processes are used to reduce or expand metal pipe into tapers, joints or sealed ends.

METAL

104

Cost	Typical Applications	Suitability
• Low to moderate tooling costs • Low to moderate unit costs	• Probes and spikes • Rigging • Telescopic poles	• One-off to high volume production

Quality	Related Processes	Speed
• High precision	• Arc welding • Tube and section bending	• Rapid cycle time, but depends on complexity of operation

INTRODUCTION

Swaging is the manipulation of metal tube, rod or wire, in a die. It is typically used to reduce cross section by drawing out the material, but pressing techniques are also capable of expanding the diameter of pipe by stretching.

Hammering techniques are a rotary operation, known as rotary swaging and tube tapering. The ideal angle of taper is no more than 30°, but it is possible to swage up to 45° where necessary. Smaller angles are simpler, quicker and more cost effective to produce. Rotary swaging is capable of producing sealed ends in pipe.

Pressing is a hydraulic operation and is often referred to as 'end forming'. The pressing action is applied from multiple angles simultaneously and can be used to expand or reduce the diameter of pipe.

TYPICAL APPLICATIONS

Applications for swaging are diverse. It is used to form large parts such as cylindrical lampposts, connectors and joints for load bearing cables, and piping for gas and water. It is also widely employed for small and precise products such as ammunition casing, electrodes for welding torches and thermometer probes.

Swaging is used a great deal to form joints. There are 2 main types of joint, friction fit and formed. Friction fit joints can be taken apart any number of times, whereas formed joints tend to be permanent. Friction fit examples include tent poles, walking frames, crutches and other telescopic parts. Formed joint examples include swaging a metal lug or terminal onto cable for rigging (sometimes known as a talurit splice), electrical cables and throttle cables. Tubes can also be swaged together to form a permanent joint.

RELATED PROCESSES

Swaging is the most rapid and widely used method for tapering and end forming. Tapered metal profiles can also be produced by ring rolling plate (see tube and section bending, page 98) and arc welding (page 282) the seam. Hydroforming is a recently developed process that is capable of forming complex hollow parts from tube. However, it is currently much more

Rotary Swaging Process

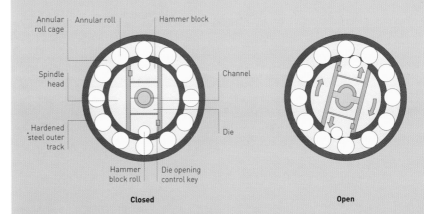

Annular roll cage — Annular roll — Hammer block — Spindle head — Channel — Hardened steel outer track — Die — Hammer block roll — Die opening control key

Closed **Open**

TECHNICAL DESCRIPTION

In rotary swaging, the metal is manipulated by a hammering action. This is generated by hammer blocks, which move in and out on the hammer block roll as the spindle head rotates rapidly.

The die opening, which is at its full extent when the spindle head rotates and the hammer block rolls fall between the annular rolls, determines the size of the tube that can be swaged. The dies are machined from tool steel (see image, below).

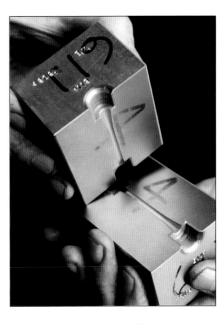

Right
Fresh dies are machined from tool steel for each application, because it is uncommon for 2 swaged products to be alike.

expensive and specialized. Applications include automotive chassis and high end bicycle frames.

QUALITY

Swaging compresses or expands metal during operation. Steel work hardens in these conditions, which improves its mechanical properties.

Components are typically produced to tolerances of 0.1 mm (0.004 in.). They can be made even more accurate, but this will increase cycle time considerably.

Ultimately, the quality is dependent on the skill of the operator. The surface finish is generally very good and can be improved with polishing (page 376).

DESIGN OPPORTUNITIES

It is possible to rotary swage tapers along the entire length of pipe. For example, if the maximum length of swage on a particular machine is 350 mm (13.78 in.), then longer tapers are made in 2 or more stages. End forming, on the other hand, is generally limited to short lengths of

→ Rotary swaging steel pipe

Rotary swaging is capable of forming not only open-ended tapers but also sealed ends in a length of pipe. This is especially useful for hollow probes and spikes.

In the first images, 25 mm (0.98 in.) diameter tube (**image 1**) is being formed into a 30° taper to a sealed end. The pipe is forced into the swaging die as it rotates at high speed (**image 2**). After 20 seconds or so, the pipe is retracted fully formed (**image 3**). The last 10 mm (0.4 in.) is likely to be solid metal.

Small diameter tube is rotary swaged using the same technique (**images 4 to 6**). In this case the finished part does not have a solid sealed end (**image 6**). The diameter of the hole at the end can be made precise by swaging the pipe over metal wire.

1

2

3

deformation due to the pressure that is required.

Precision parts, such as nozzles for MIG welding (page 282), are formed in a rotary swage die over an internal die. This ensures maximum precision on both the inside and outside measurements.

DESIGN CONSIDERATIONS
Metal is more willing to stretch than to compress, so swaging tends to draw out metal into longer lengths. Therefore thread tapping and hole making should be carried out as secondary operations.

The machinery determines the feasible diameter of metal blank. Both rotary swaging and end forming can accommodate diameters over 150 mm, (5.91 in.) such as lampposts. They can also form very fine wire, as small as 1 mm (0.04 in.) diameter.

Material thickness is also determined by the capabilities of the swager. The gauge of material affects the ability of the pipe to be formed. There is no rule of thumb as to whether thicker of thinner gauges are more suitable because both can be formed equally well. The benefit of thicker wall sections is that there is more material to manipulate, but it will take more hammering or pressing. Each design must be judged on its merits.

COMPATIBLE MATERIALS
Almost all metals can be formed in this way including steel, aluminium, copper, brass and titanium. Ductile metals will swage more easily. Tube, rod and wire are all suitable for swaging.

COSTS
Tooling is generally inexpensive, but depends on the length and complexity of the swage.

Cycle time is rapid.

Labour costs are high for manual operations because a high level of skill and experience is required to produce accurate parts.

4

5

6

ENVIRONMENTAL IMPACTS

There is very little waste made by swaging. It is not a reductive process like machining. In fact, the hammering or pressing action can strengthen the metal blank, contributing towards a longer lasting product.

Rotary swaging is generally a manually operated process. The vibration can cause 'white finger', especially in larger parts.

Featured Manufacturer

Elmill Swaging
www.elmill.co.uk

Hydraulic Swaging Process

Reducing

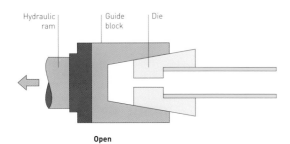

Open

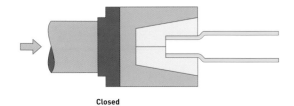

Closed

Expanding

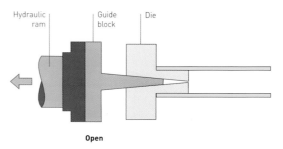

Open

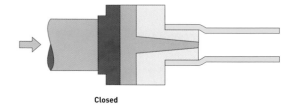

Closed

TECHNICAL DESCRIPTION

In hydraulic swaging the pressure is applied to either the inside or the outside of the pipe diameter. It is typically applied from at least 5 die segments simultaneously, which surround the pipe wall. Turning the pipe during forming ensures even deformation of the wall section.

The hydraulic action is applied along the axis of the pipe being formed. The guide block is wedge shaped and forces the die to either contract or expand. The die determines the shape and angle of swage, which can be parallel or tapered.

For parallel end forming, such as joint forming, standard tools are available for each pipe diameter. Tapered swaging requires specially designed tooling.

Case Study

→ ## Hydraulic swaging

Swaging dies are used to either expand the diameter of a pipe (**images 1**, **2** and **3**) or reduce it (**images 4**, **5** and **6**). The part is continuously rotated during operation to produce a uniform finish. Hydraulic swaging can be applied to a variety of profiles, including square and triangular sections, as well as round.

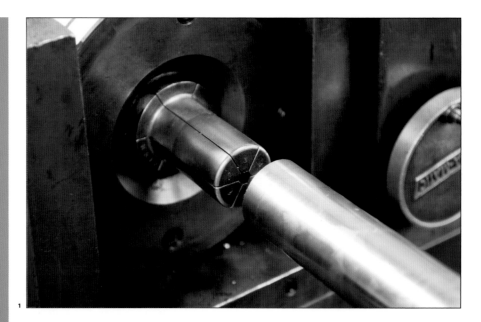

1

2

3

4

5

6

Featured Company

Pipecraft
www.pipecraft.co.uk

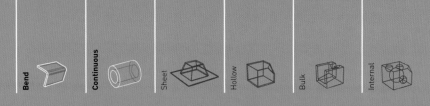

Forming Technology
Roll Forming

Sheet and strip metal can be formed into ribbed panels, channels, angles or polygons with progressive rollers. Such roll forming is a continuous process capable of producing 4,500 m (15,000 ft) of rolled steel per hour.

Costs	Typical Applications	Suitability
• High specialized tooling costs • Low to moderate unit costs	• Automotive and transportation • Construction • Enclosures for white goods	• Batch production of minimum 1,500 m (5,000 ft)

Quality	Related Processes	Speed
• Good pitch accuracy (0.125–0.25 mm/ 0.005–0.01 in.)	• Forging and extrusion • Metal stamping • Press braking	• Very rapid, but depends on the complexity of bends and length of parts • Long changeover times

INTRODUCTION

This metal sheet forming process is used to produce 2D continuous profiles with a constant wall thickness. It will, for example, bend metal strip (up to 0.5 m/ 1.6 ft wide) into continuous lengths of angle, channel, tube and polygon shapes. Roll formers (tools) are mounted alongside one another so in a continuous process they can form sheet materials into ribbed, corrugated and crimped patterns up to 1.4 m (4.6 ft) wide.

Rolling is a continuous operation that operates at speeds up to 1.25 m (4.1 ft) per second. Holes, ridges and other auxiliary patterning are carried out simultaneously to reduce cycle time.

TYPICAL APPLICATIONS

Examples of cold rolled products include roof and wall panels, lorry sidings, chassis for trains and cars, aerospace structural parts, shop fittings, structural beams for construction and brackets and enclosures for white goods.

RELATED PROCESSES

Extrusion, press braking (page 148) and roll forging (see forging, page 114) are all used to produce similar continuous profiles to roll forming. Short parts, cut to length during roll forming, can also be made by metal stamping (page 82).

Roll forming is limited to profiles with a constant wall thickness, whereas extrusion and roll forging can be used to produce profiles with varying wall thickness and parts that cannot be produced by bending a single sheet of material, such as an I-beam.

Like roll forming, press braking can produce continuous profiles. However, each bend is another operation in press braking, whereas roll forming is continuous and therefore a much more rapid production process.

QUALITY

Applying a bend to a sheet of material increases its strength; roll forming combines the ductility and strength of metal to produce parts with improved rigidity and lightness.

Cold working steel significantly improves the grain alignment and thus the strength of the material.

The roll formers do not affect the surface finish of the metal being shaped, and angles are accurate to within ±1° and between 0.125 mm and 0.25 mm (0.005–0.01 in.).

DESIGN OPPORTUNITIES

Roll forming is used to make standard and custom-made products. Many items are readily available, which eliminates tooling costs and minimum production volumes. However, new and original profiles can be designed and analysed using finite element analysis (FEA) software. Tooling is an investment that is offset by the volume of production.

This process can be used to produce a range of profiles, from ribbed panels to channels and box sections. Hollow tube

Cold Roll Forming Process

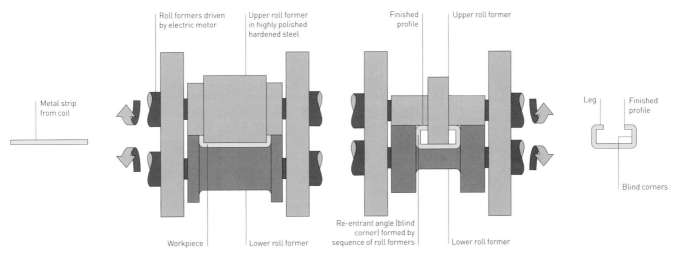

Roll formers driven by electric motor

Upper roll former in highly polished hardened steel

Metal strip from coil

Workpiece

Lower roll former

Stage 1: Progressive roller A

Finished profile

Upper roll former

Re-entrant angle (blind corner) formed by sequence of roll formers

Lower roll former

Stage 2: Progressive roller B

Leg

Finished profile

Blind corners

and box sections made in this way are seam welded because they are formed to maintain the enclosed shape.

Complex shapes with undercut features (blind corners that cannot be directly pressed by rollers), tight bends and hemmed joints can be formed with multiple roll formers in sequence.

Because roll forming is a continuous process parts of any length can be made but they are generally limited to 20 m (66 ft) for ease of handling. The profile is cut to length as it is rolled and it is possible to produce lengths as short as 0.02 m (7.87 in.) at full speed.

DESIGN CONSIDERATIONS

Tooling and changeover (set-up) is expensive in roll forming and so this process is suitable only for production runs greater than 1,500 m (5,000 ft). The most significant consideration is that bends can only be made in the line of production. They will have a radius that is equivalent to, or greater than, material thickness.

The depth of cross-section will affect the costs of tooling. Parts are generally designed so that their depth is less than the opening to the roll former, in order to ensure accurate and high quality bends in the workpiece.

The width of strip or sheet that can be processed is limited by the capabilities of the manufacturer. It is not uncommon to be restricted to less than 0.5 m (1.6 ft) wide. Multiple rolls alongside one another can form sheet materials over

TECHNICAL DESCRIPTION

As the name suggests, cold roll forming is carried out at room temperature. Sheet or strip metal is generally fed into the roll forming process from a coil, and it is bent into the desired shape by a series of progressive roll formers. However, it is also possible to form precut sheets. The bends are formed gradually over several metres (yards): each set of rollers is shaped to form the bend slightly more than the previous one. The diagram illustrates 2 roll formers in a series of 8 or more.

The roll formers are expensive because they are engineered to match precisely. They are typically machined from hardened steel and polished to a very high finish. Water-cooling and lubricants reduce tool wear and help to maintain high speed production.

The roll formers, which are fixed in place, apply an even pressure on the surface of the metal. Therefore, symmetrical designs are simple to produce and less likely to have camber or curve.

Operations including punching and notching are integrated into the production cycle. Holes that are not critical are precut into the flat metal to reduce roll forming cycle times.

Left
Highly polished roll forming tools such as these are used to manufacture continuous metal profiles at speeds up to 4,500 m (15,000 ft) per hour.

Roll forming an angle

This simple roll forming process is used to produce an angle within strip steel. A coil of metal strip is loaded onto the tool (**image 1**) and is then pulled through a series of 8 progressive roll formers (**image 2**), in order to make this single bend. The formers are powered by electric motors, which drive each roll at exactly the same speed. Tension is maintained by pressure on top of the coil.

The finished part emerges from a die at the end of the production cycle, which ensures the dimensional consistency of the part (**image 3**). The finished parts are then cut to length and stacked (**image 4**).

Varying levels of complexity, including hem, re-entrant and asymmetric profiles (**image 5**), can be achieved on thin-walled parts formed by roll forming.

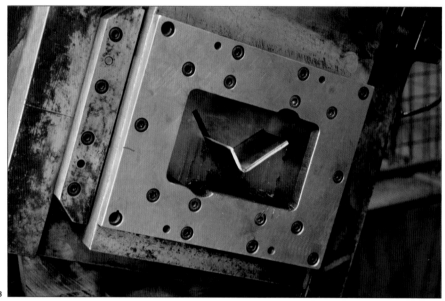

this width but are more specialized and expensive to tool up for.

Sheet thickness for roll forming is typically limited to 0.5–5 mm (0.02–0.2 in.). Bends should be at least 3 times material thickness from the edge of the strip, otherwise it may not be possible to form the bend.

COMPATIBLE MATERIALS

Almost all metals can be formed in this way. However, it is most commonly used for metals that are not feasible to extrude such as stainless steel, carbon steel and galvanized steel.

COSTS

Specialized tooling will increase the unit price considerably, but will depend on the size and complexity of the bend. Tooling costs are generally equal to progressive tooling for metal stamping.

Cycle time depends on the complexity of profile, thickness of metal and length of cut-off. Simple profiles can be formed at up to 4,500 m (15,000 ft) per hour.

ENVIRONMENTAL IMPACTS

Bending is an efficient use of materials and energy. No scrap is produced in roll forming, although it may occur during machining or finishing operations.

Featured Manufacturer

Blagg & Johnson
www.blaggs.co.uk

4

5

Forming Technology
Forging

The forming of metal by heating and hammering or pressing was traditionally performed by blacksmiths on anvils. Nowadays, forgings are made by hammering, pressing or rolling hot metal with sophisticated dies and extreme pressure.

Costs	Typical Applications	Suitability
• Moderate to high tooling costs • Moderate unit costs	• Automotive and aerospace • Hand tools and metal implements • Heavy duty machinery	• All types of production

Quality	Related Processes	Speed
• Excellent grain structure	• Casting • Machining • Tube bending (ring rolling)	• Rapid cycle time (typically less than a minute) depending on size, shape and metal

METAL

114

INTRODUCTION

Metal has been forged into high-strength tools and implements for many centuries. Examples include horseshoes, swords and axe heads, which were formed by blacksmiths hammering hot metal on anvils. There are now many different types of forging, which are used to produce an array of products from crankshafts to ice axes. The different techniques can be divided into drop forging, press forging and roll forging.

Drop forging is carried out as either an open or closed die operation. The processes forms hot metal billets through repeated hammering. The main difference between the closed and open die techniques is that open die forging is generally carried out with flat dies, where as closed die forging shapes metal by forcing it into a die cavity. Closed die forging (also referred to as impression die forging) is used to produce complex and intricate bulk shapes. Open die forging, on the other hand, is typically used to 'draw out' a billet of metal into shafts and bars and can produce parts up to 3 m (10 ft) long.

Press forging is essentially the same as drop forging, except that parts are formed by continuous hydraulic pressure, as opposed to hammering. Press forging is used to shape hot and cold metals; the temperature of the metal is determined by the material, part size and geometry.

Roll forging forms continuous metal parts through a series of metal rollers. This process is used to forge straight profiles and rings (washers), which can be up to 8.5 m (28 ft) in diameter and 3 m (10 ft) high.

Drop Forging Process

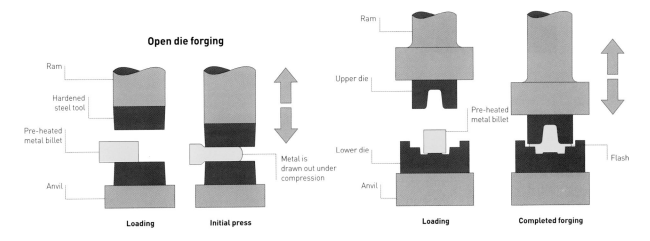

Open die forging

Ram

Hardened steel tool

Pre-heated metal billet

Anvil

Metal is drawn out under compression

Loading   Initial press

Closed die forging

Ram

Upper die

Pre-heated metal billet

Lower die

Anvil

Flash

Loading   Completed forging

TYPICAL APPLICATIONS

The high-strength nature of forged metal parts makes them ideal for demanding applications and critical components that require excellent resistance to fatigue. Forged components are found in lifting equipment, aerospace, military applications, cars and heavy machinery.

Many gears, plumbing parts, hand tools and implements are forged. Car axles and crankshafts are open die forged. The heads of nails and bolts are cold forged.

RELATED PROCESSES

Forging is suitable for one-offs and low volume production as well as mass production runs. At low volumes it is an alternative to CNC machining (page 182), and at high volumes it competes with die or investment casting (pages 124 and 130). Ring rolling as covered in tube bending (page 98) produces similar geometries to roll forging, but they require welding, whereas roll forged rings are seamless.

QUALITY

By its very nature, forging improves grain structure in the finished part. Metal billets go through plastic deformation as they are forged and as a result the metal grains align in the direction of flow. This produces exceptional strength to weight and reduces stress concentrations that tend to occur in corners and fillets. Parts

TECHNICAL DESCRIPTION

The drop forging process is carried out using either a closed or open die. An open die is typically used for drawing out (elongating a part and reducing its cross-section), upsetting (shortening a part and increasing its cross-section) and conditioning (preparing a part for closed die forging). Open die forging is also used to shape hexagonal and square section bars. The tool face is generally square or v-shaped for simple forming. The workpiece is repositioned between each drop of the hammer by the operator. This requires a great deal of operator skill and experience and is not suitable for automation. Long profiles can be forged in sections.

Closed die forging can be automated for high volume production. The tools are fabricated in tool steels (chromium-based or tungsten-based) or low alloy steels. The life expectancy of a tool is largely dependent on the shape of the part, but is also affected by the ductility of the forging material. For example, stainless steel

must be heated to over 1250°C (2282°F) and so stresses and wears the surface of the tool much more rapidly than aluminium, which is forged at only 500°C (932°F). During operation, lubricant that acts as coolant is applied to the surface of the tool to cool it down and reduce wear.

In drop forging, the ram brings the upper die into contact with the workpiece under great pressure, from 50 kg/m^2 to 10,000 kg/m^2 (362–72,330 lb/ft^2). The force is generated by gravity fed weights or by powered means (hydraulic or compression) forcing the ram down. Press forging is powered by hydraulic rams, which force the metal into the die cavity in a squeezing action. This technique can be used to process metals hot or cold. Cold metal forging is typically used for small parts, no larger than 10 kg (22 lb). The advantage of cold forging is that it can be used to produce near net shape parts that do not require secondary operations.

can be machined post-forging with no loss of quality because there are no voids or porosity in the finished article.

The tolerances range from 1 mm (0.04 in.) in small parts up to 5 mm (0.2 in.) in large parts, but vary according to requirements because reducing

tolerances increases costs. Forging is often combined with machining for improved accuracy.

DESIGN OPPORTUNITIES

Forging is suitable for low volume production and one-offs. This is

Roll Forging Process

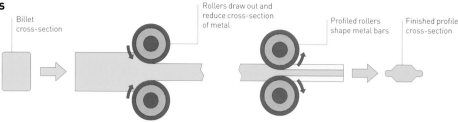

Billet cross-section

Rollers draw out and reduce cross-section of metal

Profiled rollers shape metal bars

Finished profile cross-section

Stage 1: Reducing cross-section

Stage 2: Profiling

TECHNICAL DESCRIPTION

Roll forging carried out as a continuous process. Metal bar or plate is fed into the rollers, which draw out and profile the hot metal in stages. Each stage reduces the cross-section gradually, easing the metal into the desired profile. This process is generally used to shape metal into simple continuous profiles.

Seamless rings can also be formed in this way. First, a hole is punched in the middle of a forged disk, which is then roll forged into the desired profile. This is also known as ring rolling.

because it produces parts with superior properties that cannot be manufactured in any other way. Small volumes can be machined, but will have to compensate for reduced strength resulting from random grain alignment.

Undercuts are not possible in forging. However, it is possible to form undercuts and joints with secondary forging operations. An example of this is the swivel link (see image, below). These are produced by first making 2 separate drop

forgings. They are joined together by upset forging a stopper on the shaft of the eyelet. Upset forging increases the cross-section and reduces the length of the shaft, a process similar to shaping the head of a bolt.

Typically, wall thickness should be 5 mm to 250 mm (0.2–9.84 in.). There is no restriction on step changes in wall thickness with this process.

Forging can be used to make a huge range of component sizes and geometries. Drop forgings can weigh as little as 0.25 kg (0.55 lb) or as much as 60 kg (132 lb). Roll forging can produce seamless rings weighing more than 100 tonnes (110 US tons).

DESIGN CONSIDERATIONS

Designing for forging must take into account many factors that also affect design for casting, including partition line, draft angles, ribs, radii and fillets.

Parts are formed by hammering, or pressing, which can produce surprisingly deep protrusions, up to 6 times the thickness of material. Draft angles can be minimized and even eliminated by clever design, especially in ductile materials

such as aluminium and brass. Radii, however, are very important because they encourage the flow of metal and reduce tool wear. The minimum radius increases with depth of protrusion.

COMPATIBLE MATERIALS

Most ferrous metals, including carbon, alloy and stainless steels, can be forged. Non-ferrous metals including titanium, copper and aluminium are also suitable.

COSTS

Tooling costs are moderate to high, depending on the size and geometry of the part. Closed die forging tools typically last between 50 and 5000 cycles. Tool life expectancies are affected by the complexity of forging geometry, the design of the forging cavities, sharpness of radii, the material to be forged, the temperature required to forge that material and the quality of the surface finish of the tool. Incorporating multiple cavities into a tool and pre-forming the metal billet increase the cycle time.

Cycle time is rapid. A typical forging is complete in less than a minute. However, mass production forging can produce parts in less than 15 seconds.

Labour costs are moderate to high due to the level of skill and experience required. This is a relatively dangerous process, so health and safety depend on the abilities of the workforce.

ENVIRONMENTAL IMPACTS

A great deal of energy is required to heat metal billets to working temperature and hammer or press them into shape.

No material is wasted because all scrap and off cuts can be recycled.

Left

To unite the 2 parts of the swivel link, the heated shaft of the eyelet is driven through the hole into a die that shapes the stopper.

→ Drop forging a piston end cap

In this case study a piston end cap is being drop forged in a combination of open and closed dies. The metal billet is preformed and the scale is broken off in open die forging in preparation for closed die forging.

The mild steel is delivered in 6 m (19.7 ft) lengths (**image 1**), which are cut into billets. The size of billet is determined by the weight and geometry of the part. The billets are loaded into a furnace, which heats them up to 1250°C (2282°F). Each billet takes approximately 30 minutes to heat up

sufficiently for forging. The 'cherry red' billet is removed from the furnace with a pair of tongs (**image 2**).

The metal is aligned in the open die (**image 3**) and formed by repeated blows from the hammer (**image 4**). The shape of the part is changed but the volume stays the same. The metal spraying out with impact is mill scale, which is formed as the steel surface oxidizes. This has to be removed because otherwise it will contaminate the final part.

1

2

3

4

The part is now ready for the closed die forging process (**image 5**). The floor shakes with the impact of the hammer (**image 6**). With each cycle the hot metal is forced further into the upper and lower die cavities. Once the flash has formed, it cools more rapidly than the rest of the metal because it is thinner and less ductile. Therefore, the remaining hot metal is maintained within the die cavity and is forced into the extremities of it with each repeated blow from the hammer (**images 7** and **8**).

It is a gradual, but rapid process. In this case the metal does not need to be reheated at any point and so the entire forging operation is complete within 30 seconds. Due to the depth of this forging it must be carried out while the metal is 'cherry red'. The extremities are the first to cool off, and can be seen as duller patches (**image 9**).

These lumps of metal are too heavy for a single operator to manhandle with tongs, so they have to work in pairs to move the metal between workstations (**image 10**). It requires a great deal of skill and experience to work in this environment. The flash is sheared off in a punch (**image 11**) and the part, which is still very hot, is dispensed (**image 12**).

5

6

7

8

Featured Manufacturer

W. H. Tildesley
www.whtildesley.com

9

10

11

12

Bend

Continuous

Sheet

Hollow

Bulk

Internal

Forming Technology

Sand Casting

Molten metal is here cast in expendable sand molds, which are broken apart to remove the solidified part. For one-off and low volume production this is relatively inexpensive and suitable for casting a range of ferrous metals and non-ferrous alloys.

Costs	Typical Applications	Suitability
• Low tooling costs • Moderate unit costs	• Architectural fittings • Automotive • Furniture and lighting	• One-off to medium volume production

Quality	Related Processes	Speed
• Poor surface finish and high level of porosity.	• Centrifugal casting • Die casting • Forging	• Moderate cycle time (30 minutes typically), but depends on secondary operations

INTRODUCTION

Sand casting is a manual process used to shape molten ferrous metals and non-ferrous alloys. It relies on gravity to draw the molten material into the die cavity and so produces rough parts that have to be finished by abrasive blasting, machining or polishing.

The sand casting process uses regular sand, which is bonded together with clay (green sand casting) or synthetic materials (dry sand casting) to make the molds. Synthetic sand molds are quicker to make and produce a higher quality surface finish. However, the quality of the casting is largely dependent on the skill of the operator in the foundry.

There are many variables in the casting process that the designer can do nothing about, although properly

designed parts with sufficient draft angles will result in higher quality parts.

Sand casting is most commonly used to shape metals. Glass can also be cast in this way, but because of its high viscosity it will not flow through a mold.

Evaporative pattern casting is a hybrid of the sand casting and investment casting (page 130) processes, and it can be inexpensive. Instead of a permanent pattern, evaporative pattern casting uses an expendable foam pattern (often polystyrene), which is embedded in the sand mold. The foam pattern is typically formed by CNC machining (page 182) or injection molding (page 50). The molten metal burns out the foam pattern as it is poured in. Parts made in this way have inferior surface finish, but this technique is very useful for prototyping, one-off and manufacturing very low volumes.

TYPICAL APPLICATIONS

Sand casting is used a great deal in the automotive industry to make engine block and cylinder heads, for example. Other applications include furniture, lighting and architectural fittings.

RELATED PROCESSES

Die casting (page 124), centrifugal casting (page 144) and forging (page 114) are all alternatives to sand casting metals. Sand casting is often combined with forging or CNC machining and arc welding (page 282) to create more complex parts. Die casting produces parts more accurately and rapidly and so is generally reserved for high volume production.

QUALITY

The quality of surface finish and mechanical properties depend largely on the quality of the foundry. For example, nitrogen is pumped through molten aluminium to remove hydrogen,

which causes porosity. When the parts are removed from the sand molds the surface finish is distinctive, so all cast parts are blasted with an abrasive. Varying grits are used to produce the highest quality finish. Parts can then be polished to achieve very high surface finish. Because sand casting relies on gravity to draw the molten material into the mold cavity, there will always be an element of porosity in the cast part.

DESIGN OPPORTUNITIES

There is plenty of scope for designers using this process. For low volume production it is generally less expensive than die casting and investment casting.

Draft angles are required on the sand mold to ensure removal of the pattern. The cores also require draft angles so that they can be removed from their molds. However, multiple cores can be used and they do not have to correspond to external draft angles because the mold

TECHNICAL DESCRIPTION

The sand casting process is made up of 2 main stages: moldmaking and casting. In stage 1, the mold is made in 2 halves, known as the cope and drag. A metal casting box is placed over the wooden pattern and sand is poured in and compacted down.

For dry sand casting the sand has a vinyl ester polymer coating, which is room temperature cured. The polymer coating on each sand particle helps to achieve a better finish on the cast part because it creates a film of polymer around the pattern, which forms the surface finish in the cast metal. For green sand casting, the sand is mixed with clay and water until it is sufficiently wet to be rammed into the mold and over the pattern. The clay mix is left to dry so that there is no water left in the mix. Water must be removed because otherwise it will boil and expand during casting, thereby causing pockets of air in the mold.

Meanwhile, sand cores are made in separate split molds and then placed in the drag. The runner and risers, which are covered with an insulating sleeve, will already have been built into the pattern.

Sand Casting Process

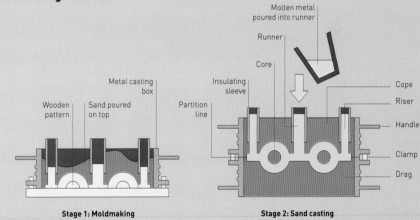

Stage 1: Moldmaking

Stage 2: Sand casting

In stage 2, the 2 halves of the mold are clamped together with the cores in place. The metal is heated up to several hundred degrees above its melting temperature. This is to ensure that during casting it remains sufficiently molten to flow around the mold. A predetermined quantity of molten metal is poured into the runner, so it fills the mold and risers. Once the mold is full an exothermic metal oxide powder is poured into the runner and risers, which burns at a very high temperature and so keeps

the metal at the top of the mold molten for longer. This means that as the metal inside the mold solidifies and shrinks as it cools it can draw on surplus molten metal in the runner and risers. Porosity on the surface of the part is therefore minimized.

is broken to remove the solidified metal parts. This means that complex internal shapes can be formed, which may not be practical for the die casting processes.

Sand casting can be used to produce much larger castings than other casting methods (up to several tonnes). The molds for castings can weigh a lot more than the casting itself. It is therefore not practical or economical to produce very large castings any other way.

DESIGN CONSIDERATIONS

Many design considerations must be taken into account for sand casting parts. These include draft angles (range from 1° to 5°, although 2° is usually adequate), ribs, recesses, mold flow and partition lines. The part must be designed to take into account all aspects of the casting process, from patternmaking to finishing. These elements affect the design and so all parties should be consulted early on in the design process for optimum results.

The wall thickness that can be cast is 2.5mm up to 130mm (1–5.12 in.). Changes in wall thickness are best avoided, although small changes can be overcome with tapers and fillets. If a large change in cross-section is required, then parts are often cast separately and assembled.

Metals cool at different rates. Therefore it is essential to know the material of choice at the design stage to ensure accurate castings in the selected material. Steel will shrink nearly twice as much as aluminium and iron, whereas brass will shrink about 50% more than aluminium. Shrinkage increases with cross-section and part size.

COMPATIBLE MATERIALS

This process can be used to cast ferrous metals and non-ferrous alloys. The most commonly sand cast materials include iron, steel, copper alloys (brass, bronze) and aluminium alloys. Magnesium is becoming increasingly popular, especially in aerospace applications because of its lightness.

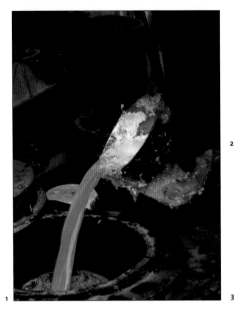

1

2

3

4

5

6

COSTS

Tooling costs are low for one-off and low volume production, the main cost being for patternmaking, which is inexpensive compared to die casting tooling. Low cost patterns can be machined in wood or aluminium. Foam patterns for the evaporative pattern method are the least expensive of all.

Cycle time is moderate but depends on the size and complexity of the part. Casting usually takes less than 30 minutes, but secondary and finishing operations increase cycle time.

Labour costs can be high, especially as the majority of this process is manually operated and the quality of casting is affected by the skill of the operator.

ENVIRONMENTAL IMPACTS

A large percentage of each casting solidifies in the runner and risers. This material can be directly recycled in most cases. The mold sand can be reused by mixing it with virgin material. In green sand casting up to 95% of the mold material can be recycled after each use.

Energy requirements for sand casting are quite high because the metal has to be raised to several hundred degrees above its melting temperature.

Case Study

→ Sand casting a lamp housing

The casting temperature of aluminium is between 730°C and 780°C (1346–1436°F), yet in preparation the aluminium is heated up to more than 900°C (1652°F) in a kiln (**image 1**). The mold is prepared by placing a metal casting box over the pattern (**image 2**). Sand coated in a film of vinyl ester is poured over the pattern (**image 3**). The box is filled incrementally and rammed down after each fill (**image 4**). The quality of the fill and so the quality of the final casting are largely dependent on the skill of the operator (**image 5**). The vinyl ester cures at room temperatures and solidifies rapidly.

In the meantime, the cores are prepared in a separate mold. They are made with the same sand coated in vinyl ester. The sand is carefully packed into the mold cavity and rammed to compact it (**image 6**). The mold can be opened almost immediately. The core can then be removed (**image 7**), but it is still very delicate at this stage and so must be handled carefully.

Once the cores have been placed into the drag (**image 8**), the mold is ready to be brought together and clamped. Hot molten metal is poured into the runner (**image 9**), where it spreads through the mold and up the risers. When the mold is full an exothermic metal oxide (aluminium oxide

in this case) is poured into the runner and risers (**image 10**). The powder burns at a very high temperature, which keeps the aluminium molten in the top of the mold for longer. This is important to minimize porosity in the top surface of the cast part.

After 15 minutes the casting boxes are separated along the partition line and the part is revealed – covered in a layer of charred sand (**image 11**). The sand is broken away from around the metal casting (**image 12**). The surface finish is relatively good and so little finishing is required. The parts are cut free from the runner system (**image 13**) and are now ready for abrasive blasting and finishing (**image 14**).

7

8

9

10

11

12

13

14

Featured Manufacturer

Luton Engineering Pattern Company
www.chilterncastingcompany.co.uk

Forming Technology

Die Casting

Die casting is a precise method of forming parts from metal. This high speed process uses pressure to force molten metal into reusable steel molds, to create intricate and complex 3-dimensional geometries.

Costs	Typical Applications	Suitability
• High tooling costs • Low unit costs	• Automotive • Furniture • Kitchenware	• High volume production

Quality	Related Processes	Speed
• Very high surface finish • Variable mechanical properties	• Forging • Investment casting • Sand casting	• Rapid cycle time that depends on size and complexity of part

INTRODUCTION

There are various techniques in die casting, including high pressure die casting, low pressure die casting and gravity die casting.

High pressure die casting is a versatile process and the most rapid way of forming non-ferrous metal parts. Molten metal is forced at high pressure into the die cavity to form the part. The high pressures mean that small parts, thin wall sections, intricate details and fine surface finishes can be achieved. The tooling and equipment is very expensive, and so this process is suitable only for high volume production.

In low pressure die casting molten material is forced into the die cavity by low pressure gas. There is very little turbulence as the material flows in and so the parts have good mechanical properties. This process is most suitable for rotationally symmetrical parts in low melt temperature alloys. A good example would be an aluminium alloy wheel.

Gravity die casting is also known as permanent mold casting; steel molds are the only feature that differentiate it from sand casting (page 120). The molds can be manually operated or automated for larger volume production. Reduced pressure means that tooling and equipment costs are lower, so gravity die casting is often used for short productions runs, which would not be economic for other die casting methods.

TYPICAL APPLICATIONS

High pressure die casting is used to produce the majority of die cast metal parts such as for the automotive industry, white goods, consumer electronics, packaging, furniture, lighting, jewelry and toys.

Low pressure methods are widely utilized in the automotive industry to make wheels and engine parts, for example. They are also used to make products for the home such as kitchenware and tableware.

RELATED PROCESSES

Forging (page 114), sand casting and investment casting (page 130) are alternative processes. However, die casting is the process of choice for large volume production of metal parts due to its versatility, speed, quality, minimum wall section, high strength to weight and repeatability.

Die casting is often compared to injection molding (page 50). The main differences are related to the material qualities. Compared to plastics, metals are more resistant to extreme temperatures, they are durable and have superior electrical properties. Therefore, die castings are more suitable for applications that demand these properties such as engine parts.

QUALITY

Die cast parts have superior surface finish, which improves with pressure. High speed injection methods cause turbulence in the flow of metal, which can lead to porosity in the casting. Voids and porosity are an inevitable part of metal casting that can be limited when engineering the product at the design stage. Mold flow simulation is used to optimize the filling of the cavity of the mould and eradicate any potential voids and porosity. Strength analysis is carried out prior to manufacture to test the mechanical properties of the part (see image, right).

Right
Strength analysis of Chair One shows the resilience of the structure. This software is used to double check that the engineering of the product is correct before manufacturing the tools for casting.

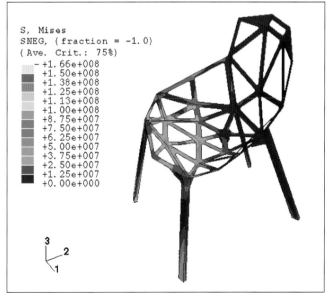

High Pressure Die Casting Process

Cold chamber method

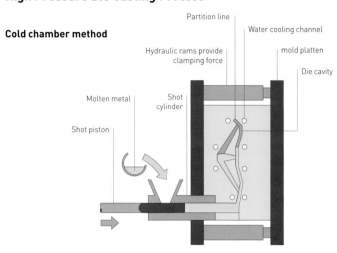

Partition line
Hydraulic rams provide clamping force
Water cooling channel
mold platten
Die cavity
Molten metal
Shot cylinder
Shot piston

Stage 1: Metal injection

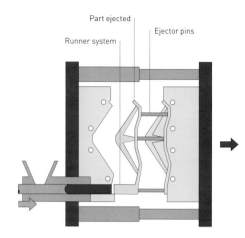

Part ejected
Ejector pins
Runner system

Stage 2: Part ejected

TECHNICAL DESCRIPTION

High pressure die casting is carried out as a hot or cold chamber process. The only difference is that in the hot chamber method the molten metal is pumped directly from the furnace into the die cavity. The machine sizes range from 500 tonnes (551 US tons) to more than 3,000 tonnes (3,307 US tons). The required clamp force is determined by the size and complexity of the part being made.

Metal ingots and scrap are melted together in a furnace that runs 24 hours a day. In stage 1 of the cold chamber method, a crucible collects a measure of molten metal and deposits it into the shot cylinder.

In the hot chamber method the shot cylinder is fed through the furnace. The hot liquid metal is forced into the die cavity by the shot piston at high pressure. The pressure is maintained until the part has solidified. Water cooling channels help to keep the mold temperature lower than the casting material and so accelerate cooling within the die cavity.

in stage 2, when the parts are sufficiently cool, which can take anything from a few seconds to several minutes depending on size, the mold halves open and the part is ejected. The flash and runner systems have to be trimmed before the part is ready for joining and finishing operations. High pressure die cast parts require very little machining or finishing, because a very high surface finish can be achieved in the mold.

DESIGN OPPORTUNITIES

There are many advantages to using die casting if the quantities justify it. Intricate or bulky fabrications can be redesigned to become more economic with improved strength and reduced weight. For example, holes in sheet material are seen as waste, whereas holes in die castings are savings because they are produced directly in the mold and so reduce material consumption.

These processes can be used to produce complex shapes with internal cores and ribs. High pressure methods will reproduce very fine details and can form thinner wall sections than other casting processes. Parts are produced to high tolerances and often require little or no machining and finishing.

External threads and inserts can be cast into the product.

DESIGN CONSIDERATIONS

This process has technical considerations similar to injection molding. These are rib design, draft angles (1.5° is usually sufficient), recesses, external features, mold flow and partition lines. The part must be designed to take into account all aspects of the casting process, from toolmaking to finishing, so all parties should be consulted early on in the design process for optimum results.

Casting is most suitable for small parts because the steel tooling becomes very large and expensive for parts above 9 kg (20 lb). Side action and cores can increase the cost of the tool considerably. However, there may be an associated benefit if the same feature reduces wall thickness, by increasing strength, for example.

Low Pressure Die Casting Process

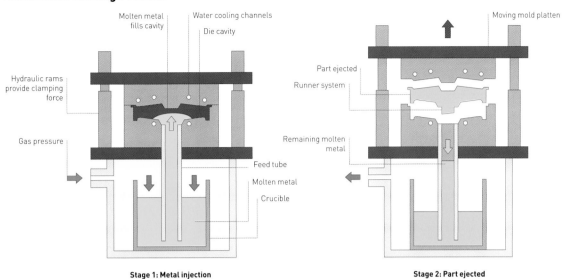

Molten metal fills cavity | Water cooling channels | Moving mold platten
Die cavity
Hydraulic rams provide clamping force
Part ejected
Runner system
Gas pressure
Remaining molten metal
Feed tube
Molten metal
Crucible

Stage 1: Metal injection   Stage 2: Part ejected

COMPATIBLE MATERIALS

Choice of material is an essential part of the design process in die casting because each material has particular properties to be exploited. Die casting is only suitable for non-ferrous metals. The melting point of ferrous metals is too high, so liquid forming is carried out by investment or sand casting, and solid state forming by forging.

Non-ferrous metals including aluminium, magnesium, zinc, copper, lead and tin are all suitable for die casting. Aluminium and magnesium are becoming popular for consumer electronics due to their high strength to weight properties. They are dimensionally stable, even at high temperatures, resistant to corrosion and can be protected and coloured by anodizing (page 360).

COSTS

Tooling costs are high because tools have to be made in steel to be able to withstand the temperature of the molten alloys. For gravity die casting, sand cores can be used for complex shapes and slight re-entrant angles.

Especially if a multiple cavity mold is used, cycle time is rapid – from a few seconds to several minutes, depending on the size of the casting.

TECHNICAL DESCRIPTION

In low pressure die casting the mold and furnace are connected by a feed tube. The mold is mounted on top of the furnace with a horizontal split line. In stage 1, the molten material is forced up the feed tube and into the die cavity by gas pressure on the surface of the metal in the furnace. Gas pressure is maintained until the part has solidified.

In stage 2, when the gas pressure is released, the molten metal still remaining in the feed tube runs back down into the crucible. The casting is left for a short time to solidify before the top half of the mold is raised and the part ejected. The nature of this process means that it is most suitable for parts that are symmetrical around a vertical axis.

Labour costs are low for automated die casting methods.

ENVIRONMENTAL IMPACTS

All the waste metal generated in casting can be directly recycled. There is no loss of strength and so the waste metal can be mixed with ingots of the same material, melted down and recast immediately.

This process uses a great deal of energy to melt the alloys and maintain them at high temperatures for casting.

→ High pressure die casting Chair One

This chair was designed by Konstantin Grcic for Magis in 2001 and took 3 years to develop from brief to production. It was designed specifically for die casting in aluminium, which is a very common material in high pressure die casting and so is used in large quantities by foundries (**image 1**). The raw material (which may come from recycled stock) is melted in the holding furnace and mixed with scrap from the casting process. A crucible for transferring the molten aluminium into the shot cylinder lowers into the holding furnace and collects a measured charge of the metal (**image 2**). It is poured into the shot cylinder (**image 3**) and forced into the die cavity by a shot piston. This machine has a clamping force of 1,300 tonnes (1,433 US tons), which is necessary for casting the 4 kg (8.82 lb) chair seat.

After 2 minutes the molds separate to reveal the solidified part, which is collected by a robotic arm (**image 4**). Once the part has been extracted the mold is steam cleaned and lubricated for the next cycle (**image 5**). Meanwhile, the part is dipped in water to cool it down and ensure that it is completely solidified (**image 6**).

The flash and runner systems are removed (**image 7**) and the scrap material is fed straight back into the holding furnace so that it can be recast. Although the stacked chair seats already have a very high surface finish (**image 8**), this is improved by polishing (page 376). Inserts are used in the tools so that different leg assemblies can be fitted (**image 9**).

1

2

3

4

5

Featured Company

Magis
www.magisdesign.com

6

7

8

9

Bend

Continuous

Sheet

Hollow

Bulk

Internal

Forming Technology

Investment Casting

Liquid metals are formed into complex and intricate shapes in this process, which uses non-permanent ceramic molds. It is also known as lost wax casting.

Costs	Typical Applications	Suitability
• Low to moderate cost wax injection tooling • Non-permanent molds • Moderate to high unit costs	• Aerospace • Construction • Consumer electronics and appliances	• Low to high volume production

Quality	Related Processes	Speed
• Very high • Complex shapes with high integrity	• Die casting • Metal injection molding • Sand casting	• Long cycle time (24 hours)

INTRODUCTION

This is a versatile metal casting process. It is more expensive than die casting (page 124), but the opportunities outweigh the price difference for many applications.

Due to its many advantages, investment casting is used to produce a wide range of products from only a few grams to more than 35 kg (77.16 lb) and less than 5 mm (0.2 in.) long up to over 0.5 m³ (17.66 ft³).

Investment casting is made up of 3 elements: expendable pattern, non-permanent ceramic mold and metal casting. The patterns are typically injection molded (page 50) wax, but other materials are also used, including rapid prototyped (page 232) models.

Both small and large volumes can be accommodated, from prototypes to mass production of 40,000 or more parts per month. It is also possible to produce very complex, intricate parts with thin and thick wall sections that cannot be cast in any other way.

TYPICAL APPLICATIONS

Applications are widespread and include products for the aerospace, automotive, construction, furniture, sculpture and jewelry industries. Parts include gears, housings, electronic chassis, covers and fascias, engine parts, turbine blades and wheels (see image, above), medical implants, brackets, levers and handles.

RELATED PROCESSES

Some parts made by investment casting are also suitable for die casting, sand casting (page 120) or metal injection molding (page 136). When volumes exceed 5,000, production may shift to die casting if the design is suitable.

QUALITY

Investment casting produces high integrity metal parts with superior metallurgical properties. The surface finish is generally very good and is determined by the quality of the expendable pattern.

Dimensions are typically accurate to within 125 microns (0.0049 in.) for every 25 mm (0.98 in.) of cast metal.

DESIGN OPPORTUNITIES

Investment casting does not have the same shape limitations as other casting techniques. This is because neither the pattern nor metal part has to be ejected at any point. The pattern and mold are both non-permanent: the wax is melted from the ceramic shell, which in turn is broken from the cast part. In other words, it is possible to cast shapes with undercuts and varying wall thickness that are not feasible with other liquid forming processes. This eliminates costly fabrication operations.

Complex internal shapes are feasible in the injection molding of the wax pattern. The molds are sometimes very complex, consisting of many parts, to reduce assembly operations later on. For precise and critical internal geometries, soluble wax or ceramic pre-formed cores

Investment Casting Process

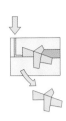

Stage 1: Wax injection

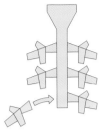

Stage 2: Wax patterns assembled onto tree

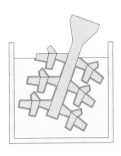

Stage 3: Assembly dipped in ceramic slurry

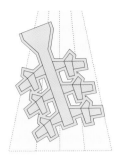

Stage 4: Ceramic stuccoed to wet surface

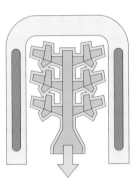

Stage 5: Wax melted from sheet mold

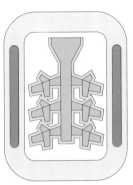

Stage 6: Sheet mold fired

Stage 7: Metal poured into hot mold

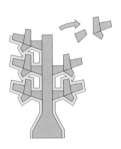

Stage 8: Casting broken out and separated from tree

TECHNICAL DESCRIPTION

There are many stages to the investment casting process, which are divided into pattern making, ceramic mold making and casting.

In stage 1, the expendable pattern is formed, which in this case is wax injection. The tooling is typically aluminium. Unlike conventional injection molding, these tools can have many parts, which are assembled are used. These are injection molded separately and then over-molded with conventional wax. When the casting is complete it is submerged in water and the soluble wax melts away.

Injection molded wax is still the most commonly used pattern material. Other materials and techniques include rapid prototyped wax (thermojet) and plastic models (quickcast), acrylic and machined or molded expanded polystyrene (EPS), known as lost foam casting. In fact, any material that can be burnt out and has a by hand. The wax is injected at low pressure, and so there is little flash, or other problems that are associated with high-pressure injection techniques.

The wax parts are molded with the gate and runner system attached. In stage 2, the whole assembly is mounted onto a central feed system. Everything is wax and so can be melted and joined together. Each assembly (tree) may hold tens or even hundreds of products, depending on the size of the parts, which increases production rates.

In stages 3 and 4, the assembly is dipped in ceramic slurry and then coated with fine grains of refractory material. The primary coating is made up of very fine particles, which ensures a good surface finish on the inside of the shell mold. The number of coats depends on the size of part and the metal being cast.

This wet dipping and dry stuccoing process, known as investing, is repeated 7 to 15 times with progressively coarser refractory materials. Between each cycle the shell is left to dry for 3 hours.

In stages 5 and 6, the wax patterns and runner system are melted out in a steam autoclave and the ceramic shell is subsequently fired at 1095˚C (2003˚F). It is removed from the kiln at between 500˚C and 1095˚C (932–2003˚F) depending on the metal being cast.

In stage 7, whilst the shell mold is still very hot, molten metal is poured in. In most cases gravity is used to fill the mold. It is also possible to pull the molten metal through using a vacuum, or force it into the mold under pressure.

Once the casting has solidified and cooled, in stage 8 it is broken out of the shell mold using impact and vibration. For delicate parts, the shell is removed using chemical dissolution or high-pressure water.

The individual parts have to be removed from the runner system and cleaned up. Machining is needed to clean up the surfaces that were in contact with the runner system. The parts are then finished with abrasive blasting (page 388), or left 'as cast', because the surface finish is generally very good.

Far left
The simulation shows potential problem areas in the original part.

Left
The part is modified in CAD. A new simulation shows that no problem areas remain.

1

2

3

sufficiently low coefficient of expansion is suitable for the pattern, although some materials will take longer than others to burn out. Wax has proved to be the most effective material for large volumes and high surface finish.

DESIGN CONSIDERATIONS
Wall sections do not have to have a uniform thickness; sometimes thicker sections are actually required to aid the flow of metal into the mold cavity. It is possible to feed metal to the cavity through multiple gates, ensuring good distribution around even complex parts.

The wall thickness depends on the alloy. The typical wall thickness for aluminium and zinc is between 2 mm and 3 mm (0.079–0.118 in.), although it is possible to cast wall sections as thin as 1.5 mm (0.059 in.). Steel and copper alloys require thicker wall sections, typically greater than 3 mm (0.118 in.), but for small areas a thickness of 2 mm (0.079 in.) is possible.

Cast parts are dimensionally accurate and minimize machining operations. However, internal threads and long blind holes can only be formed on parts as a secondary operations.

As with other liquid forming processes, cast parts are typically ribbed and reinforced to eliminate warping and distortion when cooling.

COMPATIBLE MATERIALS
Almost all ferrous and non-ferrous metal alloys can be investment cast. This process is suitable for casting metals that cannot be machined and fabricated in any other way such as superalloys.

The most commonly cast materials are carbon and low alloy steels, stainless steels, aluminium, titanium, zinc, copper alloys and precious metals. Nickel, cobalt and magnetic alloys are also cast.

COSTS
There are tooling costs for the wax injection process. These can be moderate

for complex molds with removable core segments. A simple split mold is relatively low cost and has a long life expectancy, as the wax has a low melting temperature and in not very abrasive.

Cycle time is long and generally 24 hours or more.

Investment casting is a complex process that requires skilled operators. Therefore, labour costs can be quite high.

ENVIRONMENTAL IMPACTS
Very little metal is wasted in operation and any scrap and offcuts can be directly recycled in the furnaces. The melted wax is reused in the runner systems.

The ceramic shells cannot be recycled, but their material is totally inert and non-toxic. The smoke and particles produced by casting are captured by ceramic filtration.

Case Study

→ Investment casting a window support

CAD

All parts are designed and developed on CAD. Flow simulation software is used to reduce problem areas such as porosity. The image sequence demonstrates this process. The original part (see image, far left) was gradually modified (see image, left). Areas of potential porosity are highlighted in yellow.

INVESTMENT CASTING

This is the production of a 'spider' used in the construction industry for supporting panes of glass. They are manufactured in stainless steel and often finished by electropolishing (page 384).

The expendable pattern is injection molded wax (**image 1**) in a tool very similar to that used in conventional injection molding. The process is fully automated, but molds with complex internal cores are manually operated.

The mold tool is 3% larger than the final metal casting, to allow for the wax to shrink 1% and the metal to shrink 2%.

The wax patterns, which are molded with a gate attached, are assembled onto a runner system (**image 2**). The joint interface is melted with a hot knife and held together to form a bond.

The assembly, known as a 'tree', is dipped by hand into a water-based zircon ceramic solution (**image 3**). The primary coating is always carried out by hand because it is the most critical and essential for good surface finish. The wax tree is transferred into a coating chamber, which applies fine refractory powder (**image 4**) and hung up to dry (**image 5**).

At this stage the ceramic shell is very fine and has to be reinforced with secondary coatings of a more substantial mullite ceramic. This is carried out automatically by a robot (**images 6** and **7**). The secondary coatings are applied every 3 hours.

4

5

6

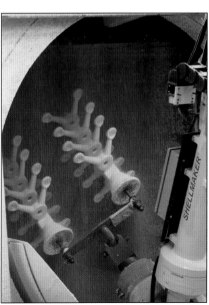

7

After at least 7 coats the ceramic shells are sufficiently robust. Metals with higher melting point and large parts require more coats. The wax is removed in a steam autoclave and recycled. The hollow ceramic shell is placed in a kiln at 1095°C (2003°F) to harden it fully and remove any residual wax. It is an exact replica of the wax pattern.

The ceramic mold is red hot when it is removed from the kiln 3 hours later (**image 8**). In the meantime, the molten stainless steel is prepared in a crucible, which is used to pour the metal into the mold (**image 9**). The metal is poured by hand into the mold (**image 10**). The ceramic molds retain a great deal of heat as the molten metal is poured in. This encourages the metal to flow through even the most complex shapes.

The mold is left to cool for approximately 3 hours (**image 11**). The ceramic shell is broken from the solidified metal with a pneumatic hammer (**image 12**).

The individual parts are removed from the runner system, ground off and polished up to produce the finished article (**image 13**).

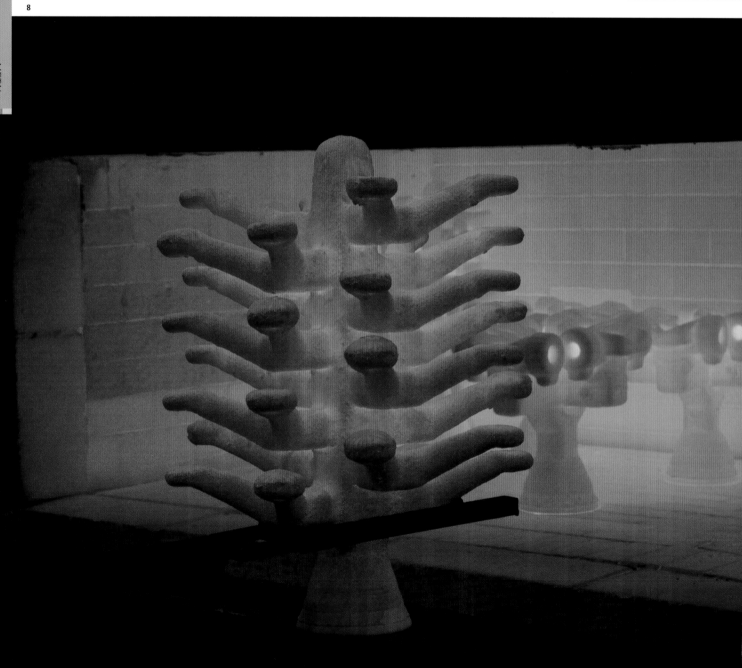

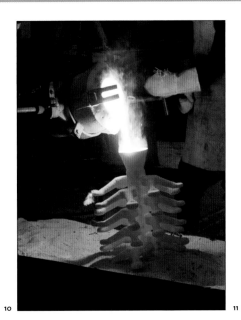

9

10

11

12

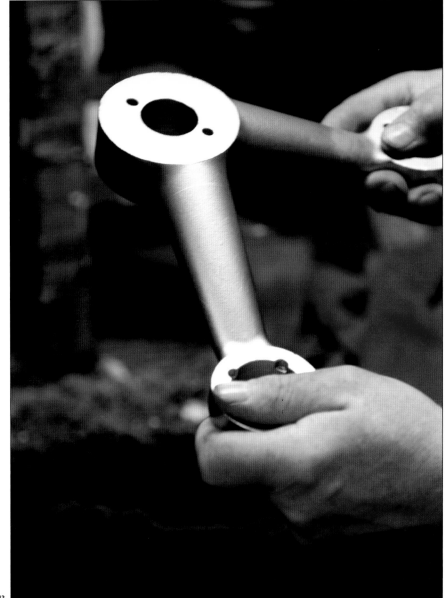

13

Featured Manufacturer

Deangroup International
www.deangroup-int.co.uk

Forming Technology
Metal Injection Molding

This process combines powder metallurgy with injection molding technology. It is suitable for the production of small parts in steel, stainless steel, magnetic alloys, bronze, nickel alloys and cobalt alloys.

Costs	Typical Applications	Suitability
• High tooling costs • Moderate to low unit costs	• Aerospace • Automotive • Consumer electronics	• High volume production • Low to medium volume production for certain applications

Quality	Related Processes	Speed
• Very high quality surface finish • High level of density	• Die casting • Forging • Investment casting	• Rapid cycle time similar to injection molding (30–60 seconds typically) • Debinding and sintering (2–3 days)

INTRODUCTION

Metal injection molding (MIM) is a powder process and a similar technique, known as powder injection molding (PIM), is used to shape ceramic materials and metal composites.

MIM combines the processing advantages of injection molding with the physical characteristics of metals. It is thereby possible to form complex shapes, with intricate surface details and precise dimensions. Parts are ductile, resilient and strong, and can be processed in the same way as other metal parts, including welding, machining, bending, polishing and electroplating.

This process is suitable for forming small components generally up to 100 g (3.53 oz). As with conventional injection molding (page 50), MIM is predominantly used for high volume production, although low to medium volumes can be viable where MIM offers technical, design or production advantages over other manufacturing processes.

TYPICAL APPLICATIONS

MIM is capable of producing a wide range of geometries and so is utilized in many industries. The accuracy and speed of the process make it ideal for manufacturing components for the aerospace, automotive and consumer electronics industries.

RELATED PROCESSES

Investment casting (page 130), die casting (page 124), forging (page 114) and CNC machining (page 182) can be used to produce similar geometries. In fact, investment casting and MIM are often interchangeable, depending on tolerances and intricacy of features. While die casting can produce similar features and tolerances to MIM, it can be used only for non-ferrous materials, not for steels and high melting point metals. Forging is generally chosen for larger components, typically those weighing more than 100 g (3.53 oz), in ferrous and non-ferrous metals, but forging cannot

produce the intricate features and accuracy of MIM.

MIM reduces, or eliminates, the need for secondary operations such as machining. It is capable of producing parts with a complexity and intricacy that may not be feasible in any other processes. Secondary operations can also work out much more expensive for large volumes of production.

QUALITY

Like injection molding, the high pressures used in this process ensure good surface finish, fine reproduction of detail and, most importantly, excellent repeatability. However, MIM has the same potential defects, including sink marks, weld lines and flash. Unlike many plastics, metal parts can be ground and polished to improve surface finish.

Careful design will eliminate the need for secondary operations. This is especially important for flash at joint lines, which becomes a metal burr

Metal Injection Molding Process

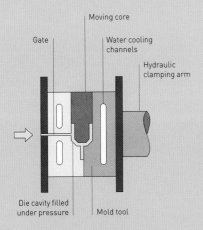

Gate | Moving core
Water cooling channels
Hydraulic clamping arm

Die cavity filled under pressure | Mold tool

Stage 1: Injection molding

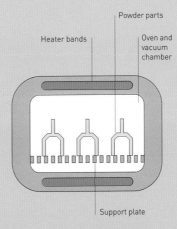

Powder parts
Heater bands | Oven and vacuum chamber

Support plate

Stage 2: Heating and sintering

after sintering. This is normally readily removed after molding, but may be problematic in surface textures and threads. Incorporating flat areas in threads and locating the partition line in non-critical areas reduce such problems.

After sintering, MIM parts are almost 100% dense and have isotropic characteristics. In other words, there is very little porosity. This ensures the structural integrity and ductility of the metal part (see artwork, above, stage 2).

DESIGN OPPORTUNITIES

Design opportunities and consideration for MIM are more closely related to injection molding than to conventional metalwork. For example, MIM can reduce the number of components and subsequent secondary operations of conventional metalworking because it can produce complex and intricate geometries in a single operation. Tools can have moving cores and unscrewing threads for example, making the inclusion of internal threads, undercuts and blind holes possible.

Complex tooling, because it comprises moving parts, may increase tool costs significantly, but if it eliminates secondary operations on the MIM parts the extra costs can be offset, particularly where large volumes are required.

TECHNICAL DESCRIPTION

Fine metal powder, typically no larger than 25 microns (0.00098 in.) in diameter, is compounded with a thermoplastic and wax binder. The spherical metal particles make up roughly 80% of the mixture. The manufacturers themselves often make up this feedstock and so the exact ingredients are likely to be a well-guarded secret.

For MIM, the injection molding machines are modified slightly in order to accommodate the plasticized powder composite material.

In stage 1, the injection cycle is much the same as for other injection molding processes, although the molded parts are roughly 20% larger in every dimension prior to heating and sintering. This is to allow for the shrinkage, which will take place as the binder is removed.

In stage 2, the 'green' parts are heated in a special debind oven, to vaporize and remove the thermoplastic and wax binder. The molded parts are now in pure metal, with all binder material removed.

The final stage is to sinter the parts in a vacuum furnace. This typically starts with a nitrogen/hydrogen mixture, (depending on material type) changing to a vacuum as the sintering temperature increases. The 'green' part will shrink roughly 15–20% during sintering, to accommodate the loss of material during debinding. The resulting metal part has very little porosity.

Certain parts, especially those with overhanging features, are supported in specially designed ceramic support plates so that they do not sag at high temperature. Parts with flat bases usually do not require additional support.

Left
These bent samples demonstrate the structural integrity and ductility of MIM material.

Left
This range of parts has been produced by the MIM process.

Without any additional costs textures and other surface details can be incorporated into the molding process, eliminating these as finishing operations.

Holes and recesses can be molded into the metal part and it is feasible to produce these with a depth to diameter ratio of up to 10 to 1.

DESIGN CONSIDERATIONS

Like conventional injection molding, when designing for MIM it is essential to involve the manufacturer in the development process to ensure that the part is designed to optimize the full advantage of the MIM process.

A significant consideration and limitation on the process is the size of part. The general size range is from 0.1 g to 100 g (0.0035–3.53 oz), or up to 150 mm (5.91 in.) in length. Such size restrictions relate to the sintering operation, which removes the plastic matrix and causes the part to shrink considerably. Large parts and thick wall sections are more likely to distort during the heating and sintering process.

Wall thickness is generally between 1 mm and 5 mm (0.04–0.2 in.). It is possible to produce wall thicknesses of 0.3 mm to 10 mm (0.012–0.4 in.), but such extreme measurements may cause

problems. Like injection molding, the wall thickness should remain constant.

To maintain a uniform wall thickness and keep material consumption to a minimum, it is necessary to core out bulk parts with recesses, blind holes and even through holes.

Internal corners are a source of stress concentration and so should have a minimum radius of 0.4 mm (0.016 in.). Outside corners can be sharp or curved, depending on the requirements of the design. Draft angles are not normally required, except for long draw faces, which makes the designer's job easier.

→ Metal injection molding a cog from a window lock mechanism

The compounded metal powder and thermoplastic and wax binder are formed into pellets for injection molding (**image 1**). The metal powder is very fine and the particles spherical to give a final dense sintered structure.

The injection molding equipment is similar to that used for plastic injection (**image 2**). The mixed materials are not as fluid as thermoplastics. The tooling is generally machined from tool steel and can have moving cores, inserts and complex ejector systems for intricate shapes (**image 3**). On release from the injection molding equipment, the parts are a dull greyish colour – which is referred to as 'green' state – and they are surprisingly heavy, due to their high level of metal content. They are replaced onto support trays, which are stacked up in the debind oven (**images 4** and **5**). This stage is often manually operated, but is suitable for automation if the volumes justify it.

The binder is completely removed in the debind oven, then high temperature sintering causes the metal particles to fuse together. The finished parts have a bright lustre (**image 6**). They are solid metal, with very little porosity. MIM parts have good mechanical properties and are stronger and less brittle than conventionally sintered components.

COMPATIBLE MATERIALS

The most common metallic materials for MIM are ferrous metals including low alloy steels, tool steels, stainless steels, magnetic alloys and bronze. Aluminium and zinc are not suitable, and parts in these materials can usually be made as die castings.

COSTS

Tooling costs are similar to plastic injection mold tools. The MIM materials are typically high cost, due to the level of processing and pretreatment necessary to make them suitable for the injection molding operation.

Injection cycle time is rapid: like plastic injection, the part is molded in 30–60 seconds. Unlike injection molding, the parts need to have the binder removed and then be sintered, which is usually done over a further 2–3 days, as the debinding time is 15–20 hours, plus several hours for sintering. Labour costs are low for injection molding because it is often fully automated. However, MIM requires sintering, which does increase manual operations and so costs.

ENVIRONMENTAL IMPACTS

Scrap material in the injection molding cycle, including feeders, can be directly recycled. Once the material has been sintered it cannot be recycled so easily, but rejects are rare at that stage as parts are accurate and repeatable.

The plastic binder is vaporized during the debinding operation and collected in waste traps, without any environmental problem.

Featured Manufacturer

Metal Injection Mouldings
www.metalinjection.co.uk

Bend | Continuous | **Sheet** | Hollow | Bulk | Internal

Forming Technology

Electroforming

Electroforming is electroplating onto non-conductive surfaces. The object being electroformed can be used as a mold, or encapsulated. Electroforming is an extremely precise technique for reproducing sheet geometries.

METAL

140

Costs	Typical Applications	Suitability
• Low tooling costs generally • High unit costs, partly dependent on electroforming material	• Architecture and interiors • Biomedical • Jewelry, silversmithing and sculpture	• One-off to batch production

Quality	Related Processes	Speed
• Very high: exact replica of the mold with relatively uniform wall thickness	• CNC machining • Investment casting • Laser cutting and engraving	• Very long cycle time (a few hours up to several weeks), depends on the materials and thickness of electroforming

INTRODUCTION

Electroforming is the same as electroplating (page 364), but it is carried out on non-conductive or non-adherent metal surfaces (such as stainless steel). It is made possible by covering the non-conductive material (mandrel) with a layer of silver particles. This forms a surface onto which metal is deposited.

The process builds up the metal layer in a gradual and precise manner. The electroformed product will be an exact replica of the mandrel. It is suitable for very small parts as well as large and complex shapes.

The electroforming process is carried out in solution and on a mandrel. Therefore, part geometry is not restricted by conventional mechanical problems such as the application of pressure. This means that a perfectly flat surface is made in exactly the same way as an undulating and intricate engraving. Adding ribs, recesses and embossed design features affects neither the cycle time nor the cost of electroforming.

TYPICAL APPLICATIONS

Due to its exceptional accuracy, electroforming is used in applications that are dimensionally precise. For example, it is popular for biomedical equipment, microsieves, filters, optical equipment and razor foils.

It is used to make tools for composite laminating (page 206), embossing and metal stamping (page 82). Electroformed nickel tooling larger than 1 m³ (35 ft³) can work out less expensive than CNC machining (page 182) steel mold tools from solid.

Decorative applications include interior fittings, sculptures, lighting, jewelry and tableware. Metal film props, such as masks and goblets, are also made in this way because electroforming can reproduce very fine details and is suitable for small batch production runs. Examples of mandrels include molded rubber and wooden carvings. Either can be used to produce electroformings or metal encapsulations.

RELATED PROCESSES

Electroforming is unmatched in its ability to reproduce the surface of a mandrel to within sub microns. It is used to replicate

Electroforming Process

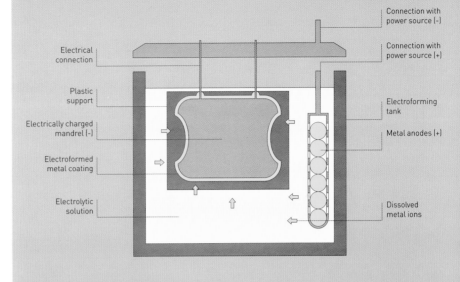

Electrical connection

Plastic support

Electrically charged mandrel (–)

Electroformed metal coating

Electrolytic solution

Connection with power source (–)

Connection with power source (+)

Electroforming tank

Metal anodes (+)

Dissolved metal ions

TECHNICAL DESCRIPTION

The difference in operation between electroforming and electroplating is that the forming process is generally carried out more slowly, for improved accuracy.

The coating on the surface of the mandrel is connected to a DC power supply. This causes suspended metal ions in the electrolytic solution to bond with it and build up a layer of pure metal.

The mandrel is generally designed to be removed and reused; it may also be permanently encapsulated within the electroforming, or dissolved away. Encapsulation can take place on almost any material, including wood, plastic and ceramics. An example of a mandrel material is silicone rubber, which is durable and reusable. They are mounted

onto a plastic support, which holds them in shape during electroforming. Rigid plastics can be molded or machined and used for mandrels for more precise parts.

As the thickness of electroplating builds up on the surface of the workpiece, the ionic content of the electrolytic solution is constantly replenished by dissolution of the metal anodes, which are suspended in the electrolyte in a perforated conductive basket.

The process is indefinite, although wall thickness is generally between 5 microns (0.00019 in.) and 25 mm (1 in.). Process time, which can take anything from a few hours to 3 weeks, is computer-controlled to ensure a high degree of accuracy and repeatability.

holograms. Similar sheet geometries can be made by investment casting (page 130), with its lost wax process, and electroforming is sometimes used to produce the molds for the wax patterns that are used for investment casting.

Electroforming mandrels are made by rapid prototyping (page 232), laser cutting and engraving (page 248), manual engraving, CNC machining, vacuum casting (page 40) and injection molding (page 50).

QUALITY

The metals used in electroforming – nickel, copper, silver and gold – are generally quite soft, with the exception of nickel, and so have to be built up in sufficient thicknesses to produce self-supporting structures. The metal content can be controlled with extreme precision: for example, silver electroformed parts are 100% silver, which is more than sterling silver parts.

The surface finish is an exact replica of the surface of the mandrel in reverse.

→ Electroforming copper

This silicone mandrel is a molding taken from a hand engraving (**image 1**). It is the negative of the product, so will make replicas of the original as electroformings.

The mandrel is coated with pure silver powder (**image 2**), which provides a base onto which the copper will deposit and grow. The whole face needs to be charged with a DC electrical current, so copper wires are connected away from the critical face of the part, on the flange of the mandrel (**image 3**). The mandrel is then submerged in the electroforming tank (**images 4** and **5**), where the electrolytic solution is supplied with fresh ions by pure copper anodes (**image 6**). The copper is deposited on the surface of the mandrel to form a uniform wall thickness.

After 48 hours this part will be fully formed (**image 7**) and have a wall thickness of 1 mm (0.04 in.). It is then removed from the mandrel, which can be cleaned up and reused. The electroformed part is cleaned, too, and trimmed prior to subsequent processing (**image 8**).

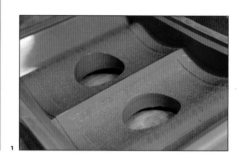

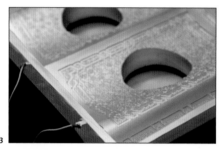

Featured Manufacturer

BJS Company
www.bjsco.com

Therefore, quality of finish is determined by the original product.

DESIGN OPPORTUNITIES
Because mandrels can be made from semi-flexible rubber, intricate shapes and re-entrant angles can be produced with a single mandrel.

Electroformed products can be manufactured in a relatively inexpensive material, such as copper or nickel, which can be deposited at faster rates than other metals. Once the process has been completed, the electroformed part can then be electroplated, to combine the cost effectiveness and strength of the base material with the decorative and hardwearing properties of another material on the surface.

Electroforming can be used to reproduce work that would take too long by hand or any other means: for example, intricate engravings can be reproduced many times with electroforming.

Metal encapsulated wood with gold

This follows the same basic electroforming process as in the case study opposite except that the mandrel is covered entirely by a uniform metal layer and so cannot be reused. The permanently encapsulated mandrel is often used as a pattern to make reusable mandrels for subsequent electroforming operations.

The original wooden carving (**image 1**) is sprayed so it has a silver conductive surface (**image 2**). The spray gun has 2 nozzles, which coat the object in water-based silver nitrate and a reducing agent. Combined, they react to leave a thin film of pure silver (**image 3**). This is the same process as the mirroring of glass.

The metal encapsulated mandrel is then ready to be coated in gold by electroforming. After processing, the part is selectively polished to highlight the carving details (**image 4**).

DESIGN CONSIDERATIONS

This process has high operating costs, which do not come down significantly with an increase in volume.

The wall thickness of an electroformed part is uniform. Depending on the requirements of the application, these are typically between 5 microns (0.00019 in.) and 25 mm (1 in.) thick.

The maximum size for electroforming is determined by the tank size. Typically this is up to 2 m² (21.5 ft²). However, there are tanks large enough to form metal parts up to 16 m² (172 ft²). At this size dimensional tolerance is reduced to 500 microns (0.019 in.). Parts electroformed in very large tanks can take up to 3 weeks to form a sufficient wall thickness for the needs of the application.

COMPATIBLE MATERIALS

Almost any material, such as wood, ceramics and plastic, can be electroformed or encapsulated with metal. Any material that cannot be used as a mandrel can be reproduced in silicone, which is also suitable for the electroforming process.

COSTS

Tooling costs depend on the complexity of the mandrel. Some mandrels are produced directly from a carved master, such as wood or metal, while others are made specially for electroforming, which can be more expensive.

Cycle time depends on the thickness of electroforming. It is also determined by the rate of deposition, temperature and electroplated metal.

Wall thickness on a part is built up at approximately 25 microns (0.00098 in.) per hour for silver and up to 250 microns (0.0098 in.) per hour for nickel. The rate of deposition will influence the quality of the electroforming: slower processes tend to produce a more precise coating thickness than faster ones.

Featured Manufacturer

BJS Company
www.bjsco.com

Labour costs are moderate to high depending on the finishing involved. Some parts require a high level of surface finish, which tends to be a manual job.

ENVIRONMENTAL IMPACTS

Electroforming is an additive rather than reductive process. In other words, only the required amount of material is used and so there is less waste.

However, plenty of hazardous chemicals are used in the process. These are carefully controlled with extraction and filtration to ensure minimal environmental impact.

Forming Technology
Centrifugal Casting

Centrifugal casting is inexpensive and suitable for casting metals, plastics, composites and glass. Because tooling costs are very low, this process is used to prototype as well as mass produce millions of parts.

Costs	Typical Applications	Suitability
• Low tooling costs • Low unit costs, but depends on chosen material	• Bathroom fittings • Jewelry • Prototyping and model making	• One-off to high volume production
Quality	**Related Processes**	**Speed**
• Very good reproduction of fine detail and surface texture	• Die casting • Investment casting • Sand casting	• Rapid cycle time (0.5–5 minutes typically)

INTRODUCTION

Centrifugal casting covers a range of spinning processes used to shape materials in their liquid state. By spinning at high speed, the material is forced to flow through the mold cavities.

The mold material – silicone or metal – differentiates the 2 main types of centrifugal casting. Small parts, from a few grams up to 1.5 kg (3.3 lb), can be cast in silicone molds. This technique is cost effective for producing any quantity of small parts from one-off to mass production because the molds are relatively inexpensive and are quick to make. Materials that can be cast include white metal, pewter, zinc and plastic.

Larger products and higher melting point materials (steel, aluminium and glass) are cast in metal molds.

Centrifugal Casting Process

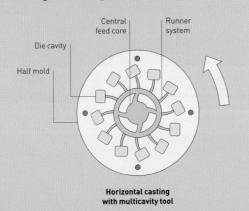

Horizontal casting
with multicavity tool

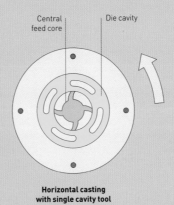

Horizontal casting
with single cavity tool

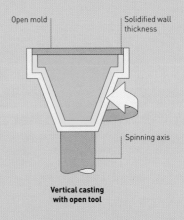

Vertical casting
with open tool

TECHNICAL DESCRIPTION

The tooling for centrifugal casting is either silicone or machined metal. The former material is used to produce small and asymmetric parts. Metals are utilized to cast high melting temperature materials and larger parts. Large molds can only produce parts that are rotationally symmetrical because of the nature of the spinning process. Silicone tooling is extremely cost effective. The pattern can be pushed into the silicone to form the die cavity, or it can be machined.

The 2 horizontal casting techniques shown above use tooling made from silicone or metal. Molten material is fed along the central feed core, which is in the upright position. As the mold spins the metal is forced along the runner system and into the die cavities. Flash forms between the meeting point of the 2 mold halves.

Runners can be integrated into the tooling to encourage air flow out of the die cavity.

Vertical casting methods are similar to rotation molding. The difference is that in centrifugal casting the mold rotates around a single axis, whereas rotation molding tools are spun around 2 axes or more. In operation, a skin of molten material forms over the inside surface of the mold to form sheet or hollow geometries.

This technique is generally chosen for products that are symmetrical around an axis of rotation and can be 1 m (39.4 in.) or more in diameter. The shape and complexity of part is limited by the nature of the process itself.

Composites and powders are also formed in metal molds. This technique is similar to rotation molding (page 36), except that the mold is spun around only 1 axis, instead of 2 axes or more.

TYPICAL APPLICATIONS

Silicone molds are used to produce prototypes, models and mass production parts for a range of industries. The largest areas of application are for jewelry, bathroom fittings and architectural models. These products are within the size limitations of silicone tooling and the low melting point metals are also sufficiently strong for centrifugal casting.

Metal molds are used to cast metal pressure vessels, flywheels, pipes and glass tableware.

RELATED PROCESSES

Regardless of the material used, this is a casting process and many comparable processes are available for shaping a spectrum of materials: for example, metal casting includes sand (page 120), die (page 124) and investment (page 130) techniques. Comparable shapes can be formed in plastic by injection molding (page 50), vacuum casting (page 40) and reaction injection molding (page 64). Also, similar glassware can be made by press molding (page 176) and glassblowing (page 152).

The quality that differentiates centrifugal casting from all these processes is that it combines relatively low costs for small parts in materials with a melting point below 500°C (932°F). The tooling costs for larger parts and higher melting point materials can be relatively small, too, because it is a low pressure process.

QUALITY

Spinning the mold at high speed forces the material to flow through small die cavities. Surface details, complex shapes and thin wall sections are reproduced very well. In small parts tolerances of 250 microns (0.00984 in.) are realistic.

The fact that centrifugal casting is a low pressure process has benefits and limitations for the material quality. For example, materials are not packed into the die cavity – as in injection molding and pressure die casting – so the parts will be more porous. This process therefore tends to be used for parts that do not have to withstand high loads. However, the benefit of low pressures is that there is less distortion as the part cools and shrinks.

Large metal parts that are formed centrifugally have a harder outer surface because a higher concentration of heavier molecules will be forced to the periphery of the die cavity.

Top
Really small parts, such as this 40:1 scale oil can for a model railway, are feasible with centrifugal casting.

Above
A 12:1 scale model of an alloy wheel in pewter,

which is part of a model Raleigh car, displays very fine detailing.

Right
This multi-cavity silicone split mold produces 35 metal parts in a single cycle.

DESIGN OPPORTUNITIES

The main opportunities for designers are associated with silicone mold centrifugal casting. This technique can produce from 1 to 100 parts in a single cycle.

Large parts are cast around the central core. Separate gates feed small parts, which are placed around the central core, so many can be cast in a single mold (see image, right).

Using semi-rigid silicone means complex details and re-entrant angles that are not suitable in hard tooling can be cast. Multiple cores can be inserted by hand for shapes that even silicone will not tolerate.

Because centrifugal casting is a low-pressure process, step changes in wall thickness do not cause significant problems. Also, parts can be fed with molten material from multiple points to overcome small wall sections in between larger sections.

As with other metal casting techniques, metal parts can be molded with inserts and removable cores. Parts can also be polished, electroplated, painted and powder coated.

DESIGN CONSIDERATIONS

In silicone molds wall thickness ranges from 0.25 mm to 12 mm (0.01–0.472 in.),

while in metal molds wall thicknesses can be much larger.

Metals produced in silicone mold centrifugal casting have a low melting point. Therefore, they will not be as strong and resilient as metals formed in other ways. To overcome this problem wall thickness can be increased and ribs added to the part.

COMPATIBLE MATERIALS

Silicone molds can be used to cast some plastics including polyurethane. Metals include white metal, pewter and zinc.

Metal molds are used to shape most other metals, powders (metal, plastic, ceramic and glass) and metal matrix composites.

COSTS

Silicone tooling costs are very low, especially for multiple cavity molds. Metal molds are more expensive, but still relatively inexpensive for their size.

The cycle time for low melting point metals and plastics ranges from 0.5 to 5 minutes. Glass products can take many days to anneal, depending on the thickness of material. Labour costs are generally low.

ENVIRONMENTAL IMPACTS

All scrap metal, thermoplastics and glass can be directly recycled. Small parts require very little energy to produce, while multicavity molds are cost effective and reduce material consumption.

White metal and pewter are alloys of lead. The exception to this rule is British Standard pewter, which contains no lead. Even though it is a naturally occurring material, in large enough quantities lead is a pollutant: for example, it causes problems with the nervous system if ingested in sufficient quantities. Therefore, lead should not be used in products designed for drinking or eating.

→ Centrifugal casting a pewter scale model

This cast pewter part is quite large (**image 1**) and so each mold can produce only 1 part at a time. The silicone mold is assembled with cores (**image 2**). These make it possible to form re-entrant angles. The 2 halves of the mold are brought together and located with pimples and matching recesses on the interface (**image 3**).

The mold is placed onto a spinning table and clamped between 2 metal discs (**image 4**). It is sealed into a box during spinning so that the flash metal can be collected. The mold is spun at high speed. Molten pewter is poured into the central feed core at more than 450°C (840°F) (**image 5**) and enters the mold, where it is pushed through the runner system by centrifugal force. It begins to cool and solidify very quickly. Once the mold is full of metal the spinning is stopped and the pewter is allowed to rest for 3 minutes.

The mold halves are separated and the cast metal part is removed (**image 6**). It is still very hot and so it is set to 1 side and allowed to cool. From molten to solid state it will shrink approximately 6%. Centrifugal casting is a low pressure process, so the part is less likely to distort as it cools.

1)

2

3

4

5

6

Featured Manufacturer

CMA Moldform
www.cmamoldform.co.uk

Forming Technology

Press Braking

This simple and versatile technique is utilized to bend sheet metal profiles for prototypes and batch production. A range of geometries can be formed including bend, continuous and sheet. It is also referred to as brake forming.

Costs	Typical Applications	Suitability
• No cost for standard tooling • Low to moderate unit costs	• Consumer electronics and appliances • Packaging • Transport and automotive	• One-off to batch production

Quality	Related Processes	Speed
• High quality and accurate bends to within ±0.1 mm (0.004 in.)	• Extrusion • Metal stamping • Roll forming	• Cycle time of up to 6 bends per minute • Machine set-up time can be long

INTRODUCTION

Press brakes are fundamental to low to medium volume metalwork. When combined with cutting and joining equipment, press brakes are capable of producing a range of products including continuous, bend and sheet geometries. They tend to be manually operated and are used for one-off, low volume and batch production up to 5,000 units.

Pressure is applied by a hydraulic ram, which forces the metal to bend along a single axis between a punch and die. There are many standard punches and dies, which are used to produce a range of bends with different angles and circumferences (see image, opposite), but always in a straight line.

TYPICAL APPLICATIONS

Press brakes are versatile machines, capable of bending both thick and thin sections up to 16 m (52 ft) long. Examples include lorry sidings, architectural metalwork, interiors, kitchens, furniture and lighting, prototypes and general structural metalwork and repairs.

RELATED PROCESSES

This process tends to be limited to volumes less than 5,000 parts. Each bend is another operation and the machine has to be set up to accommodate each bend. While this is not a lengthy process with modern computer-guided machinery, every second counts in high volume production. Thus large volumes of manufacture often justify processes with higher tooling costs provided they reduce the number of operations and cycle time.

Press Braking Process

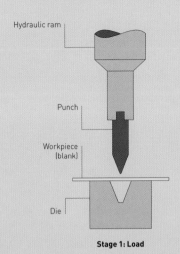

Hydraulic ram

Punch

Workpiece
(blank)

Die

Stage 1: Load

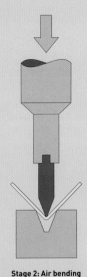

Stage 2: Air bending

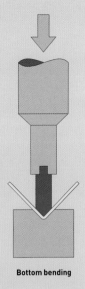

Bottom bending

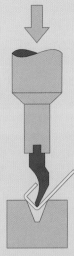

Gooseneck bending

TECHNICAL DESCRIPTION

Press brakes are driven by hydraulic rams, which apply vertical pressure. The tonnage required is determined by 4 factors: bend length, thickness, tensile strength and bend radii. Increasing the width of the lower die increases the bend radii and reduces the tonnage required to make the form. A typical press brake of 100 tonnes (110.2 US tons) will fold 3 m (10 ft) of 5 mm (0.2 in.) steel in a 32 mm (1.26 in.) wide lower die. A narrower die will require more tonnage.

In stage 1, the prepared workpiece is loaded onto the die. In stage 2, pressure is applied by the hydraulic ram so the punch forces the part to bend. Each bend takes only a few seconds. Computer-guided machines reposition themselves between bends. Alternatively, the same bend can be made on a batch of parts before the machine is repositioned for the next operation.

The geometry of the bend determines the type of punch and die that are used; there are many different types, including air bending dies, V-dies for bottom bending, gooseneck dies, acute angle dies and rotary dies. Tools can be laser hardened for improved durability.

Air bending is used for most general work such as precision metalwork, while bottom bending with matched dies (also called V-die bending or coining) is reserved for high precision metalwork because it needs more pressure to operate. Gooseneck dies are for bending re-entrant angles that cannot be accessed by a conventional tool.

Metal stamping (page 82) can produce shapes with complex undulating profiles in a single operation. Design for this process is very different from that of press braking.

Metal folding is similar to press braking: they are both used to form tight bends in thin sheet materials. Metal folding machines are often integrated into a sheet metalworking production line. They are used in the manufacture of thin walled metal enclosures, packaging and electronics housing, for example. The sorts of geometries that can be made include squares, rectangles, pentagons, hexagons and tapered shapes.

Roll forming (page 110) shapes metal sheet into continuous profiles with a constant wall section and is similar to extrusion. Roll forming has the advantage of being able to process almost any metal. Extrusion, on the other hand, is capable of producing hollow profiles with varying wall thicknesses.

Right
Many different shapes of bend can be formed without investment in tooling, using standard punches like these.

QUALITY

Applying a bend to a sheet of material increases its strength, while bending processes combine the ductility and strength of metals to produce parts with improved rigidity and lightness.

Machines are computer guided, which means they are precise to within at least 0.01 mm (0.0004 in.) and can be preprogrammed for any given part.

Even so, the aesthetic quality of press braking is largely dependent on the skill of the operator and their experience with that particular machine.

DESIGN OPPORTUNITIES

Press brakes form continuous bends in sheet material up to 8 m (26 ft) long; 16 m (52 ft) is possible by literally bolting machines together. Using an air bending die, it is possible to bend a range of angles very quickly by depressing the ram only as much as necessary. Acute-angle dies can form sheet material into acute bends down to 30°. Segmented dies can produce bends up to specific lengths, which means multiple bends can be made simultaneously.

Press brakes have the capability of producing long, tapered and segmented profiles, which are not possible with roll forming or extrusion.

DESIGN CONSIDERATIONS

The main limitation of press braking is that it can do only straight line bends. The internal radius is roughly 1 times material thickness for ductile material and 3 times material thickness for hard materials. Bend allowance (also referred to as stretch allowance) is used to calculate the dimensions of a piece of metal post-forming. Generally, it is sufficient to add the length of arc through the middle of the material thickness to calculate the dimensions of the flat net shape.

The punch has to apply pressure along the entire length of the bend. Hollow parts are produced by fabricating sheet geometries post-bending. This is also a consideration for enclosures with 2 close 90° angle bends that form an undercut feature (re-entrant), for example. Punches with overhanging features can be used; they reach into undercuts and apply pressure along the entire length of the bend.

The maximum thickness (subject to the capability of the machine) is approximately 50 mm (2 in.) if the metal

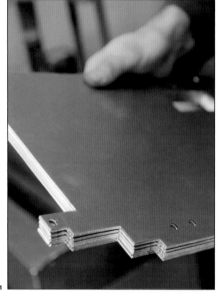

is cold formed. Beyond this, the surface of the material reaches the extent of its ductility and will tear. Introducing heat to the bend area means the process can bend higher thicknesses and the limitation is then machine capability. Maximum dimensions are generally limited by sheet size because lengths up to 16 m (52 ft) have been bent for street lighting, for example.

COMPATIBLE MATERIALS

Almost all metals can be formed using press braking, including steel, aluminium, copper and titanium. Ductile metals will bend more easily and so thicker sections can be formed.

COSTS

Standard tooling is used to produce a wide range of bent geometries: for example, an air bending die can produce a range of angles down to very acute bends. However, this is less accurate than V-dies for bottom bending. Specialized tooling will increase the unit price, depending on the size and complexity of the bend.

Cycle time is up to 6 bends per minute on modern computer guided equipment. Set-up can take a long time but is greatly reduced by a skilled operator.

Labour costs are high for manual operations because a high level of skill and experience are required to produce accurate parts.

ENVIRONMENTAL IMPACTS

Bending is an efficient use of materials and energy. There is no scrap in the bending operation, although there may be scrap produced in the preparation (of the metal blank, for example) and in subsequent finishing operations.

→ Press braking an aluminium enclosure

Aluminium blanks are prepared for press braking by turret punching (page 260) and deburring, guillotining or laser cutting (page 248). In this case study, the 3 mm (0.118 in.) aluminium blanks were cut out using a turret punch. As much of the preparatory work as possible is carried out before the bending process is done, because this is quicker and easier on a flat sheet.

The blanks are processed in batches (**image 1**) so the factory can supply its customers just-in-time (JIT).

In air bending with standard tooling, the metal blank is inserted against a computer-guided stop (**image 2**), which ensures that the part is located precisely prior to bending. The downstroke of the punch is smooth to avoid stressing the material unnecessarily and takes only a couple of seconds (**image 3**). The process is repeated to form the second 90° bend (**images 4** and **5**).

Gooseneck bending (**images 6** and **7**) is used when it would not be possible to form the second bend (**images 8** and **9**) with a conventional punch because of the length of the overhang.

The formed joints are designed in a such a way that the punch can access the joint line as easily as possible and so that bends do not run too close to the edge (**image 10**). Bends that run off at an angle, for example, may have to be formed in matched dies to maintain accuracy.

Finally the joints are TIG welded, ground and polished in preparations for spray painting (**images 11** and **12**).

6

7

8

9

10

11

12

Featured Manufacturer

Cove Industries
www.cove-industries.co.uk

Forming Technology
Glassblowing

Both decorative and functional, hollow and open-ended vessels can be created by glassblowing. The process involves blowing a bubble of air inside a mass of molten glass, which is either gathered on the end of a blowing iron or pressed into a mold.

Costs	Typical Applications	Suitability
• High tooling costs for mechanized production, low for studio glassblowing • Unit cost for mechanized production is low	• Food and beverage packaging • Pharmaceutical packaging • Tableware and cookware	• One-off to high volume production

Quality	Related Processes	Speed
• High quality and high perceived value	• Glass press molding • Plastic blow molding • Water jet cutting and scoring	• Fast cycle time if mechanized • Slow to very slow cycle time for studio glassblowing

INTRODUCTION
Glassblowing has been used to create a multitude of household and industrial products for centuries. However, many of the blown glass products that we use today are manufactured in 2 ways. The first method is known as 'studio glass' and is the production of one-off pieces, generally with artistic expression. The second method is mechanized production and can be divided into 2 main categories: machine blow and blow; and machine press and blow. Mouth glassblowing has been practised for over 20 centuries. Before the development of blowing irons, hollow containers were made using friable cores that could be scraped out once the glass had solidified. High volume mechanized production uses compressed air to blow molten glass into cooled molds, which greatly accelerates the rate of production.

TYPICAL APPLICATIONS
Glassblowing is suitable for a variety of vessels, containers and bottles, which include tableware, cookware, food and pharmaceutical packaging, storage jars and tumblers. Glassware is ideal for applications that require chemical resistance and hygienic qualities.

RELATED PROCESSES
Many items of packaging that were previously made in glass are now produced by blow molding plastic (page 22). This process is used to manufacture domestic, pharmaceutical, agricultural and industrial containers. Glass press molding (page 176) is used to produce open-mouthed and sheet geometries, while water jet cutting (page 272) and

TECHNICAL DESCRIPTION

There are many stages to glassblowing by hand, and everything that is needed for this process must be prepared in advance in order not to waste time and fuel because glassblowing is carried out at temperatures in excess of 600°C (1112°F).

The glass used in studio glassblowing is typically soda lime or crystal glass. It is maintained at more than 1120°C (2048°F) in a crucible, which is accessible through a small hole in the side of the furnace. It is only cooled down when the crucible needs to be removed and replenished, which is approximately every 18 months. The surface of the glass can develop a skin over time; this is periodically skimmed off.

In stage 1, the nose of the blowing iron is preheated in a small kiln, raising its temperature to above 600°C (1112°F). Once it is glowing red hot a small piece of coloured glass is attached to the blowing end. The coloured glass and nose of the blowing iron are dipped into the crucible of molten glass at between 30° and 45°. Molten glass is gathered onto the end of the blowing iron by making up to 4 full turns so that even coverage is achieved. The blowing iron must now be constantly turned for the duration of the process to prevent the glass from slumping.

The parison of hot glass is rolled on a 'marvering' table, which has a polished metal surface. The process of 'marvering' begins the shaping of the glass. In stage 2, air is blown in and it is intermittently inserted into the 'gloryhole' to maintain its temperature above 600°C (1112°F). This is a gas-fired chamber that is used to keep the

Studio Glassblowing Process

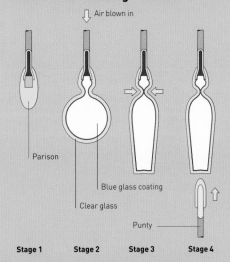

Stage 1   Stage 2   Stage 3   Stage 4   Stage 5   Stage 6   Stage 7   Finished container

glass at a working temperature because if the glass cools below 600°C (1112°F) and is reheated it may suffer thermal shock, which will cause it to shatter.

The molten glass can be marked and decorated in many ways. Coloured glass and silver foils can be rolled onto the surface and then encapsulated into the part with another layer of clear glass. Coloured threads of glass can be trailed over the surface of the parison from a separate gather. These trails can be dragged over the surface of the glass to create patterns, in a technique known as 'feathering'.

Many tools are needed to shape the hot glass when it is being blown. Blocks of wood and paper are used to prepare the parison and shape the glass. The blocks of wood are kept submerged in water so that they do not catch fire when they are held directly against the surface of the glass. Paper pads are sprinkled with water and used in the same

way. In stage 3, profiled formers or molds may be used to shape the glass accurately. Pucellas (sprung metal tongs) are used to reduce the diameter of the glass vessel. Formers are utilized to control the shaping of the glass and to achieve straight-sided vessels, for example.

In stage 4, the workpiece is transferred onto a punting iron, or 'punty'. This is a steel rod with a small glass gather, which is attached to the base of the workpiece by an assistant guided by the glassmaker. In stage 5, the glass vessel is 'cracked off' the blowing iron and in stages 6 and 7 it is shaped from the open end. The finished part is placed in an annealing kiln. Annealing is the process of cooling glass down over a prolonged period. This is essential because stress build-up occurs in glass of varying thickness as it cools at different rates and this can result in shattering. The annealing process gradually relieves these stresses.

scoring (page 276) are used to profile sheet materials.

QUALITY

Glass is a material that has a high perceived value because it combines decorative qualities with great inherent strength. Certain glass materials can withstand intense heating and cooling, and sudden temperature changes.

The structure is weakened only by surface imperfections and impurities in

the raw material. Surface imperfections can be minimized by tempering, and careful mixing and heating of batch glass with cullet (recyclate) will ensure the highest quality of finished product.

DESIGN OPPORTUNITIES

Because the studio glass processes are free from the limitations of mass production, there are many ways that glassmakers can manipulate the shape and surface of the glass to create

exciting effects. A method called 'graal' is used to create sophisticated patterns and textures on the surface of glass by etching into a layer of colour on the surface of a blown and cooled parison, which is then reheated and coated in another layer of glass to form the final product. It is possible to create a layer of cracks in the surface of glass by dipping a hot parison into warm water and then reheating it. The effect of the water is to cover the surface with cracks. Air bubbles

→ Studio glassblowing into a mold

The glass vessel being made here was designed by Peter Furlonger in 2005 and is being blown by the studio team at The National Glass Centre.

The blowing iron is preheated until it is glowing red, around 600°C (1112°F). Then a lump of coloured glass is attached to the end of the blowing iron (**image 1**). It is brought up to working temperature in a 'gloryhole' and 'marvered' on a polished steel table (**image 2**). The coloured glass is then dipped into a crucible of molten glass in a furnace, which is maintained at more than 1120°C (2048°F), and an even coating of clear glass is gathered over the coloured glass (**image 3**). The hot glass is shaped into a parison using cherry wood formers that have been soaked in water (**image 4**).

This process is repeated several times until there is plenty of glass on the end of the blowing iron for the next stage of the process.

All the time the studio glassblower is rolling the hot glass back and forth on the blowing iron. This ensures that the glass does not slump and deform.

The hot glass parison is then laid into a dish of blue powdered glass (**image 5**). Layers of colour are built up in this way to create added 'depth' in the abrasive blasted finish, which is applied later (page 388).

The gloryhole is used to maintain the temperature of the glass at above 600°C (1112°F) and around 800°C (1472°F) (**image 6**), where it feels like sticky toffee, and can be easily shaped and worked (**image 7**).

The glass parison is blown and heated and shaped until it is a suitable size with adequate wall thickness for molding (**image 8**). During blowing the temperature of the glass must be above 600°C (1112°F) so that it can be manipulated: any colder and it will become too rigid.

The mold is preheated to between 500°C and 600°C (932–1112°F) in a small kiln. The blown glass parison is placed in the mold, rotated and blown simultaneously forcing it against the relatively cooler mold walls, which starts to harden the glass (**image 9**). The glass cools down and loses its red glow. The glassblower then uses a blowtorch to produce a metallic effect on the outside of the vessel (**image 10**). This effect will be used to enhance the effects of abrasive blasting.

The parison is in its final shape. It is cut to size by 'cracking-off' the top (**image 11**). The final blown vessel (**image 12**) is annealed before it is finished with abrasive blasting.

1   2

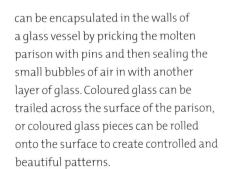

can be encapsulated in the walls of a glass vessel by pricking the molten parison with pins and then sealing the small bubbles of air in with another layer of glass. Coloured glass can be trailed across the surface of the parison, or coloured glass pieces can be rolled onto the surface to create controlled and beautiful patterns.

Mechanized methods are used only for mass production. Very precise detail can be achieved such as screw threads and embossed logos. The tolerances are fine and repeatability is good.

DESIGN CONSIDERATIONS

Studio glassblowing is limited in size only by the dimensions of the 'gloryhole' and what the glassmaker can handle. This process is nearly always carried out by 2 people and so can be expensive. A typical blown part can be made in 20 minutes or so, while more complex and sophisticated techniques will increase the operation times dramatically. The experience of the maker, however, will limit the effects that can be achieved and the rate of production.

Products for mass production glassblowing have to be designed to accommodate the production line. The

high tooling costs and set-up time mean that this process is suitable only for runs of at least 10,000 parts.

Designers must be very considerate about stress concentrations in the final product. Smoothing out the shape as much as possible will reduce stress, although tight radii can be achieved where necessary. Problems may occur in screw cap features, for example. Designers generally work to draft angles of 5°, but this is not a problem when working with round and elliptical shapes. It is advised that parts are kept symmetrical and that the neck is centralized because the product has to fit

3

4

5

6

7

8

9

10

11

12

Featured Manufacturer

The National Glass Centre
www.nationalglasscentre.com

1

2

3

4

5

6

7

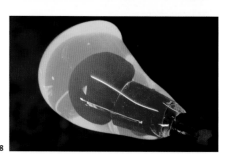

8

9

into conventional production, filling and labelling lines.

COMPATIBLE MATERIALS

Soda-lime glass is the most commonly used for high-volume production. It is made up of silica sand, soda ash, limestone and other additives. Light shades, tableware, cut glass, crystal glass and decorative objects are typically made from lead alkali glass. Borosilicate glass is used for laboratory equipment, high temperature lighting applications and cookware.

COSTS

Tooling costs are high for mechanized production methods, but are low to non-existent for studio glassblowing, where equipment costs are low. For different products, studio glassblowing often utilizes the same tools, which include metal tongs, paper formers, cherry wood paddles and cork tables.

Cycle time is governed by the preparation and annealing required in glass production. Mechanized molding cycles are extremely fast. Beatson Clark produce in excess of 15,000 glass containers every 24 hours (see case study, page 159). Studio glass, on the other hand,

→ Studio glassblowing with coloured effects

Peter Layton is renowned for his use of colour to produce dynamic and engaging works in glass. This is a relatively simple piece designed by him and blown by Layne Rowe; it shows some of the wealth of techniques used by the London Glassblowing studio team.

The 'punty' and blowing irons are preheated in a small gas-fired kiln (**image 1**). A small piece of white glass is attached to the end of a punty together with a 'gob' of clear glass (**image 2**). The white glass is heated to around 800°C (1472°F) and is worked on a 'marver' table into a long thin rod. At this point a gob of molten red and clear glass is gathered onto a punty and allowed to run over the white glass (**image 3**). Overlaying coloured glass builds up strata that add visual depth to the final piece.

The overlaid glass is worked on the marvering table and then swung, which exploits gravity to elongate it. A thin stream of molten blue glass is then trailed across the surface of the overlaid gob in a spiral (**image 4**). At this stage it is difficult to differentiate the colours, as they are all glowing red hot.

At this point, the glass gob comprises a white core, with overlaid red and clear glass and a trailed blue spiral. To enhance the pattern even further, the molten

mass of glass is wound around a punty, which coils the glass and transfers the spiralling pattern from longitudinal to helix (**image 5**). The mixed colour parison is worked on a marvering table and cherry wood formers are used to create a uniform and stable shape. The spiralling pattern within the glass can now been seen (**image 6**). The glass parison, which has now been transferred to a blowing iron, is dipped into the crucible in the furnace and a 'gather' of clear glass forms a coating over the pattern, which is worked in a wood block or former (**image 7**).

The process of gathering and forming is repeated 2 or 3 times, until there is sufficient glass on the end of the blowing iron to start blowing (**image 8**). The blowing process expands the gob of glass, magnifying the pattern as it does so. While the parison is blown it is continuously rotated and worked with damp paper pads to coax the glass into

the desired shape (**image 9**). The temperature of the glass is also maintained throughout the entire process, at more than 800°C (1472°F) in a 'gloryhole' (**image 10**).

Once blowing is complete, the parison is transferred onto a punty (**image 11**). Working on the punty enables the glassblower to develop the shape further; by forming, pulling a stem or flattening it on a cork table, for example. In this case the glassblower is making a bowl and so needs to expand the diameter of the rim. This is achieved with pucellas, which are used to draw out the hot glass (**image 12**). All the time he is moving it in and out of the gloryhole to maintain the required working temperature. Once the bowl is finished, he cracks it off the end of the punty and it is transferred to an annealing kiln for controlled cooling over a period of 36 hours (**image 13**).

GLASSBLOWING

157

10

11

12

13

Featured Manufacturer

London Glassblowing
www.londonglassblowing.co.uk

Machine Glassblowing Process

Machine blow and blow method

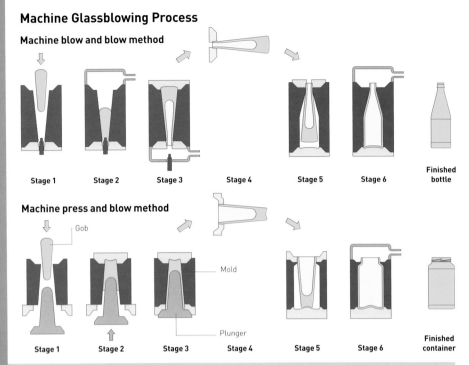

Stage 1   Stage 2   Stage 3   Stage 4   Stage 5   Stage 6   Finished bottle

Machine press and blow method

Gob

Mold

Plunger

Stage 1   Stage 2   Stage 3   Stage 4   Stage 5   Stage 6   Finished container

TECHNICAL DESCRIPTION

The mechanized glassblowing process begins in the mixing department, where the raw materials are mixed together. At this stage they are either coloured with additives, or a decolourant is added to make clear glass. They are fed into the glass-melting furnace with cullet at 1500°C (2730°F) where they fuse together to form a homogenous, molten mass. Glass is drawn through the furnace and conditioned (slowly cooled) to its working temperature of approximately 1150°C (2100°F). This process takes up to 24 hours.

The conditioned glass flows from the bottom of the forehearth and is cut into 'gobs'. These are fed into a bottle-making machine below. There are 2 different molding methods that are used, either press and blow or blow and blow. The processes are essentially the same, except that the parison (pre-form) is either pressed or blown. The press and blow method is more suited to wide-mouth jars, whereas the blow and blow technique is used for containers with a narrower neck.

In stage 1, the molten glass gob is guided along tracks into a parison mold. In stage 2 of the blow and blow method, a plunger rises and presses a neck into the molten glass, and in stage 3, air is

injected through the mold into the formed neck. In the press and blow method, all of this is done by a plunger. In stage 4, the mold opens and a partially formed vessel is released and inverted through 180°. In stage 5, the bottle is transferred to the second blow mold. In stage 6, air is injected through the neck to blow the vessel into its final shape. The glass cools against the sides of the mold before it opens and releases the part. The vessels are then passed through a 'hot end' surface treatment process to apply an external coating, which helps the glass maintain its strength during its working life. The vessels are fed by conveyor belt through the annealing lehr to remove any stress build-up. A second surface treatment is added at the 'cold end' of the lehr to improve the product's resistance to scratching and scuffing. Every container is then subject to rigorous inspections, including sidewall and base scans, pressure tests, bore tests and the flatness of the sealing surface.

is a much slower process that requires great skill and experience. Each product could take between 5 minutes and 2 hours to blow, depending on the number of stages required.

Labour costs are relatively low for mechanized methods and are relatively higher in studio glass, due to the high level of craftsmanship needed.

ENVIRONMENTAL IMPACTS

Glass is a long-lasting material. It is ideal for packaging that will be refilled, especially for food and beverages, which greatly extends a product's useful life. Successful refilling systems, such as the Finnish drinks bottles and British milk bottles, are refilled tens of times before they need to be recycled.

All scrap glass material can be recycled directly in the manufacturing process. Glass is an ideal material for recycling because it can be melted and remanufactured many times without degradation. Even so, more than 1 million tonnes of container glass still make it to landfill every year in the UK alone.

Glassblowing is energy intensive and so there have been many developments in recent years to reduce energy consumption. Improved furnace design and production techniques reduce energy usage and thus reduce the price of production. The raw ingredients of glass can affect its environmental credentials because they are mainly oxides, which will find their way into the atmosphere during production.

Mechanized glassblowing a beer bottle

A 500 ml (0.88 pint) beer bottle can be made using the blow and blow method. The main raw ingredient, silica sand, comprises about 70% of the final product and is piled up inside the factory (**image 1**). The various ingredients are mixed and melted to form molten glass, and after sufficient time in the glass-melting furnace the conditioned glass flows from the bottom of the forehearth and is cut into 'gobs' (**image 2**).

The gobs are fed along tracks to the molds, into which the molten glass settles and the neck is formed. These formed parisons (**image 3**) are then transferred to the blowing mold (**image 4**) in tandem. Robotic arms invert the parisons through 180° as the freshly blown bottles are removed (**image 5**). The split mold closes and the bottles are filled with compressed air, which forces the molten glass onto the surface of the cool mold (**image 6**).

All 16 blowing molds produce bottles continuously all year round (**image 7**) and dispense them on a conveyor belt that transfers the hot glass products (550°C/ 1022°F) to a gas-fired lehr for annealing. Having left the lehr, the bank of bottles are moved towards testing and inspection areas (**image 8**). The finished bottles leave the production line as a single stream and are fed into an automated packaging machine (**image 9**).

1

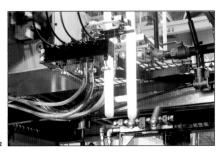

2

3

4

5

6

7

8

9

Featured Manufacturer

Beatson Clark
www.beatsonclark.co.uk

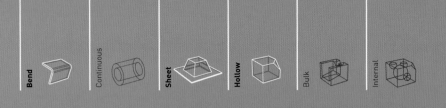
Forming Technology

Lampworking

Glass is formed into hollow shapes and vessels by lampworking, also known as 'flameworking', by a combination of intense heat and manipulation by a skilled lampworker. Products range from jewelry to complex scientific laboratory equipment.

Costs	Typical Applications	Suitability
• Typically no tooling costs • Moderate to high unit costs	• Artwork • Jewelry • Laboratory equipment	• One-off to batch production

Quality	Related Processes	Speed
• Very high quality, but depends on the skill of the lampworker	• Glass press molding • Glassblowing	• Moderate to long cycle time, depending on the size and complexity of part

INTRODUCTION

The process of lampworking has been around since the development of the Bunsen burner around 150 years ago. It is currently used to shape glass into functional and decorative objects.

This is a hot forming process: borosilicate glass is formed at 800–1200°C (1472–2192°F) and soda-lime glass is formed at 500–700°C (932–1292°F). A great deal of skill and experience are required to work hot glass intimately, and therefore this is one of the few industries that still trains apprentice workers.

Lampworking Operations

Blowing

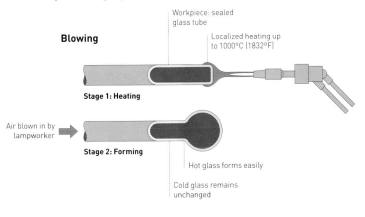

Workpiece: sealed glass tube

Localized heating up to 1000ºC (1832ºF)

Stage 1: Heating

Air blown in by lampworker →

Stage 2: Forming

Hot glass forms easily

Cold glass remains unchanged

Hole boring

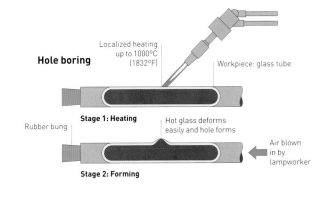

Localized heating up to 1000ºC (1832ºF)

Workpiece: glass tube

Stage 1: Heating

Rubber bung

Hot glass deforms easily and hole forms

← Air blown in by lampworker

Stage 2: Forming

Bending

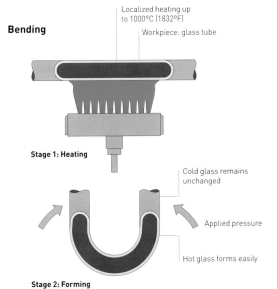

Localized heating up to 1000ºC (1832ºF)

Workpiece: glass tube

Stage 1: Heating

Cold glass remains unchanged

Applied pressure

Hot glass forms easily

Stage 2: Forming

Mandrel forming

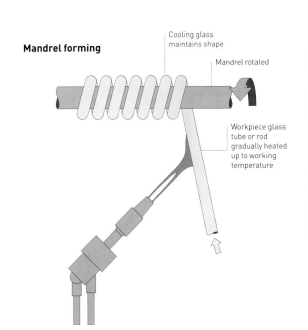

Cooling glass maintains shape

Mandrel rotated

Workpiece glass tube or rod gradually heated up to working temperature

Lampworking is carried out in 1 of 2 ways: benchwork or lathework. Benchwork can be used to create complex, intricate and asymmetric shapes, whereas lathework is suitable for parts that are symmetrical around an axis of rotation (or at least partly so).

Benchwork is generally limited to 30 mm (1.18 in.) diameter tube; it is impractical for the lampworker to hold anything larger and manipulate it at the same time. Latheworking is suitable for much larger tube diameters (up to 415 mm/16.34 in.).

TYPICAL APPLICATIONS

Lampworking is the only way to make certain specialized scientific apparatus and precision glassware. When combined with casting and grinding, lampworking

TECHNICAL DESCRIPTION

Although benchworking and latheworking are carried out in different ways, they use the same basic techniques such as blowing, bending, hole boring and mandrel forming.

A mixture of natural gas and oxygen are burnt to generate the heat required for lampworking. Alternatively, propane is used instead of natural gas, but it burns much hotter and so reduces the temperature range at which the glass can be formed. The working temperature is 800–1200°C (1472–2192°F) for borosilicate glass, at which stage it has the flexibility of softened chewing gum. All parts of the workpiece must be kept at a similar temperature to avoid cracking.

The workpiece can then be rolled, blown, twisted, bent or shaped into the desired shape – with very fine tolerances if needed. The tools used are similar to those for glassblowing: various formers shape the workpiece and 'marver' the hot glass. Tungsten tweezers are used to pull glass across a surface or to bore holes. Trails of coloured glass can be pulled across the surface or metal leaf feathered onto the hot glass.

Everything made in this way has to be annealed in a kiln. For borosilicate glass the temperature of the kiln is brought up to approximately 570°C (1058°F) and then maintained for 20 minutes, after which it is slowly cooled down to room temperature. This process is essential to relieve built up stress within the glass. Some very large pieces, such as artworks, can takes several weeks to anneal.

→ Benchworking a total condensation stillhead

A total condensation stillhead (variable takeoff pattern) is a piece of scientific apparatus used for a specific distillation process. This case study demonstrates some of the techniques used to produce it. Like any other lampworking, a stillhead starts as a series of glass tubes that are cut to length in preparation for the flamework itself.

The lampworker uses a technical drawing (BS 308) to ensure that the product is to be constructed to close tolerances (**image 1**). The first part of the stillhead is brought up to working temperature and starts to glow cherry red (**image 2**). When hot enough the glass can be pulled and manipulated like chewing gum. Tungsten picks are used to

twist and seal the end of the glass tube (**image 3**). From the cool end, air is then blown into the part, forcing the hot end of the tube into an even bubble (**image 4**). At each stage of the process the accuracy of the parts are checked against the drawing (**image 5**).

The blown tube is placed back into the flame and tungsten picks are used to form a hole (**image 6**), which is opened out with a tungsten reamer (**image 7**). In the meantime the second part of the stillhead is brought up to temperature (**image 8**) and drawn out without reducing the wall thickness (**image 9**). It is heated around its circumference and the hot glass is pushed back on itself to create a rib (**image 10**). This will help to locate it in

the previously formed part. Once in place, the 2 parts are heated simultaneously, which causes the glass to fuse and join (**image 11**). A tight bend is achieved at the neck by blowing into the tube – air pressure acts like a mandrel, which stops the walls collapsing – and by forming the bend by hand (**image 12**). This relatively simple subassembly (**image 13**) will form only a small section of the stillhead.

is used to produce everything from test tubes to complex process glassware.

Many industries utilize the versatility of lampworking in glass for prototyping and plant equipment. It is also used in the jewelry industry to make beads and other decorative pieces. Architects and artists alike have applied this process to lighting, sculptures and functional parts of their designs.

RELATED PROCESSES
Lampworking needs a great deal of skill; even simple shapes can be time-consuming to produce. Geometries that are suitable for glassblowing (page 152) or press molding (page 176) will therefore be made in that way as soon as the volumes justify the tooling costs.

QUALITY
The quality of the part is largely dependent on the skill of the lampworker. During operation, the parts have to be carefully heated and cooled for each

forming, cutting and joining action. Even tiny stresses in the part will cause the glass to shatter or crack. Annealing the parts after lampworking ensures they are stabilized and stress free.

DESIGN OPPORTUNITIES
The size, geometry and complexity of a part are limited only by the imagination of the designer. Lampworking is equally suitable for precise functional objects and decorative fluid artefacts.

DESIGN CONSIDERATIONS
Everything made in this way comes from a combination of tube and rod glass.

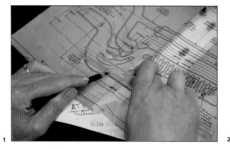
1

2

3

4

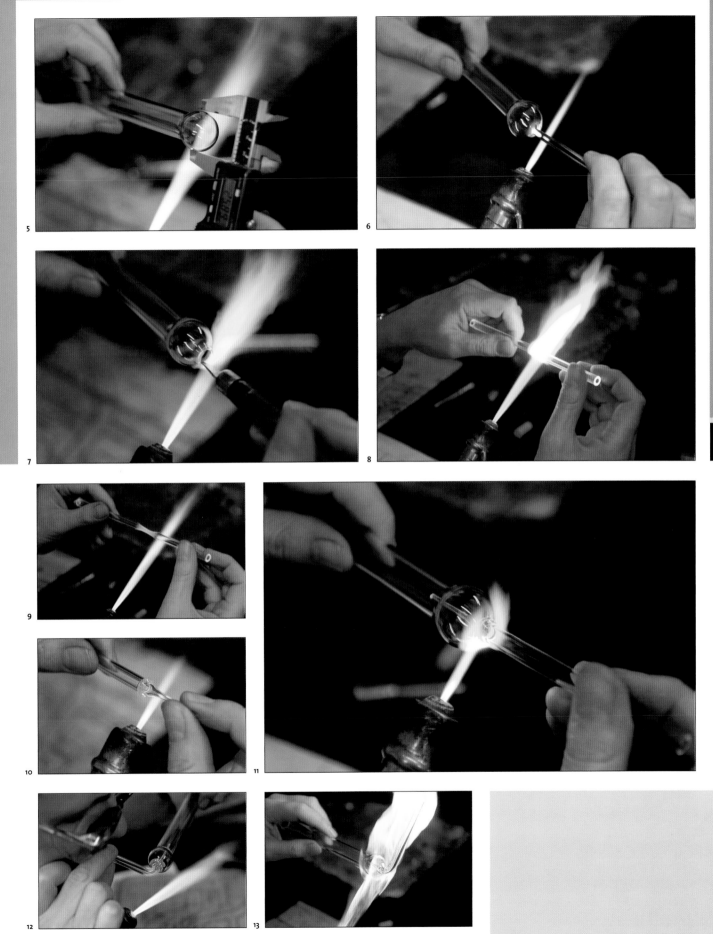

The glass must have equal coefficient of expansion (COE) to be compatible. COE is the rate at which the glass molecules expand and contract during the heating and cooling process. A part made up of glass with different COE values will crack or shatter because of the build-up of stress within the material.

The various ranges of glass are idiosyncratic: they have different colour options, operating temperatures and prices. The lampworker will advise the designer as to which is most suitable for their particular application.

COMPATIBLE MATERIALS
All types of glass can be formed by lampworking. The 2 main types are borosilicate and soda-lime glass. Borosilicate glass is a 'hard glass' known under the leading trade names Pyrex®, Duran® and Simax®. It is very resistant to chemicals, so is ideal for laboratory equipment, pharmaceutical packaging and preservation jars. Soda-lime glass, on the other hand, can be found in domestic applications such as packaging, bottles, windows and lighting. It is known as 'soft glass', because it has a lower melting temperature. Soda-lime glass is less expensive; unlike borosilicate, however, it cannot be mended or reformed once it has been annealed. One of the most common artistic glasses is Moretti Glass, manufactured in Murano, Italy.

COSTS
There are usually no tooling costs, while cycle time is moderate, but depends on the size and complexity of the part. The annealing process is usually run overnight, for up to 16 hours, but can take a lot longer depending on material thickness. Very thick glass may take several months to anneal, because in such cases the temperature is lowered by only 1.5°C (2.7°F) per day.

Labour costs are high due to the level of skill required – lampworking being complex and therefore not suitable for automation. Where possible, glassblowing and press molding are used to produce standard parts that are finished or assembled with lampworking, to reduce cycle time and labour costs.

ENVIRONMENTAL IMPACTS
Glass scrap is recycled by the supplier, so no glass is wasted during the process.

A great deal of heat is required to bring the glass up to working temperature. A mixture of natural gas and oxygen are burnt to produce heat. The combustion is non-toxic, but the glass becomes so bright when it is heated that protective eyewear must be worn. The glasses are fitted with didymium lenses, which filter the bright yellow sodium flare and prevent cataracts. They also enable the lampworker to see what he or she is doing.

A glass spiral is created using a graphite-coated mandrel. The glass must reach its optimum temperature just before it is formed over the mandrel; if it is too hot, the glass will stretch rather than bend. Such an operation requires a great deal of skill and experience. The glass tube may sag very slightly just before it is formed over the mandrel (**image 14**).

A U-bend is then worked in another tube section, using a much larger area of heat (**image 15**). Spreading the heat out makes sure that the U-bend will be even across a large diameter. As soon as the glass is up to temperature it is carefully bent by hand (**image 16**) to form the final bend (**image 17**).

The stillhead parts are then almost complete, so assembly can start. This process comprises heating and boring holes and then joining the parts together with a flame. To do this, a very small area is heated (**image 18**) and then blown, forcing the glass to stretch until it can be removed with the tungsten picks. The hole is opened out to the correct diameter with a tungsten reamer (**image 19**). The extension that will be joined to the assembly is prepared by cutting it to length with the flame (**image 20**) and then opening it out and into the profile that will neatly fit the hole (**images 21** and **22**). The 2 parts are brought together and heated until the glass fuses (**images 23** and **24**). A complete stillhead (**image 25**) takes 4½ hours to create from start to finish.

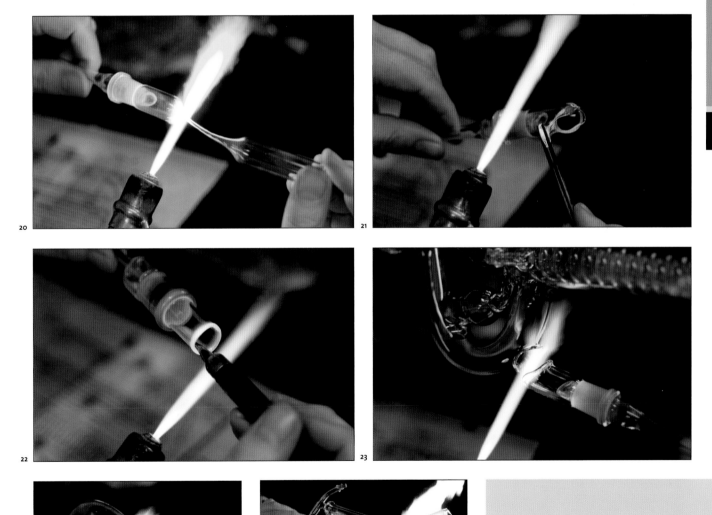

20

21

22

23

24

25

Featured Manufacturer

Dixon Glass
www.dixonglass.co.uk

1

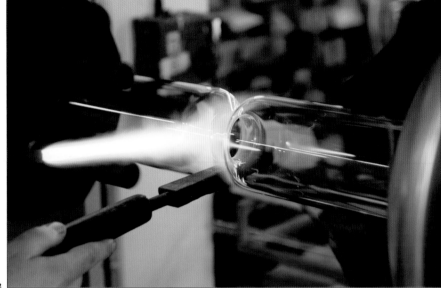

2

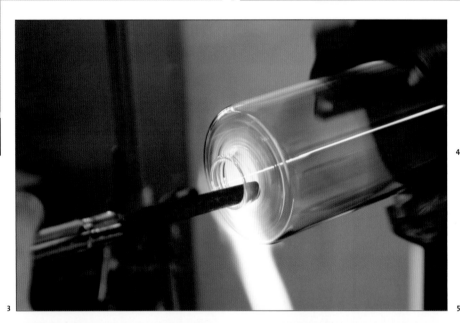

3

4

5

6

7

Featured Manufacturer

Dixon Glass
www.dixonglass.co.uk

→ Latheworking the triple walled reaction vessel

A lathe can be used to form large and small parts very accurately. The only requirement is that they are rotationally symmetrical up to a certain point (the final part may have additions that are not symmetrical).

The operation starts with a large tube section being heated gently to raise its temperature while spinning at approximately 60 rpm (**image 1**). The neck of the innermost jacket is formed with intense localized heat and a profiled carbon bar (**image 2**). The 2 halves of the tube section are separated, and

a hole is opened out using another profiled carbon bar (**image 3**). These form the first 2 reaction vessel jackets, which are assembled off the lathe (**image 4**). Throughout the process the assembly has been spinning to ensure even heat distribution.

A rod of glass is used to pick hot glass from the melt zone (**image 5**) and to bore a hole (**image 6**). The hole is opened out and the 2 parts are fused together using the profiled carbon bar (**image 7**). The process is repeated for the third and final jacket.

Once all 3 jackets of the reaction vessel are brought together, an extension tube is fused onto the bottom, using the heat of the gas torch in much the same way as benchworking (**image 8**). Some extra touches are added by boring and joining onto the outer jacket of the reaction vessel (**images 9** and **10**) to create the semi-finished product (**image 11**).

8

9

10

11

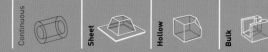

GLASS AND CERAMICS

168

Forming Technology
Clay Throwing

Ceramic products that are symmetrical around an axis of rotation can be made on a potter's wheel. The style, shape and function of each piece can be as varied as the potter who creates it, and each studio adapts and develops their own techniques.

Costs	Typical Applications	Suitability
• No tooling costs • Low to moderate unit costs	• Gardenware • Kitchenware • Tableware	• One-off and low volume production

Quality	Related Processes	Speed
• Variable, because handmade	• Ceramic slip casting • Press molding ceramics	• Moderate cycle time (15–45 minutes), depending on size and complexity • Long firing process (8–12 hours)

INTRODUCTION
Throwing (also known as turning) is used to produce sheet and hollow shapes that are rotationally symmetrical. It is often combined with other processes to make more complex products, with handles and feet, for example.

Clay throwing has been used for centuries all over the world to create a diversity of products. The process relies on the skill of the potter to create high quality and uniform pieces.

TYPICAL APPLICATIONS
Generally clay throwing is used to produce one-off and short production runs of gardenware such as pots and fountains. Kitchenware and tableware, such as pots, jugs, vases, plates and bowls, are also manufactured in this way.

RELATED PROCESSES
Slip casting ceramics (page 168) and press molded ceramics (page 176) use molds and so are more suitable for making identical parts than clay throwing.

QUALITY
Because they are handmade, thrown pots tend to have a variable quality that depends on the skill of the potter and on the material itself. Earthenware is the most commonly used pottery ceramic. It is brittle and porous and has to be sealed with a glaze to be watertight. Earthenware garden products are prone to crack in freezing conditions because they absorb water.

DESIGN OPPORTUNITIES
By the very nature of this process all

Throwing Process

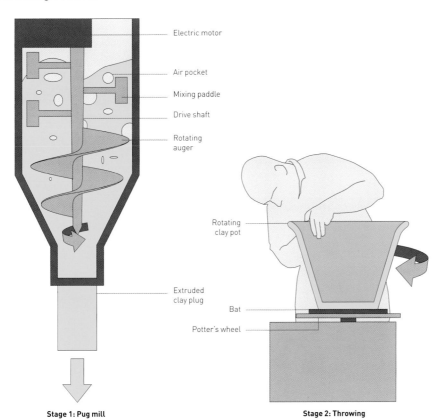

Electric motor
Air pocket
Mixing paddle
Drive shaft
Rotating auger
Extruded clay plug

Stage 1: Pug mill

Rotating clay pot
Bat
Potter's wheel

Stage 2: Throwing

parts will be rotationally symmetrical. To create asymmetric shapes, other skills such as handwork, carving and pressing are combined with clay throwing. Knobs, feet, spouts and other embellishments can be added post-throwing.

DESIGN CONSIDERATIONS

The size of part is restricted by the skill of the potter, the quality of the wheel, the wall thickness and the size of the kiln. Wall thickness ranges from 5 mm (0.2 in.) for small parts up to 25 mm (1 in.).

COMPATIBLE MATERIALS

Clay materials including earthenware, stoneware and porcelain can be thrown on a potter's wheel. Porcelain is the most difficult material to throw and earthenware the easiest, because it is more robust and forgiving (see ceramic slip casting).

COSTS

There are no tooling costs for the clay throwing process. Cycle time is moderate,

but depends on the complexity and size of the part. For example, simple shapes can be produced within 15 minutes or so, whereas tall or particularly large parts may have to be formed in stages, which prolongs cycle time. The firing time can be quite long and is determined by whether the parts are biscuit fired and then glaze fired, or are once-fired only.

Labour costs are moderate because each potter requires a high level of technical ability to throw accurate pieces.

ENVIRONMENTAL IMPACTS

There are no harmful by-products from this pottery-forming process. Any scrap produced during the throwing process can be directly recycled. Once fired, discarded pieces are not reworked into thrown clay unless special effects are required. Some studios, for example, mix 'green' and fired clay with fresh clay to produce speckled and decorative effects.

The firing process is energy intensive. Therefore the kiln is fully loaded for each firing cycle. 'Once-firing' reduces energy

TECHNICAL DESCRIPTION

In stage 1 for wheel thrown pottery, the clay body is 'pugged' in a pug mill. This process has 2 main functions: it thoroughly mixes and conditions the clay for throwing, and it removes some of the air pockets. Some pug mills are equipped with a vacuum pump to remove even more air. Pugging is frequently carried out at the beginning of the day, for 1 hour or so, in order to produce sufficient pugged clay for a day's throwing. Prior to throwing the clay is 'wedged' by hand to improve the working consistency of the material.

In stage 2, the predetermined quantity of clay is 'thrown' onto a 'bat', which is then placed on the potter's wheel. The ball of clay is centred on the spinning wheel, which is generally powered by an electric motor but can be driven by a kick wheel.

While the wheel turns, the potter gradually draws the clay vertically upwards, to create a cylinder with an even wall thickness. Clay throwing must always start in this way, to ensure even wall thickness and distribution of stress, even though the shape can subsequently be manipulated into a variety of geometries.

The bat and thrown part are then removed from the potter's wheel, and air dried together for about an hour, or until the clay is leather-hard. The weather and ambient temperature affect how long clay takes to dry out. At this point the part is trimmed, to remove any excess material, and assembled with any other parts such as feet, handles or stands.

The firing process can be carried out in the same way as with ceramic slip casting and press molding ceramics, or it can be done in a single operation, known as 'once-fired'. Parts with delicate features, such as cups with handles, are biscuit fired first to minimize the risk of breaking during glaze firing.

→ Hand throwing a garden pot

Although premixed clay is supplied to the studio (**image 1**), it still has to be pugged in a pug mill and 'wedged' by hand. The extruded pug of clay is weighed into predetermined quantities and worked by hand into the correct consistency for throwing (**image 2**). Pummelling the clay levels the density of the material and makes it supple for shaping.

The predetermined quantity of clay is then thrown down onto the bat. This action makes the clay denser at the base, which helps to give it strength. The lump of clay is rotated and 'centred' by hand, until the mass of the clay is rotationally symmetrical and can be shaped. The potter then makes a doughnut shape (**image 3**), which is drawn up into a cylinder with even wall thickness (**image 4**). Once the cylinder has reached the desired height the part can be shaped. However, if the clay pot being made is too high to be shaped

in 1 throw, as here, a second level of clay needs to be added. Potters have developed a technique that enables them to build a pot higher than 1 throw would usually allow, but it takes a great deal of skill because the lower part has to match the additional piece very accurately.

The sides of the first part are smoothed (**image 5**) and then the top is finished with a level top, achieved using a flat blade. Its height and diameter are checked (**image 6**) and it is fired with a gas torch, which hardens the clay very slightly and just enough for it to be self-supporting (**image 7**). Once the clay is sufficiently green-hard, the top is trimmed for a second time (**image 8**).

For the second part of the pot, the potter's wheel is prepared (**image 9**) and another ball of clay is thrown onto the bat. The second part is thrown in much the same way as the

first until it is the same diameter and wall thickness at the top (**image 10**). It does not, however, have a solid base. It is turned upside down and placed carefully on top of the first part (**image 11**). The clay parts are aligned and joined together. Once the bat has been removed from the base of the top part, the thrower smooths the joint line and continues to shape the pot upwards (**image 12**).

The completed pot is left to air dry until it is green-hard throughout. Because this pot is to be once-fired, the glaze is applied before the pot has been biscuit fired. The wet glaze is painted on using a small pipette (**image 13**). The pot is placed in the kiln for firing at 945°C (1730°F) for 8–12 hours (**image 14**), after which the finished pot is removed from the kiln (**image 15**).

1

2

3

4

Featured Manufacturer

S. & B. Evans & Sons
www.sandbevansandsons.com

5

6

7

8

9

10

11

12

13

14

15

Bend
 Continuous
 Sheet
 Hollow
 Bulk
 Internal

Forming Technology

Ceramic Slip Casting

Identical hollow shapes with an even wall thickness can be produced with this versatile ceramic production technique. It is used in the manufacture of many familiar household items and remains a largely manual operated process.

Costs	Typical Applications	Suitability
• Low tooling costs • Moderate to high unit costs	• Bathroom whiteware • Kitchen and tableware • Lighting	• Low volume and batch production

Quality	Related Processes	Speed
• Surface finish determined by mold, glaze and skill of the operator	• Clay throwing • Press molded ceramics	• Moderate cycle time (0.4–4 hours), depending on size and complexity • Long firing process (up to 48 hours)

INTRODUCTION

Ceramic slip casting is ideal for the manufacture of multiple identical products. A permanent plaster mold can produce up to 50 pieces before it needs replacing, and automated slip casting techniques can make many thousands of parts. Numerous household objects and tableware items are produced using this casting technique.

The ceramic slip (also known as slurry) is a finely ground (particle size around 1 micron/0.000039 in.) mixture of clay, minerals, dispersing agents and water. Traditionally, the type of ceramic slip used was determined by the location of the factory because the local clay would be incorporated in its production.

Pottery is the general term used to describe ceramic materials that are suitable for slip casting. Well-known types of pottery materials include earthenware and terracotta (characteristically reddish orange but its colouring varies from country to country),

creamware (a type of earthenware made from white Cornish clay combined with a translucent glaze), and stoneware and porcelain (fine, high quality materials that can be fired at high temperatures to enhance their shiny white and sometimes translucent qualities).

TYPICAL APPLICATIONS

Slip casting is used to form a wide variety of household items such as basins, lighting, vases, teapots, jugs, dishes, bowls, figurines and other utilitarian and decorative objects for the bathroom, kitchen and table.

RELATED PROCESSES

Clay throwing (page 168) is suitable for sheet geometries. However, it is generally associated with one-off and low volume production of idiosyncratic products. Press molding ceramics (page 176) is productive and repeatable like ceramic slip casting, but its application is also limited to sheet geometries.

QUALITY

The overall quality of the final piece is largely dependent on the skill of the operator. The materials used in slip casting are generally quite brittle and porous, which means that they are not very tough and tend to fracture rather than deform under load. Earthenware, terracotta and creamware are the most porous and so have to be glazed to be watertight. Stoneware and porcelain, on the other hand, have much better mechanical properties, even though they are still quite brittle.

DESIGN OPPORTUNITIES

This process can be used to produce a range of both simple and complex 3D sheet and hollow geometries. Simple shapes can be slip cast in a single operation, without any assembly: for example, a conical or straight-sided jug with handle and spout can be molded in 1 piece. By contrast, objects with undercuts or other intricate details may need to be molded in several pieces and assembled, or be formed in a multiple-part mold. Assembly operations are preferably avoided because of their cost, but they are sometime unavoidable. It is very important that the product is designed with consideration for the process.

DESIGN CONSIDERATIONS

The slip casting process relies on a porous plaster mold drawing moisture from the slip by capillary action. The slip must be exactly the right consistency and the plaster mold sufficiently dry so that they work in harmony.

The design and construction of the plaster mold will have an impact on the quality of the slip casting. Plaster molds are generally produced directly from the master, which can be made of clay, wood, rubber or other modelmaking material. The parting lines are worked out to optimize production and reduce assembly operations.

Shrinkage is in the region of 8%, but depends on the type of material. Draft angles are not usually a problem because the molds are inward curving.

The size of part that can be slip cast is limited for practical reasons such as weight and fragility of the material. However, large parts, such as shower trays, are feasible for ceramic slip casting.

COMPATIBLE MATERIALS

Ceramic materials such as earthenware, terracotta, creamware, stoneware and porcelain can all be slip cast. The main ingredient for all of the compatible pottery materials is clay, which is a natural material that can be dug up from the ground. It can be mixed

Slip Casting Process

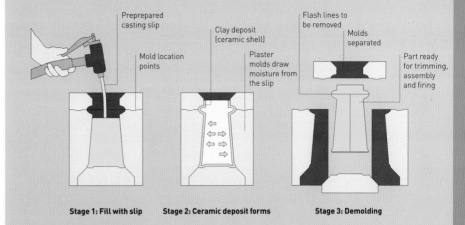

Stage 1: Fill with slip   Stage 2: Ceramic deposit forms   Stage 3: Demolding

TECHNICAL DESCRIPTION

In stage 1, the mold is prepared so it is clean and free from any previous slip contamination. For intricate molds and small details a fine dust can be used to ensure that the slip casting and molds release cleanly. The molds are fastened together with rubber bands – care being taken to ensure that the molds are secure and can withstand the appropriate internal pressure, because slip is a relatively heavy material, about twice the weight of water.

Meanwhile, the casting slip has been prepared by mixing together clay, silicate, soda ash and water. The consistency of the slip is essential for the success of the casting: it must be well blunged and free from lumps. The mold is filled with the slip and left to stand for 5 to 25 minutes. The length of time the slip is in the mold and the ambient temperature determines the wall thickness.

In stage 2, the plaster mold draws moisture from the slip, causing the clay platelets to pile up around the mold wall and create a ceramic deposit (shell). When the ideal wall thickness is achieved the slip is drained (or poured) from the mold. The mold is left to stand for anything between 1 and 24 hours to ensure that the ceramic shell is sufficiently leather-hard to be removed from the mold.

In stage 3, the 'leatherware' is carefully removed from the mold and fettled to remove any flash.

The next stages of processing depend on the surface treatment that is going to be used. The parts are left to air dry so that they are self-supporting and can withstand manhandling. The length of time is determined by the weather conditions, because hot weather will cause the clay to dry out more quickly. Following cutting, assembly and sponging, the parts are further air dried until the ceramic turns whitish in colour, known as 'greenware'. The parts are now ready to be biscuit fired, to remove all remaining moisture. This takes place in a kiln over 8 hours. The temperature of the parts is raised to 1125°C (2057°F) and retained at that level for 1 hour before cooling slowly.

Following the first firing all remaining surface decoration, such as glazing and hand painting, is carried out. The 'biscuitware' is then glaze fired using the same cycle as biscuit firing. When the finished ceramic slip castings are removed from the kiln they are watertight and rigid.

with water and a number of different minerals to create various ceramic slips. Deflocculants are added to the slip to decrease the amount of water required to make it fluid. This is achieved by aiding the suspension of the clay particles in water and reducing the porosity in the final product.

COSTS
Tooling costs are low. The plaster molds are generally produced from a rubber or clay master, which takes a great deal of skill to create. They must be engineered not only to eliminate undercuts, but also to contain the least number of pieces.

Ideally, plaster molds comprise only 2 or 3 pieces in order to maintain cycle times.

Labour costs are moderate to high due to the level of skill required, and they can be especially high in handmade pieces. This is by far the largest expense and it determines the cost of the parts.

ENVIRONMENTAL IMPACTS
During slip casting there can be up to 15% waste. The majority of this waste can be directly recycled as slip. However, if the parts have been fired then the ceramic is no longer suitable for recycling. There are no harmful by-products from these pottery-forming processes.

→ Slip casting the puzzle jug

A total of three molds are used in the production of the puzzle jug. Each mold is cleaned and prepared and the bottom of the mold and the locating points to the main body of the mold are checked (**image 1**). In turn each mold is filled to the brim with casting slip made from earthenware (**image 2**). The molds are left to stand for 15 minutes, or until a sufficient wall thickness has built up as a deposit on the inside mold wall. The remaining slip is poured from the mold into a bath, where it will be recycled for another mold (**image 3**). The slip casting is left in the mold for 45 minutes while the plaster molds continue to draw moisture from the slip and turn it leather-hard (**image 4**). The slip casting can then be demolded, but this has to be carried out very carefully to make sure that it retains its shape (**image 5**). The casting is then left to air dry before any further work is carried out (**image 6**).

The outer skin of the puzzle jug is then pierced using a profiled punch (**image 7**). The floral pattern is marked very lightly on the mold so that each jug will look very similar even though they have been handmade. A slip is used to join the parts together so they will become integral to the structure during biscuit firing. The assembly process consists of inserting and fixing the inner skin (watertight) and outer skin together at the rim, using slip (**images 8** and **9**), and placing the handle on the jug (**image 10**).

After biscuit firing a cream glaze is applied (**image 11**). The glaze is blue so that the operator can see where the glaze has already been applied. Finally the puzzle jug is placed into the kiln for glaze firing, which takes up to 8 hours (**image 12**), after which the finished jug is removed from the kiln (**image 13**).

10

11

13

12

Featured Manufacturer

Hartley Greens & Co. (Leeds Pottery)
www.hartleygreens.com

 Bend
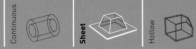 Continuous
 Sheet
 Hollow
 Bulk
Internal

Forming Technology

Press Molding Ceramics

Ram pressing, jiggering and jolleying are all techniques for manufacturing multiple replica ceramic parts with permanent molds. They are used in the production of kitchen and tableware, including pots, cups, bowls, dishes and plates.

Costs	Typical Applications	Suitability
• Low to medium tooling costs • Low to medium unit costs, depending on level of manual input	• Kitchen and tableware • Sinks and basins • Tiles	• Low to high volume production

Quality	Related Processes	Speed
• High quality finish	• Ceramic slip casting • Clay throwing	• Rapid cycle time (1–6 per minute), depending on level of automation • Long firing process (up to 48 hours)

INTRODUCTION

In this process, clay is forced into sheet geometries using permanent molds. Parts are compressed and have an even wall thickness. Press molding is often employed for the mass production of popular ceramic flatware and tiles.

The 2 main techniques used to press ceramics are jiggering (known as jolleying if the mold is in contact with the outside surface of the part rather than the inside surface) and ram pressing. Although both of these processes can be automated, jiggering and jolleying are often carried out as manual operations and can be used only on geometries that are symmetrical around an axis of rotation. Ram pressing can be utilized to make symmetrical shapes as well as oval, square, triangular and irregular ones.

TYPICAL APPLICATIONS

Uses for press molding ceramics include flatware (such as plates, bowls, cups and saucers, dishes and other kitchen and tableware vessels), sinks and basins, jewelry and tiles.

RELATED PROCESSES

Clay throwing (page 168) and ceramic slip casting (page 172) are used to make similar products and geometries. Press molding differs from them by specializing in high volume and the rapid production of identical parts.

QUALITY

The pottery materials used in press molding are brittle and porous (see ceramic slip casting), so the surface is often made vitreous by glazing, which provides a watertight seal.

A very high surface finish can be achieved with both jiggering and ram pressing techniques. When compared to ceramic slip casting and clay throwing, press molding has the advantage of producing parts that are uniform and compressed and therefore less prone to warpage.

DESIGN OPPORTUNITIES

Jiggering and ram pressing can be used to produce most shapes that can be slip cast in a 2 part mold. Using molds increases repeatability and uniformity of part, so even tall objects are easily produced with minimum skill. Manual operations, such as adding handles and spouts, tend to require a higher level of operator skill.

Ram pressing has the advantage of being able to produce shapes that are not rotationally symmetrical. Many design features (handles and decoration, for example) can be pressed directly onto the part to reduce or eliminate assembly and cutting operations. The part is pressed and cut in a single operation, thereby reducing cycle time dramatically.

DESIGN CONSIDERATIONS

The 2 part mold used in ram pressing has to be made to tight tolerances to ensure accurate and uniform reproductions. By contrast, jiggering uses a single mold

Jiggering Process

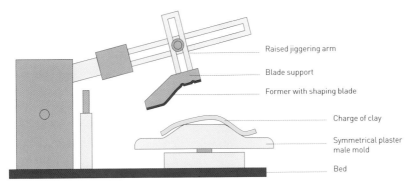

Raised jiggering arm
Blade support
Former with shaping blade
Charge of clay
Symmetrical plaster male mold
Bed

Stage 1: Open mold, loading and unloading

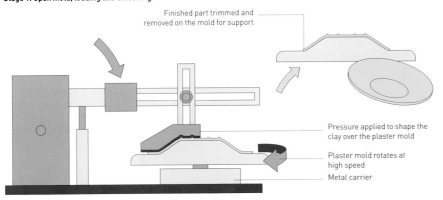

Finished part trimmed and removed on the mold for support

Pressure applied to shape the clay over the plaster mold
Plaster mold rotates at high speed
Metal carrier

Stage 2: Closed mold

and profiling tool to shape the clay. This means that jiggering is more suitable for prototyping and shot production runs.

The heat of the ram press molds leather hardens the clay during the pressing cycle, which means that the part can be removed from the mold immediately. Jiggered parts, on the other hand, have to be left to air dry on the mold. This causes a problem, in that many molds have to be used to form multiple products and these take up a large amount of storage space.

Shrinkage is in the region of 8%, but does depend on the type of material.

COMPATIBLE MATERIALS
Clay materials including earthenware, stoneware and porcelain can be pressed. Unlike ceramic slip casting, the clay is not watered down for these processes. However, it is essential that the material is sufficiently moist to flow during the pressing cycle. The clay used for ram pressing is slightly stiffer.

COSTS
Tooling costs for jiggering are relatively low because a single sided mold is used. For volume production multiple tools are required because the clay has to be left on the mold to air dry.

Ram pressing uses split molds, which are in the region of 10–20 times the cost of jiggering molds. However, they have a higher yield than jiggering molds, more rapid turnover and they last many more cycles (approximately 10,000).

Cycle time is rapid for ram pressing and slightly slower for jiggering. Automated processes accelerate production of press moldings.

Labour costs are higher for manual operations, but for automated techniques they are relatively low.

ENVIRONMENTAL IMPACTS
In all pressing operations scrap is produced at the 'green' stage and so can be directly recycled. Using molds to reproduce the parts accurately reduces

waste caused by inconsistency. There are no harmful by-products from these pottery forming processes.

The firing process is energy intensive, so therefore the kiln is fully loaded for each firing cycle.

TECHNICAL DESCRIPTION
A plaster 'male' mold is used in the jiggering process and a 'female' one for the jolleying process. In stage 1, the mold is mounted onto a metal carrier attached to an electric motor that spins at high speed. A charge of mixed clay is loaded onto the clean mold. In stage 2, as the mold and clay spin, the jiggering arm is brought down onto the clay. A profiled former with a shaping blade, which is different for each mold shape, forces the clay to take the shape of the rotationally symmetrical mold. The profiled former determines 1 side of the clay part and the mold determines the other side. The process is very rapid and takes less than a minute.

Once the final shape is complete and the edges have been trimmed, the mold and clay are removed from the metal carrier intact. The clay part is left on the mold until it is sufficiently 'green' to be removed. If the part is demolded immediately it will deform because the clay is still very supple. The length of time is determined by the weather conditions and ambient temperature: a hot environment will cause the clay to dry out more quickly. If there are secondary operations to be carried out, such as piercing or assembly, then the clay part is transferred onto a 'female' support mold.

→ Jiggering a plate

If numerous identical parts are being made with this process, then multiple plaster molds are required (**image 1**) because the parts have to be left on the molds while they harden to a 'green' state. Each mold is made individually.

To press mold each part, a charge of mixed clay is spun on a jiggering wheel so it spreads out to form an even and uniform 'pancake' of clay (**image 2**). The pancake is then transferred onto the plaster mold (**image 3**). The jiggering arm is brought down onto the clay and a profiled former shapes the outside surface while the mold shapes the inside surface (**image 4**). While the part is still spinning on the mold, the edge is trimmed, to form a perfectly symmetrical shape (**image 5**). The mold and clay part are removed intact, so that the clay can air dry and harden sufficiently before it is removed from the mold for biscuit firing (**image 6**).

In this case the piece of flatware is to be pierced and so after a few hours of air drying the plate is transferred onto a female support mold. The rim can then be pierced without forcing it out of shape (**image 7**). The part is then ready to be biscuit fired, to remove all remaining moisture. This takes place in a kiln over 8 hours. The temperature of the parts is raised and then soaked at 1125˚C (2057˚F) for 1 hour before cooling slowly.

Following the first firing, all remaining surface decoration is carried out such as glazing (**image 8**) and hand painting. The

1

2

3

4

5

6

biscuitware is then placed in specifically designed 'setters' and glaze fired using the same cycle (**image 9**). When the finished ceramic plates are removed from the kiln they are watertight and rigid (**image 10**).

7

9

8

10

Featured Manufacturer

Hartley Greens & Co. (Leeds Pottery)
www.hartleygreens.com

TECHNICAL DESCRIPTION

This is an automated process that forms parts by hydraulic action. Each product will be identical and manufactured to fixed tolerances. The molds are typically made of plaster, although these have a limited lifespan. The die casings can be made to accommodate either 1 large part or several smaller objects.

In stage 1, a charge of mixed clay is loaded into the lower mold. In stage 2, the upper and lower molds are brought together at pressures from 69 N/cm² to 276 N/cm² (100–400 psi). The pressure is evenly distributed across molds and through the clay, which produces even and uniform parts. During the pressing cycle the warm plaster molds draw moisture from the clay, to accelerate the hardening process. The perimeters of each mold cavity on the upper and lower section come together to cut the excess flash from the clay part. The ram pressing process is more rapid that jiggering and can produce up to 6 cycles every minute.

Once the pressing cycle is complete, the molds separate and the parts are instantly released by steam pressure that is forced through the porous molds. The clay parts are self-supporting and can be demolded immediately after they have been ram pressed.

Ram Press Process

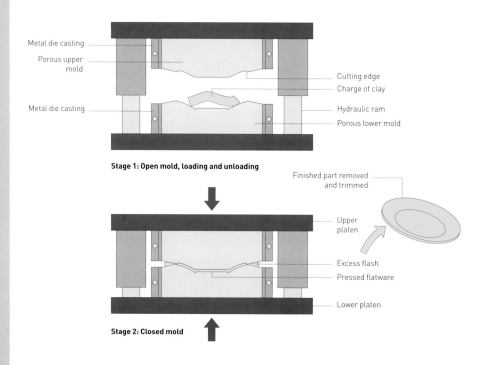

Metal die casting
Porous upper mold
Metal die casting

Cutting edge
Charge of clay
Hydraulic ram
Porous lower mold

Stage 1: Open mold, loading and unloading

Finished part removed and trimmed

Upper platen
Excess flash
Pressed flatware
Lower platen

Stage 2: Closed mold

→ Ram pressing 2 dishes

The clay used in ram pressing needs to be slightly stiffer than for jiggering. Some measured charges of mixed clay are cut from the pug roll (**image 1**). The charges of clay are placed into the lower mold cavity (**image 2**) and the 2 halves of the mold are brought together (**image 3**). The pressure forces the clay to flow plastically through the mold cavity and to squeeze out as flash around the edge (**image 4**). The flash is cut as the molds come together and the edges of the mold cavities make contact.

The molds are separated and any excess flash is quickly removed for reprocessing (**image 5**). The clay parts are instantly ejected by steam pressure, which is forced through the porous plaster mold (**image 6**). The parts are self-supporting as soon as they are demolded because the ram pressing process dehydrates the clay during the pressing cycle, and this hardens it sufficiently for immediate biscuit firing. Ram pressed clay parts take glaze and other decoration very well, as the surface is compressed and uniform.

1

2

3

4

5

6

Featured Manufacturer

Hartley Greens & Co. (Leeds Pottery)
www.hartleygreens.com

Bend
Continuous
Sheet
Hollow
Bulk
Internal

Forming Technology

CNC Machining

Using CNC machining, CAD data can be transferred directly onto the workpiece. The CNC process is carried out on a milling machine, lather or router, and results in a rapid, precise and high quality end product.

Costs	Typical Applications	Suitability
• Low tooling costs • Low unit costs	• Automotive • Furniture • Tool making	• One-off to mass production

Quality	Related Processes	Speed
• High quality finish that can be improved with grinding, sanding and polishing	• Electrical discharge machining • Electroforming • Laser cutting	• Rapid, but depends on size and number of operations

WOOD

182

INTRODUCTION

CNC machining encompasses a range of processes and operations including milling, routing, lathe turning, drilling (boring), bevelling, reaming, engraving and cutting out. It is used in many industries for shaping metal, plastic, wood, stone, composite and other materials. The terminology and use of CNC machining is related to the traditional material values of each particular industry. For example, CNC woodworking is affected by grain, greenness and warpage. By contrast, CNC metalworking is concerned with

CNC Machining Process

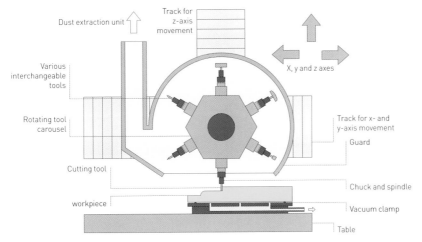

3-axis CNC with tool carousel

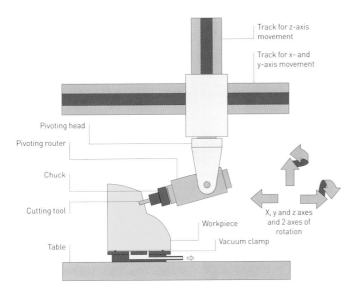

5-axis CNC with interchangeable tools

lubrication, minute tolerances and the heat-affected zone (HAZ).

The number of axes that the CNC machine operates on determines the geometries that can be cut. In other words, a 5-axis machine has a wider range of motion than 2-axis one. The type of operation can also determine the number of axes. For example, a lathe has only 2 possible axes of motion (the depth of cut and position along the length of the workpiece), whereas a router can operate on all 5 axes of possible motion (x, y and z axes, and 2 axes of rotation). The x and y axes are typically horizontal and the z axis is vertical. The 2 axes of

rotation are vertical and horizontal, to achieve 360° of possible movement.

The principles of CNC manufacturing can be applied to many other processes such as ultrasonics, fusion welding and plastic molding.

TYPICAL APPLICATIONS

Almost every factory is now equipped with some form of CNC machinery. It is an essential part of both prototyping and mass production lines. Therefore, applications are diverse and widespread across the manufacturing industry.

CNC machining is used for primary operations such as the production

TECHNICAL DESCRIPTION

Among the many different types of CNC machinery, CNC milling machines and CNC routers are essentially the same. CNC lathes, on the other hand, operate differently because the workpiece is spun rather than the tool. The woodworking and metalworking industries will probably use different names for similar tools and operations – the names and practices can be traced back to when these materials were hand worked using material-specific tools and equipment.

Most modern CNC machinery has x- and y-axis tracks (horizontal) and a z-axis track (vertical). Some older versions, or reconditioned machines, have an x- and y-axis table instead. Beyond this, there are CNC robotics in development that will occupy a space and move freely within it, rather than being fixed to a table with tracks. The new technology relies on data for each part to be preprogrammed, so that it can locate machining and assembly operations for each individual piece as it comes across them.

Many different tools are used in the cutting process, including cutters (side or face), slot drills (cutting action along the shaft as well as the tip for slotting and profiling), conical, profile, dovetail and flute drills, and ball nose cutters (with a dome head, which is ideal for 3D curved surfaces and hollowing out). By contrast, CNC lathes use single-point cutters because the workpiece is spinning.

There are several ways to change the cutting tool and here are two examples. The 3-axis CNC machine has a tool carousel with an array of cutting tools and drills. This accelerates the cutting process dramatically and tool changes are instant. The 5-axis machine has a single tool. On both CNC machines, the tools can be changed by hand, but this is rare. In most cases there is a separate magazine that is loaded with a set of tools that the CNC head will locate and use automatically.

CNC carving the Ercol Windsor chair

CNC machining can be used to cut and shape a wide range of materials. In this case study, it is used to make the various components that comprise the beech Ercol Windsor chair. Ercol is a manufacturer of traditional and contemporary wooden furniture, and they mix craftsmanship with new technology. This combination is very interesting as it illustrates the vast possibilities of CNC machining working alongside other processes.

CNC ROUTING THE SEAT

A plank of wood is prone to warping and buckling. For this reason, the planks used in the seat of the chair are first cut into small widths, which are flipped and glued back together with butt joints (**image 1**). Each section of wood in the plank will then balance the forces of its neighbours. The planks are cut to size and loaded onto the CNC machine table (**image 2**). The parts are pushed firmly onto the vacuum clamp, which is activated to hold the piece in place.

With a slot drill, the 3-axis CNC machine cuts the external profile of the seat (**image 3**), after which the tool carousel is rotated and the edge is profiled with a separate cutting tool (**image 4**). Then, material is removed from the top side of the seat by a dish-shaped cutter, to produce the ergonomic profile required (**image 5**). The process takes less then 2 minutes before the seat is removed from the vacuum clamp. A seal runs around the periphery of the vacuum clamp and it maintains the vacuum within the grooves incised across its surface (**image 6**). This is a very quick and effective method for clamping continuous runs of the same parts. Because the seat is not quite finished it is stacked up for secondary operations (**image 7**).

Holes are required for assembly, but the 3-axis CNC machine is capable only of profiling vertical holes. The seats are therefore loaded onto a 5-axis CNC, which drills holes at the correct angle for the legs and back to be assembled.

1

2

QUALITY

This process produces high quality parts with close tolerances. CNC machining is accurate across 2D and 3D curves and straight lines. Depending on the speed of operation, a CNC machine leaves behind telltale marks of the cutting process. These can be reduced or eliminated by, for example, sanding, grinding or polishing (page 376) the part.

of prototypes to minute tolerances, toolmaking and carving wood. More often, however, it is utilized for secondary operations and post-forming, including the removal of excess material and boring holes.

RELATED PROCESSES

CNC machining is versatile and widely used, competing with many other processes. For toolmaking, it is the most cost effective method of production up to approximately 1 m³ (35 ft³). Any larger and electroforming (page 140) nickel becomes more cost effective.

Laser cutting (page 248) is suitable for profiling and shaping metals and plastics as well as other materials. It is rapid and precise and more suitable for certain applications, such as profiling sheet poly methyl methacrylate acrylic (PMMA). Electrical discharge machining (EDM) (page 254) is used to profile and shape metals. It is used for concave shapes that are not practical for machining.

CNC machining is also used to prototype and manufacture low volumes of parts that can be formed by steam bending (page 198), die casting (page 124), investment casting (page 130), sand casting (page 120) and injection molding.

DESIGN OPPORTUNITIES

CNC machining can be used to produce 3D forms directly from CAD data. This is very useful in the design process, especially for prototyping and smoothing the transition between design and production of a part.

Some CNC machining facilities are large enough to accommodate a full-scale car (and larger, up to 5 x 10 x 5 m/16.5 x 33 x 16.5 ft) for prototyping purposes. Many different materials can be machined in this way, including foams and other modelmaking materials.

3

4

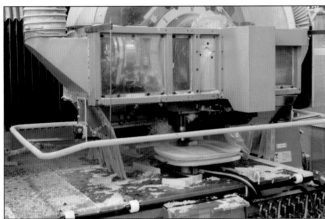

5

6

In operation, there is no difference between simple and complex shapes, and straight or curved lines. The CNC machine sees them as a series of points that need to be connected. This provides limitless design opportunity.

DESIGN CONSIDERATIONS

Most CNC machining is almost completely automated, with very little operator interference. This means that the process can run indefinitely once started, especially if the CNC machine is capable of changing tools itself. The challenge for the designers is to utilize the equipment available; it is no good designing a part that cannot be made in the identified factory.

For parts with geometries that need different cutting heads or operations, multiple machines are used in sequence. Once up and running, the CNC process will repeat a sequence of operations very accurately and rapidly. Changing the set-up is the major cost for these processes,

7

CNC LATHE TURNING THE CHAIR LEGS

The legs and spokes for the chair back are rotationally symmetrical and produced on a CNC lathe. Cut and profiled timber is loaded into the automatic feeder (**image 8**). This in turn is loaded onto the lathe centres (**image 9**), which are 'alive' and rotate with the workpiece (they are known as dead centres when they are stationary). Each workpiece is spun between the headstock and tailstock of the lathe at high speed (**image 10**). Parts with large cross-section are spun more slowly because the outside edge will be spinning proportionally faster.

The cutting action is a single, smooth arc made by the cutting head, which carves the workpiece as a continuous shaving (**image 11**). These images depict the profiling of a chair leg, which requires a second cutting operation to form a 'bolster' (shoulder) in the top for locating the leg within the seat (**image 12**).

Vertical holes are drilled for the cross bars, which make up the leg assembly (**image 13**). The profiled legs are loaded into crates for assembly in batches (**image 14**). They still have a 'handle' attached to the top end, which is waste material removed prior to assembly.

8

9

and subcontractors often charge set-up time separately from unit costs. As a rule, larger parts, complex and intricate shapes and harder materials are more expensive to machine.

COMPATIBLE MATERIALS

Almost any material can be CNC machined, including plastic, metal, wood, glass, ceramic and composites.

COSTS

Tooling costs are minimal and are limited to jigs and other clamping equipment. Some parts will be suitable for clamping in a vice, so there will be no tooling costs.

Cycle time is rapid once the machines are set up. There is very little operator involvement, so labour costs are minimal.

ENVIRONMENTAL IMPACTS

This is a reductive process, so generates waste in operation. Modern CNC systems have very sophisticated dust extraction, which collects all the waste for recycling or incinerating for heat and energy use.

The energy is directed to specific parts of the workpiece by means of the cutting tool, so very little is wasted. Dust that is generated can be hazardous, especially because certain material dusts become volatile when combined.

10

11

12

13

14

ASSEMBLING THE ERCOL WINDSOR CHAIR

Each chair is assembled by hand because wood is a 'live' material which will move and crack and so parts need to be individually inspected. Adhesive is used in all the joints and soaks into the wood grain to create an integral bond between the parts.

Adhesive is added to each joint with a cue tip (**image 15**). The legs and cross bars are then carefully put together and the joints 'hammered home' (**image 16**). The legs come right through the seat and a wedge is driven into the legs' end grain to reinforce the joint (**image 17**). Once the adhesive is dry the excess material is removed from the seat with a belt sander and a shaped sanding block (**image 18**). Sanding also improves the cut finish left by the CNC machining.

Meanwhile, the backrest for each chair will have been steam bent, the vertical spokes profiled to shape on a CNC lathe, and the relevant parts glued together.

After the legs and seat have been assembled the backrest is put in place (**image 19**), the joints being glued and reinforced in the same way as the legs. Once the legs have been cut to the same length, the assembled chair is ready for surface finishing (**image 20**).

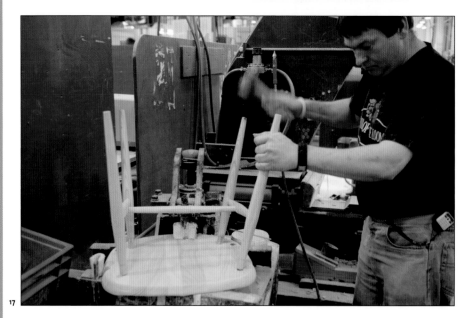

18

19

20

Featured Manufacturer

Ercol Furniture

www.ercol.com

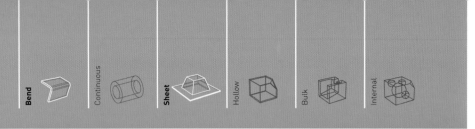

Forming Technology
Wood Laminating

Multiple sheets of veneer or solid timber are formed using molds and bonded together by very strong adhesives, to produce rigid, lightweight structures.

Costs	Typical Applications	Suitability
• Low tooling costs • Moderate unit costs	• Architecture • Engineering timber • Furniture	• One-off to medium volume production

Quality	Related Processes	Speed
• High	• CNC machining • Steam bending	• Medium to long cycle time (up to 24 hours)

INTRODUCTION

There is nothing new about the process of bonding 2 or more layers of material together to form a laminate. However, as a result of developing stronger, more water-resistant and temperature durable adhesives, lighter and more reliable structures can now be engineered in laminated wood and so greater creative opportunities have arisen in design and architecture. There are 3 main areas of wood lamination: solid wood, wood chip and veneer lamination.

Solid wood bending is a cold press process generally limited to a single axis. It consists of bending sections of wood and laminating them together with adhesive. It is typically used to form structural elements for buildings. To make a tighter bend radii possible, the wood can be kerfed, that is slots can be cut into the inside of the bend, perpendicular to the direction of bend. Laminating then acts as a means of locking the bend in place. Kerfing is a useful modelmaking technique, as bends can be formed in solid sections of wood or in plywood without high pressure or even tooling. Sheets precut with kerfing already exist for this purpose.

Wood chip techniques are often used to produce engineering timbers (page 465) for architectural applications such as load bearing beams, trusses and eaves. The products are manufactured using high pressure and penetrating adhesives to bond the wood chips permanently.

Veneer lamination is an exciting process for designers, and over the years it has been used a great deal in the furniture industry by designers such as Alvar Aalto, Walter Gropius, Marcel

Wood Laminating Processes

Kerfing

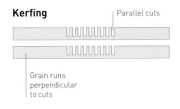

Parallel cuts

Grain runs perpendicular to cuts

Solid wood lamination

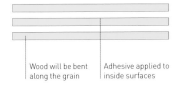

Wood will be bent along the grain

Adhesive applied to inside surfaces

Veneer lamination

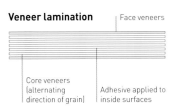

Face veneers

Core veneers (alternating direction of grain)

Adhesive applied to inside surfaces

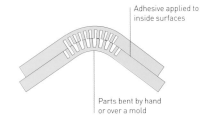

Adhesive applied to inside surfaces

Parts bent by hand or over a mold

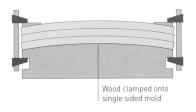

Wood clamped onto single sided mold

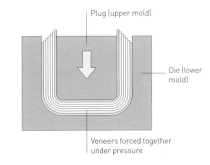

Plug (upper mold)

Die (lower mold)

Veneers forced together under pressure

Breuer and the Eames, and more recently the Azumis and BarberOsgerby. Veneers are laminated onto a single mold, with the addition of a vacuum or split mold. The adhesives are cured by low voltage heating, radiant heating, radio frequency (RF) or at room temperature.

TYPICAL APPLICATIONS

Depending on the adhesive, wood laminating processes can be used to produce articles, such as furniture and architectural products, for use both indoors and outdoors.

Wood chip and solid timber applications include engineering timber products such as trusses, beams and eaves, while veneer laminating and kerfing are used to produce an array of products that include seating, storage and room dividers.

RELATED PROCESSES

Veneer laminating has been replaced over the years by other composite constructions and metal processes. For example, light aircraft were once made from laminated veneers and are now produced in high tech composites of carbon, aramid, glass and thermosetting resins, as developments in technology have made this economically possible.

For certain applications, CNC machining (page 182) is an alternative to wood laminating. Steam bending (page 198) can be used to form solid timber. A combination of these 2 processes is frequently employed to achieve greater flexibility in manufacture.

QUALITY

The quality of a finished wood laminated article is high, although the parts often require finishing operations and sanding. The integral quality of the parts is determined by the grade of timber and strength and distribution of the adhesive. Working with timber requires a skilled workforce, and these processes are no exception.

TECHNICAL DESCRIPTION

KERFING

In this process, a series of parallel cuts are made in one side of the workpiece. This can be done with a band saw, table saw or router. The kerfs are usually between one-third and three-quarters of the depth of the workpiece. Locally reducing the thickness of material makes it more pliable, and a smooth bend can be achieved if the kerfs are cut close together. In the diagram 2 matching kerfed boards have been bent and bonded together to hide the kerfing. Kerfing can also be concealed with thick veneer.

SOLID WOOD LAMINATING

Only gradual and large radii bends can be formed using solid wood laminating. Therefore this process is generally used to strengthen a section of timber by cutting it into sections and bonding them back together in the desired shape. Laminating reduces shrinkage, twisting and warpage, which may be critical to a building project, for example. To maintain the balance between tension and compression, the sections must be inverted on one another if they are cut from the same log, to avoid warpage post-forming.

VENEER LAMINATING

This is carried out in a number of ways, such as onto a single sided mold, with the addition of a vacuum, or in a split mold. All use the same basic principle of laminating an odd number of veneers (plies) perpendicular to each other, as seen in plywood, with adhesive under pressure. The wood is the matrix for the adhesive, which determines the lamination strength.

Adhesive is applied to the face of each veneer as it is laid on top of the last. The lay-up is symmetrical, with a core made up of an uneven number of plies and face veneers of a similar material and equal thickness. This is essential to ensure that the part does not warp. Sheets are laminated this way to improve their resistance to shrinkage, warpage and twisting.

→ Veneer laminating preparation

The method of preparation is the same for all of the veneer laminating processes.

Face veneers are 'bookmatched' or 'slipmatched', depending on the grain of the wood. Slipmatching means taking veneers from the log and placing them next to one another. In bookmatching the veneers are opened next to one another (like a book) to create a symmetrical and repeating pattern. Complex and curving grain is generally bookmatched to avoid it looking too 'busy', whereas straight grained wood is often slipmatched.

In this case study the face veneers have been bookmatched (**image 1**). They are then stitched together on their reverse side with a continuous glass filament coated with a hot melt adhesive (**image 2**). This join needs to hold in place only until the veneers are permanently bonded during lamination.

The core veneers, which are often birch of a lower grade, are rotary cut (peeled) from a log. Much larger sheets can be produced this way than by slicing across the width of the log. They are cut to size on a guillotine (**image 3**). Each layer of veneer is coated with urea-formaldehyde (UF) glue (**image 4**) and then stacked in preparation for laminating.

Curved forms made by laminating have minimal spring-back, unlike those made by steam bending, and are stronger than sawn curves, because the grain of each layer is aligned to the direction of curve and not shortened.

DESIGN OPPORTUNITIES

When combined with other woodworking processes, laminating offers designers a great deal of creative freedom. But designers have to bear in mind that there are tooling costs, which will have an effect on the cost of prototyping. Simple and small molds produced in wood can be low cost, whereas high volume veneer laminating uses metal tools.

Solid wood laminating is limited to simple bends that are generally along a single axis. Veneer laminating is also generally limited to single axis bends. However, shallow dish shapes are possible with conventional veneers. In 2002, the Danish company Cinal

developed a unique bending and layering technique that makes it possible to produce much deeper dish profiles.

Cutting wood into strips, or veneers, and bonding it back together greatly improves its strength and resistance to shrinkage, twisting and warpage. Applying a simple curve to a structure further improves its strength characteristics.

DESIGN CONSIDERATIONS

Laminated veneer constructions are made up of a core section, which in some cases is covered with face veneers to give the product its final look and feel. Laminating relies on a balance between tension and compression. Therefore, its construction is crucial to the stability and strength of the formed part. Veneer laminations are made up of an odd number of veneers, with alternating grain direction, and whatever is applied to 1 side of the laminate must also be applied to the reverse. In other words, if a laminated construction is finished on 1 side with a decorative veneer, then a similar veneer must be bonded onto the opposite side for structural balance. The second veneer does not have to be the same grade, or even species in some cases, if it is out of sight.

The minimum internal radius is determined by the thickness of the individual veneers, rather than by the number of veneers or thickness of the build. This means that even parts with large wall thicknesses can be formed to tight internal radii. Each veneer is typically between 1 mm and 5 mm (0.04–0.2 in.) thick.

COMPATIBLE MATERIALS

The 2 materials that make up this composite are wood and adhesive. There are 2 main types of adhesive used, urea-formaldehyde for indoor applications and phenol-formaldehyde for exterior applications.

Any timber that is cut into veneers or solid planks can be laminated. The wood must be free from defects, such as knots, to ensure an even grain. The most flexible timbers include birch, beech, ash, oak and walnut. However, most other woods can be formed in this way. Even thin sheets of medium density fibreboard (MDF) and plywood are also suitable. Thicker sheets of material should be treated like planks, and kerfed.

Compressed wood , known as Bendywood® (page 468), can be heavily manipulated without high pressure to achieve greater freedom of design in laminating products.

COSTS

Tooling costs are low to moderate for wood laminating. Wood-based products, such as oriented strand board (OSB) and plywood, as well as solid timber can be used to produce molds, which often last for several hundred products. For high quantities the wooden molds are replaced with aluminium or steel ones.

Although cycle time can be long, it depends on the adhesive curing system. RF adhesive curing is generally between 2 and 15 minutes; radiant heat methods take between 10 minutes and an hour; and curing at room temperature takes the longest. After initial curing the products have to be left for up to 7 days to harden fully and dry out.

Labour costs are high for manual operations due to the high level of skill required to ensure consistency of quality parts. Automated processes are rapid and have much lower labour costs.

ENVIRONMENTAL IMPACTS

The various laminating processes require different amounts of energy. For example, manual laminating onto a mold at room temperature requires no energy at all, whereas laminating with RF or heat does require energy, but greatly accelerates the process.

Waste is produced as offcuts post-forming. These offcuts are incinerated and their embodied energy recovered, or they can be reused. In some cases, the offcuts are incinerated to generate steam, which is used to heat the adhesive and accelerate curing.

These processes generally have a low impact, especially if the wood is sourced locally and from renewable sources.

Featured Manufacturer

Isokon Plus
www.isokonplus.com

Case Study

→ Cold pressing the T46 table

The T46 coffee table was designed by Hein Stolle in 1946, but was not produced until 2001. It has a monocoque construction and is formed from a continuous lamination. Cut and prepared veneers of birch ply are loaded into a cold press (**image 1**). The plug (upper mold) is forced into the die (lower mold) using a manually operated screw (**image 2**). A great deal of pressure can be applied in this way, making it a very efficient method of production. As this is a cold process, the table is left in the mold for 24 hours until the adhesive has fully cured. It is then removed from the mold (**image 3**), cut out using a 5-axis CNC router and sanded. Finally the table is sprayed with a matt lacquer (**image 4**).

1

2

3

Featured Manufacturer

Isokon Plus
www.isokonplus.com

4

→ ## Cold pressing the Isokon Long Chair arms

When Marcel Breuer designed the Isokon Long Chair in 1936, he included various laminated parts. This case study describes the production of the arms, the shape of which is too complex to be formed in a single mold. In this case the birch veneers are cut into strips slightly larger than the final profile, to allow for trimming and sanding operations. The veneers are sandwiched between 2 aluminium sheets before being loaded into the cold press mold (**image 1**). The aluminium sheets ensure that the face veneers are not damaged when pressure is applied. Clamps are progressively tightened onto the veneers until the glue is oozing from the laminations (**image 2**). The mold, which comprises several parts (**image 3**) for easy disassembly, is left for 24 hours, after which the part is removed, trimmed, assembled and sanded. The final product is upholstered in red fabric (**image 4**) on a removable seat pad.

1

2

3

4

Featured Manufacturer

Isokon Plus
www.isokonplus.com

→ Radio frequency laminating the Flight Stool

The Flight Stool was designed by BarberOsgerby in 1998. It is produced by Isokon Plus in a split mold, and the adhesive curing is accelerated with RF.

The birch core veneers and walnut face veneers are prepared with adhesive. They are loaded into the metal-faced mold (**image 1**). A copper coil is inserted to connect the metal-faced mold halves before maximum pressure is applied (**image 2**). RF generation is activated, which raises the temperature of the adhesive to

approximately 70°C (158°F) by exciting the molecules. This accelerates the curing process so that the part can be removed from the mold within 10 minutes.

The part is demolded (**image 3**) and held in a jig until it has cooled down. This is to reduce spring-back. The Flight Stool is trimmed, sanded and painted. In 2005, a set of special edition Pantone colours were produced (**image 4**).

1

2

3

4

Featured Manufacturer

Isokon Plus
www.isokonplus.com

→ Bag pressing the Donkey3 storage unit

This process, which uses a vacuum to force the part onto a single sided mold, reduces costs and increases flexibility. However, only shallow geometries like the base of Donkey3 can be formed in this way. This product was designed by Shin and Tomoko Azumi in 2003 and is a development of the original Isokon Penguin Donkey designed by Egon Riss in 1939.

The birch veneers are prepared and a film of adhesive is applied to each surface. The veneers are then laid on a single sided mold (**image 1**). The rubber seal is drawn over the parts (**image 2**) and a vacuum forces the lamination to take the shape of the mold. A heater is introduced to raise the temperature on the mold to 60°C (140°F) and decrease cycle time. After 20 minutes the adhesive is fully cured and the parts can be removed from the mold (**image 3**). The final product is lacquered (**image 4**).

1

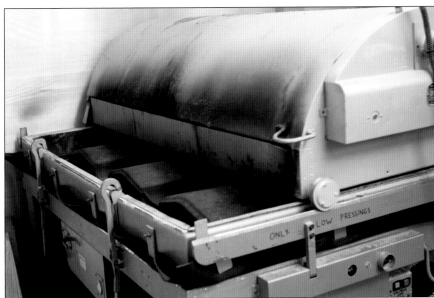

2

3

4

Featured Manufacturer

Isokon Plus
www.isokonplus.com

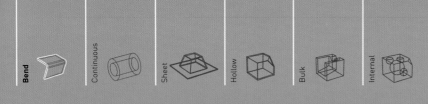

Steam Bending

Certain woods are suitable for bending over a shaped former when they are steamed and softened. This process, which combines industrial techniques with traditional craft, is used to produce tight and multi-axis bends in solid wood.

Costs	Typical Applications	Suitability
• Low tooling costs • Moderate to high unit costs	• Boat building • Furniture • Musical instruments	• One-off to high volume production

Quality	Related Processes	Speed
• Good quality and high strength due to grain alignment	• CNC machining • Wood laminating	• Slow cycle time (up to 3 days)

INTRODUCTION

Bentwood production was industrialized by Michael Thonet during the 1850s, with the production of chair type No. 14 (now known as No. 214). At the time many other manufacturing techniques were going through a similar transition from manual labour to mechanized production. This period, instigated by the invention of the steam engine in the 18th century, is known as the Industrial Revolution. Michael Thonet was a pioneer of mass production; more than 50 million No. 14 chairs were sold within the first 50 years of production.

Steam bending is still used extensively in furnituremaking, boat building and construction. It remains the most effective technique for bending solid wood into both single axis and multiple axis bends. Other techniques, such as laminating, do not require such high quality timber because they are a composite: the wood merely provides a matrix for the adhesive.

TYPICAL APPLICATIONS

Steam bending is commonly used to produce furniture, boats and a range of musical instruments.

RELATED PROCESSES

While CNC machining (page 182) and wood laminating (page 190) can be used to produce similar geometries, steam bending is specified for applications that require the aesthetic appeal and structural benefits of solid wood.

QUALITY

The primary attribute of bentwood is that its grain runs continuously along its entire length. If the bend is made across more than 1 axis, the timber will be twisted to align the grain. In other words, the grain will remain on the same face along the entire length of timber. This gives added strength to the bend and minimizes springback. By contrast, a sawn timber profile will have had its lengths of grain cut through, thus

shortening them and weakening the structure as a whole.

High quality timber is essential for steam bending. Solid timber will not delaminate, although it may split. Splitting will usually occur during bending only as a result of defects such as knots, rot or uneven grain in the wood.

Even though the same jigs are used time and again, no 2 pieces of bentwood will be the same. Therefore, steam bending is not suitable for applications that demand precision.

DESIGN OPPORTUNITIES

The main advantage of bentwood is its strength, which means parts are lighter.

Design in wood requires both an understanding of production technology and the material itself. Wood is a forgiving material. Unlike many plastic, metal and glass processes, which require expensive and complex equipment, wood can be prototyped and even batch produced in a workshop. Steam bending

Steam Bending Process

Circle bending

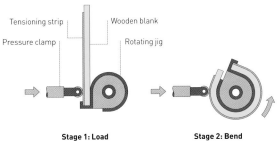

Tensioning strip
Pressure clamp
Wooden blank
Rotating jig

Stage 1: Load   Stage 2: Bend

Open bending

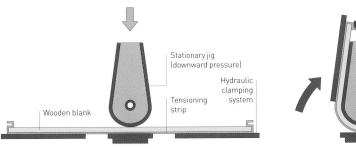

Stationary jig (downward pressure)
Hydraulic clamping system
Tensioning strip
Wooden blank

Stage 1: Load   Stage 2: Bend

TECHNICAL DESCRIPTION
There are essentially 3 main bending techniques. The first is manually operated for bending along multiple axes, and this can take many different forms. The other 2 are power assisted and are used for single axis bends: circle bending is designed for forming enclosed rings, such as seat frames and armrests, whereas open bending is used for open-ended bend profiles such as a backrest.

All of the processes work on the same principle: the wooden blank is steamed and softened. In stage 1 it is clamped in position onto a jig and in stage 2 it is formed by the jig.

Wood is a natural composite made up of lignin and cellulose. To make the wood adequately pliable the lignin, which bonds the cellulose chains together, must be softened (plasticized) to reduce its strength. This is achieved by thermo-mechanical steam treatment. Once the lignin is sufficiently flexible, the wood can be manipulated. The wooden blank is locked in place and the lignin is encouraged to harden in a drying chamber. The whole process can take several days, depending on the dimensions and type of wood.

All of the processes rely on the tensioning strip, which is held on the outside edge of the wooden blank in order to minimize stretching along that edge of the bend. The bend is therefore formed by compression.

can be done with relatively simple equipment: a plastic pipe, a metal water container and a stove. This means that it is possible to explore design ideas quickly and inexpensively.

For complex bends and continuous forms, sections are shaped individually and then joined. This makes it possible to produce almost any shape. The only limiting factor is cost.

DESIGN CONSIDERATIONS

Modern production techniques have minimized springback and maximized consistency. However, wood is a variable material and no 2 pieces will be alike. Therefore, consideration must be given to ensure that the assembly can accommodate some flexibility. Steam bent parts often need to be anchored; otherwise they will gradually unwind over time. For example, an umbrella handle will gradually straighten out because it is a tight bend and the end is not secured.

All of the profiling (shaping of the timber length) must be carried out prior to bending because it is much easier and more cost effective than shaping a piece that has already been steam bent. Simple shape changes and joint profiles, however, are carried out post-bending.

Bentwood profiles tend to be square or circular in cross-section, and these are the easiest profiles to bend. Complex shapes and undulations in the profiles are weak points and are generally avoided. However, tapers are often incorporated and can be used to aid the bending process. For example, thin sections are preferable around tight bends and thicker sections may be required for joints.

The minimum bend radius depends on the dimensions and type of wood.

COMPATIBLE MATERIALS

Hardwoods are more suitable than softwoods for steam bending, and some hardwoods are more pliable than others.

Beech and ash are common in furnituremaking, as they produce tightly packed grain and have good flexibility. Oak is common in construction because it is durable, tough and suitable for outdoor use. Elm, ash and willow are traditionally used in boatbuilding, as they have interlocking grain, are lightweight and have good resistance to water. Maple is suitable for musical instruments because is it decorative, durable and bends well.

Other materials that are suitable for steam bending include birch, hickory, larch, iroko and poplar.

COSTS

Tooling costs are low. The cycle time for steam bending is quite slow, due to the length of each stage in the process:

Steam bending the Thonet No. 214 chair

The classic Thonet No. 214 chair was designed by Michael Thonet in the 1850s and was the first mass produced chair by the company. There were 2 versions, with and without arms (**image 1**), and to date more of these chairs have been manufactured than any other piece of furniture.

The chair is made with beech purchased from within a 113 km (70 mile) radius of the manufacturer's premises. Once the timber has been delivered, it is air dried until it reaches about 25% water content – a state known as 'green' (**image 2**). In this state the timber is sawn into lengths and profiled on a lathe in preparation for bending (**image 3**) – green timber being considerably easier to cut than timber that has been kiln dried.

The profiled square or circular parts are submerged in a bath of softened water at 60°C (140°F) for 24 hours. Then they are steamed at 104°C (219°F) in a pressure chamber (at 0.6 bars/8.7 psi) for 1–3 hours (depending on the size of timber) (**images 4** and **5**). Such a thermo-mechanical process ensures the lignin structure is plasticized sufficiently for bending.

Wood will compress when the lignin becomes plastic, but it will tear very quickly if it is stretched more than 1% of its length. Therefore, the legs and backrest parts are each manually bent by 2 operators and secured in a tensioning strip with clamps (**image 6**). This strip has 2 functions: it stops the outside fibres stretching and helps to control the twist of the wood. The craftsmen are synchronized in their movements as they clamp the bentwood into the metal jig (**image 7**). The part in its jig is loaded into the drying chamber, set to 80°C (176°F), for up to 2 days. The length of time needed to dry the timber to approximately 8% moisture content depends on the size of cross-section.

The parts are removed from the jig so that it can be reused (**image 8**). The pencil mark, which can still be seen, indicates the centre of the length of wood and is used to align it on the jig during bending. This is very important because the part has been tapered already and misalignment would make it a write-off during assembly into the final chair.

In addition to the manual bending illustrated above, there are 2 different methods of power assisted bending: circle and open bending. Circle bending is used to form the chair seat and arms for the No. 214 chair. Like the manual bending, the wood for the arms is held under tension to avoid the outside fibres stretching (**images 9–11**). Once bent and locked into the jig, the part is moved into the drying chamber for up to 2 days, after which it is removed from the jig and retains its bent shape (**image 12**).

9

10

11

12

soaking (24 hours), steaming (1–3 hours) and final drying (24–48 hours).

Labour costs are moderate to high, due to the level of craftsmanship needed.

ENVIRONMENTAL IMPACTS

Steam bending is a low impact process. Often timber is locally sourced: for example, Thonet buy all their timber from within a 113 km (70 mile) radius of the processing plant. Less waste is produced in solid wood bending than in laminating and machining.

Featured Manufacturer
Thonet
www.thonet.de

Forming Technology
Paper Pulp Molding

Molded pulp packaging is made entirely of waste material from the paper industry. The process by which it is manufactured is environmentally friendly as the recycled material needs only the addition of water.

Costs	Typical Applications	Suitability
• Low to moderate tooling costs • Low to moderate unit costs	• Biodegradable flowerpots • Packaging	• Batch and mass production

Quality	Related Processes	Speed
• Variable	• Die cutting • Molded expanded polystyrene • Thermoforming	• Rapid cycle time (5–10 cycles per minute) • Drying time 15 minutes

WOOD

202

INTRODUCTION

Molded paper pulp is a familiar material. Although it has become associated mainly with egg boxes, it is also used extensively for packaging fruit, medical products, bottles, electronic equipment and other products. It is utilized as an enclosure and as a lining, to protect and separate the contents within a box (see image, left).

Pulp packaging provides a useful outlet for industrial and post-consumer paper waste. Shredded paper and card are mixed with water, pulped and molded. It can be reused or recycled and is completely biodegradable.

Air dried paper pulp moldings have a smooth surface on 1 side only. The surface quality is improved by wet pressing or hot pressing after molding.

TYPICAL APPLICATIONS

The primary use of paper pulp molding is for packaging. There are a couple of exceptions, including biodegradable flowerpots and even a lampshade.

RELATED PROCESSES

Molded expanded polystyrene (page 432), thermoforming (page 30) and die-cut corrugated card (page 266) are all used for similar applications. As well as its environmental benefits, paper pulp has many other advantages: it is lightweight, cushioning and protective, anti-static, molded (increases versatility) and relatively inexpensive.

QUALITY

Paper pulp is made up of natural ingredients and so by its very nature is a variable material. Even so, paper

Paper Pulp Molding Process

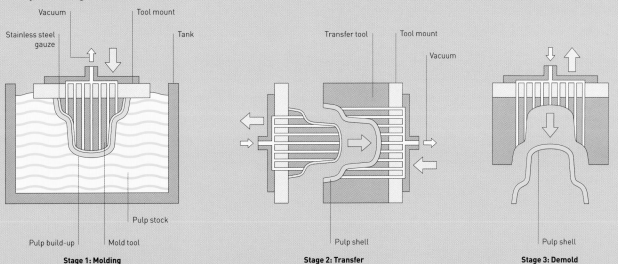

Stainless steel gauze — **Vacuum** — **Tool mount** — **Tank**

Pulp build-up — Mold tool — Pulp stock

Stage 1: Molding

Transfer tool — **Tool mount** — **Vacuum**

Pulp shell

Stage 2: Transfer

Pulp shell

Stage 3: Demold

TECHNICAL DESCRIPTION

The mold tool is typically machined aluminium, and it is covered with a fine stainless steel mesh (see image, top) which acts like a sieve and separates the water from the wood fibres. The tool is covered with holes, roughly 10 mm (0.4 in.) apart, which provide channels for the water to be siphoned up (see image, middle).

The pulp stock is a mixture of 1.4% pulped paper and water. In stage 1, it is stirred constantly to maintain an even mixture as the tool is dipped in it. The molding is formed over the tool by a vacuum, which draws water from the pulp stock, and in doing so begins to dry out the pulp shell that forms over it. The tool is dipped in the stock long enough for it to build up adequate wall thickness, typically 2–3 mm (0.79–0.118 in.). The vacuum is constant, and in stage 2 it holds the pulp onto the surface of the tool as it emerges from the tank.

In stage 3, the pulp shell is demolded into a transfer tool, which continues to apply a vacuum. Typically this tool is polyurethane, which is less expensive to machine than aluminium. The transfer tool places the pulp shell, which is now only 75% water and self-supporting, onto a conveyor belt. The shell is then warmed and dried in an oven for about 15 minutes. Alternatively, it is pressed into a hot tool that dries it out rapidly.

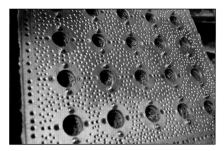

pulp moldings are generally accurate to within 0.1 mm (0.004 in.). They can also be made resistant to grease and water with an additive known as 'beeswax'.

The smoothness of the molding is improved by wet pressing or hot pressing in the drying cycle. With such techniques logos and fine text can be embossed into the molding. Alternatively, parts can be hot pressed in the final stage of air drying. This technique is a cost effective alternative to conventional hot pressing, which requires up to 4 sets of tools to accommodate the progressively shrinking pulp molding.

DESIGN OPPORTUNITIES

The main advantage of paper pulp molding compared to the related processes is its environmental impacts.

Essentially, paper pulp packaging is board formed into sheet geometries. It is therefore typically not an added value material. However, it can be coloured and printed, foil blocked and blind embossed. It is naturally grey or brown,

Top
The face of this tool is covered with stainless steel mesh.

Middle
The tool is covered with holes, for water to be drawn through it and a pulp molding to form.

Above
Cull-UN packaging is UN certified for the transportation of hazardous chemicals. The contents, in a cardboard box, can be dropped onto a steel plate from 1.4 m (4.6 ft), without breaking.

→ Molding paper pulp packaging

The corrugated card waste produced in 1 side of Cullen Packaging's facility is raw material in the paper pulp production plant (**image 1**). A predetermined quantity is loaded into a vat, mixed with water and pulped gradually with a rotating paddle. The process is carried out slowly to avoid shortening the fibre length of the pulp.

The pulp stock, which is 4% water, is fed through a series of vats that remove contamination and condition it in preparation for molding (**image 2**).

The tools are mounted on a rotating arm, and usually work in tandem (**image 3**). It is important that the tools all require roughly the same mass of pulp to maintain even distribution of pulp in the stock. The tools are dipped into the stock for about 1 second (**image 4**), after which they emerge coated with a rich brown layer of pulp (**image 5**). The partially dry

pulp shell is demolded into a transfer tool. The role of the transfer tool is to place the fragile pulp shell carefully onto the conveyer (**image 6**).

The pulp moldings are fed into a gas heated oven (**image 7**). Hot air (200°C/392°F) is gently circulated around the moldings to dry them out. After 15 minutes or so, the completed pulp shells are stacked onto a pallet for shipping (**image 8**).

1

2

depending on the stock. Brown pulp is typically made up of high quality craft paper, whereas grey pulp is composed of recycled newsprint and is not as sturdy or resilient as brown pulp.

Air dried moldings will have 1 smooth side (the tool side) and 1 rough one. The rough side will not show fine details that have been reproduced on the smooth side. Wet pressing improves surface finish. And hot pressing improves surface finish and reduces permeability. Logos, instructions and other fine details can be pressed into a partially dry molding.

DESIGN CONSIDERATIONS

Paper pulp is made up of cellulose wood fibres bonded together with lignin (the 2 main ingredients of wood itself). No additional adhesives, binders or other ingredients are added to strengthen this natural composite material. Therefore, the wall thickness is increased and ribs are designed in for added strength. The ribbing provides added strength

for the part when it is both wet and dry. This is especially important for air dried moldings, which have to be self-supporting within seconds of forming, because they are deposited onto a conveyor belt while still having at least 75% water content.

When fully dry, any ribbing provides added support for the package contents, including cushioning, strength and a friction fit.

Material thickness can range from 1 mm to 5 mm (0.04–0.2 in.), depending on the design needs. Pulp moldings can be up to 400 x 1,700 mm (15.75 x 66.92 in.) and 165 mm (6.5 in.) deep. Vertical dipping machines can make products deeper than 165 mm (6.5 in.).

Draft angles are especially important for this process. The tools come together along a single axis and there is no room for adjustment. The molded part must therefore be extracted along that axis. Draft angles are usually 5° on each side, but will depend on the design of the molding. It must be remembered that, when extracted from the tool, the molded part comprises up of 75% water and so is very supple.

The parts are typically stacked onto a pallet and, because of the draft angle, the parts can be packed more densely, saving on costly transportation expenses.

COMPATIBLE MATERIALS

This process is specifically designed for molding paper pulp. It can be industrial or post-consumer waste, or a mixture of both of these.

It is possible to mix in other fibres, such as flax, with the waste paper material, to reduce further the environmental impact of the product. Additional fibres will affect the aesthetics and strength of the part.

COSTS

Tooling costs are generally low but do depend on the complexity of the molding. Wet pressing and hot pressing

3

4

5

6

7

require additional tooling. As the pulp molding dries it shrinks, and so for hot pressing several sets of tools are sometimes required. Each set of tooling roughly doubles the initial costs.

Cycle time is rapid. Some machines can operate at up to 10 cycles per minute. However, it is more common to mold around 5 parts per minute. The drying time extends the cycle time, adding roughly 15 minutes to the process. Hot pressing is quicker.

Labour costs are relatively low because most operations are automated.

ENVIRONMENTAL IMPACTS

This process for disposable packaging has admirable environmental credentials, because it uses 100% recycled material, which can be recycled again and is fully biodegradable. In the above case study from Cullen Packaging, the paper pulp is made up of by-product from their corrugated card production facility next door, diverting it from landfill.

Waste produced by the process can be directly recycled and averages 3%. Water used in production is continuously recycled to reduce energy consumption.

8

Featured Manufacturer

Cullen Packaging
www.cullen.co.uk

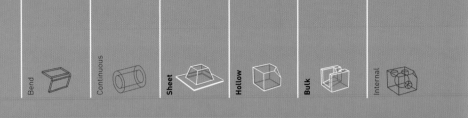

Bend
Continuous
Sheet
Hollow
Bulk
Internal

Forming Technology

Composite Laminating

Strong fibres and rigid plastics can be amalgamated to form ultra lightweight and robust products, using composite laminating. A range of material combinations is used to produce parts that are suitable for the demands of high performance applications.

Costs	Typical Applications	Suitability
• Moderate to high tooling costs • Moderate to high unit costs, determined by surface area, complexity and performance	• Aerospace • Furniture • Racing cars	• One-off to batch production

Quality	Related Processes	Speed
• High performance lightweight products	• DMC and SMC molding • Injection molding • Thermoforming	• Long cycle time (1–150 hours), depending on complexity and size of part

INTRODUCTION

Composite laminating is an exciting range of processes that are used in the construction of racing cars, aeroplanes and sailing boats alike.

There are 3 main types of laminating: wet lay-up, pre-preg (short term for pre-impregnated with resin) and resin transfer molding (RTM) (also known as resin infusion). The resins used are typically thermosetting and so will cure at room temperature. In wet lay-up they are soaked into woven mats of fibre reinforcement, which is draped into a mold. For precision products, autoclaves

are used to apply heat and pressure during the curing phase.

Pre-preg is the most expensive composite laminating process, and therefore is limited to applications where performance is critical. It is made up of a mat of woven reinforcement with resin impregnated between the fibres. The quantity of resin is precise, so each layer of the lamination provides optimum performance. This reduces the weight and increases the strength of the product, which is cured under pressure in an autoclave.

RTM is used for larger volume manufacturing. Unit price is reduced by accelerated production as a result of using split molds, heat and pressure and dividing the labour force into specialized areas of production.

TYPICAL APPLICATIONS
Until recently composite laminating was limited to low volume production. However, since the development of RTM it is now possible to make thousands of identical products. These processes are finding application in the production of automotive parts and bodywork and in the aerospace industry.

Manual laminating processes are generally limited to high performance products, due to high costs. However, designers have long craved the immense potential of carbon fibre and other composites, and so occasionally it is used for domestic products such as furniture. Fibreglass is much cheaper and often the material of choice for such applications.

High performance applications include the production of racing cars, boat hulls and sailing equipment,

structural framework in aeroplanes, satellite dishes, heat shielding, bicycle frames, motorcycle parts, climbing equipment and canoes.

RELATED PROCESSES
Composite laminating is unequalled in its versatility and performance. Processes that can produce similar geometries include dough molding compound (DMC) and sheet molding compound (SMC) (page 218), injection molding (page 50) and thermoforming (page 30). Metalworking processes that are used to produce the same geometries include panel beating, superforming (page 92) and metal stamping (page 82).

DMC and SMC bridge the gap between injection molding and composite laminating.

QUALITY
The mechanical properties of the product are determined by the combination of materials and lay-up method. All of them produce products with long strand fibre reinforcement. The resins used include polyester, vinylester, epoxy, phenolic and cyanate. They are all thermosetting and so have cross-links in their molecular structure. This means they have high resistance to heat and chemicals as well as very high fatigue strength, impact

Top left
Nomex® honeycomb core is used to increase the strength and bending stiffness of composite laminates.

Above left
Nomex® honeycomb core is laminated into Kevlar® aramid fibre epoxy for exceptional

strength to weight properties.

Above
A lightweight aerodynamic racing car spoiler is made by CNC machining PUR foam and encasing it in carbon fibre reinforced epoxy resin.

resistance and rigidity. Laminates cured under pressure have the least porosity.

Processes other than RTM use single sided molds, which produce a gloss finish on only 1 side. However, by joining two 3D sheet geometries together a 3D hollow part is made with an all over molded finish. The surface finish of wet lay-up and RTM products is improved by using a gel coat.

DESIGN OPPORTUNITIES
These versatile processes can be used for general-purpose glass fibre products or high performance application with carbon and aramid fibre. The materials and method of lamination are selected according to the budget and application, which makes these processes suitable for a wide range of prototyping and production applications.

Fibre reinforcement is typically glass, carbon, aramid or a combination. Various weaves are available to provide different

Lay-Up Processes

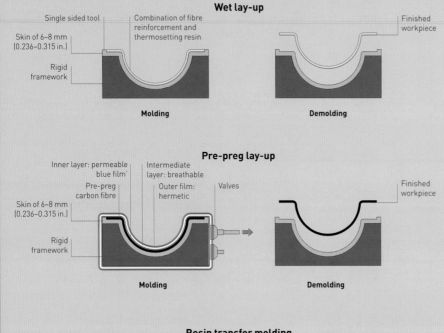

Wet lay-up

Single sided tool · Combination of fibre reinforcement and thermosetting resin · Finished workpiece

Skin of 6–8 mm (0.236–0.315 in.)

Rigid framework

Molding · Demolding

Pre-preg lay-up

Inner layer: permeable blue film' · Intermediate layer: breathable · Finished workpiece

Pre-preg carbon fibre · Outer film: hermetic · Valves

Skin of 6–8 mm (0.236–0.315 in.)

Rigid framework

Molding · Demolding

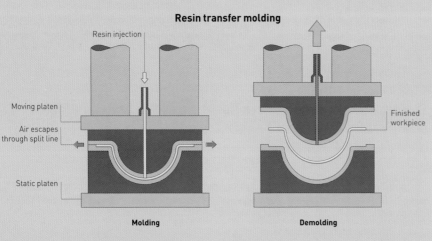

Resin transfer molding

Resin injection

Moving platen

Air escapes through split line · Finished workpiece

Static platen

Molding · Demolding

The 3 main types of composite laminating are wet lay-up, pre-preg and resin transfer molding (RTM).

All types of weave and thermosetting resin can be applied by wet lay-up, which is the least precise of all the laminating methods. The mold is single sided and is made up of a skin of the composite material supported by a rigid framework. It is essential that the mold is not only strong and supportive during lay-up, but is also sufficiently flexible to allow the molding to be removed post-curing.

Wet lay-up is typically started with a gel coat. The gel coat is a thermosetting resin (the same as in lamination), which is painted or sprayed onto the surface of the mold prior to lamination. Gel coats are anaerobic; in other words, they cure when not in the presence of oxygen, which is ideal for the mold face.

Mats of woven fibre reinforcement are laid onto the gel coat, and then thermosetting resin is painted or sprayed

areas that need density of material, such as surrounding a racing driver's head.

Surface area is limited to 16 m² (172 ft²), although it is possible to make larger products in more than 1 piece. Monocoque boat hulls, for example, are made in stages – each area being laminated and cured one at a time.

Manual lay-up methods are labour intensive and expensive. Each product may have between 1 and 10 layers of fibre reinforcement, and each layer is applied by hand. A complex product may be constructed from 10s of parts, which makes it very expensive.

Molds should be made from the same materials as the part to be laminated. This will ensure that the molds have the same coefficient of expansion as the materials. Wall thickness is typically 6 mm to 8 mm (0.236–0.315 in.).

When the mold parts are joined together, the fibre reinforcement is overlapped. The case studies illustrate 2 methods for doing this skilful and

strength characteristics. The direction of weave will affect the mechanical properties of the part. For high performance products this is calculated using finite element analysis (FEA) prior to manufacture. Certain weaves have better drape and so can be formed into deeper profiles. However, fibre alignment is critical; just 5° of movement will reduce its strength by 20%.

Core materials are used to increase the depth of the parts and thus increase torsional strength and bending stiffness. The role of the core material is to maintain the integrity of the composite skin. Examples of core material include

DuPont™ Nomex® honeycomb (their trademark for aramid sheet), foam and aluminium honeycomb (see images, page 207). Cores make step changes in wall thickness possible.

DESIGN CONSIDERATIONS

Carbon and aramid fibres are very expensive, so every effort should be taken to minimize material consumption while maximizing strength. Wall thickness is limited to 0.25–10 mm (0.01–0.4 in.) (any thicker and the exothermic reaction can be too dangerous). Carbon fibre is generally between 0.5 mm and 0.75 mm (0.02–0.03 in.) and is only ever built up in onto it. It is important to achieve the right balance of resin to fibre reinforcement. Rollers are used to remove porosity.

Pre-preg lay-up is more time-consuming, precise and expensive. It is most commonly used to form carbon fibre. No gel coat is needed because this would increase weight; instead, the carbon fibre is cut to predetermined patterns, which are laid into the pre-preg mold. Because the fibres are sticky, they can be rubbed together to form a lamination that is free of porosity.

After lay-up the whole mold is covered with 3 layers of material. The first is a blue film, which is permeable, while the intermediate layer is a breathable membrane. These 2 are sealed in with a hermetic film, and a vacuum is applied. These layers ensure that an even vacuum can be applied to the whole surface area, because if a vacuum was applied under 1 layer of film it would stick and air pockets would be left behind.

The pre-preg lay-up is placed into an autoclave, which is raised to a pressure of 4.14 bar (60 psi) and a temperature of 120°C (248°F) for 2 hours. The pressure and temperature are lowered when core materials are used.

RTM uses matched molds to produce parts with a high quality finish on both sides, known as 'double A side'. The molds are typically made from metal. Small molds are machined from solid. Molds larger than 1 m³ (35.31 ft³) are typically electroformed because at this size this process is less expensive than machining and is capable of producing parts with a surface area up to 16 m² (172 ft²).

The molds are preheated, then the fibre reinforcement is laid into the open mold. When closed, resin is injected under pressure. Alternatively, the resin can be drawn through the mold under vacuum (resin infusion) or it can simply be poured in prior to molding.

Because RTM is basically a wet lay-up process, gel coats are required for glass fibre products. For high volume production a thermoplastic–thermosetting combination is used, which produces a very high quality finish because as the thermosetting resin cools and shrinks the thermoplastic takes its place and forms a 'low profile' surface finish.

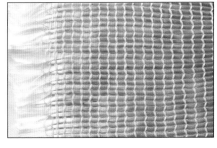

Left above
General purpose glass fibre chop strand mat is used in wet lay-up and resin transfer molding.

Left below
Carbon fibre twill is a high performance weave material.

Right above
Unidirectional glass fibre weave has specialized uses.

Right below
Kevlar® aramid and epoxy composite has very good resistance to high temperatures.

time-consuming process. To maximize strength, each layer of the lamination is overlapped at a different point to create a staggered lap joint.

COMPATIBLE MATERIALS

Fibre reinforcement materials include fibreglass, carbon and aramid.

Glass fibre is a general purpose laminating material that is heat resistant, durable and has good tensile strength. It is relatively inexpensive and can be used for a range of applications. Non-woven materials are the least expensive and known as chop strand mat (see image, above left). Weaves include plain (known as 0–90), twill and specialist (see image, above right). For large surface areas chopping and spraying the glass fibres directly onto the mold's surface produces a similar material to chop strand mat.

Carbon fibre has higher heat resistance, tensile strength and durability than glass fibre. When combined with a precise amount of thermosetting plastic it has an exceptional strength to weight ratio, which is superior to steel. Carbon fibre twill (see image, below left) is the most common weave.

Aramid fibre (see image, below right) is commonly referred to by the DuPont™ trademark name Kevlar®. Aramid is available only as spun fibres or sheet material because there is no other practical way to make it. It has very high resistance to abrasion and cutting, very high strength to weight and superior temperature resistance.

Since composite laminating has become more important in the automotive industry there have been many significant improvements in materials such as a material that is made up of glass fibres and polypropylene (PP) woven together. The composites are loaded into a heated mold and pressed, which causes the PP to melt and flow

around the glass fibre reinforcement (see images, page 216, above).

COSTS

Tooling costs are moderate to high, as moldmaking is a labour intensive process. Cost depends on the size and complexity of the product. The tooling is done in the same way as the laminated product. Therefore, a master (pattern) has to be made. However, it is possible to form the mold from almost any material and then fill, spray paint and polish the surface to produce the required finish.

Cycle time depends on the complexity of the part. A small part might take an hour or so, whereas a large one with lots of undercut features and cores may require as many as 150 hours.

Labour costs are high because composite laminating is a skilful and labour intensive process.

ENVIRONMENTAL IMPACTS

Laminated composites reduce the weight of products and so minimize fuel consumption. However, harmful chemicals are used in their production and it is not possible to recycle any of the offcuts or scrap material.

Operators must wear protective clothing in order to avoid too much contact with the materials and potential health hazards.

Material developments are reducing the environmental impact of the process. For example, hemp is being researched as an alternative to glass fibre, and in some cases thermoplastics are replacing thermosetting materials.

Case Study

→ Wet lay-up for the Ribbon chair

The Ribbon chair (**image 1**) was designed by Ansel Thompson in 2002. It is constructed with vinylester, glass and aramid reinforcement, and a polyurethane foam core, in a lengthy process that takes approximately 1 day. The sequence of events are: mold preparation, release agent, gel coat, lamination, cure time, demolding and then finishing.

The mold is prepared before each molding (**image 2**). Because the chair is a complex shape, the mold is designed to come apart to make demolding easier. Particular care is taken in preparing the surface finish because it will be reflected on the outside surface of the part. Once the parts have been assembled, a wax release agent is applied with a soft cloth (**image 3**).

This agent stops the gel coat bonding with the surface of the tool.

The flanges are taped to enable the mold to be closed (**image 4**), then the gel coat is applied and the masking tape removed (**images 5** and **6**).

1

2

3

4

5

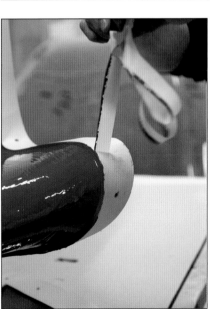

6

The glass and aramid fibre mat is cut into patterns to fit the chair, using 1 of 2 different types of fibre reinforcement (**image 7**). Sheets of fibre are laid onto the gel coat, and vinylester is applied to the back (**image 8**). Air is then removed with a paddle roller (**image 9**).

The layers are gradually built up, and care is taken to ensure that joints do not overlap and cause weaknesses (**image 10**). An area of aramid is then applied to the inside of the seat, to increase resilience and strength (**image 11**).

7

8

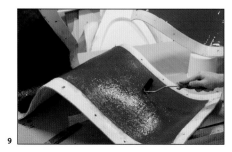
9

10

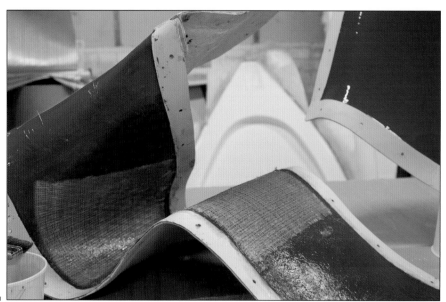
11

The molds are then brought together and bolted along the flange (**image 12**). Excess resin squeezes out as the bolts are tightened. The design of the mold means that an overlap of material forms a strong joint. While clamped shut an expanding polyurethane (PUR) foam is injected into the mold cavity. This forces the lamination against the surface of the mold, in order to improve the surface finish. It also increases the strength of the chair by supporting the thin composite wall sections.

After about 45 minutes the vinylester is fully cured and the mold is separated (**image 13**). It is trimmed and polished to complete the process (**image 14**).

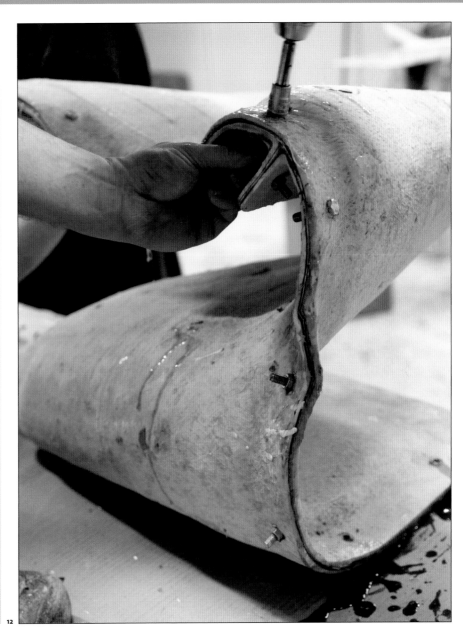

12

13

14

Featured Manufacturer

Radcor
www.radcor.co.uk

→ Racing car design with carbon fibre

Pre-preg carbon fibre is used in the production of high performance racing cars such as the Lola B05/30 Formula 3 car (**image 1**), which is manufactured by Lola Cars. Construction (**image 2**) is closely tied into design and engineering. High performance products have to be engineered to take the maximum load, yet be as lightweight as possible. It is the role of a carbon fibre engineer to push carbon fibre to its limits.

The process of designing and producing a car takes approximately 8 months. By using detailed design, FEA and controlled testing, parts can be produced directly from the CAD drawing. In the CAD drawing of the monocoque chassis on the car (**image 3**), FEA software is used to determine the stresses and strains on the structure (**image 4**). This helps the engineer to calculate the optimal structure within set parameters.

Critical parts are molded and tested to ensure that the calculations are correct. A crash test is simulated on the nose cone of the car – the average deceleration being 25G during impact (**image 5**). The aim of this part is to deaccelerate the car in a head-on collision. The nose cone is coloured yellow and is attached to the monocoque structure.

Success is measured in energy absorption per gram (0.035 oz) of carbon fibre. The image shows how the carbon has shattered into fragments. Each of the fragments is absorbing a small amount of the impact; so the smaller the fragments, then the more successful the engineering design.

Exact replicas of the cars, including carbon fibre wheels, are tested at 50% scale in a wind tunnel (**image 6**). This helps to solve issues of aerodynamics, balance and tuning.

COMPOSITES

214

1

2

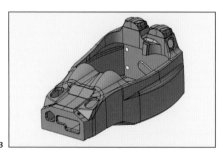

3

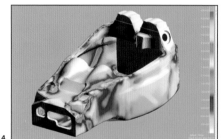

4

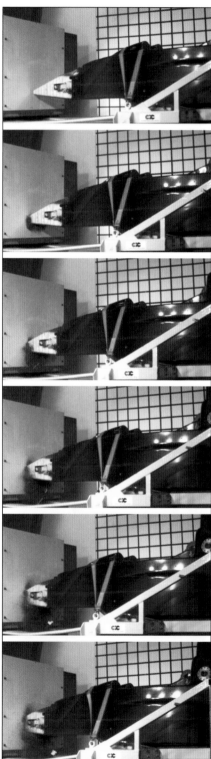

6

5

Featured Manufacturer

Lola Cars International
www.lolacars.com

→ Pre-preg carbon fibre lay-up for the roll hoop trailing edge

This case study demonstrates the production of a small piece of the Intersport Lola B05/40 racing car (**image 1**). In total, this car is made up of hundreds of carbon fibre components.

Production starts when the designers complete the 'lay-up handbook', which outlines the production requirements of each part, including pattern profile, number of laminations and sequence of production. This case study covers the production of the roll hoop trailing edge, which is made with a simple split mold, so is relatively simple. Parts such as the monocoque structure (see opposite) comprise many different pieces, cured at different stages and incorporates core material. The lay-up handbook ensures that each part is made according to the design requirements.

The patterns of carbon fibre are fed through to a kit cutter, which functions much like an x–y plotter (**image 2**). The carbon fibre is coated with plastic film on either side. This is peeled off just prior to lay-up (**image 3**). The mold is in 2 parts (**image 4**), which are laid up separately. The carbon fibre patterns are aligned on the part with the rubbed side downwards (**image 5**). Each 1 is trimmed (**image 6**), leaving a small overlap to make a stronger joint interface, before the next layer is applied.

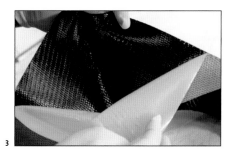

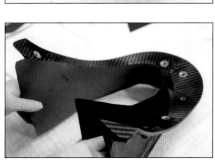

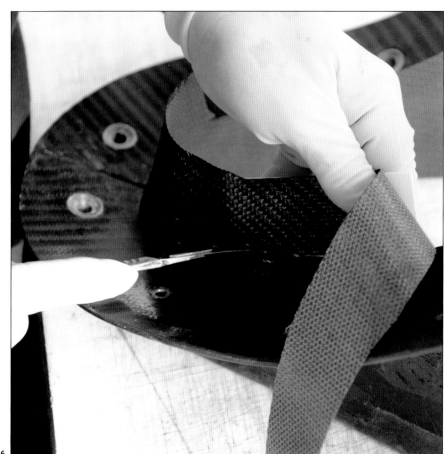

The second part of the mold is laid up in exactly the same way. Each pattern of carbon fibre is cut to fit the mold like a dress (**image 7**). Profiled 'dobbers' are used to rub the weave into tight corners (**image 8**). Three layers of carbon fibre are applied to each surface, and the 2 halves of the mold are brought together (**image 9**). The layers are carefully overlapped in the joint with another dobber (**image 10**).

When all the layers are in place the entire molding and tool are covered with a permeable blue film (**image 11**). This is overlaid with a layer of breathable membrane (**image 12**). The mold is then placed inside a pale pink hermetic film (**image 13**). The resulting 'sandwich' enables a vacuum to be applied on the mold, which forces the laminate onto its surface (**image 14**).

The vacuumed mold is then placed in an autoclave (**image 15**), which cures the resin with heat and pressure. The size of the autoclave limits the size of part – larger parts are restricted to oven or room temperature curing methods.

Many molds are placed in the autoclave at 1 time because curing is a 2 hour process. When finished, the part is removed (**image 16**). At the Lola Cars workshop (**image 17**), up to 20 laminators may be working at 1 time to produce the complete car. All of the parts are finally brought together and assembled to make up the complete car (**image 18**).

7

8

9

10

11

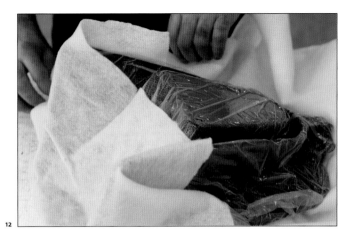

12

13

14

15

17

16

18

Featured Manufacturer

Lola Cars International

www.lolacars.com

 Bend
 Continuous
 Sheet
 Hollow
 Bulk
 Internal

Forming Technology

DMC and SMC Molding

Compression molding is used to form dough molding compound (DMC) and sheet molding compound (SMC) into structural and lightweight parts. These processes bridge the gap between composite laminating and injection molding.

Costs	Typical Applications	Suitability
• Moderate tooling costs • Low unit costs (3 to 4 times material cost)	• Automotive • Building and construction • Electrical and telecommunication	• Medium to high volume production

Quality	Related Processes	Speed
• High strength parts with long fibre length	• Composite laminating • Compression molding • Injection molding	• Cycle time between 2–5 minutes

INTRODUCTION

Combined with compression molding, DMC and SMC have been used to replace steel and aluminium in applications such as automotive bodywork and structural electronic enclosures.

DMC and SMC are typically glass reinforced thermosetting materials such as polyester, epoxy and phenolic resin. DMC is also referred to as BMC (bulk molding compound). There are thermoplastic alternatives such as GMT (glass mat thermoplastic), which is a fibre reinforced polypropylene (PP). Other types of reinforcement include aramid and carbon fibre. In recent years natural materials with suitable properties, such as hemp, have been trialled in an attempt to reduce the environmental impact of these materials.

DMC is compression molded to form bulk shapes. In contrast, SMC is compression molded to form lightweight sheet components with a uniform wall thickness. Fibre length is longer in

SMC: the raw materials are available with different weave patterns for different structural applications. This has significant mechanical advantages, similar to those of laminated composites (page 206).

A similar process, called pultrusion, is used to produce continuous lengths of fibre-reinforced plastic. Instead of compressing the materials, pultrusion draws fibre through a bath of thermosetting resin, which cures to form a continuous profile similar to extrusion. Pultrusion is the bridge between filament winding (page 222) and extrusion for continuous profiles.

TYPICAL APPLICATIONS

These materials can be mass-produced and so are suitable for application in a range of industries.

Thermosetting composite materials have high resilience to dielectric vibration, high mechanical strength and are resistant to corrosion and

fatigue. These qualities make DMC and SMC very useful in electrical and telecommunication applications. Some examples include enclosures, insulating panels and discs.

The high volume, high strength and lightweight characteristics of DMC and SMC make them suitable for automotive parts. Examples include body panels, seal frames and shells, engine covers and structural beams. Electric cars are making use of these materials because they combine mechanical strength with high dielectric strength.

In the building and construction sector, DMC and SMC are suitable for door panels, flooring and roof covering materials. They are very durable and so are suitable for public space furniture and signage.

Compression Molding DMC Process

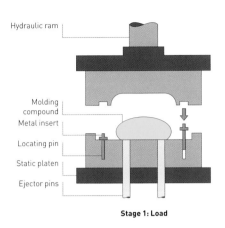

Stage 1: Load

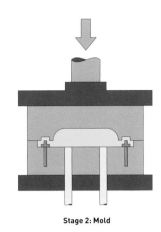

Stage 2: Mold

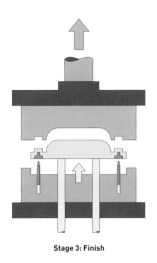

Stage 3: Finish

RELATED PROCESSES

Similar products can be made by injection molding (page 50) and composite laminating. Reinforced injection molded products are structurally inferior because the length of the fibre is considerably shorter.

QUALITY

This is a very high quality process. Many of the characteristics can be attributed to the materials such as heat resistant and electrically insulating polyester or phenolic. Thermosetting plastics are more crystalline and as a result are more resistant not only to heat, but also to acids and other chemicals.

Surface finish and reproduction of detail is very good. The compression, rather than injection, of material in the die cavity produces parts with reduced stress that are less prone to distortion.

DESIGN OPPORTUNITIES

The type of fibre reinforcement and length of strand can be modified to suit the requirements of the application. This helps to reduce weight and maximize the efficiency of the process.

Metal inserts and electrical components can be over-molded. This reduces secondary operations.

Step changes in wall thickness are not a problem with DMC molding.

TECHNICAL DESCRIPTION

The diagram illustrates compression molding DMC. SMC is a similar process, except that it is used for sheet profiles as opposed to bulk shapes. The molding compound is a mixture of fibre reinforcement and thermosetting resin: DMC is made up of chopped fibre reinforcement, whereas SMC contains sheets of woven fibre reinforcement.

The sequence of operation is the same for both and includes loading, molding and de-molding. In stage 1, a measure of DMC or SMC is loaded into the die cavity in the lower tool. Metal inserts with locating pins are loaded into slots. They are in line with the direction of ejection because otherwise the part would not release from the mold.

In stage 2, the upper tool is gradually forced into the die cavity. It is a steady process that ensures even distribution of material throughout the die cavity.

Thermosetting material plasticizes at approximately 115°C (239°F) and is cured when it reaches 150°C (302°F), which takes 2–5 minutes. In stage 3, the parts of the mold separate in sequence. If necessary, the part is relieved from the lower or upper tool with ejector pins.

It is a simple operation but is suitable for the production of complex parts. It operates at high pressure, ranging from 40 to 400 tonnes (44–441 US tons), although 150 tonnes (165 US tons) is generally the limit. The size and shape of the part will affect the amount of pressure required. Greater pressures will ensure better surface finish and reproduction of detail.

DESIGN CONSIDERATIONS

Colours are applied by spray painting (page 350) because the thermosetting resins in DMC and SMC have a limited colour range.

As with injection molding, there are many design considerations that need to be taken into account when working with compression molding. Draft angles can be reduced to less than 0.5°, if both the tool and the ejector system are designed carefully.

The size of the part can be 0.1–8 kg (0.22– 17.64 lb) on a 400 tonne (441 US ton) press. The dimensions are limited by the pressure that can be applied across the surface area, which is affected by part geometry and design. Another major factor that affects part size is venting gases from the thermosetting material as it cures and heats up. This plays an important role in tool design, which aims to get rid of gasses with the use of vents and clever rib design.

1

2

3

4

Wall thickness can range from 1 mm to 50 mm (0.04–1.97 in.). It is limited by the exothermic nature of the thermosetting reaction because thick wall sections are prone to blistering and other defects as a direct result of the catalytic reaction. It is generally better to reduce wall thickness and minimize material consumption, so bulky parts are hollowed out or inserts are added. However, some applications require thick wall sections such as parts that have to withstand high levels of dielectric vibration.

COMPATIBLE MATERIALS

Thermosetting materials molded in this way include polyester, epoxy and phenolic resin. Thermoplastic composites are generally polypropylene; the molding process for these is different because they have to be plasticized and formed.

Fibre reinforcement can be glass, aramid, carbon or a natural fibre such as hemp, jute, cotton, rag and flax. Other fillers include talc and wood. The various fibre reinforcement and filler materials are used to increase the strength, durability, resistance to cracking, dielectric resilience and insulating properties of the part.

COSTS

Tooling is generally less expensive than for injection molding because less pressure is applied and the tooling tends to be simpler.

Cycle time is 2–5 minutes, depending on the size of the part and length of time it takes to cure.

Labour costs are moderate because the process requires a great deal of manual input.

ENVIRONMENTAL IMPACTS

The main environmental impacts arise as a result of the materials used. Thermosetting plastics require higher

→ Compression molding 8-pin rings

This is a DMC compression molding operation. In preparation, the metal inserts are screwed onto locating pins (**image 1**). The surface of the metal inserts are knurled to increase the strength of the over-molding. The pins are loaded into the upper tool (**image 2**). Inserts are also loaded into the lower tool. A pre-determined weight of polyester and glass based DMC is loaded into the die (**images 3** and **4**).

The 2 halves of the die come together to force the polyester to fill the die cavity.

After curing, the molds separate to reveal the fully formed part (**image 5**). The locating pins attached to the metal inserts remain on the part as it is removed (**image 6**).

The locating pins are unscrewed from the metal inserts (**image 7**) and the workpiece is de-flashed by hand (**image 8**). The finished moldings are stacked (**image 9**).

5

6

7

8

molding temperatures, typically between 170°C and 180°C (338–356°F). It is not possible to recycle them directly due to their molecular structure, which is cross-linked. This means that any scrap produced, such as flash and offcuts, has to be disposed of.

Featured Manufacturer

Cromwell Plastics
www.cromwell-plastics.co.uk

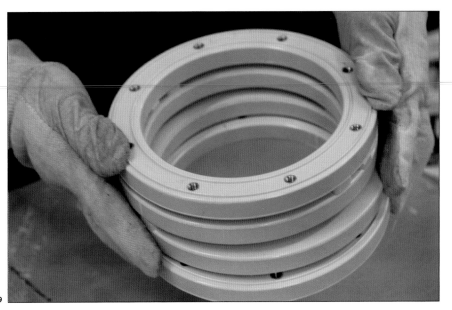

9

Bend

Continuous

Sheet

Hollow

Bulk

Internal

Forming Technology
Filament Winding

Layers of carbon fibre monofilaments, coated in epoxy resin, are wound onto a shaped mandrel to give the ultimate strength characteristics. The mandrel is either removed and reused, or permanently encapsulated by the carbon fibres.

COMPOSITES

222

Costs	Typical Applications	Suitability
• Low to moderate tooling costs, depending on size • Moderate to high unit costs	• Aerospace • Automotive • Deep sea submersibles	• One-off to batch production

Quality	Related Processes	Speed
• High gloss surface finish • High performance, lightweight products	• 3D thermal laminating • Composite laminating • DMC and SMC molding	• Moderate cycle time for small parts (20–120 minutes); large parts may take several weeks

INTRODUCTION

Filament winding is used to produce continuous, sheet and hollow profiles for applications that demand the high performance characteristics of fibre reinforced composites. They are produced by winding a continuous length of fibre around a mandrel. The fibre is coated with thermosetting resin to produce high strength and lightweight shapes.

The fibre (typically glass, carbon or aramid) is applied as a tow that contains a set number of fibre monofilaments. For example, 12k tow has 12,000 strands, while 24k has 24,000 strands.

There are 2 main types of winding, 'wet' and 'pre-preg'. Wet winding draws the fibre tow through an epoxy bath prior to application. Pre-preg winding uses carbon fibre tow pre-impregnated with epoxy resin, which can be applied directly to the mandrel without any other preparation.

TYPICAL APPLICATIONS

Filament wound products can be found in high performance applications in the aerospace, deep sea and automotive industries. Some examples include blades for wind turbines and helicopters, pressure vessels, deep sea submersibles, suspension systems, torsional drive shafts and structural framework for aerospace applications.

RELATED PROCESSES

Filament winding is used to produce low volumes of cylindrical parts. Higher volumes can be produced by DMC and SMC molding (page 218). Composite laminating (page 206) is used to produce similar profiles. The benefit of filament winding is that the direction of the strand can be adjusted precisely throughout production from almost 0° up to 90°. In composite laminating, filaments are woven into mats with intertwined warp and weft.

The 3D thermal laminating (3DL, page 228) developed by North Sails is similar to filament winding. The difference is that filaments are laid onto a static mold in a process known as tape laying. In this case the molds can be very large and up to 400 m² (4,305 ft²). 3DL is best suited to sheet parts, where as filament winding is more suited to hollow parts.

Above
This pressure vessel has an aluminium liner filament wound with carbon fibre composite.

Left
The gloss surface finish is a gel coat resistant to heat or chemical attack.

Top
CNC machining after filament winding produces very accurate surface dimensions. Channels, recesses and tapers on the outside surface are applied in this way.

QUALITY

There are 4 main types of finish. These are 'as wound', taped, machined and epoxy gel coat. 'As wound' finishes have no special treatment. Taped finishes produce a smooth surface finish. The principle is the same as vacuum bagging in composite laminating, but the surface finish is more controllable. Surfaces are machined where precise tolerances are required (see image, top). Smooth, glossy finishes can be achieved with epoxy gel coat (see image, left).

Stiffness is determined by the thickness of lay-up and tube diameter.

Winding is computer guided, making it precise to 100 microns (0.0039 in.). The angle of application will determine whether the layer is providing longitudinal, torsional (twist) or circumferential (hoop) strength. Layers are built up to provide the required mechanical properties. Products cannot be interchanged between applications because their properties will be designed specifically for each.

DESIGN OPPORTUNITIES

Opportunities for designers are limited to cylindrical and hollow parts. But they need not be rotationally symmetrical;

Filament Winding Process

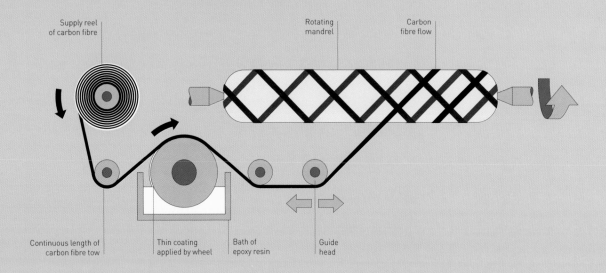

Supply reel
of carbon fibre

Rotating
mandrel

Carbon
fibre flow

Continuous length of
carbon fibre tow

Thin coating
applied by wheel

Bath of
epoxy resin

Guide
head

TECHNICAL DESCRIPTION

The carbon fibre tow is applied to the rotating mandrel by a guide head. The head moves up and down along the mandrel as it rotates, and guides the filament into the geodesic overlapping pattern.

The width of the tow is chosen according to the material being used and the layer thickness required.

The fibre is continuous and only broken when a new supply reel of fibre

reinforcement is loaded. This is the wet lay-up process; the fibre reinforcement is coated with an epoxy resin by a wheel rotating in a bath of the resin.

A complete circuit is made when the guide head has travelled from 1 end of the mandrel to the other and back to the starting point. The speed of the head relative to the speed of mandrel rotation will determine the angle of the fibre. A single circuit may have

several different angles of tow, depending on the requirements.

Bulges can be made by concentrating the tow in a small area. These are used either to create localized areas of strength or to build up larger diameters that can be machined for accuracy.

oval, elliptical, sharp edged and flat-sided profiles can be filament wound (see image above).

Parallel-sided and conical mandrels can be removed and reused, while 3-dimensional hollow products that are closed at both ends can be made by winding the filament tow over a hollow liner, which remains as part of the final product (see image, page 227, above). This technique is known as bottle winding and is used to produce pressure vessels, housing and suspension systems.

Other than shape, the benefits of winding over a liner include forming a water, air and gas tight skin.

DESIGN CONSIDERATIONS

This is a high cost process for low volumes. Increasing the volumes reduces unit cost. Even so, the materials are expensive and so every effort is made to reduce material consumption while maximizing strength.

Parallel-sided parts can be made in long lengths and cut to size. Conventional filament winding is typically limited to 3 m (10 ft) long and up to 1 m (3.3 ft) in diameter. However, much larger forms are produced by filament winding, such as space rockets, which may take several weeks to make.

Winding over a liner will increase the cost of the process because a new liner is manufactured for each cycle.

These products are typically used for demanding applications, so the liners are typically produced by electron beam welding (page 288) high-grade aluminium or titanium.

Winding over a hollow liner produces parts that have inward and outward facing corners such as a bottleneck profile. Even though it is possible, it is generally not recommended to wind over outward facing corners with a radius of less than 20 mm (0.8 in.) because product performance will be affected. There is no lower limit on the radius for inward facing corners.

COMPATIBLE MATERIALS

Types of fibre reinforcement include glass, carbon and aramid. An outline of their particular qualities is given under composite laminating.

Resins are typically thermosetting and include polyester, vinylester, epoxy and phenolics. Thermosetting plastics have cross-links in their molecular structure, which means they have high resistance to heat and chemicals. Combined with carbon fibres these materials have very high fatigue strength, impact resistance and rigidity.

COSTS

Tooling costs are low to moderate. Encapsulating the mandrel will increase the cost.

Filament winding cycle time is 20–120 minutes for small parts, but can take several weeks for very large parts. Curing time is typically 4–8 hours, depending on the resin system.

The winding process is computer-guided. Even so, there is a high level of manual input, so labour costs are moderate to high.

ENVIRONMENTAL IMPACTS

Laminated composites reduce the weight of products and so reduce fuel consumption.

Operators have to wear protective clothing to avoid the potential health hazards of contact with the materials.

Thermosetting materials cannot be recycled, so scrap and offcuts have to be disposed of. However, new thermoplastic systems are being developed that will reduce the environmental impacts of the process.

Top
Non-circular profiles are suitable for filament winding. Examples include the blades for wind turbines and helicopters.

Above
This flywheel is made up of 3 parts: the inner lining is glass fibre and the other 2 layers are carbon fibre.

→ Filament winding a racing propshaft

These are used in cars to deliver power from the engine to the wheels. Traditionally these components are made from metal, but when carbon composite is used it is possible to make weight savings of up to 65%. Finite element analysis is used to determine the stress and strain on the product prior to manufacture. The results, seen in the computer generated images, indicate the required angle of tow and number of layers.

They are a parallel section tube and so are made on removable and reusable mandrels (**image 1**). Carbon fibre tow is supplied from a spool (**image 2**). Many different spools can be running simultaneously for multifibre composites or if more than 1 mandrel is being wound at the same time. The fibres are coated with epoxy resin as they pass over the wheel (**image 3**).

The angle of application ranges from 90° to almost 0°. The mandrel ends are tapered so that adequate tension can be pulled on the tow without it slipping along the mandrel (**image 4**) at shallow angles; this is not a problem when tow is laid at or close to 90° (**image 5**). In this case the carbon fibre is sealed onto the mandrel with a plastic tape (**image 6**). The tape squeezes excess epoxy resin from the carbon and ensures a high quality, smooth finish. The excess is cut off while it is uncured (**image 7**). This is so that the mandrel can be removed once the composite has cured.

The filament wound assemblies are placed into an oven, which cures the resin at up to 200°C (392°F) for 4 hours. The whole curing process takes 8 hours because the temperature inside the oven is ramped up and down gradually. Small droplets of resin form on the surface of the tape during curing (**image 8**). This is removed when the tape is peeled off.

The cured composite is removed from the mandrel (**image 9**). It is possible to remove long cylinders from mandrels because metal expands and shrinks more than carbon fibre, so is slightly smaller once it has cooled. This provides just enough space to remove the mandrel.

The ends of the shaft are cut off to precise tolerances, and machined metalwork is assembled onto the ends (**image 10**). These are bonded in with adhesives, which are heat cured.

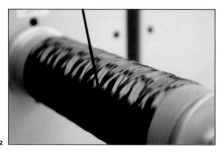

1

2

3

4

5

6

7

8

9

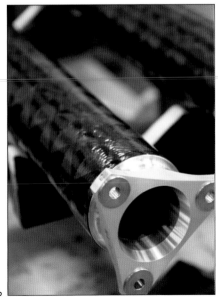

10

Featured Manufacturer

Crompton Technology Group
www.ctgltd.com

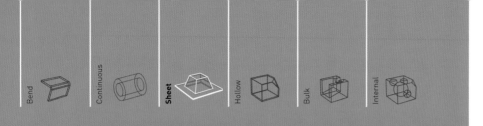
Forming Technology
3D Thermal Laminating

This technology was developed by North Sails to make ultra lightweight, seamless 3-dimensional sails. The fibre reinforcement is continuous over the entire surface of the sail, replacing traditional methods of cutting, stitching and gluing.

Costs	Typical Applications	Suitability
• Very high tooling costs • Very high unit costs	• Sailing	• One-off and small batch production
Quality	**Related Processes**	**Speed**
• Very high strength to weight properties	• Composite laminating • Filament winding • Stitching	• Cycle time up to 5 days

INTRODUCTION

The first 3-dimensional laminating (3DL) sail was constructed by Luc Dubois and J. P. Baudet at North Sails in 1990. Since then the process has been adopted by nearly every racing boat in the Volvo Ocean Race and America's Cup – 2 major sailing grand prix.

Sails have a 3D 'flying shape'. In conventional sailmaking, cut patterns are stitched or glued together to produce the optimum shape. In 3DL, the sails are molded in the optimum flying shape and so eliminate cutting and joining. This produces sails that are up to 20% lighter than conventional equivalents. They are also stronger and seamless.

The 3DL process combines the benefits of composite laminating (page 206) with filament winding (page 222). The 3-dimensional sheet geometries are formed over computer-guided molds. The fibre reinforcements (aramid and carbon) are laid down individually along pre-determined lines of stress. They are sandwiched between thin sheets of polyethylene terephthalate (PET) coated with a specially developed adhesive.

3DL manufacturing is a long process, and demand within the industry is very high. Therefore, 3-dimensional rotary laminating (3Dr) was developed as a method for continuous production. Instead of many large 3-dimensional molds, 3Dr sailmaking is carried out on a single rotating drum. The shape of the drum is manipulated as the sail is constructed over it. The process is a very recent development and so is still undergoing refinement.

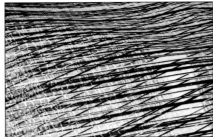
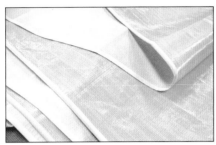

Fibre layouts
Any combination of performance fibres can be laminated by the 3DL and 3Dr processes. The quantity, type and direction of fibre is adjusted to suit the level of performance demanded by the application. Types of fibre include aramid, carbon, polyester and other advanced yarns. Above, from left to right, 600 series, 800 series and 800 series. Left, from left to right, 900 series and TF1 series.

TYPICAL APPLICATIONS

Although these laminating technologies are currently limited to sailmaking, there is potential for them to be developed for application in a variety of products such as high altitude balloons, dirigibles, tension structures, temporary structures and inflatable structures.

RELATED PROCESSES

The only processes capable of producing very large and seamless shapes are 3DL and 3Dr. Circular, conical, elliptical and similar shapes, however, can be produced by filament winding, composite laminating and stitching (see upholstery, page 338). Filament winding is generally limited to structures no larger than 3 m (10 ft) in length and 1 m (3.3 ft) in diameter, but this is not always the case.

QUALITY

These sails are built to perform a very specific function. They are required to hold their shape, even in strong winds, be durable and lightweight and produce minimum resistance to the flow of air over the surface.

Aerospace vacuum and pressure techniques are used to cure the laminates on the mold. This ensures that the composite will not delaminate, even when extreme loads are applied.

The thermally formed PET maintains the position of the fibres, which are laid down in alignment with the direction of stress across the sail.

DESIGN OPPORTUNITIES

At present, opportunities for designers are limited because these processes are restricted to sailmaking. In the future there is potential for application of these techniques in other industries.

Specific fibre layouts and composites provide varying levels of performance (see images, above). The 600 series is made up of high-modulus aramid fibres and laid down along the lines of stress to maintain shape and durability. The flowing shape produced in the catenary curves joins areas of stress with smooth, flowing lines of fibre reinforcement. The 800 and 900 series combine carbon fibre with the aramid to reduce weight and stretch. Different films can be used on the outside surfaces to protect against tearing and chafing and provide UV stability. Plain weave polyester, known as taffeta, is used on the outside surface of the TF1 series to increase durability.

DESIGN CONSIDERATIONS

The materials and manufacturing processes used in 3DL are expensive and therefore are limited to high performance applications.

Sails are designed to be strong under tension. Therefore, this process is not suitable for use in applications that require resistance to compression.

The sails are not molded with a uniform wall thickness; the thickness varies with the intended application. Minimum wall thickness is limited by the thickness of PET film and fibre reinforcement.

COMPATIBLE MATERIALS

The 2 outer films are PET coated with a special adhesive. The fibre reinforcement is of carbon, aramid, polyester or a combination of the 3 materials.

COSTS

Tooling costs are high. The male molds are computer guided and adjusted to the flying shape of sail being made. They are also very large, up to 400 m² (4,306 ft²).

Cycle time is up to 5 days because of the lengthy curing process.

Labour costs are high due to the level of skill required to make reliable products for high performance applications.

ENVIRONMENTAL IMPACTS

These sails minimize material consumption and reduce weight.

→ 3DL sails on a TP 52 type racing boat

The TP 52 type racing boat (**image 1**) has here been fitted with North Sails 3DL composite sails. The process starts with a bespoke sail design (flying shape), because each boat requires a slightly different shape. There are a range of sizes possible from 10 m² to 500 m² (108–5,382 ft²). The shape of the molds is computer guided, so they can be adjusted for many different sail shapes (**images 2** and **3**).

The PET film is laid across the mold (**image 2**) and pulled tight with tensioning straps. The 6-axis computer controlled gantry lays down the strands of carbon fibre (**image 4**). An operator in a hang-gliding harness inspects the sail (**image 5**).

The PET and fibre are vacuum bagged and held under pressure. An operator then moves over the sail applying heat to thermally form the composite over the mold surface, using either a conductive heating system in contact with the laminate surface (**image 6**), or an infrared heating system that travels above the laminate surface (**image 7**), depending on the fibre in the sail. The finished laminate is inspected after curing and left for a further 5 days to ensure that the adhesive has reached its full bond strength.

When the sail has fully cured, corner reinforcements, batten pockets, eyelets, and other details are applied by sail makers using traditional cutting and sewing techniques.

1

2

3

4

5

Featured Manufacturer

North Sails
www.northsails.com

6

7

Three-Dimensional Laminating Process (3DL)

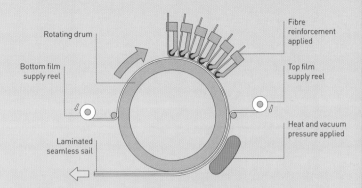

- Supply reel
- Composite lay-up
- Computer-guided head
- Computer manipulated mold
- Pneumatic rams

Three-Dimensional Rotary Laminating Process (3Dr)

- Rotating drum
- Bottom film supply reel
- Laminated seamless sail
- Fibre reinforcement applied
- Top film supply reel
- Heat and vacuum pressure applied

TECHNICAL DESCRIPTION

Each sail has an individual 3D shape. The designer produces a CAD file, which is transferred onto the surface of the mold by a series of pneumatic rams. The shape of the mold will determine the 'flying shape' of the sail when it is made.

A PET film is laid over the mold and put under tension. It is coated with a specially developed adhesive, designed for North Sails by its industrial partners for the 3DL process. Strands of fibre reinforcement are applied by a fibre placement head guided by a computer controlled overhead gantry. It operates on 6 axes, so can follow the profile of the mold exactly. The fibres follow the anticipated 'load lines'. Adhesive is applied to the fibres as they are laid down onto the mold, which helps to keep them in place.

A second layer of film is laid on top of the fibres and forced to laminate under pressure at approximately 8.6 N/cm^2 (12.5 psi) from a vacuum bag. Heat is then applied with a blanket to cure the adhesive. After curing, the composite sail is removed from the mold and brought to the curing floor for a further 5 days to ensure the adhesive is fully cured and will not delaminate.

The 3Dr process is carried out on a rotating drum as opposed to the static horizontal mold used in 3DL. This enables the sails to be put together much more quickly. The composite sail is formed as the drum rotates and leaves the drum completed. The shape of the drum is computer controlled to produce 2-directional curvature: the flying shape of the sail.

Case Study

→ 3Dr sails on a Melges 24

A more recent development at North Sails has been to make sails using rotary molds (3Dr), which is a more rapid and cost effective process for sailmaking. The Melges 24 racing boat is fitted with a 3Dr laminated sail (**image 1**).

The rotating drum contains about 2000 actuators, which shape the surface into complex curvatures describing the flying shape of a sail. The shape of the drum is manipulated to accommodate the change in shape along the length of the sail. The drum surface is shaped to curve in 2 directions simultaneously, using the principals of Gaussian curvature (**images 2–4**).

Fibre reinforcement is applied to the bottom PET film, which rotates on the drum (**image 5**). The sail leaves the drum fully formed in 3 dimensions. Finishing touches are applied by skilled sail makers.

1

2

3

4

5

Featured Manufacturer

North Sails
www.northsails.com

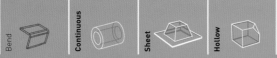

Forming Technology

Rapid Prototyping

These layer building processes can be used to prototype one-offs or to direct manufacture low volume production runs from CAD data. There is no tooling involved, which not only helps to reduce cost, but also has many advantages for the designer.

Costs	Typical Applications	Suitability
• No tooling costs • SLS is the cheapest and DMLS the most expensive process	• Automotive, F1 and aerospace • Product development and testing • Tooling	• One-offs, prototypes and low volume production

Quality	Related Processes	Speed
• High definition of detail and surface finish	• CNC machining • Electrical discharge machining • Investment casting	• Long process, but turnaround is rapid because there is no tooling and data is taken directly from CAD file

INTRODUCTION

Rapid prototyping is used to construct simple and complex geometries by fusing together very fine layers of powder or liquid. The process starts with a CAD model sliced into cross-sections. Each cross-section is mapped onto the surface of the rapid prototyping material by a laser, which fuses or cures the particles together. Many materials such as polymers, ceramics, wax, metals and even paper can be formed in this way.

This description will concentrate on 3 main processes: stereolithography (SLA), selective laser sintering (SLS) and direct

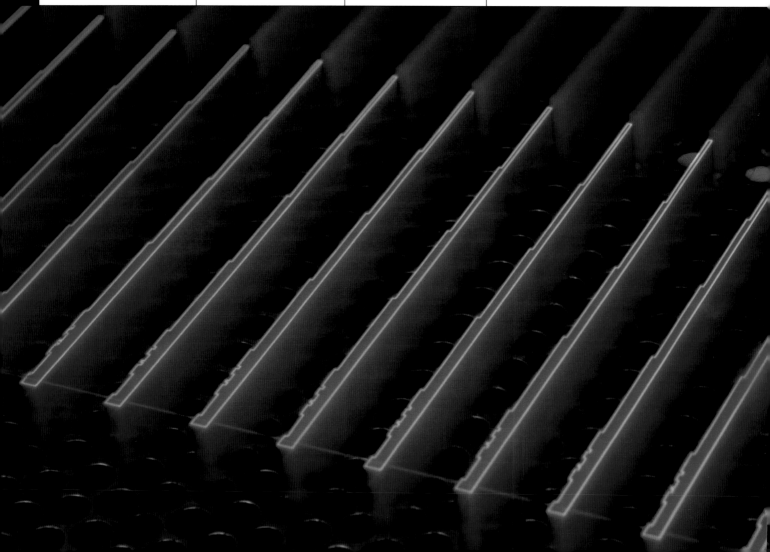

metal laser sintering (DMLS). SLA is the most widely used rapid prototyping technique and produces the finest surface finish and dimensional accuracy.

These processes are utilized mainly for design development and prototyping, in order to reduce the time it takes to get a product to market. However, they can also be used to direct manufacture products that have to be precisely reproduced with very accurate tolerances.

TYPICAL APPLICATIONS

The high tolerances of the SLA process mean that it is ideal for producing fit-form prototypes that are used to test products before committing to the chosen method of production.

The SLS technique is often selected to produce functional prototypes and test models because the materials have similar physical characteristics to injection molded parts.

DMLS is suitable for injection and blow molding tooling, wax injection and press tools. This process is typically used to produce functional metal prototypes and low volume production runs of parts for the automotive, F1, jewelry, medical and nuclear industries.

RELATED PROCESSES

Conventional CNC machining (page 182) operations remove material, whereas rapid prototyping builds only what is necessary. Machining produces very accurate parts, but is a slow, energy intensive process. Rapid prototyping can produce complex internal shapes by undercutting, which would be time-consuming and expensive to machine.

Electrical discharge machining (page 254) is another reductive process. Material is removed by sparks that are generated between the tool and workpiece. This process is mainly used to form concave profiles that would be impractical to CNC machine.

Investment casting (page 130) has many of the geometry advantages of rapid prototyping because the ceramic molds are expendable.'

QUALITY

SLA techniques produce the highest surface finish and dimensional accuracy of all rapid prototyping ones. All these layer-building processes work to fine tolerances: SLS builds in 0.1 mm (0.004 in.) layers and is accurate to ±0.15 mm (0.006 in.); SLA forms in 0.05–0.1 mm (0.002–0.004 in.) layers and is accurate to ±0.15 mm (0.006 in.); and DMLS builds in 0.02–0.06 mm (0.008–0.0024 in.) layers and is accurate to ±0.05 mm (0.002 in.).

Top
PP mimic part with live hinges and snap fits has been produced by stereolithography.

Above left
Direct metal laser sintering is suitable for making nickel–bronze

electric seat adjustment cogs for use in the automotive industry.

Above
This carbon strand filled nylon impeller has been manufactured by selective laser sintering.

As a result of manufacturing 3D forms in layers, contours are visible on the surface of acute angles. All of the parts therefore require finishing when they come out of the machine. It is possible to acquire a very high 'glass-like' surface finish on the water clear SLA epoxy resin, by polishing.

Micro modelling can be used to produce intricate and precise parts (up to 77 x 61 x 230 mm/3.03 x 2.40 x 9.05 in.). This process builds in 25 micron (0.00098 in.) layers, which are almost invisible to the naked eye and so eliminates surface finishing operations.

Stereolithography Process

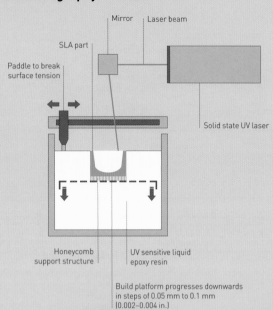

Mirror   Laser beam

SLA part

Paddle to break
surface tension

Solid state UV laser

Honeycomb
support structure

UV sensitive liquid
epoxy resin

Build platform progresses downwards
in steps of 0.05 mm to 0.1 mm
(0.002–0.004 in.)

Selective Laser Sintering Process

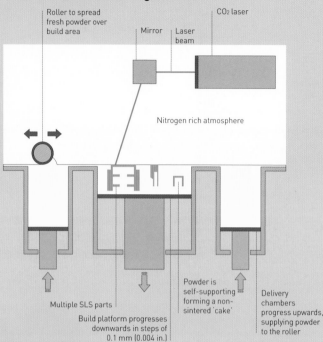

Roller to spread
fresh powder over
build area

Mirror   Laser
beam

CO_2 laser

Nitrogen rich atmosphere

Multiple SLS parts

Build platform progresses
downwards in steps of
0.1 mm (0.004 in.)

Powder is
self-supporting
forming a non-
sintered 'cake'

Delivery
chambers
progress upwards,
supplying powder
to the roller

TECHNICAL DESCRIPTION

STEREOLITHOGRAPHY

All of these rapid prototyping processes start with a CAD drawing sliced into cross-sections. Each cross-section represents a layer in the build, generally between 0.05 mm and 0.1 mm (0.002–0.004 in.) thick for SLA modelling. The model is built 1 layer at a time by an UV laser beam directed by a computer-guided mirror onto the surface of the UV sensitive liquid epoxy resin. The UV light precisely solidifies the resin it touches. Each layer is applied by submersion of the build platform into the resin. The paddle sweeps across the surface of the resin with each step downwards, to break the surface tension of the liquid and control layer thickness. The part gradually develops below the surface of the liquid and is kept off the build platform by a support structure. This is made in the same incremental way, prior to building the first layer of the part.

SELECTIVE LASER SINTERING

In this layer-additive manufacturing process, a CO_2 laser fuses fine nylon powder in 0.1 mm (0.004 in.) layers, directed by a computer-guided mirror. The build platform progresses downwards in layer thickness steps. The delivery chambers alternately rise to provide the roller with a fresh charge of powder to spread accurately over the surface of the build area. Non-sintered powder forms a 'cake', which encapsulates and supports the model as the build progresses. The whole process takes place in an inert nitrogen atmosphere at less than 1% oxygen to stop the nylon oxydizing when heated by the laser beam.

DESIGN OPPORTUNITIES

There are many advantages to using rapid prototyping technology such as reducing time to market and lowering product development costs. However, the most desirable qualities of this process for designers are not cost savings, they are that complex, intricate and previously impossible geometries can be built with these processes to fine tolerances and precise dimensions. There is no tooling and so changes to the design cost nothing to implement. The combination of these qualities provides limitless scope for design exploration and opportunity.

The SLS technique forms parts with physical characteristics similar to that of injection molded polymers. Because parts can be made with live hinges and snap fits, this process is ideal for functional prototypes.

The SLA technique is suitable for water clear, translucent and opaque parts. The surface finish can be improved with polishing and painting. SLA materials mimic conventional thermoplastics, which means SLA is good for making parts with the visual and physical characteristics of the final product.

DMLS provides an alternative to machining aluminium parts. The advantage of DMLS is that it produces very accurate parts (±0.05 mm/0.002 in.) with fine surface definition (0.02 mm/0.0008 in. layers), and the final product is 98% dense metal. As a result, DMLS parts have good mechanical strength and so are suitable for tooling and functional metal prototypes.

DESIGN CONSIDERATIONS

The main restriction for these processes is the size of the machine: SLS is limited to parts of 350 x 380 x 700 mm (13.78 x 14.96 x 27.56 in.); SLA is restricted to parts of 500 x 500 x 500 mm (19.69 x 19.69 x 19.69 in.); while the DMLS process is

DIRECT METAL LASER SINTERING

A considerable amount of heat is generated during this process because a 250 watt CO_2 laser is used to sinter the metal alloy powders. An expendable first layer of the part is anchored to the steel plate to stop distortion caused by differing rates of contraction. Such a layer also means that the part is easier to remove from the steel plate when the build is complete. During the sintering process, the delivery chamber rises to dispense powder in the path of the paddle, which spreads a precise layer over the build area. The build platform is incrementally lowered as each layer of metal alloy is sintered onto the surface of the part. The whole process takes place in an inert nitrogen atmosphere at less than 1% oxygen to prevent oxydization of the metal powder during the build.

Direct Metal Laser Sintering Process

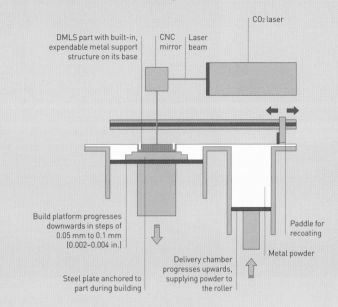

confined to parts of 250 x 250 x 185 mm (9.84 x 9.84 x 7.28 in.).

The orientation of the part can affect its mechanical properties – strand filled materials having better strength in certain geometries. The orientation of SLS parts in regular nylon powders has to be considered, to maintain accuracy. For example, a tube is built vertically to keep it round; if built horizontally the tube would be very slightly oval. A large flat plane should be manufactured at an incline because if it is built flat there will be too much stress in the part, which will result in warpage.

A live hinge is always constructed so that the hinge is on the horizontal plane, for strength; if it was built vertically the layers would be too short to withstand the stress of opening and closing, and would fail.

In the SLS system, multiple parts can be built simultaneously and on different planes because the non-sintered powder supports the sintered parts. By contrast, parts made by SLA and DMLS processes need to be supported and undercuts must be tied into the build platform. This means that fewer parts can be formed

at the same time and fine undercuts are more difficult to achieve.

COMPATIBLE MATERIALS

The SLS process is compatible with a variety of nylon-based powders. The tough Nylon 11 is heat resistant up to 150°C (302°F) and so can be used to produce functional prototypes that are suitable for working situations. Carbon filled and glass filled materials have been developed to build structural parts. The carbon filled material Windform™ XT was designed specifically for use in wind tunnels. It has good resistance to wind load and vibration, superior mechanical properties and high surface finish, which makes this an ideal material for F1 and aerospace applications. Recent material developments for the SLS process include powders with rubber-like qualities and those with integral colour, to reduce post-processing operations.

The SLA process uses liquid epoxy resin polymers that are categorized by the thermoplastics that they are designed to mimic. Some typical materials include acrylonitirile butadiene styrene (ABS) mimic,

polypropylene/polyethylene (PP/PE) mimic and water clear polybutylene terephthalate (PBT)/ABS mimic. Materials that can withstand temperatures up to 200°C (392°F) have been developed for the SLA process.

The DMLS process is compatible with specially developed metal alloys: 2 examples are nickel–bronze, which is slightly harder wearing than aluminium tooling, and steel alloy, which has similar characteristics to mild steel.

COSTS

There are no tooling costs.

The cost of rapid prototyping is largely dependent on build time. Cycle time is slow, but these processes reduce the need for any preparation or further processing and so turnaround is very rapid. Individual part cost is reduced if multiple products are manufactured simultaneously. The SLS powders are self-supporting and so large numbers of components can be built around and inside one another to reduce cost.

The build times are affected by the choice of process and layer thickness: typically SLS machines build 2 mm to

3 mm (0.079–0.118 in.) per hour in 0.1 mm (0.004 in.) layers; SLA machines build 1.2 mm to 12 mm (0.047–0.47 in.) per hour; and DMLS machines build at a rate of 2–12 mm³ (0.00012–0.00073 in.³) per hour. One drawback with SLS is that parts have to be left to cool, which can increase cycle time by up to 50%.

Labour costs are moderate, although they depend on finishing required. SLS parts are generally less expensive than SLA ones because they need less post-building processing. DMLS parts are typically cut from the steel plate by EDM wire cutting and are then polished. This can be a lengthy process, but it depends on the product and application.

ENVIRONMENTAL IMPACTS
All the scrap material created during rapid prototyping can be recycled, except the carbon filled powders. These processes are an efficient use of energy and material, as they direct thermal energy to the precise point where it is required.

Case Study

→ Building an SLA part of PE mimic

The rapid prototyping machine works automatically, overnight. The CAD data from a .stl file guides the UV laser (**image 1**). The SLA parts appear as ghost-like forms in the clear epoxy resin (**image 2**). Each pass of the laser fuses another 0.05 mm to 0.1 mm (0.002–0.004 in.) to the preceding layer by constructing the 'skin' (**image 3**) and then filling in the 'core' material (**image 4**). The build platform moves down 1 step, and the paddle sweeps across the top of the build tank to break the viscous surface tension and ensure that the correct depth of layer is in place to be fused (**image 5**). The finished parts are left to drain and then removed from the build tank along with any uncured epoxy resin residue (**image 6**). The parts are separated from the build platform (**image 7**) and the support structure that separated them is carefully detached (**image 8**). An alcohol-based chemical (isopropinol alcohol) is used to clean off the uncured resin liquid and any other contamination (**image 9**) and the parts are then fully cured under intensive UV light for 1 minute (**image 10**). The build strata are just visible in the finished part (**image 11**) and can be removed with abrasive blasting, polishing or painting.

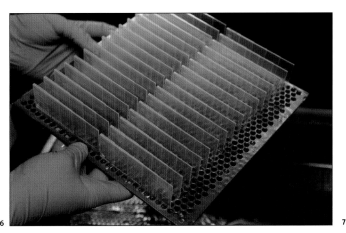

6

7

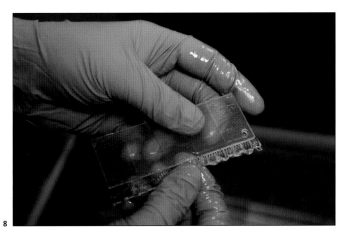

8

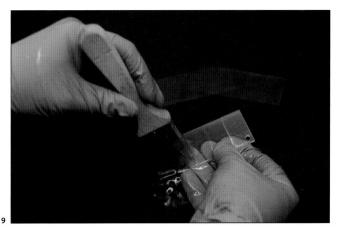

9

Featured Manufacturer

CRDM
www.crdm.co.uk

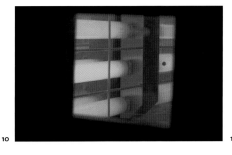

10

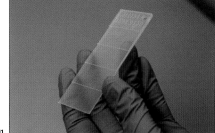

11

→ Building an SLS part

The SLS process takes place in a sealed, nitrogen rich atmosphere that contains less than 1% oxygen. The temperature inside the building chamber is maintained at 170°C (338°F), just below the melting point of the polymer powder, so that as soon as the laser makes contact with the surface particles they are instantly fused (**image 1**) by the 12°C (22°F) rise in temperature. Following the sintering process the delivery chamber moves up to deliver powder to the roller, which spreads it across the surface of the build area (**image 2**), coating the part with an even layer of powder (**image 3**).

The building process can take anything from 1 hour to 24 hours. Once it is complete, the build platform is raised (**image 4**), pushing the mixture of non-sintered powder and sintered parts into a clear acrylic container. The block of powder is disposed of in a clean-up booth and work begins to excavate the parts (**image 5**). The non-sintered 'cake' encapsulates the parts and has to be carefully brushed away so that individual parts can be removed for cleaning (**image 6**). Once most of the excess powder has been removed (**image 7**), the parts are blasted with a fine abrasive powder (**image 8**). The final part is an exact replica of the computer model, accurate to 150 microns (0.0059 in.) (**image 9**).

Featured Manufacturer

CRDM
www.crdm.co.uk

1

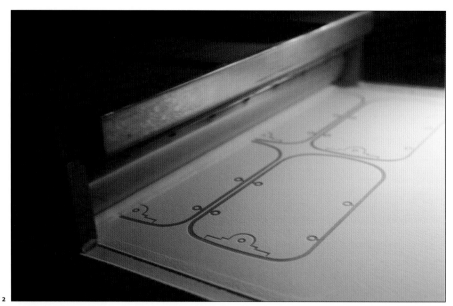

2

3

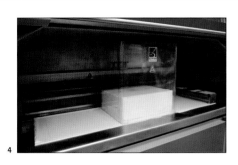

4

5

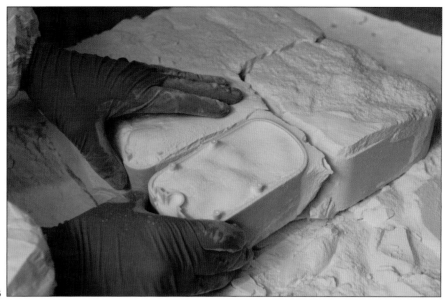

6

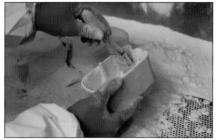

7

8

9

→ Building a DMLS part

The DMLS process builds a metal part from data within a .stl file. To make the process more efficient, each layer of the build is not completely filled in by the laser. The part is broken up into 3 main elements, which are the outer skin, the inner skin and the core. For every 3 layers of metal powder that are spread the outer skin is sintered 3 times, the inner skin is sintered twice and the core is sintered only once. A cross-section of a typical DMLS part shows the outer skin, inner skin and core, visible in the different tones (**image 1**).

A steel build plate is set precisely inside the build chamber (**image 2**). Its thickness is 13–45 mm (0.51–0.77 in.), depending on the depth of part to be built on it.

The fine metal powder used to form this part comprises spheres of nickel–bronze alloy, 20 microns (0.00078 in.) in diameter. The powder is sieved into the delivery chamber and then spread evenly across the build area in preparation for the first pass of the laser (**image 3**). When the build area is ready for sintering to begin (**image 4**), the CO_2 laser

is guided across the surface layer by a CNC mirror (**image 5**). After each pass of the laser, a new layer of powder is spread over the build area (**image 6**).

Once building is complete, the build platform is raised (**image 7**), the excess powder is brushed away (**image 8**) and the steel plate removed with the part still attached (**image 9**). The part is eventually removed from the steel build plate by EDM wire cutting (see electrical discharge machining, page 254).

1

2

3

4

This part will be used as an insert in an injection molding tool. Some 20,000–30,000 components can be produced with the part before any significant wear occurs. Harder metal alloy powders will produce 100,000–200,000 parts from a single impression tool.

5

6

7

8

9

Featured Manufacturer

CRDM
www.crdm.co.uk

Part

2

Cutting Technology

Cutting Technology
Photochemical Machining

Unprotected metal is chemically dissolved in photochemical machining. Masks are designed so that components are cut out and engraved simultaneously. The results are both decorative and functional.

Costs	Typical Applications	Suitability
• Very low tooling cost • Moderate to high unit costs	• Aerospace • Automotive • Electronics	• Prototype to mass production

Quality	Related Processes	Speed
• High: accurate to within 10% of material thickness	• Abrasive blasting • CNC machining and CNC engraving • Laser cutting	• Moderate cycle time (50–100 microns/ 0.002–0.004 in. per hour)

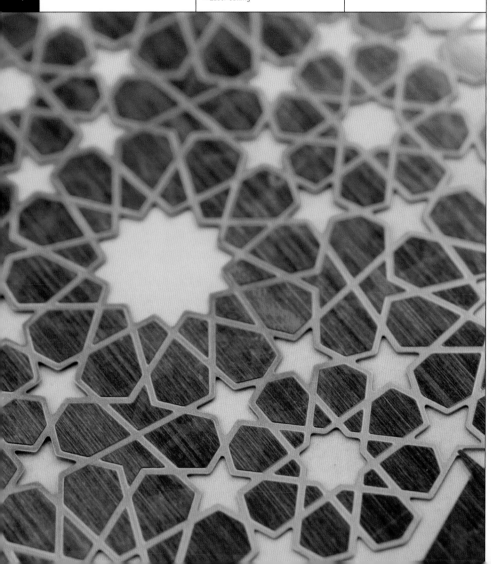

INTRODUCTION

This chemical cutting process, which is used predominantly to mill and machine thin sheet metals, is also known as chemical blanking and photofabrication. Decorative chemical cutting is known as photo etching (page 392).

Photochemical machining has 3 main functions: weight reduction, scoring and cutting out (known as profiling). It can chemically remove surface material, and can mark lines on most metals. The cutting action is precise to within 10% of material thickness and so is suitable for technical application. Profiling is achieved by attacking the material from both sides simultaneously. Therefore, this process is limited to foils and thin sheet metal between 0.1 mm and 1 mm (0.004–0.04 in.) thick. However, accuracy can be maintained only in sheet materials up to 0.7 mm (0.028 in.) thick.

TYPICAL APPLICATIONS

The technical aspects of this technology are utilized in the aerospace, automotive and electronics industries.

Circuit boards are made by coating the plastic (typically polycarbonate) board with a thin layer of copper. Areas of the metal are then chemically removed in order to create the positive image of the circuit board.

Other products include modelmaking nets, control panels, grills, grids, meshes, electronic parts, micro metal components and jewelry.

RELATED PROCESSES

Laser cutting (page 248) and engraving are used to produce similar products. However, lasers heat up the workpiece,

Photo Etching Process

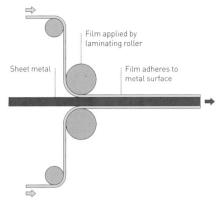

Stage 1: Applying photosensitive resist film

Film applied by laminating roller

Sheet metal

Film adheres to metal surface

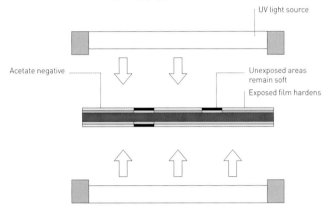

UV light source

Acetate negative

Unexposed areas remain soft

Exposed film hardens

Stage 2: UV exposure

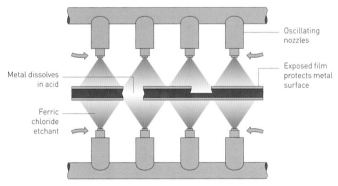

Oscillating nozzles

Metal dissolves in acid

Exposed film protects metal surface

Ferric chloride etchant

Stage 3: Chemical cutting

TECHNICAL DESCRIPTION

In stage 1, the metal workpiece is carefully prepared, because it is essential that it is clean and grease free to ensure good adhesion between the film and the metal surface. The photosensitive polymer film is applied by hot roll laminating or dip coating (page 68). In this case it is being hot roll laminated onto the metal. The coating is applied to both sides of the workpiece because every surface will be exposed to the chemical machining process.

In stage 2, acetate negatives (the tool) are prepared in advance and are printed from CAD or graphics software files. The negatives are applied to either side of the workpiece, then both sides of the combined workpiece – resist film and negative – are exposed to UV. The patterns are different on each side, so where they do not meet up, the cut will only go half way through the sheet. The soft and unexposed photosensitive resist film is chemically developed away. This process exposes the areas of the metal that are to be etched.

In stage 3, the metal sheet is passed under a series of oscillating nozzles that apply the chemical etch. The oscillation ensures that plenty of oxygen is mixed with the acid to accelerate the process. Finally, the protective polymer film is removed from the metalwork in a caustic soda mix to reveal the finished etching.

which can cause distortion in very thin metals. Lasers are typically suitable for thicker materials, which are impractical for photochemical machining.

CNC machining (page 182) and CNC engraving (page 396) also produce a heat-affected zone (HAZ), which can result in distortion in the workpiece.

Photochemical machined engravings have the same finish as an abrasive blasted surface (page 388).

QUALITY

This process produces an edge finish free from burrs, and it is accurate to within 10% of the material thickness. Another advantage of photochemical machining is that there is no heat, pressure or tool contact, and so the process is less likely to cause distortion, and the final shape is free from manufacturing stresses.

The chemical process does not affect the ductility, hardness or grain of the

metal structure. The surface finish is matt, but can be polished (page 376).

DESIGN OPPORTUNITIES

Photochemical machining is used to score lines, mill holes, remove surface material and profile entire parts (blanking). These different functions are applied simultaneously during machining. The type of cut is determined by the design of the protective mask,

Photochemical machining a brass screen

The design for the negatives is prepared using graphics software (**image 1**), which shows how the chemical machining process works. The left hand drawing is the reverse and the right hand drawing is the front of the workpiece. The 2 sides are etched simultaneously, so the areas that are blacked out on both sides will be cut through (profiled). Areas that are black only on 1 side will be half etched.

The negatives are printed onto acetate (**image 2**). Then a 0.7 mm (0.028 in.)

brass sheet is degreased in a bath of 10% hydrochloric acid. It is washed and dried, and a photosensitive film is laminated onto both sides (**image 3**). The process takes place in a dark room, to protect the film.

The coated workpiece is placed into an acrylic jig (**image 4**). The negatives are aligned and mounted onto each side of the jig. The assembly is placed under a vacuum and exposed to UV light on both sides (**image 5**). Unexposed photosensitive film is washed off in a developing process (**image 6**).

After the first stage of chemical machining (**image 7**), the process is repeated until the chemical has etched through the entire thickness of material (**image 8**). This can take up to 25 minutes for 0.7 mm (0.028 in.) brass. Ferrous metals will take longer.

Finally, the brass screen is removed from the brass sheet and the tabs that connected the workpiece are trimmed (**image 9**).

that is, by the photosensitive polymer resist film.

Score lines can be machined to act as fold lines. Changing the width of score will determine the angle of fold, and this is particularly useful for modelmaking and other secondary operations.

Profiling entire parts is achieved by chemically cutting the sheet from both sides. Lines that do not match up on both sides will become surface markings on 1 side. Therefore, sheets can be profiled and decorated (or scored) in a single operation.

Photochemical machining is suitable for prototyping and high volume production. Tooling costs are minimal: the negatives can be produced directly from CAD drawings or artwork and will last for many thousands of cycles. This means that small changes are inexpensive and adjustments can be made to the design. For these reasons this process is suitable for experimentation during design.

DESIGN CONSIDERATIONS
The intricacy of a pattern is restricted by the thickness of the material: any configuration can be machined, as long as the smallest details are larger than the material thickness. The minimum dimension for internal and external

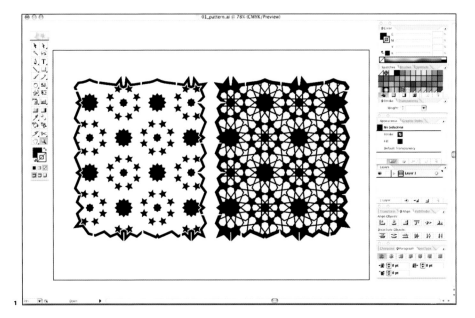

radii, holes, slots and bars is 1.5 times the thickness of the material.

Although any thickness of sheet material can be photo etched, only sheets and foils between 0.1 mm and 1 mm (0.004–0.04 in.) can be profiled using photochemical machining. Materials more than 1 mm (0.04 in.) thick will have a visible concave or convex edge profile left by the chemical process.

Tabs are an essential part of the design for profiling. They connect the parts to the workpiece and hold the parts in place as they are being cut out. Tabs make handling easy, especially if the parts are very small or in secondary

operations such as electroplating (page 364) and for flat sheet packaging. The shape of the tabs – V-shape or parallel – is determined by the secondary operations. A V-shape tab is for manually breaking out, whereas a parallel tab is for parts that are punched or machined. V-shaped tabs can be sunk within the profile of the part so that when removed there is no burr. The tabs can be half the material thickness and a minimum of 0.15 mm (0.006 in.) each.

COMPATIBLE MATERIALS
Most metals can be photo etched including stainless steel, mild steel,

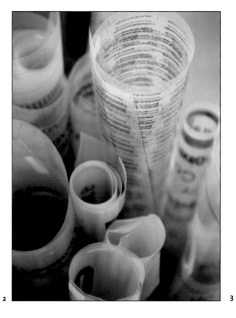

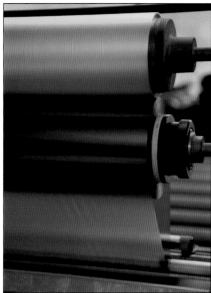

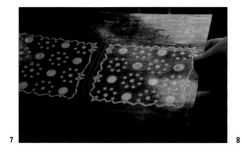

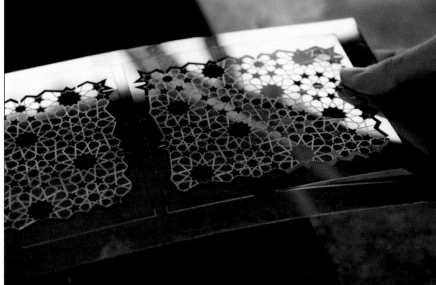

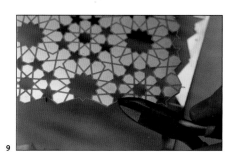

aluminium, copper, brass, nickel, tin and silver. Of these, aluminium is the easiest, and stainless steel is the hardest and so takes longer to etch.

Glass, mirror, porcelain and ceramic are also suitable for photo etching, although different types of photo resist and etching chemical are required.

COSTS

The only tooling required is a negative that can be printed directly from CAD data or a graphics software file.

Cycle time is moderate. Processing multiple parts on the same sheet reduces cycle time considerably.

Labour costs are moderate and depend on the complexity and duration of photochemical machining process.

ENVIRONMENTAL IMPACTS

During operation, metal that is removed from the workpiece is dissolved in the chemical etchant. However, offcuts and other waste can be recycled. There are very few rejects because photochemical machining is a slow, controllable process.

The chemical used to etch the metal is one-third ferric chloride. Caustic soda is needed to remove spent protective film. Both of these chemicals are harmful, and operators must wear protective clothing.

Featured Manufacturer

Mercury Engraving
www.mengr.com

Laser Cutting

This is a high precision CNC process that can be used to cut, etch, engrave and mark a variety of sheet materials including metal, plastic, wood, textiles, glass, ceramic and leather.

Costs	Typical Applications	Suitability
• No tooling costs • Medium to high unit cost	• Consumer electronics • Furniture • Model making	• One-offs to high volume

Quality	Related Processes	Speed
• High quality finish • Precision process	• CNC machining and engraving • Punching and blanking • Water jet cutting	• Rapid cycle time

INTRODUCTION

The 2 main types of laser used for this process are CO_2 and Nd:YAG. Both work by focusing thermal energy on a spot 0.1 mm to 1 mm (0.0004– 0.004 in.) wide to melt or vaporize the material. Both operate at very high speeds to precise tolerances and produce accurate parts with very high edge finish. The main difference between them is that CO_2 lasers produce a 10 micron (0.00039 in.) infrared wavelength and Nd:YAG lasers produce a more versatile 1 micron (0.000039 in.) infrared wavelength.

TYPICAL APPLICATIONS

Applications are diverse and include modelmaking, furniture, consumer electronics, fashion, signs and trophies, point of sale, film and television sets, and exhibition pieces.

RELATED PROCESSES

CNC machining (page 182), water jet cutting (page 272) and punching and blanking (page 260) can all be used to produce the same effect in certain materials. However, the benefit of the laser cutting process is that it cuts thermoplastics so well that they require no finishing; the cutting process leaves a polished edge. Laser cutting can also be used to score and engrave, so competes with CNC engraving (page 396), abrasive blasting (page 388) and photo etching (page 392) for some applications.

QUALITY

The choice of material will determine the quality of the cut. Certain materials, like thermoplastics, have a very high surface finish when cut in this way. Laser processes produce perpendicular, smooth, clean and cuts with a narrow kerf in most materials.

DESIGN OPPORTUNITIES

These processes do not stress the workpiece, like blade cutting, so small and intricate details can be produced without reducing strength or distorting the part. Therefore very thin and delicate materials can be cut in this way.

Raster-engraving methods can be used to produce logos, pictures and fonts on the surface of materials with cuts of various depths. Certain systems are very flexible and can be used to engrave from a variety of file formats.

DESIGN CONSIDERATIONS

These are vector-based cutting systems: the lasers follow a series of lines from point to point. The files used are taken directly from CAD data, which is divided up into layers that determine the depth of each cut. It is important that all lines are 'pedited' (joined together) so that the laser cuts on a continuous path. Replica lines also cause problems because the laser will treat each line as another cut and so increase process time.

Compatible file formats include .DXF and .DWG. Any other file formats may need to be converted.

Laser Cutting Process

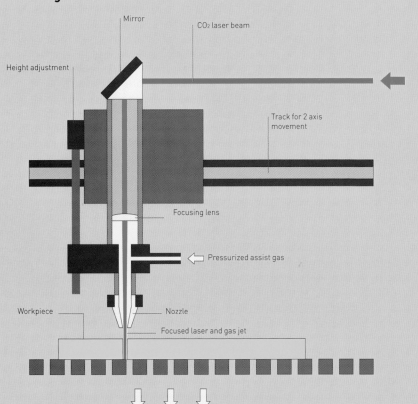

Mirror

CO₂ laser beam

Height adjustment

Track for 2 axis movement

Focusing lens

Pressurized assist gas

Workpiece

Nozzle

Focused laser and gas jet

Vacuum bed

TECHNICAL DESCRIPTION

CO_2 and Nd:YAG laser beams are guided to the cutting nozzle by a series of fixed mirrors. Due to their shorter wavelength, Nd:YAG laser beams can also be guided to the cutting nozzle with flexible fibre optic cores. This means that they can cut along 5 axes because the head is free to rotate in any direction.

The laser beam is focused through a lens that concentrates the beam to a fine spot, between 0.1mm and 1mm (0.0004–0.004 in). The height of the lens can be adjusted to focus the laser on the surface of the material. The high concentration beam melts or vaporizes the material on contact. The pressurized assist gas that blows along the path of the laser beam removes the cutting debris from the kerf.

This process is ideally suited to cutting thin sheet materials down to 0.2 mm (0.0079 in.); it is possible to cut sheets up to 40 mm (1.57 in.), but thicker materials greatly reduce processing speed. Different laser powers are required for different operations. For example, lower powered lasers (150 watts) are more suitable for cutting plastics because they leave a polished edge. High-powered lasers (1 to 2 kilowatts) are required to cut metals, especially reflective and conductive alloys.

COMPATIBLE MATERIALS

These processes can be used to cut a multitude of materials including timber, veneers, paper and card, synthetic marble, flexible magnets, textiles and fleeces, rubber and certain glasses and ceramics. Compatible plastics include polypropylene (PP), poly methyl methacrylate (PMMA), polycarbonate (PC), polyethylene terephthalate glycol (PETG), carbon fibre, polyamide (PA), polyoxymethylene (POM) and polystyrene (PS). Of the metals, steels cut better than aluminium and copper alloys, for example, because they are not as reflective to light and thermal energy.

COSTS

There are no tooling costs for this process. Data is transmitted directly from a CAD file to the laser cutting machine.

Cycle time is rapid but dependent on material thickness. Thicker materials take considerably longer to cut.

The process requires very little labour. However, suitable CAD files must be generated for the laser cutting machine, which may increase initial costs.

ENVIRONMENTAL IMPACTS

Careful planning will ensure minimal waste, but it is impossible to avoid offcuts that are not suitable for reuse. Thermoplastic scrap, paper and metal can be recycled, but not directly.

Above
Laser cutting produces a polished edge on certain thermoplastics and so eliminates finishing operations.

→ Laser cutting, raster engraving and scoring

LASER CUTTING PLASTIC

The pattern was designed by Ansel Thompson for Vexed Generation in 2005. This series of samples demonstrate the versatility of the laser cutting process. The strength and depth of the CO_2 laser can be controlled to produce a variety of finishes. The laser is used to cut translucent 3 mm (0.118 in.) thick poly methyl methacrylate (PMMA) (**image 1**). This is a relatively thin material, so the laser can work rapidly even though it is a complex and intricate profile. The result (**image 2**) took only 12 minutes to complete. The laser leaves a polished edge on PMMA materials and so no finishing operations are required.

LASER RASTER ENGRAVING

Adjusting the strength and depth of the laser cutter can produce interesting finishes. Raster engraving uses only a small percentage of the laser's power and produces engravings up to 40 microns (0.0016 in.) deep (**image 3**). This form of engraving can be carried out on a variety of materials and surfaces. For example, anodized aluminium may be raster engraved to reveal the bare aluminium beneath, instead of being printed. The sample took approximately 25 minutes to complete, which is considerably longer than for a simple cutting operation.

More powerful lasers can engrave deeper channels in materials. However, when the depth exceeds the width, problems with material removal result in a poor quality finish on the cut edge.

LASER SCORING

In this case the laser is being used at only 3% of its potential. Scoring produces 'edge glow' effect in the cut detail (**image 4**). This is caused by light picked up on the surface of the material being transmitted out through the edges. The scoring acts like an edge and so lights up in the same way.

LASER CUTTING WOOD

In this example 1 mm (0.04 in.) thick birch plywood is being cut and scored to form part of an architectural model. The first pass scores surface details laid out in the top layer of the CAD file. Secondly, the laser cuts internal shapes and finally the outside profile (**image 5**). The entire cutting and scoring process takes only 8 minutes. The parts are removed and assembled to form a building façade in relief (**image 6**).

1

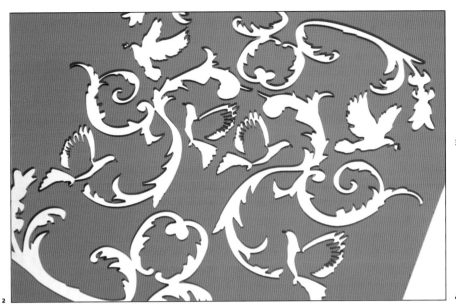

2

3

4

5

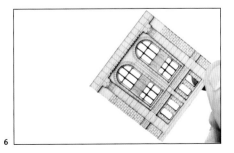

6

Featured Manufacturer

Zone Creations
www.zone-creations.co.uk

1

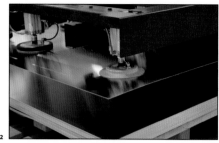

2

3

4

5

Featured Manufacturer

Luceplan
www.luceplan.com

→ Laser cutting the Queen Titania

Alberto Meda and Paolo Rizzatto designed the Queen Titania for Luceplan in 2005 (**image 1**). It is a modified (1.4 m/4.6 ft long) version of the Titania, which was first manufactured by Luceplan in 1989.

The structure of the light is reminiscent of an aeroplane wing. The parts are laser cut from a sheet of aluminium and assembled to form a lightweight and rigid structure.

The process starts by loading a sheet of aluminium into the laser-cutting machine (**image 2**). This is an Nd:YAG and oxygen cutting process operating at 600 watts. The machine is capable of cutting sheet steel up to 25 mm (0.98 in.) thick and aluminium up to 5 mm (0.2 in.) thick (**image 3**). Aluminium produces a larger HAZ (heat-affected zone) and so is trickier to laser cut.

The laser cutting process is rapid: each sheet takes only 8 minutes to process

(**image 4**). The parts on this sheet do not necessarily make up an entire light. Many lights are made in each batch, and so the parts are nested to optimize production and reduce waste.

The cut parts do not have to be de-burred or treated in any other way, which is a major advantage of laser cutting. Each part is removed from the sheet and hung up ready for assembly (**images 5** and **6**).

The light is assembled in 2 halves, which are joined by riveting (**images 7** and **8**).

The assembled structure is anodized (page 360) before the electronic and lighting components are added. The reason for anodizing is to ensure longevity of the finish of the light. The surface of untreated aluminium is less stable and more likely to change over time. The surface finish is grey, and light filters produce the vibrant colours.

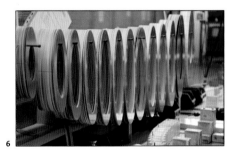

6

7

8

 External Internal Channel Surface

Cutting Technology
Electrical Discharge Machining

High voltage sparks erode the surface of the workpiece or cut a profile by vaporizing the material, making this a precise method of machining metals. Material is removed from the workpiece and a texture is applied simultaneously.

THERMAL

254

Costs	Typical Applications	Suitability
• Low tooling costs for EDM; no tooling costs for wire EDM; very high equipment costs • Moderate to high unit costs	• Precision metalwork in the aerospace and electronics industries • Tool making	• One-offs and modifications to existing metalwork • Low volume production

Quality	Related Processes	Speed
• Very high control of surface finish and texture	• CNC machining • Laser cutting • Water jet cutting	• Long cycle time

INTRODUCTION
Electrical discharge machining (EDM) has revolutionized toolmaking and metal prototyping. It is extremely precise and can be used to texture metal surfaces as well as machine them. This makes it ideal for toolmaking for injection molding (page 50) and other plastic forming processes, which is the largest area of application for this process.

There are 2 versions of the EDM process: die sink EDM (also known as spark erosion) and wire EDM (also known as wire erosion). Die sink EDM can be used to bore holes, texture surfaces

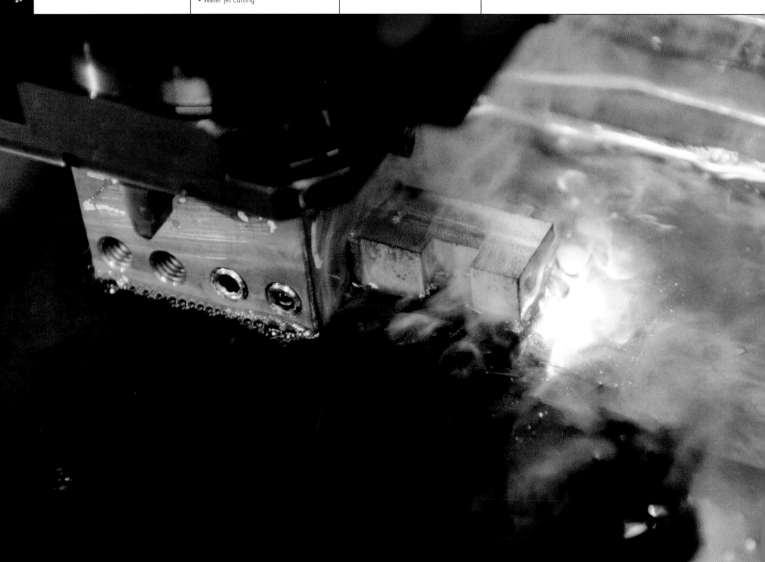

and die-sink complex geometries deep into metal parts. Wire EDM is similar in principle to hot wire cutting polymer foams and is used to create internal and external profiles with either parallel or tapering sides. Combined, these processes provide metalworkers with unlimited versatility.

TYPICAL APPLICATIONS

EDM equipment is very expensive, so its use is limited to applications that demand the very high levels of precision and the ability to work hardened steels and other metals that are impractical for CNC machining (page 182). It has been widely adopted by the toolmaking industry for injection molding, metal casting and forging. It is also used for modelmaking, prototyping and low volume production of typically no more than 10 parts.

RELATED PROCESSES

EDM is generally used in conjunction with CNC machining. The electrodes (tools) for die sink EDM are machined by conventional metalworking techniques (see image, top). However, complex and intricate tools are sometimes cut using wire EDM, which combines the 2 processes (see image, above right).

Die sink EDM has replaced CNC machining in applications that require accurate and intricate internal features. EDM is used because it can produce geometries that are not feasible with any other process. Internal features, especially in hard metals, are not practical for CNC machining because they would require very fine cutting tools, which would wear out rapidly.

Alternatives to wire EDM include water jet cutting (page 272) , laser cutting (page 248) and the power beam processes (page 288) for certain profiling and drilling procedures. All of these processes have high equipment costs, so the choice is often determined by available facilities. Wire EDM is suitable for parts up to 200 mm (7.87 in.) thick that require high dimensional accuracy, parallel kerf walls and a controlled surface texture.

QUALITY

The quality of EDM parts is so high they can be used to manufacture tooling for injection molding without any finishing operations. Surface finish is measured on the Association of German Engineers' VDI scale (see image, above left). The VDI scale is comparable with roughness average (Ra) 0.32–18 microns (0.000013–0.00071 in.). Many of the plastic products surrounding us today have been molded in tools shaped and finished by EDM.

Top
On these die sink EDM tools, the patination of black vaporized metal shows the cutting area.

Above left
The VDI scale is used to measure the texture of the surface.

Above right
This very small die sink EDM tool was cut out using the wire EDM process. It is too small and intricate to produce by conventional machining.

The quality of the finish and resulting texture are determined by the cutting speed and voltage. High voltage and high cutting speed produce a rough texture. Lower voltage, slower cutting speeds and more passes produce finer surface textures. Parts can be manufactured accurate to 5 microns (0.00019 in.).

DESIGN OPPORTUNITIES

A major advantage for designers is the ability of this process to cut metals and apply a texture simultaneously. Textures, including matt and gloss, are usually applied after machining operations using abrasive blasting or photo etching techniques. EDM produces accurate

TECHNICAL DESCRIPTION

Die sink EDM takes place with the electrode (tool) and workpiece submerged in light oil, similar to paraffin. This fluid is continuously running and so both maintains the temperature of the workpiece and flushes out vaporized material. It is also dielectric, which insulates the working area and sustains the static discharge within in it.

The copper electrode (tool) and metal workpiece are brought within close proximity, which initiates the spark erosion process. High voltage sparks leap from the electrode to the workpiece and vaporize the surface of the metal. The electrical discharges jump between the closest points on the electrode and the workpiece, forming continuous and even surface material removal.

There is no pressure required in operation. The electrode is lowered into the workpiece very gradually; the speed of the process is dependent on the finish required and can range from 2 mm³ (0.00012 in.³) per minute to over 400 mm³ (0.024 in.³) per minute for very rough finishes. As the tool is sunk into the workpiece it is continuously agitated and descends in a spiral. This action flushes vaporized material away from the cutting area and ensures even and efficient material removal.

Die sink EDM Process

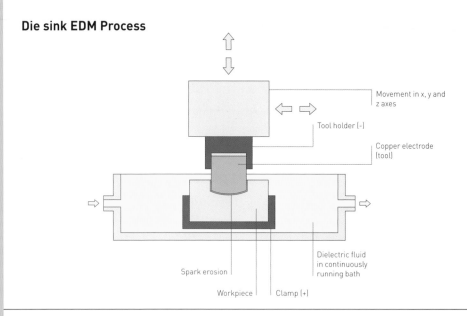

Movement in x, y and z axes

Tool holder (-)

Copper electrode (tool)

Dielectric fluid in continuously running bath

Spark erosion

Workpiece   Clamp (+)

---

textures, which are determined by the machine settings, eliminating the need for further finishing operations.

These processes can be used to produce geometries that are not possible with conventional machining. For instance, the guide heads on many wire EDM machines can move independently during the cutting process. The advantage is that complex tapers, up to 30°, can be cut with extreme precision, which is not possible with any other machining technique.

Die sink EDM can be used to produce internal geometries on parts that are not possible with conventional machining.

This is because a negative copper electrode can be machined into shapes that are not suitable for cavities. The negative electrode (tool) is reproduced in the workpiece, creating sharp corners and complex features that would otherwise be impractical. The erosion of the copper electrode (tool) is considerably slower than the erosion of the workpiece (0.1%), so small internal radii and complex features are reproduced to the same precision as simple geometries.

Wire EDM can be used in much the same way as hot wire cutting polymer foam, although it is considerably more precise and much slower. The equipment requires a high level of skill to operate.

Stress is not applied to the electrode (tool or wire) or workpiece during processing because the metal is not being shaped by force, but instead by high voltage sparks that vaporize the surface. Other than the obvious, this has many processing advantages. For example, multiple thin-walled parts can be stacked up and cut by wire EDM to reduce processing time.

DESIGN CONSIDERATIONS

Although these processes can produce internal radii as small as 30 microns (0.0012 in.), it is always better to use larger radii to avoid stress concentration.

However, these do not need to be any larger than 500 microns (0.02 in.) in most applications. The minimum internal radius is also affected by the thickness of the wire electrode, which is typically 50 microns (0.0020 in.) to 300 microns (0.012 in.) in diameter.

The thickness of material that can be cut by wire EDM ranges from 0.1 mm (0.004 in.) up to 200 mm (7.87 in.), but depends on the capabilities of the equipment. The depth that can be produced by die sink EDM is limited by the ease with which vaporized metal (black powder) can be flushed away, so maximum depth is affected by the depth to diameter ratio. Very deep profiles are possible, but may require extra flushing, which will increase cycle time.

COMPATIBLE MATERIALS

Many metals can be shaped by EDM techniques. The hardness of the material does not affect whether it can be processed in this way. Metals including stainless steel, tool steel, aluminium, titanium, brass and copper are commonly shaped in this way.

COSTS

Wire EDM does not require tooling. However, it does consume wire electrode continuously, which must be replaced.

→ Die sink EDM

This process is widely used for tool-making. As well as machining entire cavities for injection molding, it is often used to modify existing tools. It can also be used to bore holes or emboss surface textures and graphics. In this case study, Hymid Multi-Shot are forming a cavity directly into the surface of high-carbon steel (**image 1**). It is not practical to machine complex and intricate cavities into hard metals like this, other than by using EDM.

The copper alloy electrode (tool) is inserted into a tool holder that is registered with the computer guided tool head (**image 2**). The EDM machines are programmed to the settings required for the tool. The black areas show the parts of the tool that have been exposed to spark erosion. Each tool may last for only 5 uses before it has to be replaced. If extreme levels of precision, to within 5 microns (0.00019 in.), are required, new tools are machined for each EDM operation,

including a tool for roughing out and a separate tool for finishing.

The tool and workpiece are inserted and submerged in a dielectric fluid, which is similar to paraffin (**image 3**). In fact, paraffin was once used as the insulating fluid, until this oil was developed.

Sparks and fumes are given off during rough cutting (**image 4**). The copper electrode is charged with an electric current, which jumps to the oppositely charged workpiece when they come into very close proximity. There are several thousand sparks per second. Each spark vaporizes a small piece of surface material. The arcs will jump across the shortest distance between the electrode and workpiece, which ensures evenly distributed surface removal. In this case the process is vaporizing around 400 mm³ (0.024 in.³) per minute. The resulting surface finish is very rough (**image 5**).

The second stage of machining is much slower, in this case as low as 50 mm³ (0.003 in.³) per minute (**image 6**). This produces a much finer surface texture (**image 7**). This is a very shallow impression; die sink EDM can also be used to form very deep cavities.

1

2

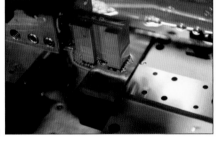

3

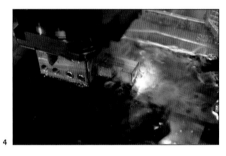

4

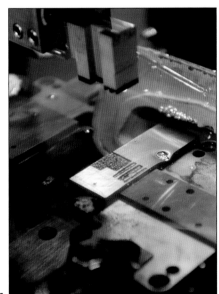

5

6

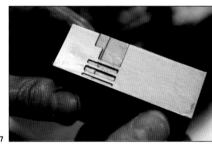

7

Featured Manufacturer

Hymid Multi-Shot
www.hymid.co.uk

TECHNICAL DESCRIPTION

In this process the wire electrode, which is usually copper or brass, is fed between the supply spool and take up spool. It is charged with a high voltage, which discharges as the wire progresses through the workpiece. Similar to die sink EDM, the spark occurs in the smallest gap between the metals. There are many thousand sparks per second, which vaporize very small amounts of the metal surfaces. The wire electrode is not recycled; instead it is replaced continuously, which maintains the accuracy of the process.

This process is submerged in deionized water, which is maintained at 20ºC (68ºF). The water is continuously running to flush spent material and recycled through a filtration system.

The upper wire guide can be moved along x and y axes to enable cutting angles up to 30º.

Wire EDM Process

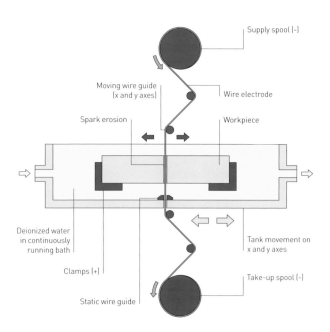

Supply spool (-)

Moving wire guide (x and y axes)

Wire electrode

Spark erosion

Workpiece

Deionized water in continuously running bath

Clamps (+)

Static wire guide

Tank movement on x and y axes

Take-up spool (-)

Die sink EDM tools are typically made from a copper alloy, which can be shaped by conventional machining processes or wire EDM. The tooling is relatively inexpensive, but for precision parts new tools are required for each operation.

Different factors affect cycle time for wire and die sink EDM. In wire EDM applications, cycle time is affected by the thickness of the material. Thicker materials require higher power and consume more wire electrode. As a rough guide, a piece of 36 mm (1.42 in.) hardened steel can be cut at 1.5 mm (0.059in.) per minute. This will produce an even, matt finish. Slowing the process down, or using multiple passes, will produce a finer texture. Speeding the process up will produce a rougher surface finish.

Similarly, the cycle time of die sink EDM is determined by the size of the cutting area and desired finish. Typically, internal geometries are roughed out at between 200 mm³ and 400 mm³

(0.012– 0.024 in.³) per minute. As the term suggests, this will produce a rough finish. After this a new tool is used at a much slower cutting rate, which can be as low as 2 mm³ (0.00012 in.³) per minute, to produce a very fine finish. Therefore, very fine surface textures are best suited to small surface areas.

Labour costs are moderate due to the high level of operator skill required and the length of the process.

ENVIRONMENTAL IMPACTS

This process requires a great deal of energy to vaporize the metal workpiece. However, it does eliminate the need for any further processing, such as abrasive blasting (page 388) or photo etching (page 392).

The electrically insulating fluids are continually recycled for reuse. The metal electrodes are also suitable for recycling. Fumes are given off during operation, which can be hazardous.

→ Wire EDM

Unlike die sink EDM, which is used to form internal relief cavities, wire EDM is used to cut internal and external profiles. The wire is held under tension to cut straight lines. The guide heads move in tandem along x and y axes to produce profiles and independently along x and y axes to produce tapers. The wire acts like the copper tool used in die sink EDM and is usually a copper alloy.

There are many cutting operations in this case study. This collection of images illustrates one of those operations, the principles of which can be applied to all other wire EDM operations. The partly machined high-carbon steel workpiece (**image 1**) is loaded into a jig and clamped in place. A small hole is drilled in preparation, which the wire electrode is automatically fed through (**image 2**).

The guide heads are brought close to the workpiece for precision and both the workpiece and wire electrode are submerged in deionized water, which acts as an insulator (**image 3**). Once submerged, the cutting process begins (**image 4**). It is a long process; in this case the 36 mm (1.42 in.) workpiece is cut at 1.5 mm (0.059 in.) per minute. This will produce the desired finish.

After cutting, which takes approximately 2 hours, the part is removed and cleaned (**image 5**). The accuracy of the part is measured on a micrometer (**image 6**). On the finished article you can just make out the kerf where the wire has cut from the pre-drilled hole to the cutting profile (**image 7**): on the left are the parts prior to cutting, in the centre is the finished workpiece and on the right the material that has been removed.

2

3

1

4

5

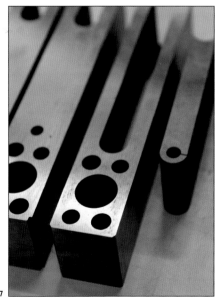

6

7

Featured Manufacturer

Hymid Multi-Shot
www.hymid.co.uk

External | Internal | Channel | Surface

Cutting Technology

Punching and Blanking

Circular, square and profiled holes can be cut from sheet materials using a hardened steel punch. Tooling is either dedicated or interchangeable, depending on the geometry and complexity of design.

Costs	Typical Applications	Suitability
• Low to moderate tooling cost • Low to moderate unit costs	• Automotive and transportation • Consumer electronics and appliances • Kitchenware	• One-off to mass production

Quality	Related Processes	Speed
• High quality and precise, but edges require de-burring	• CNC machining • Laser cutting • Water jet cutting	• Rapid cycle time (1–100 per minute) • Tooling changeover is time consuming

INTRODUCTION

Punching and blanking are shearing processes used in metalwork. They are essentially the same, but the names indicate different uses: punching refers to cutting an internal shape (see images, below and opposite) and blanking is cutting an external shape in a single operation.

Cutting out large parts (blanking), or removing large areas of material from the centre of the workpiece (punching) becomes impractical with a single tool above 85 mm (3.35 in.) diameter because it would be too expensive. In such cases

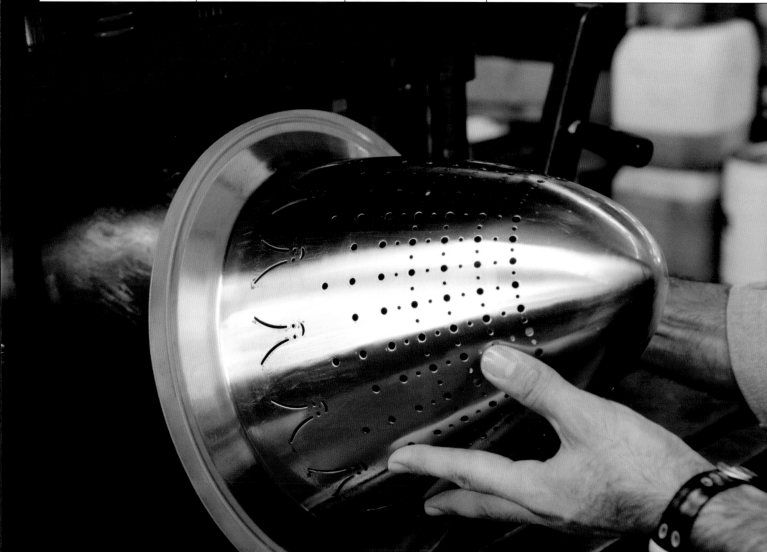

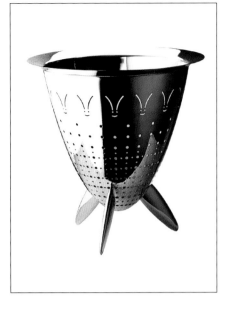

multiple punches around the perimeter of the cut string together in a process known as 'nibbling'.

Another operation, known as 'notching', is the removal of material from the outside of the workpiece.

These processes are generically referred to as punching. They are often combined in a production line such as progressive dies in metal stamping (page 82). In other applications, punching may be the core process, in which case the operations are carried out on a punch press, or turret punch. A punch press uses a single tool and is typically set up for a repetitive operation. A turret punch is computer controlled, and is loaded with multiple tools that can accommodate complex and varied operations, including nibbling. Flypressing is a manually operated spinning press.

TYPICAL APPLICATIONS
Some typical products include kitchenware, such as Alessi colanders, bowls and plates, consumer electronic and appliance enclosures, filters, washers, hinges, separators, general metalwork and automotive body parts.

These processes are used a great deal in low volume and prototype production of parts that will ultimately be shaped by processes such as stamping.

RELATED PROCESSES
These processes can only be used for thin sheet materials up to 5 mm (0.2 in.). Water jet cutting (page 272) and laser cutting (page 248) are suitable for a wide range of materials and thicknesses. Small cuts can be made more quickly by punching because each cycle removes a chunk of material, whereas laser and water jet have to trace the outline.

CNC machining (page 182) of holes and profiles is more precise, but consequently more expensive. A machining operation, such as drilling or milling, will produce perpendicular walls without burr, but cycle time is considerably longer.

QUALITY
The shearing action forms 'roll over' on the cut edge and fractures the edges of the material. This results in burrs, which are sharp and have to be removed by grinding and polishing (page 376). Problems are minimized by careful tool design and machine set up.

DESIGN OPPORTUNITIES
It is not just flat sheet that can be punched. Stamped, deep drawn (page 88), roll formed (page 110) and extruded metal parts are also suitable. Specialized tooling is required for 3D parts, which

Above left
These figures are punched from Alessi stainless steel products and leave a hole of exactly the same shape.

Above
The perforation on the Max Le Chinois colander is punched in 2 stages on a rotating die to accommodate its profile.

will increase investment cost. It may also have to be carried out in several stages because punching can only occur in the vertical position. Thus, Max Le Chinois (see image, above), the Alessi colander designed by Philippe Starck, requires 2 punches: 1 for the perforations on the side and 1 for the base. The product is punched and rotated in stages until the full circumference is complete.

Turret punches can be used to prototype parts without investing in expensive tooling. A typical metalworking factory will have many tools for cutting a range of profiles guided by computer.

Punching tends to be used to add function to a metal part such as perforations or fixing points. It is also used for decorative applications because a punched hole, for example, does not have to be circular or square.

Selective material removal will reduce weight but may not reduce strength too much. Therefore, it is sometimes practical to carry out functional punching and selective material reduction on an item in the same operation.

DESIGN CONSIDERATIONS

The first consideration is the width of material being removed or left behind. Punch diameter or width should be not smaller than material thickness. Also, the width of material left behind should be greater than the thickness of material. Fixing points should be inset as far as possible for maximum strength.

Parts that are cut out have to be tied into the sheet of material by tabs. These ensure that the part does not move about during production, but are sufficiently fragile to allow the parts to be broken out by hand. The burr left behind will have to be polished off, unless it is concealed in a recess.

COMPATIBLE MATERIALS

Almost all metals can be processed in this way. It is most commonly used to cut carbon steel, stainless steel and aluminium and copper alloys.

Other materials, including leather, textiles, plastic, paper and card, can also be punched, but metalworking machinery is not suitable. These materials are softer and easier to cut than metal. Punching in this case is typically referred to as die cutting (page 266), which is the process of making all the cuts in a single stroke. Thin metal sheet is also suitable for die cutting.

COSTS

Standard and small tools are inexpensive. Specialized and 3D tools can be more expensive but are suitable for small batch production.

Cycle time is rapid. Between 1 and 100 punches can be made every minute, but this depends on loading time and continuity of production. Tooling changeover time can be expensive.

Labour costs are moderate to high because these processes require skilled operators to ensure high quality cuts. Machine set up determines the quality and speed of cut.

Punching and Blanking Process

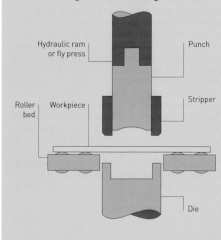

Stage 1: Loading

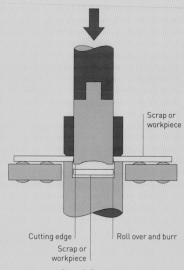

Stage 2: Punching

TECHNICAL DESCRIPTION

The operation is the same whether it is carried out on a turret punch, punch press or flypress. It is possible to punch a single hole, multiple holes simultaneously, or many holes with the same punch.

In stage 1, the workpiece is loaded onto the roller bed. In stage 2, the stripper and die clamp the workpiece. The hardened punch stamps through it, causing the metal to fracture between the circumferences of the punch and die. The die is slightly larger than the punch to allow for roll over and burr caused by the shearing process. The offset is determined by the type and thickness of material and ranges from 0.25 mm to 0.75 mm (0.01– 0.03 in.). Therefore, the sides of the hole are not exactly perpendicular to the face.

Once cut, the punch retracts and the stripper ensures that the metal comes free. Either the punched material or the surrounding material is scrap, depending on whether it is a punching or blanking operation. In both cases the scrap is collected and recycled.

Dedicated tooling may be made up of many punches joined together. They operate simultaneously in a punch press. Complex and intricate shapes are more easily achieved with these tools such as in the Alessi case study.

ENVIRONMENTAL IMPACTS

Parts can be nested very efficiently on a sheet to minimize scrap. Any scrap is collected and separated for recycling, so there is very little wasted material.

The de-burring process is abrasive and so produces some waste.

→ Punch pressing the Alessi Tralcio Muto tray

This case study illustrates a stage in the production of the Tralcio Muto tray (**image 1**), which was designed for Alessi by Marta Sansoni in 2000.

The tray is stamped in stainless steel; the edges are trimmed, de-burred and rolled. The part is coated in a thin film of oil (**image 2**). Trays are loaded into the tool (**image 3**) one at a time (**image 4**). The design is such that the punching operation is carried out in 2 stages. There is too much cutting for a single stroke on this 150 tonne (165.35 US ton) press.

The first punch is made (**images 5** and **6**) and the product is rotated through 90° and punched again to produce the complete pattern (**images 7** and **8**).

The tray is punched from the back so that the burrs are on the flat front surface. This is to ensure that the tray can be polished to a very high finish. If it were punched from the front, the edges of the holes would have a slight radius from the impact and so could not be polished to such a high level.

The final part is removed from the tool and taken to be polished (**image 9**).

1

2

3

4

5

6

7

8

9

Featured Manufacturer

Alessi
www.alessi.com

→ Turret punching an aluminium blank

The turret punch is loaded with a series of matching punches and dies (**image 1**). The orange surround on the punch is the stripper, which makes sure the punch can retract from the workpiece.

The die set is loaded into the turret and their location programmed into the computer guiding software (**image 2**). Optimization software ensures that the punch selected is the most efficient for each cut. Many different tools may be used for a single part, all of which are pre-loaded into the turret. Changeover time can be lengthy, and all the while the machine is not working it is losing money.

Parts are nested together on a sheet to minimize scrap. The punching operation takes less than 2 minutes for these parts (**image 3**). The sheet is moved around on the roller bed and the turret rotates to select the die set for each operation.

Each cycle produces a small piece of scrap, which is siphoned into a collection basket for recycling (**image 4**). In a blanking operation, these pieces of metal are the workpiece. Many of the pieces, such as squares and circles, are from straightforward punching operations. The crescent shaped part was produced by a nibbling operation, in which a circular punch makes a larger diameter hole with many overlapping cuts.

The cut out part is inspected and cleaned up (**image 5**). Tabs maintain the blank in the sheet until it is further processed (**image 6**). In this case, the metal blanks are formed into enclosures by press braking (page 148).

1

Featured Manufacturer

Cove Industries
www.cove-industries.co.uk

2

Cutting Technology

Die Cutting

Using die cutting, thin sheet materials can be cut through, kiss cut, perforated and scored in a single operation. It is a rapid process and is utilized a great deal in the packaging industry for mass producing cartons, boxes and trays.

Costs	Typical Applications	Suitability
• Low tooling costs • Low unit costs	• Packaging • Promotional material • Stationery and labels	• Low to high volumes

Quality	Related Processes	Speed
• High quality edge finish	• Laser cutting • Punching • Water jet cutting	• Very rapid cycle time (up to 4,000 per hour)

INTRODUCTION

This process, which is also referred to as blanking, is a stamping process in which shapes are cut from sheet material using steel knives mounted on wooden tools. It is the opposite of punching (page 260), which is the removal of material from the workpiece by a similar action.

Die cutting, which is carried out as a linear or rotary operation, is the most cost effective way to cut complex net shapes from most non-metallic sheet materials, including plastic, paper, card, felt and foam. It combines many operations into a single stamping action.

The depth and shape of each steel rule determines whether it cuts through, kiss cuts (in which the top layer is cut through and the support layer is left intact), scores or perforates the material.

TYPICAL APPLICATIONS

This process is primarily used in the packaging industry. The number and angle of lines does not affect the operating cost, so it is ideal for packaging systems that have intricate details or complex shapes and are not based on right angles. Many boxes, cartons and trays are made in this way.

A major area of application is cutting out labels, including pressure sensitive and adhesive backed ones. Other stationery applications are plastic and paper folders, envelopes and promotional material. Lighting products are also made using die cutting.

This process lends itself to low volume production, as low as 50 parts, because of its relatively inexpensive tooling.

RELATED PROCESSES

Die cutting is the process of choice for most non-metallic and non-glass flat sheet cutting applications. Sheet metals are cut into similar shapes by punching, water jet cutting (page 272), laser cutting (page 248) or simply with a guillotine. Sheet glass is typically scored (page 276) or water jet cut.

QUALITY

The tools used in die cutting are typically laser cut and so precise. The steel knives wear very slowly and have a long lifespan. The quality of the cut is therefore very high and repeatable

Top
This is the steel rule wooden die used for the Libellule light (page 269).

Above
The foam pads that surround the sharp blades eject the cut or perforated material.

Right
These steel rule wooden dies are used to mass produce cardboard boxes and trays.

over large volumes. The cutting action is between a sharp cutting rule and a steel cutting plate, and therefore results in clean, accurate cuts. Scoring, kiss cutting and perforating are equally precise, and the depth of cut is accurate to within 50 microns (0.0019 in.). Accuracy also depends on the material being die cut: for example, corrugated materials have a unidirectional core pattern that may affect scored bends.

Some materials resist certain cutting techniques, especially kiss cutting. Cutting such materials may result in a halo around the perimeter, or cause laminated materials to delaminate.

Testing is essential to eliminate such aesthetic defects.

DESIGN OPPORTUNITIES

The complexity of a design does not affect the cost of die cutting – unlike many mass production techniques in the packaging industry such as the printing and rotary slotting machines used to produce cardboard boxes. Therefore, designers are free to explore innovative structures such as closures, handles and display windows in packaging without fear of ramping up the cost.

The size of the product is also insignificant because multiple small

TECHNICAL DESCRIPTION

Die cutting consists of 3 main elements: the steel rule wooden die, the press either side of it and the sheet material to be cut.

Steel rules are mounted in a wooden die. The slots for the rules are typically laser cut. If the blades wear out then they can be easily replaced. The steel rules pass right through the wooden die and press against the steel support plate. This ensures that all of the energy produced in the hydraulic rams is directed into the cutting or scoring action.

The pressure required for cutting is determined by the thickness and type of material. It is also sometimes possible to cut through multiple sheets simultaneously. Generally, the pressure required to die cut is between 5 and 15 tonnes (5.5–16.5 US tons), but some die cutters are capable of 400 tonnes (441 US tons) of pressure.

In stage 1, the sheet is loaded onto the cutting plate, which rises up to meet the steel rule wooden die. In stage 2, the sharp steel rules cut right through the sheet material, while foam and rubber pads either side of each cutting rule apply pressure to the sheet material to prevent it from jamming. The cutting action is instantaneous, each sheet being processed within a few seconds. However, particularly tight and complex geometries can clog up the die, in which case the operator may have to remove material manually.

It is possible to score certain materials, such as corrugated card and plastic, using different techniques, including perforating and creasing, and by adding a ribbed strip on the cutting plate. The type of steel rule used to score the material is determined by the angle and depth of score required.

Die Cutting Process

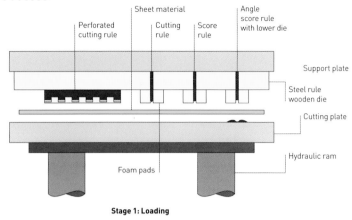

Stage 1: Loading

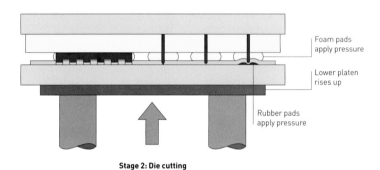

Stage 2: Die cutting

shapes can be mounted onto a single die, maximizing output. Cutting beds can typically accommodate sheet materials up to 1.5 x 2.5 m (5 x 8 ft).

A multitude of materials can be die cut. This means that an entire packaging system can be shaped in this way, including the card enclosure and foam padding, for example.

Multiple operations are performed simultaneously. Materials can be cut to specific depths in increments only microns apart, such as in kiss cutting, which is used for the production of self-adhesive labels on a backing film. Because these operations have to be extremely precise (within 50 microns/0.0019 in.), they take longer to set up and require skilled operators to control the tools.

Prototyping in sheet materials is generally an inexpensive process. Most cutting operations are carried out on a CNC x- and y-axis cutter, which transfers vector based CAD data to a cutting head operating on x and y axes. This same process is used to cut fabric patterns for upholstery (page 338) and fibre reinforcement for composite laminating (page 206).

DESIGN CONSIDERATIONS

The type of material will determine the thickness that can be cut.

Possible shape, complexity and intricacy are determined by the production of the steel rule dies. Steel rule blades can be bent down to a 5 mm (0.2 in.) radius. Holes with a smaller radius are cut out using a profiled punch. Sharp bends are produced by joining 2 steel rules together at the correct angle.

The width between the blades is limited to 5 mm (0.2 in.); any smaller and ejection of material becomes a problem. Thin sections should be tied into thicker sections to minimize ejection problems.

Nesting parts is essential to reduce material consumption, cycle time, scrap material and costs in general.

→ Die cutting the Libellule light

This Libellule pendent light was designed by Black & Blum (**image 1**). Because the volumes for this product do not justify fully automated production, it is made on a manually operated die cutting machine (**image 2**). To reduce weight and maximize production efficiency, 2 shades are produced on each piece of polypropylene sheet, which is cut to size (**image 3**).

Each cycle requires the operator to load and unload the cut sheet (**image 4**), after which the cut shapes are easily removed from the sheet (**images 5** and **6**), which becomes scrap material (**image 7**).

1

2

3

4

5

6

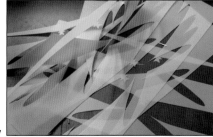

7

Featured Manufacturer

PFS Design & Packaging
www.pfs-design.co.uk

→ Die cutting and assembling a crashlock box

This crashlock cardboard box (**image 1**) is produced in very large numbers and so production is fully automated. It is called a 'crashlock' because the base is glued together in such a way that it unfolds as the box is opened up.

Die cutting is the most effective method of manufacturing the box because there are angled cuts and scores that cannot be produced on printing and slotting machines. Very simple boxes that are scored and cut at right angles can be produced economically with rotary cutters on the end of the print line. However, as soon as there is an angle to score, such as on the base of the crashlock box, die cutting techniques are used.

Large quantities of card are needed for this production process that requires 4,000 feeds per hour (**image 2**). The card is die cut in an enclosed production line, after which is it fed onto the folding production line (**image 3**). This is an incredible and rapid process, which is hypnotic to stand and watch.

The combination of levers, arms, rollers and glue dispensers gradually assemble the boxes (**images 4–7**). The finished boxes emerge from the press,

with the adhesive fully cured, flat and ready to ship (**image 8**).

This particular set-up is not dedicated, and most of the mechanisms are computer controlled. Therefore at the click of a button they move into place ready for the next sequence of events. However, setting up the process for the first time is still a long and highly skilled job.

COMPATIBLE MATERIALS

Many materials can be cut by die cutting, including corrugated plastic, plastic sheet, plastic film, self-adhesive films, cardboard, corrugated card, board, foam, rubber, leather, veneer, felt, textile, very thin metals, fibreglass and other fibre reinforcement, flexible magnets, cork, vinyl and DuPont™ Tyvek®.

COSTS

Tooling costs are low. Tools are usually wooden based and wear out very slowly.

Cycle time is very rapid. Automated systems can operate at up to 4,000 feeds per hour. Each feed may produce 4 or

more parts, which means a cycle time of 16,000 parts per hour.

Labour costs are low in automated systems. Manually operated production is restricted to low volume parts and is slightly more expensive, but operation is rapid and needs little adjustment or maintenance. Set-up can be expensive in both automated and manually operated versions, due to the machine down time.

ENVIRONMENTAL IMPACTS

Die cutting does produce offcuts. However, scrap can be minimized by nesting the shapes together on a sheet. Most scrap can be recycled by

the material supplier. In some cases the offcuts can be recycled by the manufacturers themselves, such as in the case of Cullen Packaging, who use the scrap material from their cardboard packaging facility as raw material in their paper pulp production plant (page 204).

3

4

5

6

7

8

Featured Manufacturer

Cullen Packaging

www.cullen.co.uk

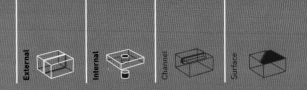

Cutting Technology
Water Jet Cutting

A high-pressure jet of water, which is typically mixed with abrasives, produces the cutting action. It will cut through almost any sheet material from soft foam to titanium, and is capable of cutting stainless steel up to 60 mm (2.36 in.) thick.

Costs	Typical Applications	Suitability
• No tooling cost • Moderate unit cost	• Aerospace components • Automotive • Scientific apparatus	• One-off to medium volumes

Quality	Related Processes	Speed
• Good quality	• Die cutting • Laser cutting • Punching and blanking	• Moderate cycle time, depending on the type and thickness of material

INTRODUCTION

This is a versatile process for cold cutting sheet materials. It has been used for commercial industrial applications since the 1970s and has continued to develop at a rapid pace. It is carried out as either water only cutting or abrasive water jet cutting. Water only cutting is a high-velocity jet of water at pressures up to 4,000 bar (60,000 psi). The supersonic jet of pure water erodes the materials and produces the cutting action. In abrasive water jet cutting small particles of sharp material are suspended in the high-velocity jet of water to aid the cutting process in hard materials. In this case the cutting action is executed by the abrasive particles as opposed to the water. Both are very accurate and work to tolerances of less than 500 microns (0.02 in.).

TYPICAL APPLICATIONS

Applications for this process are diverse. As with most new technology, the aerospace and advanced automotive industries were the early adopters. However, this process is now an essential part of many factories. Specific applications include cutting intricate glass profiles for scientific apparatus, titanium for aerospace and carbon reinforced plastic for motor racing.

RELATED PROCESSES

This process competes with a diversity of machining operations as a result of the diversity of compatible materials. Laser cutting (page 248), die cutting (page 266), punching and blanking (page 260) and glass scoring (page 276) are all alternatives for profiling sheet materials. Laser cutting is suitable for a range of materials, but produces a heat-affected zone (HAZ). Die cutting and punching and blanking are used to cut a wide range of thin sheet materials. Glass scoring is only suitable for thin materials.

QUALITY

One of the main advantages of water jet technology is that it is a cold process; therefore it does not produce a heat-affected zone (HAZ), which is most critical in metals. This also means that there is no discolouration along the cut edge and pre-printed or coated materials can be cut this way.

Pure water jet produces a much cleaner cut than abrasive systems. In both operations, there is no contact between the tool and workpiece and so there is no edge deformation. However, the flow of water drags in deep materials due to reduced pressure and so produces a rougher finish. The process is slowed down to accommodate this in harder materials, which increases cycle time.

DESIGN OPPORTUNITIES

This process cuts most sheet materials between 0.5 mm and 100 mm (0.02–3.94 in.) thick. The hardness of the material will determine the maximum

Abrasive Water Jet Process

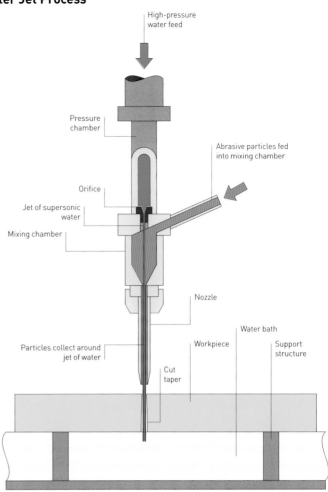

thickness. For example, polymer foam 100mm (3.94 in.) thick will cut with very little drag, but the maximum thickness for stainless steel is 60 mm (2.36 in.) and the cycle time is considerably longer. One-off to medium volumes can be accommodated because there are no tooling costs. Also, the scale of the workpiece does not dramatically increase costs. Therefore, this process is ideal for prototyping and experimentation. Materials can be changed and tested without any start-up cost because the main cost factor is time.

External and internal profiles can be cut in the same operation. Entry holes are unnecessary, except in materials that are likely to delaminate or shatter on impact. The process does not create stresses in the workpiece, so small, intricate and complex profiles are possible.

DESIGN CONSIDERATIONS

Reducing cycle time reduces the cost of the water jet process. Sharp corners and tight radii slow down the process; the water jet cutter will slow down to avoid drag, which inadvertently increases cut taper on curves. Also, holes with a diameter smaller than the depth of material should be drilled.

The sharp particles used in abrasive water jet cutting vary in size much like sandpaper (120, 80 and 50). Different grit sizes affect the quality of the surface finish; finer grit (higher number) is slower and produces a higher quality surface finish.

The final accuracy of the part is determined by a combination of many factors including material stability, thickness and hardness, accuracy of

TECHNICAL DESCRIPTION

This is only a small part of the equipment required to produce water jet cutting. Tap water is supplied to the cutting nozzle at very high pressure from a pump and intensifier combination. In the pressure chamber the water reaches up to 4,000 bar (60,000 psi). At this high pressure it is forced through a small opening in the 'orifice' (0.1 to 0.25mm in diameter). This is also sometimes called a jewel because it is made of diamond, sapphire or ruby.

In abrasive jet water cutting the sharp mineral particles (often garnet) are fed into the mixing chamber and come into contact with the supersonic water, which propels them at very high speed towards the workpiece. The abrasive particles create a beam 1 mm (0.04 in.) in diameter, which produces the cutting action. The taper left by the cutting process can be

decreased by reducing the cutting speed or increasing the water pressure. The same techniques will reduce drag and other cutting defects, which are especially prominent on internal and external radii.

The difference between this process and pure water jet cutting is the addition of abrasive particles. Without the mineral particles it is the water alone that erodes the workpiece.

The high velocity jet is dissipated by the bath of water below the workpiece. This water is continuously sieved, cleaned and recycled.

→ Water jet cutting glass

The water jet cutter is CNC, so every operation is programmed into the machine from a CAD file (**image 1**). The discs are being cut from 25mm (0.98 in.) plate glass. The cutting nozzle progresses slowly around the workpiece to achieve a clean cutting action. As it progresses, the operator inserts wedges to support the part as it is being cut (**image 2**). This image clearly shows the drag on the high-velocity jet of water as it erodes the material. The faster the cutting process, the greater the effect of drag and subsequent drop in quality.

After cutting, the part is carefully removed. The cutting programme is designed to overrun at the start and end of the process to ensure a perfectly symmetrical disc (**image 3**). The surface finish that is left by the abrasive water jet cutting action (**image 4**) is improved by polishing, which takes place on a diamond encrusted polishing wheel (**image 5**). Finally, a small chamfer is ground onto the cut edge (**image 6**).

1

2

3

the cutting bed, consistency of water pressure and speed of cut.

Very thin materials may break before the cutting process has finished due to the weight of the part on the uncut material. Tabs can be designed in to avoid this, or wedges inserted to support the part. Tabs mean secondary operations because they will have to be removed afterwards.

COMPATIBLE MATERIALS

There are very few materials that cannot be cut in this way. The first water jet cutting was developed to machine wood. Very little wood is now cut in this way,

but it is possible to cut wood and other natural materials.

Mild steel, stainless steel and tool steel can all be cut with high accuracy. Titanium, aluminium, copper and brass are suitable for complex profiles and can be cut rapidly with this process compared to other cutting technologies.

Marble, ceramic, glass and stone can be cut into intricate profiles even though they are quite brittle.

Laminates and composite materials (including carbon reinforced plastic) can be cut very effectively. Laminated parts tend to delaminate under stressful cutting conditions, and the water jet

process does not cause these problems. Even printed and coated materials can be cut without any detrimental effects to the surface.

COSTS

There are no tooling costs.

Cycle time can be quite slow but depends on the thickness of material and quality of cut. Thin sheets of material can be stacked up and cut in a single operation to reduce cycle time.

Labour costs are moderate due to the skilled workforce that is required.

4

5

6

ENVIRONMENTAL IMPACTS

The kerf is quite narrow and so very little
material is wasted during operation.
Offcuts in most materials can be recycled
or re-used. Computer software improves
efficiency by nesting parts together to
produce minimal waste. Water jet does
not produce a HAZ or distortion, so the
parts can be nested relatively closely and
as little as 2 mm (0.079 in.) apart.

There are no hazardous materials
created in the process or dangerous
vapours off-gassed. The water is usually
tapped from the mains and is cleaned
and recycled for continuous use.

Featured Manufacturer

Instrument Glasses
www.instrument-glasses.co.uk

MECHANICAL

276

Cutting Technology

Glass Scoring

This is a precise method for cutting sheet glass materials. The cutting wheel is guided along x and y axes at high speed from a CAD file to produce one-offs or high volumes of parts.

Costs	Typical Applications	Suitability
• No tooling costs • Low unit costs	• Furniture • Glass panes and tiles • Stained glass	• One-off to high volume production

Quality	Related Processes	Speed
• Good quality cut edge, but with some lateral cracking	• Laser cutting • Water jet cutting	• Rapid cycle time (roughly 100 m/328 ft per minute)

INTRODUCTION

Glass scoring is the most widely used technique for shaping and sizing sheet glass from 0.5 mm to 20 mm (0.02–0.8 in.) thick. It is used for both industrial and decorative applications: manufacturers and craftsmen alike utilize this process for general sheet cutting operations.

Carried out on a computer guided x–y plotter, this is a high speed and precise cutting technique. Handheld scoring tools are also widely used, especially for simple, long and straight cuts.

Glass Scoring Process

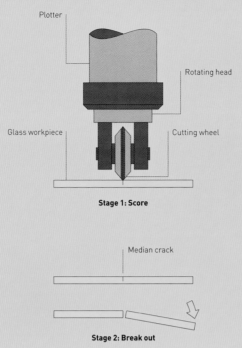

Stage 1: Score

Stage 2: Break out

(Labels in diagram: Plotter, Rotating head, Glass workpiece, Cutting wheel, Median crack)

TECHNICAL DESCRIPTION

Glass scoring can be carried out by hand or on a computer-guided x–y plotter. This diagram shows computer-guided scoring.

Attached to the rotating head is a tungsten carbide cutting wheel. Low pressure is applied to the cutting wheel as it runs across the surface of the glass. This produces a flaw that forms as a crack just ahead of the cutting wheel. This crack is known as a median or vent crack and is typically less than 1 mm (0.04 in.) deep.

Once the operation is complete, the glass sheet is removed from the cutting bed. Applying pressure to the glass forces the shallow crack to extend through its depth and the parts are broken out.

This method produces a good quality cut edge. The median crack caused by the cutting wheel is a different texture to the break (see image, opposite).

TYPICAL APPLICATIONS

Display screens, tabletops, lenses, filters, protective glass covers and glass signage are just a few of the wide range of products made in this way.

This is the main technique for cutting out stained glass designs.

RELATED PROCESSES

Water jet cutting (page 272) and laser cutting (page 248) are also used to profile sheet glass materials. They are both slower than scoring for simple external profiles, but are capable of cutting through a wide variety of materials and thicknesses. Glass scoring is suitable for materials up to 20 mm (0.8 in.) thick. By contrast, water jet cutting is capable of cutting sheet glass up to 70 mm (2.75 in.) thick. Laser cutting produces a very high surface finish on the cut edge with less defects than other cutting methods.

QUALITY

The scoring technique produces a clean break. However, if the cut edge is exposed in application, it should be polished.

DESIGN OPPORTUNITIES

The main advantage for designers is the speed and versatility of this process. Designs can be prototyped freehand, with a compass, or around a profile. In production, the design is cut from sheet by a computer guided cutting wheel.

Small shapes are possible in thin materials. For example, disks down to 5 mm (0.2 in.) in diameter are feasible.

DESIGN CONSIDERATIONS

This process is limited by what can be broken out of the glass sheet. It is possible to cut simple straight or curved lines of any length, but internal shapes cannot be made, only external profiles.

Each score must run from edge to edge, or as a continuous shape. Therefore it is ideal for cutting out disks, rectangles and simple irregular shapes. But designs with indents in the profile, such as a crescent shape, cannot be easily made. Also, acute angles and complex shapes are not practical because it may not be possible to break them out of the sheet in a single piece.

COMPATIBLE MATERIALS

All types of sheet glass can be cut such as float glass, textured, coloured, mirrored and dichroic glass.

COSTS

There are no tooling costs.

Cycle time is rapid, and cutting speeds of 100 m (328 ft) per minute are typical.

Labour costs are low for automated operations and high for manual processes such as stained glass.

ENVIRONMENTAL IMPACTS

No glass is wasted in this operation: all offcuts can be recycled. This is a very low impact process, which requires very little energy in operation.

Unlike water jet cutting, this process does not remove a kerf width from the material. Therefore, parts can be nested closer together, further reducing material consumption.

1

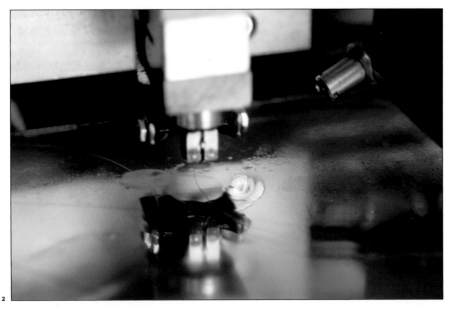

2

3

→ Glass scoring dichroic lenses

The discs are cut out on a computer guided x–y plotter. A small amount of lubricant is sprayed onto the cutting wheel and glass prior to cutting (**image 1**).

A small median crack forms under the tungsten carbide cutting wheel (**image 2**). The amount of pressure is adjusted according to the thickness of material. For example, 20 mm (0.8 in.) thick glass needs a relatively high level of pressure to form a median crack large enough to enable the shape to be broken out afterwards. At this point the crack is very shallow. Gentle pressure is applied, which causes it to 'run'. Straight edges can be broken out by hand (**image 3**).

Circles and other shapes are broken out using 'running pliers' (**image 4**). These break the glass outside of the desired shape, causing it to fail along the median crack. The shallow cracking caused by the wheel is visible in contrast with the 'run' fracture (**image 5**). Scoring causes small amounts of lateral cracking. To give a sense of scale, this sample is only 3 mm (0.118 in.) thick.

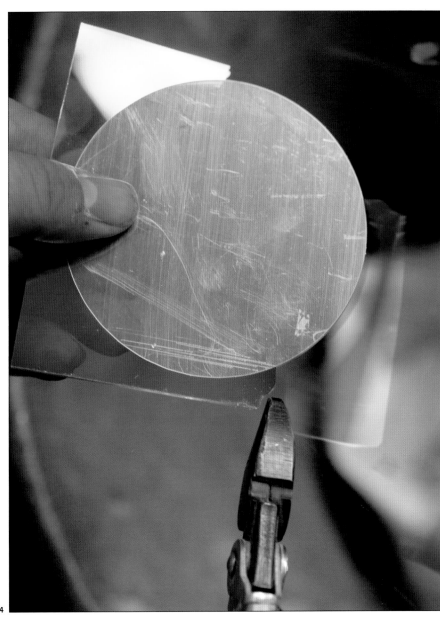

4

5

Featured Manufacturer

Instrument Glasses
www.instrument-glasses.co.uk

Part

3

Joining Technology

Arc Welding

Arc welding encompasses a range of fusion welding processes. These processes can only be used to join metals because they rely on the formation of an electric arc between the workpiece and electrode to produce heat.

Costs		Typical Applications		Suitability	
• No tooling costs • Low unit costs		• Containers • Fabrications • Structures		• One-off to mass production	
Quality		**Related Processes**		**Speed**	
• High quality		• Friction welding • Power beam welding • Resistance welding		• Slow to rapid cycle time	

INTRODUCTION

The most common types of arc welding are manual metal arc (MMA), metal inert gas (MIG) and tungsten inert gas (TIG), which are respectively called shielded metal arc welding (SMAW), gas metal arc welding (GMAW) and gas tungsten arc welding (GTAW) in the US. Arc welding also includes submerged-arc welding (SAW) and plasma welding (PW).

The joint interface and electrode (in some cases) melt to form a weld pool, which rapidly solidifies to form a bead of weld metal. A shielding gas and layer of slag (in some cases) protects the molten weld pool from the atmosphere and encourages formation of a 'sound' joint.

MMA, MIG and TIG welding can all be operated manually. PW and SAW are more suited to mechanized systems due to the bulkiness and complexity of the equipment, but there is a microplasma variant that is suited to manual work. MIG welding is also very well suited to automation because it uses continuous-feed consumable electrode and separate shielding gas. TIG welding is only suitable for certain automated applications such as orbital welding of pipes.

TYPICAL APPLICATIONS

Arc welding is an essential part of the fabrication process, used extensively in the metalworking industries. The various processes are suited to different uses, determined by volumes, material type and thickness, speed and location.

MMA welding is the most portable of the processes, requiring relatively little equipment; it is used a great deal in the construction industry and for other site applications such as repair work.

Manual Metal Arc Welding Process

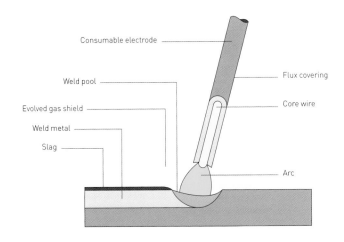

Consumable electrode

Flux covering

Weld pool

Core wire

Evolved gas shield

Weld metal

Slag

Arc

TECHNICAL DESCRIPTION

MMA welding, also known as stick welding, has been in use since the late 1800s, but has seen major development in the last 60 years. Modern welding techniques use a coated electrode. The coating (flux) melts during welding to form the protective gas shield and slag. There are several different types of flux and electrode, which improve the versatility and quality of the weld.

MMA welding can only produce short lengths of weld before the electrode needs to be replaced. This is time consuming and stresses the welded joint because the temperature is uneven. The slag that builds up on top of the weld bead has to be removed before another welding pass can take place. It is not suitable for automation or large volume production, and the quality of the weld is largely dependent on the skill of the operator.

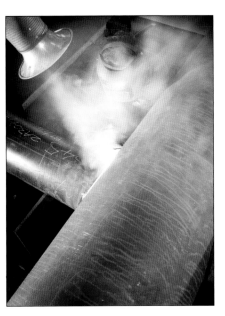

Above left
An operator uses MMA welding to join steel plate.

Left
Joining steel pipes with MMA welding.

Above
An arc is formed between the coated electrode and workpiece.

Top
A steel joint that has been MMA welded.

Above
Coated electrodes are used in MMA welding.

Featured Manufacturer

TWI
www.twi.co.uk

TECHNICAL DESCRIPTION

MIG welding is also referred to as semi-automated welding. The principle is the same as MMA welding; the weld is made by forming an arc between the electrode and the workpiece and is protected by a plume of inert gas. The difference is that MIG welding uses an electrode continuously fed from a spool and the shielding gas is supplied separately. Therefore, MIG welding distinguishes itself from MMA welding through higher productivity rates, greater flexibility and suitability for automation.

The gas shield performs a number of functions, including aiding the formation of the arc plasma, stabilizing the arc on the workpiece and encouraging the transfer of molten electrode to the weld pool. Generally the gas is a mixture of argon, oxygen and carbon dioxide. MIG is also known as metal active gas (MAG) in the UK when CO_2 or O_2 are present in the shielding gas. Non-ferrous metals require an inert argon-helium mixture.

Metal Inert Gas Welding Process

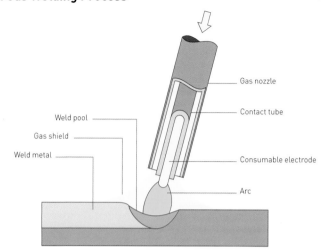

Gas nozzle

Contact tube

Weld pool

Gas shield

Weld metal

Consumable electrode

Arc

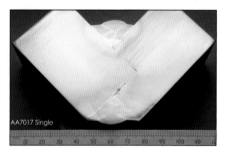

AA7017 Single

Top
An operator uses MIG welding to seal the end of a steel pipe.

Above left
Example of a MIG welded tee joint profile.

Above
Cross-section of a MIG welded overlapping joint profile.

Opposite
MIG consumable electrode, which also acts as a filler wire during welding.

Modified MMA technology can be used for wet underwater welding of pipes and other offshore structures.

MIG welding accounts for about half of all welding operations and is used in many industries. It is used extensively for car assembly because it is rapid and produces clean welds.

The quality and slower speeds of TIG welding make it ideally suited to precise and demanding applications. It may be manual or fully automated.

There are 3 main types of PW, which have different applications. Microplasma welding is suited to thin sheet and mesh materials. Medium current welding

provides an alternative to conventional TIG welding; the equipment is bulkier but the weld is more penetrating. Keyhole welding has the advantage of deep penetration and high welding speeds. This means it is suitable for sheet material up to 10mm (0.39 in).

Due to the nature of SAW it is only suitable for horizontal welding positions.

RELATED PROCESSES

More recent fusion welding technologies, including laser and electron beam welding (page 288), are providing an alternative to the conventional processes, especially for thicker materials that require deep penetration.

Other welding processes, such as friction welding (page 294), resistance welding (page 308) and gas welding, are alternatives for certain applications. Adhesive technologies are becoming more sophisticated, providing increased joint strength with no heat-affected zone (HAZ), which is sometimes preferable.

QUALITY

The quality of manual arc welding is largely dependent on operator skill. Precise and clean weld beads can be formed with MMA, MIG and TIG welding.

Automated welding operations produce the cleanest and most precise

welded joints. MIG welding produces a neat and continuous weld bead, as can be seen on aluminium alloy metal ship decks. Steel ship decks are SAW.

DESIGN OPPORTUNITIES

As well as being suitable for batch and mass production of actual parts, these processes make it possible to realize complex metal assemblies without expensive tooling costs. Combined with profiling operations such as gas, plasma, or laser cutting (page 248), welding can be used to fabricate accurate prototypes. MMA is especially useful for quickly fabricating steel structures, whereas TIG welding can be used to produce intricate and delicate metal fabrications.

The equipment for MMA, MIG and TIG welding can be used in horizontal, vertical and inverted positions, making these versatile for manual operations.

DESIGN CONSIDERATIONS

Welding is a line-of-sight process. Therefore, parts need to be designed so that a welder or robot can get access to the joint. The geometry of the joint will affect the speed of welding. Deposition rate and duty cycle (how much welding is achieved in an average hour, allowing for set up, cooling, jigging and so on) are used as a measure of the efficiency of

a process. Deposition rate is measured as kg/hour (lb/hr), and duty cycle as a percentage. MMA has a deposition rate of approximately 2 kg/hr (4.4 lb/hr) and duty cycle of 15–30%; TIG has a deposition rate of approximately 0.5 kg/hr (1.1 lb/hr) and duty cycle of 15–20%; MIG has a deposition rate of approximately 4 kg/hr (8.8 lb/hr) and a duty cycle of 20–40%; and SAW has a deposition rate of approximately 12.5 kg/hr (27.5 lb/hr) and duty cycle of 50–90%.

Certain metals, such as titanium, can only be welded to like metals. In contrast, carbon steel can be welded to stainless steel or nickel alloys, and aluminium alloys to magnesium alloys.

MIG welding is limited to materials between 1 mm and 5 mm (0.039–0.2 in.) thick. PW can accommodate the largest range of materials, from 0.1 mm up to 10 mm (0.0039–0.39 in.) thick.

As with all thermal metal fabrication processes, arc welding produces a HAZ. All welding produces residual stress in HAZ weld area and beyond. The problems are caused by a change in structure leading to change in properties and susceptibility to cracking.

All of these processes require the workpiece to be connected to an earth return to the power source, so that the electrical current passes through the electrode and creates an arc with the weld pool. If the metalwork is painted or coated in any way, this will insulate the metal contact and make welding difficult or impossible. Also, assemblies and fabrications may contain other elements that could be damaged by the electrical current, so it is important to mount the metal contact close to the weld zone to create a path of least resistance avoiding the vulnerable parts of the assembly.

COMPATIBLE MATERIALS

MMA welding is generally limited to steel, iron and nickel alloys. There are also electrodes for welding copper alloys. MIG and TIG welding, PW and SAW can be used on ferrous and non-ferrous metals.

Tungsten Inert Gas Welding Process

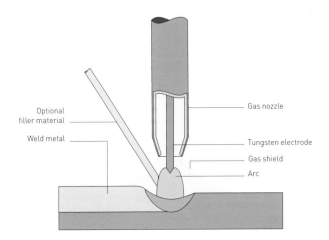

Optional filler material

Weld metal

Gas nozzle

Tungsten electrode

Gas shield

Arc

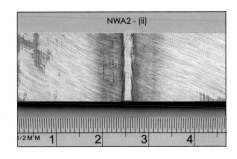

NWA2 - (ii)

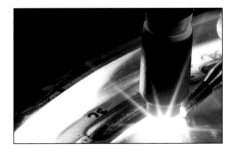

TECHNICAL DESCRIPTION

TIG welding is a precise and high quality welding process. It is ideal for thin sheet materials and precise and intricate work. The main distinction is that TIG welding does not use a consumable electrode; instead, it has a pointed tungsten electrode. The weld pool is protected by a shielding gas and filler material can be used to increase deposition rates and for thicker materials.

The shielding gas is usually helium, argon or a mixture of both. Argon is the most common and is used for TIG welding steels, aluminium and titanium. Small amounts of hydrogen can be added to produce a cleaner weld that has less surface oxidization. Hydrogen makes the arc burn hotter and so facilitates high welding speeds, however, it can also increase weld porosity. Helium is more expensive, but helps the arc burn hotter and so facilitates higher production rates.

Top
Detail of TIG welded butt joint profile.

Middle
An arc is formed between the tungsten electrode and workpiece, and the filler material melts into the weld pool.

Above
Automated TIG welding in operation.

Below
Joining metal pipe by TIG welding.

TIG is widely used on carbon steel, stainless steel and aluminium, and it is the main process for joining titanium. MIG welding is commonly used to join steels, aluminium and magnesium.

Certain PW technologies can be used on very thin sheets, meshes and gauzes.

COSTS

There are no tooling costs unless special jigs or clamps are required.

Cycle time is slow for MMA welding, especially because the electrode needs to be replaced frequently. MIG welding is rapid, especially if automated, and TIG welding falls somewhere in the middle.

Labour costs are high for manual operations, due to the level of skill required. Some automated versions can run almost completely unattended and so labour costs are dramatically reduced.

ENVIRONMENTAL IMPACTS

A constant flow of electricity generates a great deal of heat during the welding process, and there is very little heat insulation, so it is relatively inefficient. SAW insulates the arc in a layer of flux, producing 60% thermal efficiency, compared with 25% for MMA welding.

This process produces very little waste material; slag must be removed.

Featured Manufacturer

TWI
www.twi.co.uk

TECHNICAL DESCRIPTION

Plasma welding is very similar to TIG welding. This means that the process is versatile and capable of welding everything from meshes and thin sheets to thick materials in a single pass.

There are 3 main categories of plasma welding, which are microplasma, medium current and keyhole plasma. Microplasma is operated with relatively low electrical current, so is suitable for materials as thin as 0.1mm (0.0039 in.) and meshes. Medium current techniques are a direct alternative to TIG welding but with deeper penetration. Keyhole welding can be achieved with plasma because of its powerful, small diameter. The temperature of the plasma is estimated to be between 3000°C and 6000°C (5432–10832°F). This is capable of penetrating material up to 6mm (0.236 in.) in a single pass.

Plasma Welding Process

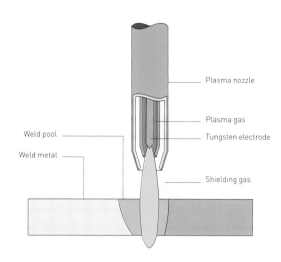

- Plasma nozzle
- Plasma gas
- Tungsten electrode
- Weld pool
- Weld metal
- Shielding gas

Submerged Arc Welding Process

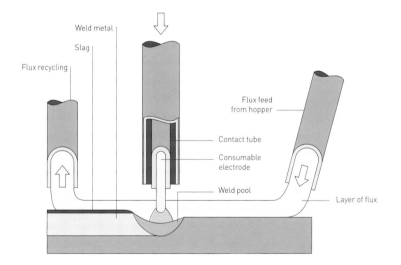

- Weld metal
- Slag
- Flux recycling
- Flux feed from hopper
- Contact tube
- Consumable electrode
- Weld pool
- Layer of flux

Right
In submerged arc welding a solid flux is used, which limits the process to horizontal joints including lap, butt and tee profiles.

TECHNICAL DESCRIPTION

SAW, like MIG welding, forms the weld by creating an arc between the continuously fed electrode and workpiece. However, there is no need for a shielding gas because the electrode is submerged in a layer of flux fed to the joint from a hopper. There are many associated benefits of submerging the arc, including the absence of a visible arc light, spatter-free weld area and very little heat loss.

As the welding torch passes along the joint, the flux is laid down in front and collected for recycling behind. The weld pool is not visible and it is a complex operation, so it is generally mechanized. The flux will affect the depth of penetration and metal deposition rate, so the workpiece must be horizontal to maintain the layer of flux. However, overlapping, butt and tee joint profiles are all possible and frequently used.

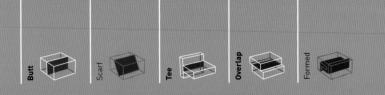

Power Beam Welding

Of the powerful welding processes, electron beam welding is capable of joining steels up to 150 mm (5.91 in.) thick and aluminium up to 450 mm (17.72 in.) thick, while laser beam welding is mainly used for materials thinner than 15 mm (0.6 in.).

Costs	Typical Applications	Suitability
• No tooling costs	• Aerospace	• Specialist to mass production
• Very high equipment costs	• Automotive	
• High unit costs	• Construction	

Quality	Related Processes	Speed
• High strength joint	• Arc welding	• Rapid cycle time
	• Ultrasonic welding	

INTRODUCTION

Like arc welding, power beam welding joins materials by heating and melting the joint interface, which solidifies to form a high integrity weld. The difference is that power beam processes do not rely on the formation of an electric arc between the welding electrode and workpiece to generate heat. Heat is produced by the concentration of energy in the power beam.

Both laser and electron beam technologies are capable of establishing a 'keyhole' in the workpiece to deliver heat deep into, or through, the joint. Indeed, this ability means they can be used to cut, machine and drill materials as well as weld them together.

Lasers produce light waves that are focused to achieve power densities above 100 W/mm² (1 mm² = 0.039 in.²). At this power most engineering materials melt or vaporize, and laser beam welding (LBW) is therefore suitable for welding a range of materials, including most

metals and thermoplastics from 1 mm to 15 mm (0.04–0.59 in.). The many different types of laser technologies include CO_2, direct diode and Nd:YAG, as well as Yb disc and Yb fibre lasers. Besides welding, lasers are used to cut and engrave (see laser cutting, page 248), and fuse layers of particles (see rapid prototyping, page 232).

Electron beams are generated in a high vacuum. Electrons are energized using a cathode heated to above 2000°C (3632°F) and up to two-thirds the speed of light for normal industrial electron beam guns. The high velocity electrons bombard the surface of the workpiece, causing it to heat up and melt, or vaporize in an instant. The focused electrons create power densities as high as 30,000 W/mm² (1 mm² = 0.039 in.²) in a localized spot. Thus electron beam welding (EBW) produces rapid, clean and precise welds in materials between 0.1 mm and 450 mm (0.004–17.72 in.) thick, depending on the material.

TYPICAL APPLICATIONS

LBW is used on a wide range of industrial applications, including aluminium car frames (such as the Audi A2), construction, shipbuilding and airframe structures. Since the 1970s, it has been developed for plastics and utilized for joining thermoplastic films, injection molded parts, textiles and transparent components in the automotive, packaging and medical industries.

EBW has been in development since the 1960s and has been used for many years in Japan for welding heavy duty offshore and underground piping. The ability of this process to penetrate thick materials makes it ideal for construction and nuclear applications.

RELATED PROCESSES

Although they are specialized processes, LBW and EBW are gradually becoming more widespread for joining parts in a single pass, which would previously have been almost impossible to achieve. There are no processes that compete with them in this area.

Arc welding (page 282), however, is an alternative to power beam welding for thin sheet applications. For plastics, ultrasonic welding (page 302) is an alternative to LBW.

LBW can be combined with arc welding to increase opportunities for production. For example, large structures often have varying joint gaps that cannot be avoided. Because power beam technologies are not suitable unless accurate joint configurations are lined up precisely with the laser or electron beam, laser-arc hybrids developed for shipbuilding have been used instead. These can accommodate variations in joint gap. The combination of high powered laser welding with versatile metal inert gas (MIG) welding increases the quality of joints and improves production rates.

QUALITY

Both LBW and EBW processes form rapid and uniform high integrity welds. They have superior penetration and so produce a relatively smaller heat-affected zone (HAZ) than arc welding. A major advantage of power beam welding is its ability to make deep and narrow welds. However, this means there is a huge temperature differential between the melt zone and parent material, which can cause cracking and other problems in certain metals, especially high carbon steels.

LBW in plastics produces extremely high quality joints. In overlapping geometries the laser penetrates the top part and affects only the joint interface, leaving absolutely no trace of the process on either surface.

DESIGN OPPORTUNITIES

Power beam welding has many advantages such as increased production rate, narrow HAZ and low distortion. Another major advantage for designers is the ability to join dissimilar materials: for example, EBW can be used to fuse a range of different metals. As well as joining dissimilar metals, LBW can weld a range of different thermoplastics to one another. Factors that affect material choice include relative melting point and reflectivity.

It is now possible to weld clear plastics and textiles using a technique known as Clearweld®. The Clearweld® process was invented and patented by TWI, and is being commercialized by Gentex Corporation. This process is particularly useful in applications that demand high surface finish (see image, above). The principle is that an infrared absorbing medium is applied to the joint region, either printed or film. This causes the joint to heat up on exposure to the laser beam. Without the infrared absorbing medium the laser would pass right through clear materials and textiles. Hermetic seals can be achieved in plastics and fabrics, so this technology

Above
A clear plastic printer ink cartridge laser welded using Clearweld®.

has potential applications in waterproof clothing, for example.

EBW is carried out in a vacuum, which until recently has meant that the workpiece size is restricted by the size of the vacuum chamber. However, mobile EBM has been made possible by reduced pressure techniques developed in the 1990s. It is now feasible to apply a local vacuum in a small EBW unit that travels around the joint. This has many advantages, including on-site welding of parts too large or unsuitable for the vacuum chamber.

DESIGN CONSIDERATIONS

These processes require expensive equipment and are costly to run, especially EBW which is carried out in a vacuum. Both EBW and LBW are automated, and need careful programming prior to any welding. The cost and complexity of power beam processes means that they are suitable only for specialist applications and

TECHNICAL DESCRIPTION

In LBW, the CO_2, Nd:YAG and fibre laser beams are guided to the workpiece by a series of fixed mirrors. Nd:YAG and fibre laser beams can also be guided to the welding or cutting head by fibre optics, which has many advantages. These laser technologies typically weld at 7 kW, which is suitable for fusion welding 8 mm (0.315 in.) carbon steel in a single pass at up to 1.5 m (5 ft) per minute. These processes can be used for many different welding operations, including profile, rotary and spot welding. Spot welding processes have been recorded at rates of up to 120,000 welds per hour.

A shielding gas is used to protect the melt zone from oxidization and contamination. Because of their shorter wavelength, Nd:YAG laser beams can also be guided with fibre optics. This means that they can cut along 5 axes, as the head is free to rotate in any direction.

The laser beam is focused through a lens that concentrates the beam to a fine spot, between 0.1 mm and 1 mm (0.004–0.04 in.) wide. The height of the lens is adjusted to focus the laser on the surface of the workpiece. The high concentration beam is capable of melting the workpiece on contact, which resolidifies to form a homogenous bond.

Laser Beam Welding Process

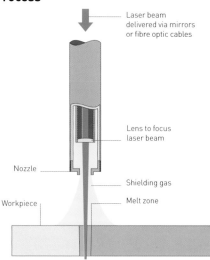

Laser beam delivered via mirrors or fibre optic cables

Lens to focus laser beam

Nozzle

Shielding gas

Workpiece

Melt zone

mass production. The payoff is that they are very rapid: EBW can produce high integrity welds at up to 10 m/minute (33 ft/minute) in thick sections.

For an effective butt weld the joints have to fit tight tolerances and align with the beam very accurately. As a result, preparation can also be costly and time-consuming, especially for large and cumbersome parts.

Material thickness is much greater for power beam welding than with arc welding. High power EBW can be used to join steel from 1 mm up to 150 mm (0.04–5.91 in.) thick, aluminium up to 450 mm (17.72 in.) thick and copper up

to 100 mm (3.94 in.) thick. Nor are these processes limited to very thick materials; they are also used to join thin sheet and film materials with extreme precision.

COMPATIBLE MATERIALS

Many different ferrous and non-ferrous metals can be joined by power beam welding. The most commonly welded metals include steels, copper, aluminium, magnesium and titanium. EBW, however, is limited to metallic materials because the workpiece must be electrically conductive.

Aluminium is highly reflective but can be welded with both lasers and electron beams. Certain alloys are more suitable than others for power beam welding, such as 5000 and 6000 series.

Titanium can be power beam welded, but it is highly reactive to oxygen and nitrogen. This means that it is more difficult and therefore more costly to weld. Tungsten inert gas (TIG) welding is used as an alternative but is up to 10 times slower than laser welding.

LBW can be used to join most thermoplastics, including polypropylene (PP), polyethylene (PE), acrylonitrile butadiene styrene (ABS), polyoxymethylene (POM) acetal and poly methyl methacrylate (PMMA).

COSTS

There are no tooling costs, but jigs and clamps are necessary because of the precision required for the welding process to be successful. Equipment costs are extremely high, especially for processes that involve EBW.

Cycle time is rapid, but set-up will increase cycle time, especially for large and complex parts. For EBW each part has to be loaded into a vacuum chamber and a considerable vacuum applied, which can take up to 30 minutes. Mobile reduced pressure techniques have lowered set-up time and increased cycle time considerably.

Labour costs are high.

ENVIRONMENTAL IMPACTS

Power beam welding efficiently transmits heat to the workpiece. EBM typically requires a vacuum, which is energy intensive to create. Recent developments have made it possible to weld steels up to 40 mm (1.57 in.) thick. However, weld quality and width to depth ratio are diminished.

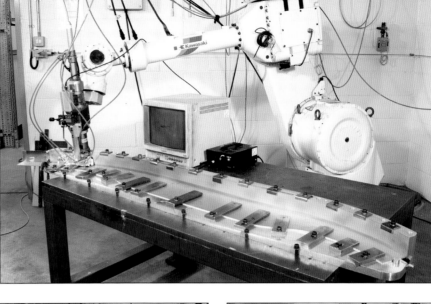

Above
An operator checks the CO2 LBW equipment.

Above right
A robot manipulates the Nd:YAG lasers.

Right
The process of Nd:YAG laser beam welding commences.

Far right
This detail shows the Nd:YAG laser beam welding head while it is in operation.

Below
Detail of the finished Nd:YAG laser weld.

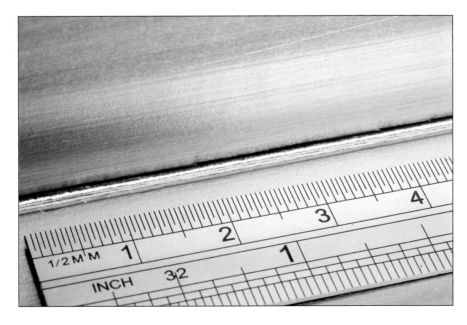

Featured Manufacturer

TWI
www.twi.co.uk

Electron Beam Welding Process

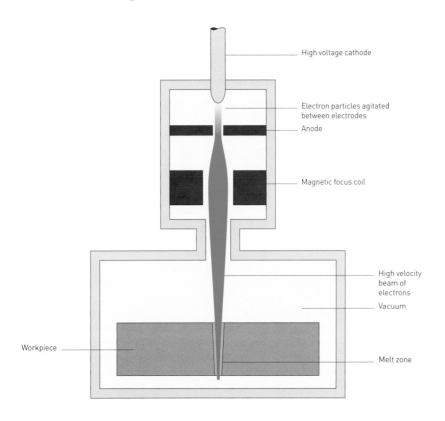

High voltage cathode

Electron particles agitated between electrodes

Anode

Magnetic focus coil

High velocity beam of electrons

Vacuum

Workpiece

Melt zone

TECHNICAL DESCRIPTION

Electrons are emitted from a heated tungsten cathode. They are pulled and accelerated into the gun column at up to two-thirds the speed of light. Magnetic focusing coils concentrate the beam into a fine stream, which impacts on the surface of the workpiece. The electrons' velocity is transmitted into heat energy on contact, which vaporizes a localized area that rapidly penetrates deep into the workpiece.

EBW is carried out in a vacuum of 10^{-5}–10^{-2} mbar (0.000000145–0.000145 psi). The level of vacuum determines the quality of the weld because the atmosphere can dissipate the electron beam, causing it to lose velocity. Reduced pressure EBW is carried out at 1–10 mbar (0.0145–0.145 psi), which requires considerably less energy and time. As the fusion welding is carried out in a vacuum there is very little contamination of the melt zone, which ensures high integrity welds.

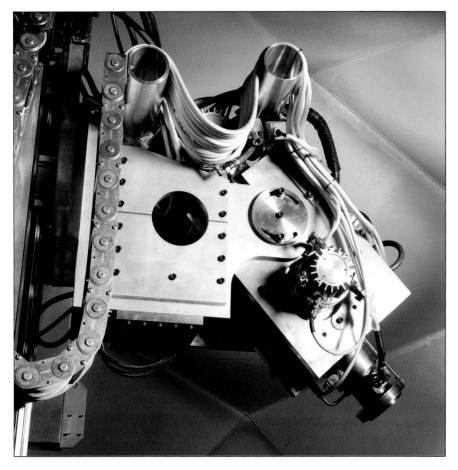

Left
A piece of electron beam welding equipment.

Above
Electron beam welding in progress.

Above
The scale of this electron beam welded gear is shown by the ruler beneath it.

Right
This close-up is of the electron beam welded gear shown left.

Below
This cross-section of electron beam welded joint demonstrates the integrity of a deep weld.

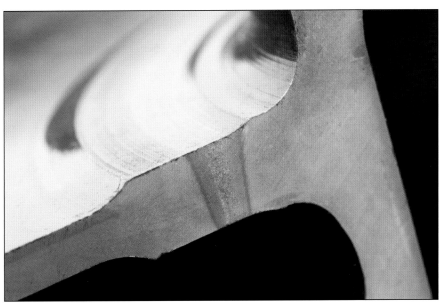

Featured Manufacturer

TWI
www.twi.co.uk

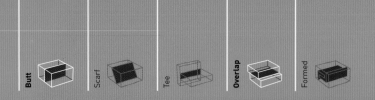
Joining Technology
Friction Welding

Forge welding processes are used to form permanent joints in metals. There are 4 main techniques: rotary friction welding (RFW), linear friction welding (LFW), orbital friction welding (OFW) and friction stir welding (FSW).

THERMAL

294

Costs	Typical Applications	Suitability
• RFW, LFW and OFW: No tooling costs • FSW: Inexpensive tooling costs • Low to moderate unit costs	• Aerospace • Automotive and transportation • Shipbuilding	• High volume production
Quality	**Related Processes**	**Speed**
• High integrity hermetic seal • High strength joint that can have similar characteristics to base material	• Arc welding • Power beam welding • Resistance welding	• Rapid cycle time that depends on size of joint

INTRODUCTION

The 4 main friction welding techniques can be separated into 2 groups: conventional techniques including LFW, OFW and RFW processes; and a recent derivative, FSW.

LFW, OFW and RFW operations weld materials with frictional heat generated by rubbing the joint interface together. The joint plasticizes and axial pressure is applied, forcing the materials to coalesce. The rotary technique (see main image) was the earliest and is the most common of the friction welding techniques.

In FSW the weld is formed by a rotating non-consumable probe (tool), which progresses along the joint mixing the material at the interface (see image, page 396, above left).

TYPICAL APPLICATIONS

Application of these processes is concentrated in the automotive, transportation, shipbuilding and aerospace industries.

In the automotive industry, RFW is used for critical parts including drive shafts, axles and gears. LFW is utilized to join engine parts, brake discs and wheel rims. OFW has not yet found commercial application in metals, but is utilized in the plastics industry (see vibration welding, page 298). FSW is used to join flat panels, formed sheets, alloy wheels, fuel tanks and space frames.

The first commercial application of FSW was in the shipbuilding industry, for welding extruded aluminium profiles into large structural panels. This has benefits for many industries: for example, the railway one, in the construction of prefabricated structural components in train carriages (see images, page 297, above, top right and above right). FSW is suitable as it causes very little distortion in the welded parts, even across long joints in thin sections.

Recently, FSW has been introduced into the consumer electronics industry, such as in the fabrication of Bang &

Olufsen aluminium speakers (see images, page 297, above left and left).

RELATED PROCESSES

Even though the welds are of similar quality, friction welding is not as widely used as arc welding (page 282) and resistance welding (page 308). This is mainly because it is a more recent development: for example, TWI patented friction stir welding only in 1991. It is also because friction welding is a specialized technique, and the equipment costs are extremely high.

Friction welding does not take the weld zone above melting point, so this process can join metals that are not suitable for fusion welding by arc or power beam welding (page 288).

Friction welding plastics is known as vibration welding (page 298).

QUALITY

Friction welding produces high integrity welds. Butt joints are fused across the

Friction Welding Processes

Rotary friction welding

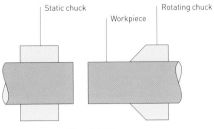

Static chuck | Workpiece | Rotating chuck

Stage 1: Load

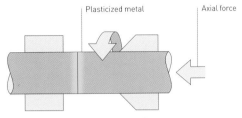

Plasticized metal | Axial force

Stage 2: Friction through rotation

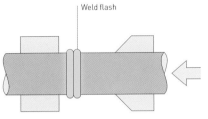

Weld flash

Stage 3: Axial pressure

Linear friction welding

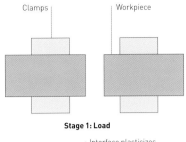

Clamps | Workpiece

Stage 1: Load

Interface plasticizes | Reciprocating motion

Stage 2: Friction through rubbing | Axial pressure

Weld flash

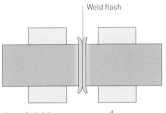

Stage 3: Axial pressure | Axial pressure

Above
This titanium blisk (1-piece bladed disc) for jet engines is made by linear friction welding.

Right
These beech blocks have been joined by linear friction welding.

TECHNICAL DESCRIPTION

The use of RFW is limited to parts where at least 1 part is symmetrical around an axis of rotation. In stage 1, the 2 workpieces are secured onto the chucks. In stage 2, 1 of the parts is spun while the other is stationary. They are forced together, and friction between the 2 faces causes the metal to heat up and plasticize. In stage 3, after a specified time – a minute or so – the spinning stops and the axial force is increased to 20 tonnes or so. Flash forms around the circumference of the joint and can be polished off afterwards.

Another technique uses stored energy and is known as inertia friction welding. In this process the spinning workpiece is attached to a flywheel. Once up to speed, the flywheel is left to spin freely. The parts are brought together and the stored energy in the flywheel spins the parts sufficiently for a weld to form. The process is similar to RFW, except that a mandrel is used to maintain the internal diameter of the pipe.

LFW and OFW are based on the same principles as RFW. However, instead of rotation, the parts are oscillated against one another. In stage 1, the 2 workpieces are clamped. In stage 2, the joint interface is heated by rubbing the parts together. In stage 3, axial force is increased and maintained until the joint has formed.

LFW typically operates at frequencies up to 75 Hz and amplitudes of ±3 mm (0.118 in.). These processes were developed for parts that are not suitable for rotary techniques.

TECHNICAL DESCRIPTION

This process is similar to RFW, LFW and OFW because the metals are welded by mixing the joint interface. Likewise, there is no consumable wire, shielding gas or flux. However, FSW differs from the other friction welding techniques because it uses a non-consumable probe to mix the metals.

The probe is rotated in a chuck at high speed, pushed into the joint line and progresses along the joint interface, plasticizing material that comes into contact with it. The probe softens the material, while the shoulder and backing bar prevent the plasticized metal from spreading. As the probe progresses, the mixed material cools and solidifies, to produce a high integrity weld.

Even though FSW tools are non-consumable, they can run for only up to 1000 m (1094 yd.) before they need to be replaced. Some techniques run more than 1 tool in parallel, for a broader weld or for a weld from either side to form a deeper joint.

Friction Stir Welding Process

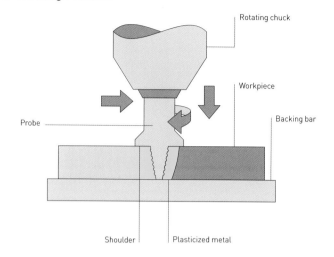

Rotating chuck

Workpiece

Backing bar

Probe

Shoulder | Plasticized metal

Above left
This aluminium butt joint profile is being made using friction stir welding.

Above right
The tool (probe) and butt joint have been formed in aluminium by friction stir welding.

entire joint interface. These are solid state welding processes. In other words, they weld metals below their melting point. The heat-affected zone (HAZ) is relatively small and there is very little shrinkage, and thus distortion, even in long welds.

DESIGN OPPORTUNITIES

As the friction welding techniques are relatively new processes, especially FSW, many commercial opportunities have yet to be fully understood.

Friction welding offers important advantages such as the ability to join dissimilar materials without loss of weld

integrity. Useful material combinations include aluminium and copper, and aluminium and stainless steel.

It is also possible to join materials of dissimilar thickness. Multiple layers stacked up can be welded in a single pass with FSW.

Use of RFW, LFW and OFW is limited to parts that can be moved relative to one another along the same joint plane. FSW, on the other hand, can be computer-guided around circumferences, complex 3D joint profiles and at any angle of operation. Combined, these processes can fuse almost any joint configuration and part geometry. FSW is also capable of

producing butt, lap, tee and corner joints. It is particularly useful in the fabrication of parts that cannot be cast or extruded such as large structural panels made up of several extrusions.

These processes are not affected by gravity, and so can be carried out upside down if necessary.

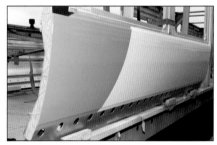

Left above and left
Bang & Olufsen BeoLab aluminium speakers were designed by David Lewis and launched during 2002.

Above, top right and above right
Lightweight and structural aluminium panels, made by friction stir welding, are needed in the production of train carriages.

DESIGN CONSIDERATIONS

Because they are comparatively new, the equipment costs for friction welding are still very high. Therefore it is expensive to develop products with these processes in mind unless there are high volumes of production to justify the investment. This is partly why friction welding is used extensively in the automotive and shipbuilding industries and only to a limited extent elsewhere.

FSW is suitable for materials ranging from 1.2 mm up to 50 mm (0.047–1.97 in.) in non-ferrous metals. It is possible to weld joints up to 150 mm (5.91 in.) deep if welded from both sided simultaneously. Recently, microfriction welding techniques have been developed that are capable of welding materials down to 0.3 mm (0.012 in.).

Joint size in RFW, LFW and OFW techniques is limited to less than 2,000 mm² (3.1 in.²).

COMPATIBLE MATERIALS

Most ferrous and non-ferrous metals can be joined in this way, including low carbon steel, stainless steel, aluminium alloys, copper, lead, titanium, magnesium and zinc. It is even possible to weld metal matrix composites.

Pipes can be joined in a process known as radial friction welding, which is a development on RFW. The difference is that in radial friction welding an internal mandrel is used to support the weld area.

Recent developments in LFW have made it possible to join certain woods, including oak and beech (see image, page 295, right). Although TWI assessed the process in 2005, it is still in the very early stages of development. In the future this process has the potential to replace conventional wood joining techniques.

COSTS

FSW requires tooling, the cost of which depends on the thickness and type of material; even so, it is inexpensive. The cost of this process is largely dependent on the equipment and development for the application.

Cycle time is rapid for RFW, LFW and OFW. FSW is the slowest of the friction welding processes because it runs along the entire length of the joint. Using FSW, 5 mm (0.2 in.) aluminium can be welded at approximately 12 mm (0.472 in.) per second. Thick materials will take considerably longer.

ENVIRONMENTAL IMPACTS

Friction welding is an energy efficient process for joining metals; there are no materials added to the joint during welding such as flux, filler wire or shielding gas. This process generates no waste; the exception is run-offs in FSW, where the weld runs from edge to edge.

Featured Manufacturer
TWI
www.twi.co.uk

Joining Technology
Vibration Welding

Homogenous bonds in plastic parts can be created using vibration welding. Rapid linear or orbital displacement generates heat at the interface, and this melts the joint material and forms the weld.

Costs	Typical Applications	Suitability
• Low to moderate tooling costs • Low unit costs	• Automotive • Consumer electronics • Packaging	• Medium to high volume production

Quality	Related Processes	Speed
• High strength and homogenous bond • Hermetic seals are possible	• Friction welding • Hot plate welding • Ultrasonic welding	• Very rapid cycle time (up to 30 seconds)

INTRODUCTION
Vibration welding is based on friction welding (page 294) principles. The parts are rubbed together to generate frictional heat, which plasticizes the joint interface. The process is carried out as linear or orbital vibration welding. Linear vibration welding uses transverse, reciprocating motion – the vibration occurs in only 1 axis. Orbital vibration welding uses constant velocity motion – a non-rotating offset circular motion in all directions. The vibration motion occurs equally in both the x and y axes and all axes in between.

TYPICAL APPLICATIONS
Although this process is typically utilized in the automotive industry, it is now steadily becoming more widespread in, for example, medical, consumer electronics and white good appliances.

Vibration Welding Process

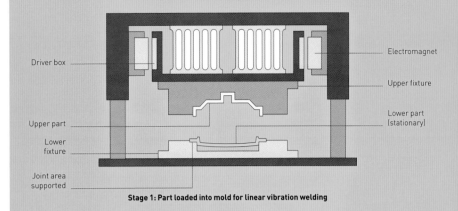

Driver box
Upper part
Lower fixture
Joint area supported

Electromagnet
Upper fixture
Lower part (stationary)

Stage 1: Part loaded into mold for linear vibration welding

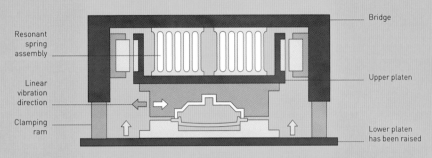

Resonant spring assembly
Linear vibration direction
Clamping ram

Bridge
Upper platen
Lower platen has been raised

Stage 2: Closed mold, welding and clamping

RELATED PROCESSES

Linear friction welding metal and linear vibration welding plastic are the same process; both join parts by rubbing them together using linear motion, under axial force. Ultrasonic welding (page 302) and certain adhesive technologies are frequently an alternative, especially for thin walled and delicate parts.

Ultrasonic welding transfers energy through the part to the joint interface. By contrast, the vibration process transfers energy directly to the joint interface, so does not rely on the ability of the part material to transmit energy. Fillers, regrind, colour additives and contamination all affect a material's ability to transmit energy and so reduce the effectiveness of ultrasonic welding but not vibration welding.

Recent developments include heating the joint interface prior to welding, in a process similar to hot plate welding (page 320). Combined, these processes create neat and high strength welds with much less debris, especially important for such applications as filter housings.

QUALITY

Vibration welding produces strong bonds. It is possible to form hermetic seals for automotive and packaging applications. When certain material combinations are joined in this way a complete mixing of the materials occurs, resulting in a homogenous bond.

Further variables to influence weldability include moisture content and the addition of resin modifiers.

TECHNICAL DESCRIPTION

In stage 1 of linear vibration, 1 part is placed in the lower platen and the part to which it is to be joined is positioned in the upper platen. In stage 2, the heat necessary to melt the plastic is generated by pressing the parts together and vibrating them through a small relative displacement from 0.7 mm to 1.8 mm (0.028–0.071 in.) at 240 Hz or a 4 mm (0.016 in.) peak to peak at 100 Hz in the plane of the joint under a force of 1–2N/mm² (0.69–1.37 psi).

Heat generated by the resulting friction melts the plastic at the interface within 2–3 seconds. Vibration motion is then stopped and the parts are automatically aligned. Pressure is maintained until the plastic solidifies to bond the parts permanently with a strength approaching that of the parent material itself.

During orbital vibration welding, each point on the joint surface of the moving part orbits a different and distinct point on the joint surface of a stationary part.

The orbit is continuous and of constant speed, identical for all points on the joint surface. The movement's orbital character is produced by three electromagnets arranged horizontally at relative angles of 120°. These magnets influence the movement of the fixture and generate a harmonic movement. Compensation for the decreasing force of one electromagnet is provided by attraction of the next electromagnet, together with a change in direction. This causes rotary vibration of the fixture, which ensures a controlled relative movement in a frequency between 190 Hz and 220 Hz and an amplitude range between 0.25 mm and 1.5 mm (0.1–0.59 in.) peak to peak.

Linear vibration welding an automotive light

In this case study Branson Ultrasonics are producing a short production run of automotive sidelights. A pair of right- and left-hand sidelights is being molded simultaneously. The red reflectors (**image 1**) are placed into the lower platen (**image 2**), then the sidelight housings (**image 3**) are positioned into the upper platen, where they are held by vacuum (**image 4**).

During the welding cycle, the lower platen holding the stationary part (in this case the red reflectors) rises to meet the upper platen (**image 5**), forcing the parts together at a pressure of 103–207 N/cm² (150–300 psi) (**image 6**). Either hydraulic or electromagnetic drivers are used to create the vibration in the upper platen. This is mounted onto a resonant spring assembly to ensure that the parts align accurately once the vibration sequence concludes and the polymers resolidify. The welding process lasts no more than 2–15 seconds, after which the parts are held under a measured clamping force for another 2 to 15 seconds and the melted polymer in the joint interface cools to form a continuous weld. Finally the mold platens part to reveal the welded parts held by vacuum in the upper platen. The vacuum is released, dispensing the sidelights (**image 7**). These are quality checked (**image 8**) and packed for shipping.

1

2

DESIGN OPPORTUNITIES

Both linear and orbital vibration welding provide a fast, controlled and repeatable method of achieving a permanent strong bond between 2 plastic parts. They also eliminate the need for mechanical fastenings and adhesives.

These processes can accommodate parts with complex geometry, whether large or small, and can be used to create perimeter and internal welds. As long as the parts can be moved relative to each other along the same joint plane, vibration welding can be used.

DESIGN CONSIDERATIONS

Vibration welding is most suitable for injection molded and extruded parts. However, 2 essential requirements must be taken into account when considering the process. Firstly, the design must allow for the required movement between the parts to generate sufficient frictional heat; the plane of vibration must be flat, or at least within 10°. Secondly, the parts must be designed so that they can be gripped adequately to ensure sufficient energy transmission to the joint.

Linear vibration welding can be used to weld parts up to 500 x 1,500 mm (20 x 59 in.), but it is not recommended for products with thin wall sections or long unsupported walls because they will flex unless supported by the tooling.

Orbital technology, on the other hand, requires less clamping pressure than linear methods and so can be used for more delicate products. It is currently suitable for relatively small parts, up to 250 mm (9.84 in.) in diameter.

Vibration welding is not suitable for applications where the workpiece cannot be vibrated, for instance, as products with moving parts in the top section.

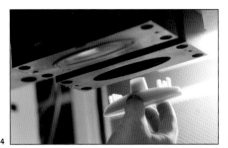

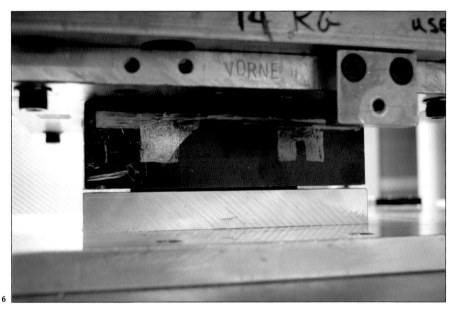

COMPATIBLE MATERIALS

Thermoplastics (including amorphous and semi-crystalline) are suitable for this process, and some polymer based composite materials, thermoplastic films and fabrics can also be welded in this way. Vibration welding is particularly useful for materials that are not easily joined by ultrasonic welding or adhesive bonding such as polyoxymethylene (POM) acetal, polyethylene (PE), polyamide (PA) nylon, and polypropylene (PP). Plastic welding is limited to joining like materials. There are a couple of exceptions to the rule such as acrylonitrile butadiene styrene

(ABS), poly methyl methacrylate (PMMA) acrylic and polycarbonate (PC), which can be welded to one another.

COSTS

Tooling has to be designed and built specifically for each part, which increases the start-up costs of the process, especially if the product is made up of multiple parts that need to be joined in this way.

Cycle time is rapid: welding is typically between 2 and 15 seconds and clamping time is the same. Welding multiple parts simultaneously can reduce cycle time. Labour costs are relatively low, especially

when fully automated or integrated into a production line.

ENVIRONMENTAL IMPACTS

There are no materials added to the joint to form the weld, which means that this process does not generate any waste.

Featured Manufacturer

Branson Ultrasonics
www.branson-plasticsjoin.com

| Butt | | Scarf | | Tee | | Overlap | | Formed | |

Joining Technology

Ultrasonic Welding

This process forms permanent joints using ultrasonic waves in the form of high-energy vibration. It is the least expensive and fastest plastic welding process and so is the first to be considered in many welding applications.

THERMAL

302

Costs	**Typical Applications**	**Suitability**
• Low tooling costs • Low unit costs	• Consumer electronics and appliances • Medical • Packaging	• Batch and mass production
Quality	**Related Processes**	**Speed**
• High quality permanent bond • Hermetic seals are possible	• Hot plate welding • Power beam welding • Vibration welding	• Rapid cycle time (less than 1 second) • Automated and continuous processes can form many welds very rapidly

INTRODUCTION

Ultrasonic technology includes welding, swaging, staking (page 316), inserting, spot welding, cutting, and textile and film sealing. Welding is the most widely used ultrasonic process and is employed by a range of industries, including consumer electronics and appliances, automotive, medical, packaging, stationery and toys.

It is fast, repeatable and very controllable. The conversion of electrical energy into mechanical vibration is an efficient use of energy. No fillers are required, and there is no waste or contamination. It is safe to use for toys, medical applications and food preparation. Ultrasonic cutting is used to cut cakes, sandwiches and other foods without squashing them: the high-energy vibrations effortlessly cut through both hard and soft foods. It is also used to seal drinks cartons, sausage skins and other food packaging.

TYPICAL APPLICATIONS

Ultrasonic welding in used for a vast range of products in many different industries. The packaging industry uses ultrasonics for joining and sealing cartons of juice and milk, ice cream tubs, toothpaste tubes, blister packs, caps and enclosures.

Applications in consumer electronics include mobile phone covers and screens, watch bodies, hairdryers, shavers and printer cartridges. Applications in the textile and non-woven industries include nappies, seat belts, filters and curtains.

RELATED PROCESSES

Ultrasonic welding is often the first joining process to be considered. If it is not suitable then hot plate welding (page 320, vibration welding (page 298), laser welding (see power beam welding, page 288) may be used.

QUALITY

Ultrasonic welding produces homogenous bonds between plastic parts. Joint strength is high and hermetic seals are possible.

A small amount of flash is produced, but this can be minimized with careful design such as housing the joint out of sight or with the provision of flash traps.

DESIGN OPPORTUNITIES

The advantage of ultrasonic technology is the range of welding processes and techniques that are available. For example, as well as conventional joining ultrasonic welders can spot weld, stake, embed metal components (known as inserting), swage and form joints.

This diversity gives greater design freedom. For example, products like mobile phones and MP3 players are made up of material combinations that may not be possible with multi-shot injection molding (page 50), or other processes. Parts that cannot be molded in a single operation can be ultrasonically joined to produce complex, intricate and otherwise impossible geometries.

DESIGN CONSIDERATIONS

The main considerations for designers are type of material (ease of welding) and geometry of the part. There are 3 main types of joint design: straightforward overlapping material, energy director and shear joints (see images, page 304).

The difference is that energy director and shear joint techniques have molded details to aid the ultrasonic process. Overlapping joints, such as extruded tube packaging sealed at one end, are the least strong, but are suitably effective for the application (see image, above left).

Top and opposite
The TSM6 mobile phone by Product Partners uses ultrasonic welding on the front assembly.

Above left
Ultrasonic welding produces hermetic seals in extruded plastic tube, for instance, those for cosmetic packaging.

Above
Triangular beads, known as energy directors, are molded onto each side of the joint interface, perpendicular to one another, to maximize the efficiency of the welding process.

The energy director is raised bead of material on one side of the joint. It is generally triangular and is molded in line with the joint or perpendicular to it (see image, above). The role of the triangular bead is to minimize contact between the 2 surfaces and therefore maximize the energy transferred to the joint. This reduces the amount of energy needed to produce the weld, reduces flash and minimizes cycle time.

The alternative method, shear joint, works on the same principle of reducing surface area in the joint to maximize efficiency. The difference is that instead

Ultrasonic Welding Process

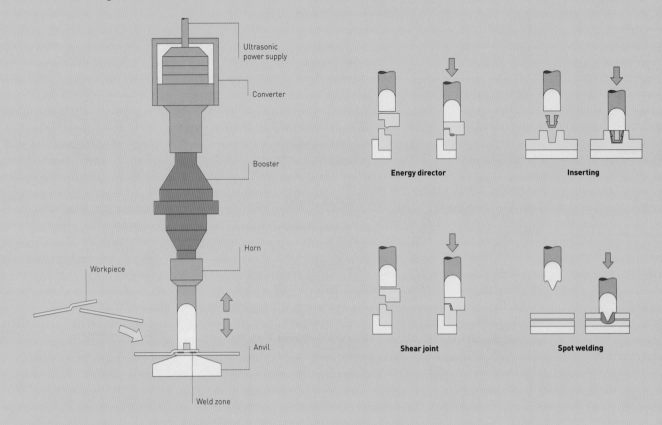

Ultrasonic power supply

Converter

Booster

Horn

Workpiece

Anvil

Weld zone

Energy director

Inserting

Shear joint

Spot welding

of a triangular bead, the joints are designed with a shoulder or step. This detail concentrates the vibratory energy to a thin line of contact. As the parts are pressed together, the shoulder progresses down the joint until the entire interface is welded.

Horn (tool) size and shape is limited because the horn has to resonate at the correct frequency and withstand high-vibration energy. Horns are typically no larger than 300 mm (11.81 in.). Longer welds can be made in more than a single step, with multiple heads or as a continuous process.

Joints must be designed with consideration for transmitting sufficient energy to the interface. The tool has to be within a certain distance of the joint. Generally, amorphous materials dampen the vibrations less and so can be far field welded, meaning the joint is more than 5 mm (0.2 in.) from the horn contact point. However, semi-crystalline and low stiffness parts have to be near

field welded, meaning the joint is within 5 mm (0.2 in.) of the contact point.

COMPATIBLE MATERIALS

All thermoplastics can be joined in this way. Many amorphous thermoplastics such as acrylonitrile butadiene styrene (ABS), poly methyl methacrylate (PMMA), polycarbonate (PC) and polystyrene (PS) can be joined to themselves and in some cases to one another. Semi-crystalline thermoplastics such as polyamide (PA), cellulose acetate (CA), polyoxymethylene (POM), polyethylene terephthalate (PET), polyethylene (PE) and polypropylene (PP) can only be joined to themselves.

There are many factors that affect the ability of a material to weld to itself or another plastic. For example, a material has to be sufficiently stiff to transfer vibratory energy to the joint interface.

Amorphous materials are more suited to energy directing joint designs, whereas semi-crystalline materials are best welded with shear joints.

Textile and non-woven materials, including thermoplastic fabrics, composite materials, coated paper and mixed fabrics, can be joined.

Some metals can be joined in this way. However, the technology is more specialized and less widespread. It is also known as 'cold welding' because the parts are joined below their melting point. It is similar in principle to diffusion bonding, which is a form of brazing (page 312). However, no flux or other surface preparation is required as long as the joint fits properly.

COSTS

Horns have to be made from high-grade aerospace aluminium or titanium. Even so, tooling costs are generally low.

Cycle time is very rapid. Welding time is typically less than 1 second. Loading and unloading will increase cycle time slightly. Automated and continuous operations are very rapid and can produce several welds per second.

TECHNICAL DESCRIPTION

Ultrasonic welding works on the principle that electrical energy can be converted into high-energy vibration by means of piezoelectric discs. Electricity is converted from mains supply (50 Hz in Europe or 60 Hz in North America) into 15 kHz, 20 kHz, 30 kHz or 40 kHz operating frequency. The frequency is determined by the application; 20 kHz is the most commonly used frequency because it has a wide range of application.

The converter consists of a series of piezoelectric discs, which have resistance to 15, 20, 30 or 40 kHz frequencies. The crystals that make up the discs expand and contract when electrically charged. In doing so, they convert electrical energy into mechanical energy with 95% efficiency.

The mechanical energy is transferred to the booster, which modifies the amplitude into vibrations suitable for welding. The horn transfers the vibrations to the workpiece. The size and length of the horn are limited because it has to resonate correctly.

The ultrasonic vibrations are transferred to the joint by the workpiece. This generates frictional heat at the interface, which causes the material to plasticize. Pressure is applied, which encourages material at the joint interface to mix. When the vibrations stop, the material solidifies to form a strong, homogenous bond.

The horn in the diagram is shaped to apply vibratory energy to selected areas of the joint. Other joining techniques include energy director, inserting, shear joint and spot welding. In each case, the red area indicates the weld zone. This is barely visible from the surface, but clear in a microscopic view (see image, below).

Labour costs are generally low. Tooling changeover is quick and the process is highly repeatable without the need for operator intervention.

ENVIRONMENTAL IMPACTS

Ultrasonic welding is an efficient use of energy; almost all of the electrical energy is converted into vibrations at the joint interface, so there is very little heat radiation. It is rapid and there are no other materials added to the joint. There is no risk of contamination, which makes this process suitable for food packaging, toys and medical products.

Welding reduces mixing materials, weight and cost by eliminating the need for mechanical fasteners or adhesives.

This is a permanent joining method, which means parts cannot be easily disassembled for recycling. However, if only like materials are joined, this is not a problem.

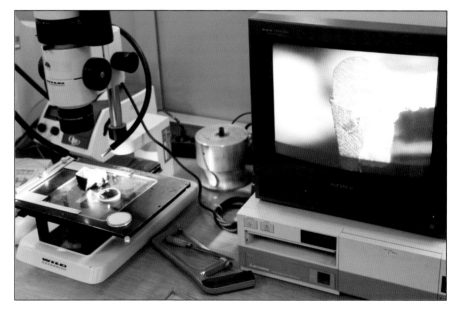

Top
Each application requires a new horn to be machined from high-grade aerospace aluminium or titanium.

Above
Ultrasonic welding is a precise and neat joining process, as can be seen in this microscopic image of a shear joint.

→ Ultrasonic welding a shear joint

This case study illustrates the equipment used in ultrasonic joining and a simple shear joint in a small impellor.

This product is suitable for ultrasonic joining because a shear joint is used, housed in the lower part. This provides a clean finish with very little flash. It is a 3-dimensional shape, but tooling is made for every application, so this does not increase costs.

Boosters are anodized aluminium and coloured to indicate the frequency at which they operate (**image 1**), which is 15, 20, 30 or 40 kHz. This application requires 30 kHz. The converter, booster and horn are screwed together and inserted in a vertical pillar assembly (**images 2** and **3**).

The parts that are going to be joined (**image 4**) make up a small impellor. There is a small step on the blades of the impellor, which provide the interference for the shear joint. The lower part is placed in the anvil, which provides support (**image 5**). The upper part, which is over-molded onto a metal bar, locates on top and is self-aligning.

The horn is brought into contact with the part and the welding process is completed in less than a second (**image 6**). The finished part is removed from the anvil with a permanent hermetic joint (**image 7**).

1

2

3

4

5

6

7

Featured Manufacturer

Branson Ultrasonics
www.branson-plasticsjoin.com

Butt
 Scarf
 Tee
Overlap
 Formed

Joining Technology
Resistance Welding

These are rapid techniques used to form welds between 2 sheets of metal. Spot and projection welding are used for assembly operations, and seam welding is used to produce a series of overlapping weld nuggets to form a hermetic seal.

Costs	Typical Applications	Suitability
• Low tooling costs, if any • Low unit costs	• Automotive • Furniture and appliances • Prototypes	• One-off to mass production

Quality	Related Processes	Speed
• High shear strength, low peel strength • Hermetic joints possible with seam welding	• Arc welding • Friction welding • Riveting	• Rapid cycle time

INTRODUCTION

All of the resistance welding techniques are based on the same principle: a high voltage current is passed through 2 sheets of metal, causing them to melt and fuse together.

As the name suggests, these processes rely on metal's resistance to conducting electricity. High voltage, concentrated between 2 electrodes, causes the metal to heat and plasticize. Pressure applied during operation causes the melt zone to coalesce and subsequently form a weld.

The 3 main resistance welding processes are projection, spot and seam.

Projection Welding Process

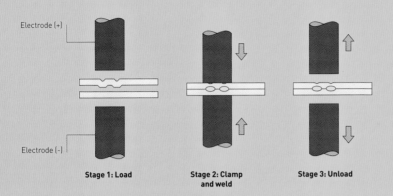

Electrode (+)

Electrode (-)

Stage 1: Load   **Stage 2: Clamp and weld**   **Stage 3: Unload**

TECHNICAL DESCRIPTION

In projection welding, the weld zone is localized. This can be done in 2 ways: either projections are embossed onto 1 side of the joint, or a metal insert is used.

This process is capable of producing multiple welds simultaneously because unlike spot welding the voltage is directed by the projection or insert. The electrodes do not determine the size and shape of the weld. Therefore, they can have large surface area that will not wear as rapidly as spot welding electrodes.

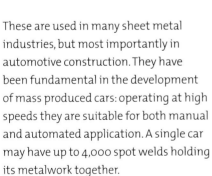

These are used in many sheet metal industries, but most importantly in automotive construction. They have been fundamental in the development of mass produced cars: operating at high speeds they are suitable for both manual and automated application. A single car may have up to 4,000 spot welds holding its metalwork together.

TYPICAL APPLICATIONS

Applications are widespread, including the automotive, construction, furniture, appliance and consumer electronic industries. These processes are used for prototypes as well as mass production.

Above left
The ring is placed onto a lower electrode, which locates it to ensure repeatable joints.

Above
The second part has protrusions that localize the voltage.

It is clamped into the upper electrode.

Right
Projection welding takes a second or so. Both welds are formed simultaneously and are full strength almost immediately.

Spot and projection welded products include car chassis and bodywork, appliance housing, electronic circuitry, mesh and wire assembly.

Seam welding is used on products such as radiators, gas and water tanks, cans, drums and fuel tanks.

Spot Welding Process

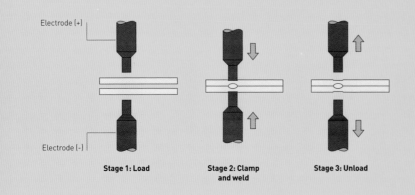

Electrode (+)

Electrode (-)

Stage 1: Load

Stage 2: Clamp and weld

Stage 3: Unload

TECHNICAL DESCRIPTION

Spot welding is the most versatile of the resistance welding processes. The weld zone is concentrated between 2 electrodes that are clamped onto the surface of the metal joint. Very high voltage plasticizes the metal and pressure is applied, forcing it to coalesce and subsequently form a weld nugget.

Because the weld is concentrated between the electrodes only 1 weld can be produced with each operation. Multiple welds are produced in sequence.

Equipment is generally not dedicated, so this process is the least expensive and the most suitable for prototyping and low volume sheet metal work.

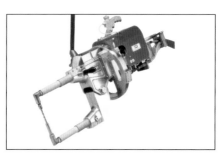

Left
Spot welding stainless steel mesh with a hand held welder and larger, static lower electrode.

Below left
The welded mesh covers a paper pulp molding tool (page 202).

Above
Heavy duty handheld welding guns like this are used for general sheet metal assembly work. Sophisticated computer-guided robotic systems are used for high volume welding operations.

RELATED PROCESSES

Resistance welding stands alone in the welding techniques: it is simple, consistent, low cost and de-skilled. Arc welding (page 282) and power beam welding (page 288) require a high level of skill to operate with such efficiency.

Formed and riveted joints require preparation. Resistance welding can be applied with very little surface preparation. Weld nuggets are formed easily in most metals, regardless of their hardness. Slightly higher clamping pressures are required when welding harder materials to achieve the same level of coalescence.

Seam Welding Process

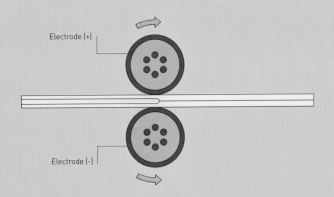

Electrode (+)

Electrode (−)

TECHNICAL DESCRIPTION
Seam welding is a continuous process in which the welding occurs between rolling electrodes. It is possible to form hermetic joints in sheet metal using this technique.

An alternative technique is used to apply multiple spot welds in a line. This is done by replacing the rolling electrodes with wheels that have a single electrode on the perimeter, which produces a spot weld when the upper and lower electrodes are aligned. This technique is used for high volume production of metal radiators, for example.

QUALITY

Weld quality is consistently high. Joints have high shear strength, but peel strength can be limited with small and localized weld nuggets.

DESIGN OPPORTUNITIES

These are simple and versatile processes. Spot welding is widely available for prototyping sheet metalwork; it is relatively inexpensive and can be used on a range of materials.

Dissimilar metals can be joined together, although the strength of the joint may be compromised. Also, different thickness of material can be joined together, or multiple sheet materials laid on top of each other. Assemblies up to 10 mm (0.4 in.) deep can be joined with a single spot weld, but thickness is generally limited to between 0.5 mm and 5 mm (0.02–0.2 in.).

DESIGN CONSIDERATIONS

The position of the weld is limited by 2 factors: the reach of the welding equipment and the shape of the electrode. The form of a typical hand held spot welding gun (see image, opposite, above) makes clear the constraints.

The weld point must be accessible from the edge of the sheet. Welding equipment typically operates on a vertical axis, so weld points must be accessible from above and below. Even so, it is possible to weld in confined spaces with offset or double bent electrodes.

COMPATIBLE MATERIALS

Most metals can be joined by resistance welding, including carbon steels, stainless steels, nickels, aluminium, titanium and copper alloys.

COSTS

Tooling costs are not always a consideration: many joints can be welded with inexpensive standard tooling. Specially designed tooling may be required to weld contoured surfaces, but is generally small and not expensive.

Most assemblies can be welded with standard electrodes (tooling), which minimizes costs. However, certain applications, including contoured surfaces, require dedicated electrodes.

Cycle time is rapid. Spot welding can only produce 1 weld at a time and so is limited to about 1 weld per second. Projection welding can produce multiple welds simultaneously, so is more rapid. The speed is affected by changes in material thickness, welding through coatings and variability in part fit up.

ENVIRONMENTAL IMPACTS

No consumables (such as flux, filler or shielding gas) are required for most resistance welding processes. Water is sometimes used to cool the copper electrodes, but this is usually recycled continuously without waste.

There is no waste generated by the process itself, and spot welding does not require any preparatory or secondary operations that create waste.

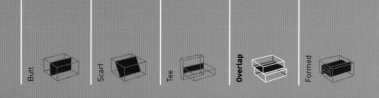

Joining Technology

Soldering and Brazing

Both these processes form permanent joints by melting a filler material between adjacent parts. The difference between the processes is the melting point of the filler materials, which is lower for soldering than it is for brazing.

Costs	Typical Applications	Suitability
• No tooling costs, but may require jigs • Low unit costs	• Electronics • Jewelry • Kitchenware	• One-off to mass production

Quality	Related Processes	Speed
• High strength bond, close to strength of parent material	• Arc welding • Resistance welding	• Rapid cycle time (1–10 minutes depending on technique and size of joint)

INTRODUCTION

These processes have been used in metalworking for centuries to join lead frames in stained glass windows and copper sculptures, for example. More recently, filler materials have been developed that are suitable for joining many metallic and ceramics materials. The most well known application is soldering printed circuit boards (PCBs). Traditionally, lead-based solders were used for many applications, including PCBs. As a result of the negative environmental impacts of lead and other heavy metals, tin, silver and copper filler materials are now more commonly used.

The melting temperature of the filler material determines whether the process is known as soldering, which is below 450°C (840°F), or brazing, which is above. But the melting point of the workpiece is always higher than the filler material, so filler materials are non-ferrous.

There are 3 main elements to the process: heating, flux and filler material.

There are several techniques that employ different combinations of these elements to achieve a range of design opportunities. For example, the filler material is drawn into the joint by capillary action, dipped, or inserted as a pre-form. Heating methods include induction, furnace and flame. Fluxes protect the surface of the joint from oxidizing. An alternative is to carry out the operation in a vacuum or in the presence of an inert gas (such as hydrogen brazing).

TYPICAL APPLICATIONS

These are widely used processes. The most common use of soldering is joining electronic components, such as surface, or through hole mounting onto PCBs. It is ideal for this because solder is conductive and is suitable for joining delicate components. Many electronic parts are assembled by hand using a soldering iron, especially for low volume parts. High volume production is carried out in

a process known as 'wave soldering'. In operation, the PCBs are assembled with all the relevant components, preheated and dipped in a tank of flux, followed by a tank of solder. The solder wets the unmasked metal connections and forms a permanent, conductive joint when it is baked.

An alternative method for mass-producing soldered joins is known as 'reflow soldering'. In this case, a paste of solder and flux is screen printed onto the part and then baked. The advantage is that hundreds, or even thousands of joints can be coated in a single pass of the squeegee.

Soldering is also used a great deal in jewelry, domestic plumbing (hermetic seals in copper pipes), silverware and food preparation items, restoration and repair work (re-soldering).

Brazing is generally stronger than soldering because the filler material has a higher melting point than solder. Typical applications include industrial

Soldering and Brazing Processes

Conduction method

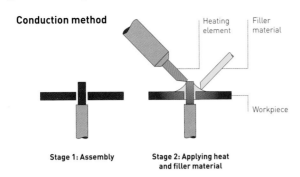

Heating element
Filler material

Workpiece

Stage 1: Assembly

Stage 2: Applying heat and filler material

Torch method

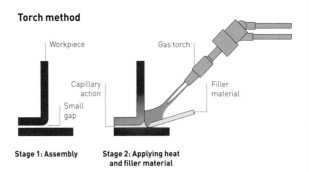

Workpiece

Gas torch

Capillary action

Filler material

Small gap

Stage 1: Assembly

Stage 2: Applying heat and filler material

Furnace method

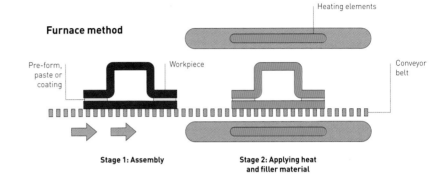

Heating elements

Pre-form, paste or coating

Workpiece

Conveyor belt

Stage 1: Assembly

Stage 2: Applying heat and filler material

TECHNICAL DESCRIPTION

Soldering and brazing is made up of the following main elements: joint preparation, flux, filler material and heating. These 3 diagrams illustrate the most common methods for heating a joint.

There are many different techniques, but the basic principle of soldering and brazing is that the workpiece is heated to above the melting point of the filler material. At this point the filler becomes molten and is drawn into the joint by capillary action. The liquid metal filler forms a metallurgical bond with the workpiece to create a joint that is as strong as the filler material itself.

The filler material is typically an alloy of silver, brass, tin, copper or nickel, or a combination of these. The choice of filler material is determined by the workpiece material because they have to be metallurgically compatible.

Fluxes are an essential part of the process. The type of flux is determined by the filler material. The role of flux is to provide a clean, oxide-free surface, so the filler can flow around the joint and form a strong bond. This is also referred to as surface 'wetting'.

Wetting is the coverage achieved by the filler material. It is hindered by surface contamination and surface tension between the materials. In some instances, such as reflow soldering, fluxes are incorporated into the filler to aid wetting. To encourage wetting on the surface of ceramic materials they are coated with metal by electroplating (page 364), or vacuum metalizing for non-conductive materials (page 372) .

CONDUCTION HEATING

Conduction heating is typically carried out with a soldering iron. The diagram depicts a through joint, which is common on circuit boards. The alternative is surface mounting. In this case the filler material, solder, is fed into the joint separately as a rod or wire. Applying a fine coat of filler and flux to the workpieces prior to assembly is similarly effective; on heating, the filler material coalesces to form the joint. On heating, the filler material coalesces to form the joint.

TORCH HEATING

The next method of heating is using a gas torch, usually oxyacetylene. However, soldering can be carried out with a cooler burning gas such as propane. As with the previous technique, the joint is heated to the desired temperature and the filler material is added. A small gap of approximately 0.05 mm (0.02 in.) is necessary to facilitate capillary action. The gap can be larger, but this will affect the strength of the joint.

This is not restricted to manual operations. Mass production methods may have many torches fixed in place and pointing at the workpiece such as in the production of certain bicycle frames.

The filler may be added as a pre-form, or coating, which becomes molten under the intense heat of the flames.

FURNACE HEATING

Both of the previous methods have been based on heating localized areas of the workpiece. Furnaces heat up the entire workpiece to the melting point of the filler material. This process is ideal for making multiple joints simultaneously, for mass production and for materials that are brazed in a protective atmosphere such as titanium.

The filler material can be added as a pre-form, or as in the case of reflow soldering, the flux and filler are screen printed onto the joints. The baking process can take longer than other methods, but many parts can be joined simultaneously.

Instead of a furnace, it is possible to apply heat using a shaped element, which surrounds the part without touching it, to elevate the temperature of the joint area. This technique is more commonly used for large volume production brazing.

pipework, bicycle frames, jewelry and watches. Brazing is also used for joining components in engines, heating elements, and parts for the aerospace and power generation industries.

RELATED PROCESSES

Like arc welding (page 282), brazing uses thermal energy to melt the filler material. But soldering and brazing do not melt the parent material, which is beneficial for thin, delicate and sensitive components. Resistance welding (page 308) is used for similar applications but with different joint geometry.

QUALITY

Soldering and brazing do not heat the joint above the melting temperature of the parent material and so there is minimal metallurgical change, or heat-affected zone (HAZ). The finish of the joint bead is usually satisfactory without the need for significant grinding. And even though brazing is usually carried

out with brass filler, the colour can be adjusted to suit the parent material.

DESIGN OPPORTUNITIES

The main advantage of these processes is that they do not affect the metallurgy of the parent material. Even so, an integral metallurgical bond is made at the interface by the molten filler material.

Soldering and brazing are simple processes, but there are many variations, which make them versatile and suitable for many different materials and joint geometries. The range of techniques cover manual and automated production, making these processes equally suitable for one-off and mass production. One-off and small batch production is typically carried out with a handheld torch and filler rod, whereas mass produced items are fed through a furnace on a conveyor belt.

Filler material is drawn into the joint by capillary action. Therefore, the joints do not have to fit exactly; the filler

material will bridge the gap. This also means very complex and intricate joints can be made because the molten filler is drawn right through.

Alternatively, the filler material can be inserted into the joint as a pre-form. In such cases, it is possible to join multiple products, or multiple joints simultaneously.

DESIGN CONSIDERATIONS

The filler material determines the optimum service temperature and strength of soldered and brazed joints.

Butt, tee and scarf joint types are not generally suitable for these processes because joint interface must be as large as possible. Therefore, whatever the configuration, the joint interface is designed to provide substantial surface area. This may affect the decorative aspects of the product.

In furnace and vacuum methods the size of the products is limited by chamber size.

→ # Brazing the Alessi Bombé milk jug

This is the Bombé milk jug (**image 1**). It was designed by Carlo Alessi in 1945 and is still in production today. It was originally made by metal spinning brass (page 78), but is now deep drawn stainless steel (page 88). Brazing is used to join the spout and handle even though it would be quicker by resistance welding. Brazing is still used to maintain the integrity of the original design.

First of all, the joint is checked and coated with flux paste (**image 2**). An overlap has been created between the spout and body to maximize the surface area to be joined.

They are mounted into a jig (**image 3**). The brazing process is very rapid, lasting 30 seconds or so (**image 4**). The craftsman first warms up the join area with an oxyacetylene torch. When it is up to temperature, the filler material is added. It flows into the joint interface and forms an even bead.

There is very little finishing required. The brazed part is removed from the jig complete (**image 5**). It is lightly polished and cleaned before being packed (**image 6**).

1

2

COMPATIBLE MATERIALS

Most metals and ceramics can be joined with these techniques. Metals include aluminium, copper, carbon steel, stainless steel, nickel, titanium and metal matrix composites.

Ceramics can be joined together and to metals. Many ceramics can be joined, but brazing is generally reserved for engineering materials due to the sophistication of the process.

A similar process to brazing, known as diffusion bonding, is suitable for joining ceramic, glass and composite materials. This process joins materials in a vacuum chamber, using a small amount of pressure and very thin film of filler coated onto the joint interface. When the temperature is raised a very small amount of pressure is applied, which causes the molecules in the joint to mix and form a strong bond. Dissimilar materials, such as metals and ceramics, can be joined in this way.

COSTS

There are no tooling costs. But jigs may be necessary to support the assembly. However, joints can be designed to locate and therefore do not need to be jigged.

Cycle time is rapid and ranges from 1–10 minutes for most torch applications. The cycle time for furnace techniques may be longer, but is offset because multiple products can be joined simultaneously.

Labour costs are generally low.

ENVIRONMENTAL IMPACTS

Soldering and brazing operate at lower temperatures than are welding. There are very few rejects because faulty parts can be dismantled and reassembled.

3

4

5

6

Featured Manufacturer

Alessi
www.alessi.com

Joining Technology
Staking

Thermoplastic studs are heated and formed into permanent joints. This process is suitable for joining dissimilar materials such as plastic to metal. There are 2 main techniques: hot air and ultrasonic staking.

Costs	Typical Applications	Suitability
• Low tooling costs • Low unit costs	• Appliances • Automotive • Consumer electronics	• Medium to high volume production

Quality	Related Processes	Speed
• High strength joints with variable appearance	• Hot plate welding • Ultrasonic welding • Vibration welding	• Rapid cycle time (0.5–15 seconds)

INTRODUCTION

These are clean and efficient processes used to assemble injection molded thermoplastic parts with other materials. They utilize the ability of thermoplastics to be heated and reformed without any loss of strength.

The joint configurations resemble rivets. A plastic stud, injection molded onto the component, is heated and formed into a tight fitting joint. Dissimilar materials can be joined, as long as 1 is thermoplastic; staking is used extensively to join metal circuitry into plastic housing.

TYPICAL APPLICATIONS

The largest areas of application are in the consumer electronics and automotive industries. In many instances, staking has replaced mechanical methods such as screws and clips.

Applications in the automotive industry include control panels, dashboards and door linings. Staking is ideal for joining electrical components into plastic housing because the studs are insulating.

RELATED PROCESSES

If 2 similar materials are being joined then ultrasonic (page 302), hot plate (page 320), vibration (page 298) or other welding techniques can be used. Staking stands alone in its ability to join plastics to metals without additional fixings.

QUALITY

These are permanent joints. The strength of the joint is determined by the diameter of the stud and mechanical properties of the parent material.

Staking Process

Hot air staking

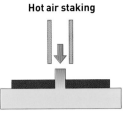

Stage 1: Applying hot air

Stage 2: Applying cold stake

Ultrasonic staking

Stage 1: Assembly

Stage 2: Ultrasonic staking

Alternative profiles

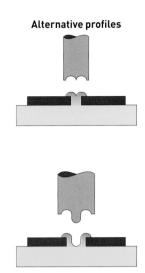

DESIGN OPPORTUNITIES

This is a simple process that eliminates consumables such as screws and rivets. The studs are injection molded into the part and add no cost to the forming process.

An alternative to staking is molded snap fits. The advantage of staking is that multiple fixing points can be molded in, as long as they are in the same line of draw (tool action). Snap fits are more difficult to integrate onto a product's surface without complex tool actions.

Multiple studs can be heated and formed simultaneously to form large joint areas. It is possible to produce watertight seals by sandwiching a rubber seal in the assembly.

There are no restrictions on the layout, pattern or number of studs. Pressure is applied during operation and so the joints are tight and free from vibration.

Low pressures are sufficient to form the joints, so thin walled and delicate parts that may not be able to stand up to vibration can be joined in this way.

DESIGN CONSIDERATIONS

The principal consideration is that the studs should align in a vertical direction, otherwise they cannot be formed in a single stroke.

Staking is limited to injection molded parts (page 50) and so large volumes.

These joints are usually internal and so appearance is not a major consideration. Due to the nature of the process joints are neat and clean, but the finish is not controllable.

COMPATIBLE MATERIALS

This process is limited to injection molded parts. Suitable materials include polypropylene (PP), polyethylene (PE), acrylonitrile butadiene styrene (ABS) and poly methyl methacrylate (PMMA).

COSTS

Tooling costs are low. Aluminium jigs are required to support the part during assembly. Ultrasonic staking uses a horn (tool). Hot air staking uses standard tooling, which is laid out to suit the orientation of the studs.

Cycle time is rapid. Ultrasonic staking can form joints in 0.5–2 seconds. Hot air staking cycle times are 5–15 seconds.

Labour costs are low because these processes are generally automated.

ENVIRONMENTAL IMPACTS

Staking eliminates consumables such as rivets, screws and clips.

TECHNICAL DESCRIPTION

The thermoplastic stud is softened with hot air or ultrasonic vibration. It is then formed into a domed, knurled, split or hollow head. The shape of the tool is adjusted to fit the requirements of the application and stud diameter.

Round studs are typically 0.5 mm to 5 mm (0.02–0.2 in.). Rectangular and hollow studs can be much larger, as long as the wall thickness is thin enough for staking.

Hot air staking is a 2 stage process. In stage 1, hot air is directed at the stud. The temperature of the air is determined by the material's plasticizing point. In stage 2, a cold stake with a profiled head presses down onto the hot stud. This simultaneously forms and cools the stud and joins the materials.

Ultrasonic staking is more rapid and takes only a few seconds. It is a single stage operation in which the stud is heated up by ultrasonic vibrations. As the stud is heated up it softens and is formed by pressure applied by the tool.

Ultrasonic welding is suitable for a range of other operations including welding, sealing and cutting.

→ Hot air staking

This case study demonstrates a typical hot air staking application. The parts being assembled are the lamp housing and electrical contacts.

The lamp housing is placed into a tool and the pressed metal contacts are placed over the studs (**images 1** and **2**).

A stream of hot air is directed at each stud for a few seconds (**image 3**). Cold dome-headed tools form the hot plastic studs (**image 4**). The stakes retract and the assembly operation is complete (**images 5** and **6**).

THERMAL

318

1

2

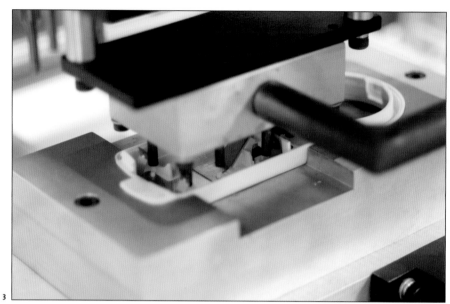

3

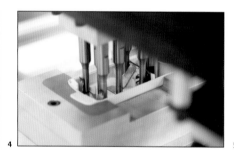

4

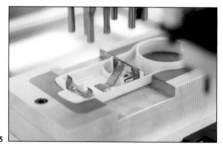

5

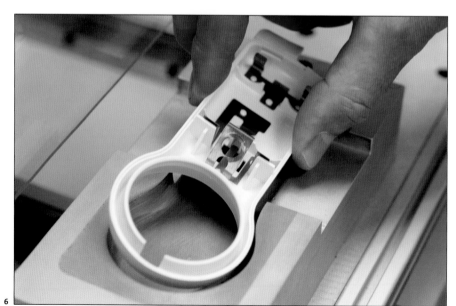

6

Featured Manufacturer

Branson Ultrasonics
www.branson-plasticsjoin.com

→ Ultrasonic staking

In this case, ultrasonic staking is used to join a rubber seal onto an injection molded part. The 2 parts are assembled on a jig (**image 1**). The studs are rectangular to provide a larger joint area.

The ultrasonic horns are mounted onto a single booster and work simultaneously. The inside of the horn forms the stud into the desired shape (**image 2**).

The ultrasonic horns compress onto the parts and apply vibrations to heat up the studs (**image 3**). The ultrasonic vibrations form the studs very quickly. After a couple of seconds the joint is complete and the horns retract (**image 4**).

1

2

3

4

Featured Manufacturer

Branson Ultrasonics
www.branson-plasticsjoin.com

Joining Technology

Hot Plate Welding

This is a simple and versatile process used to join materials. The joint interface is heated to above its melting point, causing it to plasticize. The parts are then clamped together to form the weld.

THERMAL

320

Costs	Typical Applications	Suitability
• Moderate tooling costs • Low unit costs	• Automotive • Packaging • Pharmaceutical	• Batch to mass production

Quality	Related Processes	Speed
• High quality homogenous bonds • Hermetic seals possible	• Ultrasonic welding • Vibration welding	• Variable cycle time (30 seconds to 10 minutes)

INTRODUCTION

Hot plate welding is used to form joints in extruded and injection molded (page 50) thermoplastic parts. It is a very simple process: the joint interface is heated until it plasticizes and then pressed together until it solidifies. It is also a versatile process: size is restricted only by the size of the heating platen, and the joint profile may be flat or 3D. The equipment can be either portable or fixed to a production line.

TYPICAL APPLICATIONS

The automotive industry is the largest user of hot plate welding. The process is also used for some packaging and pharmaceutical products.

RELATED PROCESSES

Similar joint profiles are suitable for vibration (page 298) and ultrasonic (page 302) welding. This process is selected because only small levels of pressure are

Hot Plate Welding

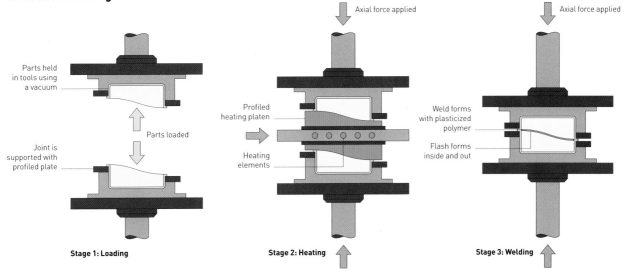

Parts held
in tools using
a vacuum

Parts loaded

Joint is
supported with
profiled plate

Stage 1: Loading

Axial force applied

Profiled
heating platen

Heating
elements

Stage 2: Heating

Axial force applied

Weld forms
with plasticized
polymer

Flash forms
inside and out

Stage 3: Welding

required to form the joint, unlike these other processes, which means small and delicate parts can be welded.

QUALITY

The strength of the weld is affected by the design of the part and type of material. Strength is therefore very difficult to quantify because it varies according to the application. Flash produced by pressure in the welding operation is often left untrimmed. It is possible to conceal the joint flash with a flange around the weld area.

Heating and plasticizing is localized, up to 1 mm (0.04 in.) on either side, and so the process does not affect the structure of the workpiece.

DESIGN OPPORTUNITIES

There are few restrictions on part size and geometry; the joint interface can be very complex and have both internal and external welds. The only requirement is that it can be accessed by a hot plate along one axis. Joints on different axes must be welded in a second operation.

This process is not specific to any thermoplastics material. In other words, different materials can be welded with the same tooling. All that need to be adjusted is the temperature of the heating platen.

TECHNICAL DESCRIPTION

This process is made up of several stages, which can be divided into 3 main operations: loading, heating and welding.

In stage 1, the parts are loaded into the tools, which generally operate on a vertical axis. They are held in place by a small vacuum. The tools align the joint interface, so that it can be heated and welded very accurately.

A pre-heated platen is located between the parts. In stage 2, the parts are brought into contact with the heated platen, which raises the temperature of the joint interface and plasticizes the outer layers of material. The temperature is adjusted according to the plastic parts being welded. Generally, it is set 50°C to 100°C (90–180°F) above the melting temperature of the polymer because it has to warm the plastic up enough for it not to fall below melting point before the weld has been made. Maximum temperature is around 500°C (932°F).

The hot plate is coated with a thin film of polytetrafluoroethylene (PTFE), which prevents the melting plastics from sticking during heating. For particularly sticky plastics, heat is applied without contact (convection rather than conduction), but this can cause problems. The benefit of contact with the tool is that pressure can be applied for uniform heating. Non-contact heating is less effective,

and impractical for complex and undulating joint profiles. PTFE coatings start to degrade above 270°C (518°F) and give off toxic fumes. Therefore, they can only be used for low melt temperature plastics such as polypropylene (PP) and polyethylene (PE).

The hot plate is generally a flat plate of aluminium, but can be profiled to accommodate parts with a 3D joint interface (as shown in the diagram).

In stage 3, the parts separate from the heated platen, which is withdrawn to allow the parts to be brought together. Axial pressure is applied and the plasticized joint interfaces mix to form a homogenous bond. A pre-determined amount of material is displaced by the pressure, which produces small beads of flash.

The parts are held in place until the polymer has solidified and cooled sufficiently so that it can be removed. Small parts are typically heated for around 10 seconds and then clamped (welded) for a further 10 seconds. Therefore, cycle time is up to 30 seconds. Large applications may take considerably longer.

→ Hot plate welding an automotive part

This product is a typical application for hot plate welding. It is part of a water-cooling system for an automotive under-the-bonnet application. This process is ideal because both internal and external welds can be made simultaneously. Only a small amount of pressure is required to form the weld, which is suitable for thin wall sections.

The part is made of 2 injection molded halves (**image 1**). The top and bottom half are loaded into their respective jigs and are held in place by a small vacuum (**images 2** and **3**). The required pressure, heating and welding time have been established through tests and so the process is run automatically to ensure accurate repeatability; the operator sets the programme for these particular parts (**image 4**).

Heating takes place on a heated platen (**image 5**), which raises the temperature of the material to more than 50°C (90°F) higher than its melting point. This ensures that there is sufficient heat build up in the joint for welding to take place. After only a few seconds, the tools separate and the heating platen

is withdrawn. The parts are then brought together and held under pressure until the joint interface has mixed and solidified (**image 6**). The whole process takes no more than 25 seconds.

The tools part and leave the welded product in the bottom half (**image 7**). The part is removed and checked. A bead of flash is typical with this process (**image 8**); it builds up around the joint as the pressure is applied.

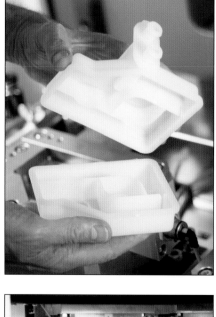

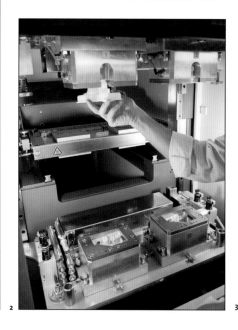

DESIGN CONSIDERATIONS

This process is most commonly used to weld butt joints. Lap joints are possible in simple parts and extrusions, but this is a much less common configuration.

The strength of the joint is improved by increasing the surface area of the weld. Incorporating a tee-shape or right angle at the joint interface, for example, will increase the weld area.

The maximum dimension of part that can be welded is limited by the size of equipment, which can be up to 620 mm by 540 mm (24.41 x 21.26 in.) on the heating platen and up to 350 mm (13.78 in.) high.

COMPATIBLE MATERIALS

Most thermoplastics can be joined in this way, although it is limited to injection molded and extruded parts. Some materials, including polyamide (PA), oxidize when they are heated to melting point, which can decrease the strength of the weld.

COSTS

Tooling costs are moderately expensive because they are required to support the part accurately during welding. Parts are often supported with a vacuum, which further increases cost.

Cycle time is generally rapid; around 10 seconds. However, complex and large welds can take considerably longer; up to 10 minutes.

Hot plate welding is either partially or fully automated, so labour costs are relatively low.

ENVIRONMENTAL IMPACTS

This process does not add any material to the joint and there is no waste produced during welding. Hot plate welding has a low environmental impact.

4

5

6

7

8

Featured Manufacturer

Branson Ultrasonics
www.branson-plasticsjoin.com

Joining Technology
Joinery

Contemporary furniture is constructed with both handmade and machine made joints. There are many different types, and it is up to the joiner to select the strongest and most visually pleasing for each application.

Costs	Typical Applications	Suitability
• No tooling costs; jigs may be necessary • Moderate to high unit costs depending on the complexity	• Construction • Furniture and cabinet making • Interiors	• One-off to high volume production

Quality	Related Processes	Speed
• High quality seamless and strong joints	• Friction welding • Timber frame construction	• Cycle time depends on complexity

MECHANICAL

324

INTRODUCTION

Joinery remains an essential part of furniture and cabinet making. Craft and industry have combined over the years and as a result a standard selection of joint configurations have emerged. These include butt, lap, mitre, housing, mortise and tenon, M-joint, scarf, tongue and groove, comb, finger and dovetail. Additionally, butt joints are strengthened with dowels or biscuits.

All of these joints can be further reinforced with screws and nails, but this section is dedicated to joints that are secured only with adhesive. (For metal fixtures in timber frame construction, see pages 344–7.)

Joints have to work both functionally and decoratively. The art of joinery is using wood to its strengths. It is anisotropic and is stronger along the length of its grain. It is also prone to shrinking or expanding as it dries or absorbs water from the atmosphere. Therefore, the joint design must be sympathetic to the strengths and instabilities of this natural material.

Joints can be seamless and almost invisible, or they can provide contrast to emphasize the joint. These are decisions made by the designer and made possible by a skilled joiner.

TYPICAL APPLICATIONS

Joinery is used in woodworking industries, including furniture and cabinet making, construction, interiors, boat building and patternmaking.

Typical furniture includes tables, chairs, desks, cabinets and shelves. Joinery is used in construction for timber roof trusses, gable ends, doors and window frames. Interior applications for this process include floors, walls, structures and stairwells.

RELATED PROCESSES

Joinery and timber frame structures (page 344) overlap. Simple joinery is used a great deal in timber frame construction, especially if the joints are on display. But timber frame construction tends to be concerned with speed of operation and reliability of joint. Metal fixtures are often used, rather than spending time cutting a complex joint.

The advantage of gluing is that it spreads the load over the entire joint and is not visible from outside.

In the future, linear friction welding (page 294) may compete with conventional gluing techniques.

QUALITY

The quality of joint is very much dependent on skill. There is very little room for error, but mistakes are inevitable. Skilled cabinet makers differentiate themselves by their ability to repair mistakes imperceptibly.

Products made from wood have unique characteristics associated with visual patterns (growth rings), smell, touch, sound and warmth. High quality woodwork is often left exposed.

Joint Configurations

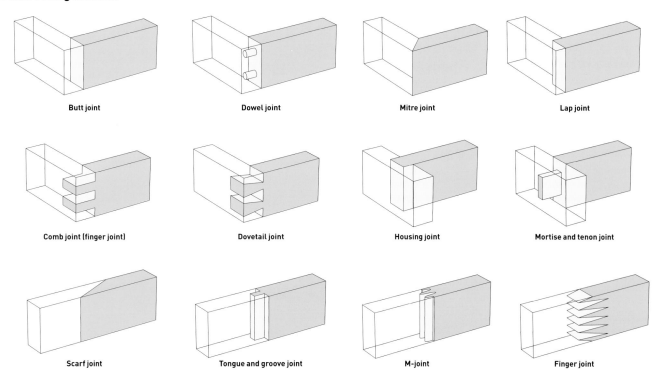

Butt joint

Dowel joint

Mitre joint

Lap joint

Comb joint (finger joint)

Dovetail joint

Housing joint

Mortise and tenon joint

Scarf joint

Tongue and groove joint

M-joint

Finger joint

TECHNICAL DESCRIPTION

The diagram illustrates the most common joint types. They include handmade and machine made configurations, which are used in furniture construction, house building and interior structures.

There are 4 main types of adhesive, which are urea polyvinyl acetate (PVA), formaldehyde (UF), 2-part epoxies and polyurethane (PUR). PVA and UF resins are the least expensive and most widely used. PVA is water based and non-toxic, and excess can be cleaned with a wet cloth. PUR and 2-part epoxies can be used to join wood to other materials, such as metal, plastic or ceramic, and are waterproof and suitable for exterior use. They are rigid, so restrict the movement of the joint more than PVA.

Butt joints are the simplest form of joinery. They are inexpensive to prepare because the 2 planks are simply cut to length. However, they are also the weakest because there is a relatively small interface for gluing and 1 face is end grain. End grain is the least strong face for all fixing types, including adhesives, nails and screws.

Butt joints are reinforced with dowels, biscuits or metal fixings. The benefit of

these is that they are concealed within the joint interface. These are widely used in high volume production and flat pack furniture.

A mitre joint is a simple, neat method of joining 2 planks at right angles. It is more aesthetically pleasing than a butt joint because it ensures continuity of long grain and avoids exposed end grain.

A lap joint increases the size of the interface to include gluing to long grain on both sides. This increases the strength of the joint and requires more preparation.

A comb joint (also called a finger joint) provides a much larger gluing area and is a common joint for boxes or drawers in tables and cabinets. It is made by a series of spaced cutters on a spindle molder, a spinning cutter for profiling lengths or ends of wood.

Dovetail joints are modified finger joints. They are cut with re-entrant angles, which increase the strength of the joint in certain directions. They are especially useful for drawers, which are repeatedly pulled and pushed from the front.

A housing joint is a lap joint in the middle of a workpiece. It is common for shelves and cabinet making.

Mortise and tenon joints are used to join perpendicular lengths of wood. The leg is the tenon and the mortise the hole. The tenon is usually cut on a band saw, and the mortise on a mortiser, which is a machine with a drill bit inside a chisel that cuts square holes in wood. Alternatively, the mortise can be cut by hand. There are many different types including haunch, shoulder, blind, through (illustrated) and pegged.

Scarf joints, tongue and groove, M-joints and finger joints are primarily used to join planks of wood to make larger boards. They increase the gluing surface between 2 planks in a butt joint configuration.

M-joints are machine-made and suitable for mass production. They are cut on a high-speed spindle molder.

The finger joint is machine-made and designed to integrate lengths of timber into a continuous profile. It is designed to maximize gluing area and joint strength.

There are many variations on each of the above joints, including the angle of interception and whether the joint detail penetrates right through the wood or is blind (not right through).

Mortise and tenon and M-joints in the Home Table

This is the Home Dining Table, which was designed by BarberOsgerby in 2000. It is produced from solid oak with a range of different joints (**image 1**).

The legs are mitred together (**image 2**). The legs and table frame are joined with a mortise and tenon, the traditional and strongest joint for this application (**images 3–5**). The top is made by joining solid planks of oak with M-joints (**images 6** and **7**).

On the assembled table the mitre joint is almost invisible (**image 7**). The grain merges on the front of the leg because the 2 halves of the leg are cut from the same plank of wood.

Materials like medium density fibreboard (MDF) tend to be concealed under several layers of paint. A very high level of finish can be achieved on these surfaces with repeated painting and polishing.

1

DESIGN OPPORTUNITIES

Joints have several primary functions, which include lengthening or widening, change of grain direction and joining shapes that cannot be made in a single process or from a single piece.

The style of joint should minimize the amount that wood will naturally twist and buckle, reduce weight and maximize gluing area.

Joinery maximizes the opportunities of designing with timber. It is possible to make large flat surfaces, such as tabletops, of planks joined together with M-joints or tongue and grooves and with 'bread board' ends of wood fixed across the ends of the planks to prevent warping. Long and continuous lengths can be made with M-joints and finger joints. Shelving and cabinets can be assembled with simple housing, lap or butt joints reinforced with dowels or biscuits. Simple boxes can be made with butt, mitred and lap joints, or more decorative comb and dovetail joints.

Change of grain direction is especially useful for load bearing applications.

For example, the framework and legs of a table are joined together, partly because the grain should run in different directions for optimum strength. The most common joint for this application is a mortise and tenon because it helps to prevent twisting.

Joinery is often combined with CNC machining (page 182), laminating (page 190), steam bending (page 198), timber frame construction and upholstery (page 338). The joints may be exposed and decorative or purely functional.

DESIGN CONSIDERATIONS

There is no limit on the size of plank or length of wood that can be joined. However, the type of joint and its size relative to the wood are important considerations. There are guidelines to maximize retained strength in the timber that is cut to form the joint.

For example, mortises (mortise and tenon joint) and grooves (tongue and groove joint) should be no wider than

one-third the thickness of the workpiece. This will ensure that there is sufficient material around the joint to support it.

The choice of joint is determined by a balance between functional and decorative requirements, and economic factors. There are types of joint that have been used traditionally for joining legs to tables and panels to doorframes or making shelves and drawers, for example. For each application the joints are modified and adjusted. The case studies illustrate some of the most common joint types and how they are used.

Aesthetics play an important role for exposed joints. For example, a mitred joint will cover up end grain and provide a neat edge detail with continuity of surface grain, whereas comb and dovetail joints express craftsmanship and add perceived value.

Economic factors, such as cycle time and labour costs, often have a role to play. For example, a part that is being made

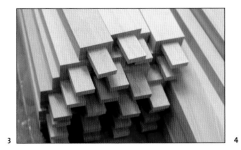

using butt joints will be less expensive than finger or dovetail types. However, glued butt joints may not provide such a large and strong joint interface as other methods. Therefore, very strong adhesives, dowels and biscuits are used to maximize strength.

COMPATIBLE MATERIALS

The most suitable wood for joinery is solid timbers, including oak, ash, beech, pine, maple, walnut and birch.

Joinery is not limited to wood. These methods can be applied to any material as long as it is sufficiently hard to cut a joint into. Synthetic materials will not require such complex profiles because they are more stable than wood.

COSTS

Most applications do not require tooling. Some machine made joints may need tools cut for spindle molders or routers, but these are generally inexpensive.

Cycle time is totally dependent on the complexity of the job and the skill of the operator. A single joint may take only 5 minutes to cut and assemble, but there might be 15 different joints on the product. Generally, the same or similar joints will be used as much as possible on a product to minimize set up and changeover time on the machines.

Labour costs tend to be quite high due to the level of skill required.

ENVIRONMENTAL IMPACTS

Wood has many environmental benefits, especially if it is sourced from renewable forests. Timber is biodegradable, can be reused or recycled and does not cause any pollution.

Joints tend to be formed by cutting, so waste is unavoidable. Dust, shavings and wood chips are often burnt to reclaim energy in the form of heating.

Featured Manufacturer

Isokon Plus
www.isokonplos.com

→ Mitred and biscuit reinforced butt joints in a bedside table

This is a simple, veneered bedside table. The 4 sides of the product are lipped with solid oak edges, veneered, cut to size and mitred. Grooves are cut for the biscuit reinforced butt joints.

To join the mitred joints, the parts are laid face up (joint down) on the table. The faces are taped (**image 1**) to keep the joint tight during assembly. Prior to assembly, the biscuits (**image 2**) are placed into pre-cut grooves on the butt joint, and glue is applied to all of the joints (**image 3**).

The 4 sides are assembled with the tape still in place (**image 4**). This keeps the joint tight and acts like a clamp (**image 5**).

1

2

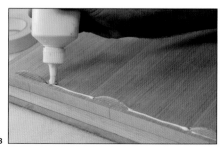

3

4

5

Featured Manufacturer

Windmill Furniture
www.windmillfurniture.com

→ Dowelled butt joints in a table drawer

This case study illustrates dowels used to strengthen butt joint configurations. The parts are drilled with the holes set apart to exact measurements (**image 1**). This is often carried out on a drill with 2 heads, which are set exactly the same distance apart for both sides of the joint.

The dowels are made from beech or birch, because they are suitably hard materials (**image 2**). They are inserted into the joint with glue and hammered into place (**image 3**).

1

2

3

Featured Manufacturer

Windmill Furniture
www.windmillfurniture.com

→ Comb jointed tray

Comb joints are traditionally used to join the sides of trays and drawers. The joint is cut by spindle molder with a set of cutters separated by matched spacers. Both sides are cut with the same set up to ensure a perfect fit (**image 1**), which can be assembled by hand (**image 2**).

The base of the tray, which is located in a housing joint between the 4 sides, holds it all square, and the finished product is lacquered (**image 3**).

1

2

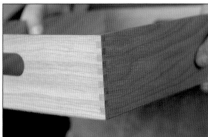

3

Featured Manufacturer

Windmill Furniture
www.windmillfurniture.com

→ Housing joints in the Donkey

This version of the Donkey was designed by Egon Riss in 1939. It is made from birch plywood (**image 1**). Housing joints are a simple and a strong way to fix the shelves into the end caps (**image 2**). This use of this joint means that the product can be assembled and glued in a single operation. The end caps are clamped together, which applies even pressure to all the joints.

1

2

Featured Manufacturer

Isokon Plus
www.isokonplus.com

Case Study

→ **Decorative inlay**

This is a simple form of wood inlay, used visually to separate 2 veneers on a tabletop (**image 1**). The inner veneer is bird's eye maple and the outer veneer is plain maple. This type of decorative inlay is made up of layers of exotic hardwoods and fruitwoods, which are cut into strips (**image 2**). The groove is cut by router, and the strips of inlay are bonded in with UF adhesive (**image 3**).

1

2

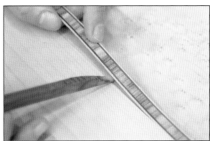

3

Featured Manufacturer

Windmill Furniture
www.windmillfurniture.com

Joining Technology

Weaving

Weaving is the process of passing strands or strips of material over and under each other to form an intertwined structure. Fibre strength and alignment can be adjusted specifically for each application, reducing weight and material consumption.

Costs	Typical Applications	Suitability
• There are no tooling costs • Low unit costs, but dependent on the raw material	• Furniture • Interiors • Storage	• One-off to high volume production

Quality	Related Processes	Speed
• Depends on raw material	• Steam bending • Upholstery • Wood laminating	• Machine weaving is rapid • Hand weaving is moderate to slow, but depends on size and complexity of part

INTRODUCTION

Weaving is used across a wide range of industries, including textiles, rug making, sail making and architecture. This section focuses on rigid textiles, which are used in the construction and upholstery of furniture, baskets, fences, screens and mats. Rigid textiles can be made not only as flat panels, but also as 3D, self-supporting structures.

There are 3 main types of rigid textile weaving: plain (see image, opposite, above middle), twill (see image, opposite, top) and satin (in which either the warp or the weft bridges 5 perpendicular strands or more). Satin has a slightly less stable textile, a warp- or weft-rich surface and a more densely packed weave. Other types of weaving, such as machine-made tri-axial 'strand caning' (see image, opposite, above) and 'basket weaving' (see image, below), are combinations of these techniques.

Machine made textiles are woven as panels, which are then secured to a structural framework. To make 3D profiles they are draped over molds and coated in adhesive to retain their shape.

Hand weaving techniques date back thousands of years and are very similar today because many are not suitable for mechanized mass-production. Nevertheless, there is a substantial market for hand woven products, which are made in large quantities in countries with enough labour to make production economically attractive.

Hand weaving is carried out either on a loom (in which case it can be mechanized), between rigid elements (which do not have to be parallel as they are in machine weaving), or as a 3D self-supporting structure. There are many techniques including plaiting (general weaving), hand caning, lashing, splint seat weaving and coiling.

TYPICAL APPLICATIONS

Weaving is used in many different areas of furniture construction. Typical products include stools, chairs, tables, sofas, beds, lights, storage boxes, blinds and screens.

Other products using similar techniques and materials include baskets, fences and wall and floor panels.

Top
Twill, or herringbone, is a diagonal pattern created by overlapping 2 strands or more at a time.

Above middle
Plain weave is a simple '1 up, 1 down' pattern.

Above
Cane weave. This is the most popular method of caning, based on the traditional 7-step manual technique to form an octagonal pattern

Above
The Lloyd Loom Nemo chair, designed by Studio Dillon in 1998, is made up of a single steam bent ring, onto which the woven material is fixed.

RELATED PROCESSES

Alternatives to rigid textiles in furniture include upholstery (page 338) with leather or 'soft' textiles, wood laminating (page 190) and composite laminating (page 206). Similar to upholstery, woven 'rigid' textiles are often fixed wonto a steam bent (page 198) or CNC machined (page 182) wooden support structure.

Wood and composite laminates rely on adhesives to hold the layers together. By contrast, woven materials are maintained by friction. This means woven structures tend to be more flexible and deform permanently. The advantage is that woven structures can be shaped, for example onto a mold in the case of the Lloyd Loom Nemo (see image, above right).

QUALITY

The quality of weave is determined by the combination of raw material

Loom Weaving Process

Independent
heddle bar

Beater

Intertwined weft

New weft fed
into place

Continuous
supply of warp

Eyelet

1

2

TECHNICAL DESCRIPTION

Weaving rigid textiles on a loom consists of 3 movements repeated many times: raising and lowering the heddle bars, feeding the weft and beating.

Each strand of warp is fed through an eyelet in the heddle bar. The heddle bars are operated individually or as a set, and are computer-guided or moved by depressing a foot pedal. Moving them up and down determines whether the warp or weft will be visible from the top side. This is how patterns are made, and they can be very intricate. In the diagram the heddles are separated into 2 sets, which creates a basket weave pattern.

A weft is fed into the space between the fibres and in front of the beater. The beater is a series of blunt blades that sit between each fibre. They are used to 'beat' each weft tightly into the overlapping warp.

The weft is held in place by the beater while the lower heddle bar moves up and the upper heddle bar moves down, which locks the weft between the warps. The process is repeated to form the next run.

and pattern. Nearly all contemporary weaving is carried out on computer-controlled looms, which produce high quality, repeatable materials. The quality of handmade weaves depends on the skill of the weaver.

Rigid textiles tend not to be as tightly woven as 'soft' textiles. Material is used only where necessary and there is give between the fibres, which makes them lightweight and durable structures. They are breathable, which is advantageous for applications such as beds and chairs, especially in hot and humid climates.

DESIGN OPPORTUNITIES

Woven structures tend to be multifunctional, and there are many opportunities associated with construction and application.

Each type of weave has a different appearance, drape and robustness. Combined with different materials, an endless number of structural properties can be achieved. This makes woven structures suitable for both self-supporting and reinforced applications. For example, food parcels dropped in World War II were made by placing 1 woven basket inside another for protection: this was durable enough

to drop without a parachute, saving valuable materials.

Colours and patterns can be woven into rigid textiles. Like 'soft' textiles, they are treated as repeating modules. Alternatively, they can be printed on, or dyed in a solid colour.

Different types and thickness of material are combined to produce structures with specific load-bearing capabilities. For example, the weft can bind a structural warp together, as in the Lloyd Loom basket weave.

Handmade weaves illustrate a major advantage, which is that a woven structure can be designed and constructed to suit a specific load bearing application.

The continuity and direction of fibres directly affect the strength of a woven product. Laminated composites, filament winding (page 222) and 3D thermal lamination (page 228) exploit this property to huge advantages with 'soft' fibres reinforced with rigid adhesive.

Case Study

→ Weaving upholstery

This case study demonstrates weaving and the subsequent upholstery of woven material onto a steam bent wooden structure. The example being made is the Lloyd Loom Burghley chair (**image 1**).

Lloyd Loom manufacture their own paper-based weaving material. The warp is made by twisting strips of Kraft paper into tight fibres (**images 2** and **3**). It is coated with a small amount of adhesive to lock it in place. The weft has a metal filament along its centre.

The metal is concealed in paper (**image 4**) and is the structural element of the basket weave.

Each of the looms is loaded with 664 bobbins of twisted paper warp (**image 5**). The looms produce flat and continuous woven material 2 m (6.6 ft) wide (**image 6**). The wire wefts are folded along the edge to lock the warps in place (**image 7**). Otherwise they would spread out and fall off the end of the structural wefts.

The woven material is transferred onto a steam bent structural framework (**image 8**). The edge is stapled with a braid of twisted paper to secure the strands and prevent any fraying (**image 9**). Finally, the completed chair is sprayed with a protective coating, which ensures the longevity of the material.

3

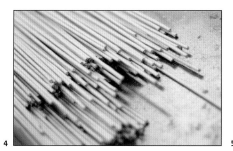

4

5

6

9

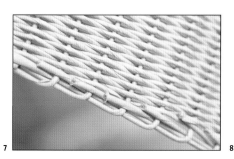

7

8

Featured Manufacturer

Lloyd Loom
www.lloydloom.com

The benefits of weaving 'rigid textiles' according to strength requirements are weight reduction and an aesthetic that is directly related to function.

DESIGN CONSIDERATIONS

Hand weaving techniques, which include rattan and wicker furniture, are time consuming and labour intensive. Machine made weaves tend to be a standard pattern. Even so, there is a wide range to choose from. They are supplied as a flat panel. This can be bent, but typically only tightly in 1 direction, or gently over a shaped mold. Forcing it into complex profiles will push the fibres out of alignment and affect the strength and aesthetics of the product.

The size of bend is determined by the type of material. Some materials, such as cane, can be bent to very tight angles. Other materials are not so pliable and may split. These shapes have to be woven into a 3D shape by hand, or by joining panels of material.

Many of the materials used in rigid weaving are natural and so need to be protected from the elements. This is often done with a clear or coloured spray coating (page 350).

COMPATIBLE MATERIALS

Woven furniture was traditionally handmade using natural fibres such as rattan, willow and bamboo. Mass production methods can also produce continuous woven materials from metal, paper, plastic and wood.

COSTS

There are no tooling costs unless the weave is formed over a mold. Even then the tooling costs tend to be low.

Cycle time depends on size, shape and the complexity of the weave or 3D product.

Labour costs tend to be high because a high level of skill is required. Weaves made by machine and then upholstered onto products require less labour, but still require highly trained operators.

1

2

3

ENVIRONMENTAL IMPACTS

This process creates products with minimal materials. Hand weaving methods traditionally use locally grown materials and so do not require the transportation of raw materials over large distances. However, these practices are becoming less common.

The mechanical join is formed by intertwining materials. Therefore, there are no chemicals, toxins or other hazards associated with melting, fusing, or otherwise altering materials to join and shape them.

Most waste is either biodegradable or capable of being recycled.

→ Strand caning

The S32 was designed by Marcel Breuer and production at Thonet began in 1929 (**image 1**). It is among most highly mass-produced tubular steel chairs in history.

The sheets of strand cane are pre-woven on a loom (**image 2**). There are many different types and patterns, but this octagonal pattern is the most popular. It is a reproduction of the traditional 7-step hand caning technique.

First of all the cane is cut into strips and soaked for at least 35 minutes. This is to ensure that it is sufficiently pliable to be formed. The seat back, which is the product being made here, is steam bent and a groove is cut around the front into which the cane is fixed.

The cane is hammered into the groove using a specially shaped tool (**image 3**). A spline of cane is then pressed into the groove to secure the weave (**image 4**). They are both locked in place with adhesive.

The excess material is trimmed off and the spline finished off (**images 5** and **6**). The assembly is placed into a warm press, which applies even pressure (**image 7**) and the finished seat backs are stacked ready for assembly onto the tubular metal frame (**image 8**).

4

5

6

7

8

Featured Manufacturer

Thonet
www.thonet.com

Joining Technology

Upholstery

Upholstery is a highly skilled process and the quality of craftsmanship can set products apart. It is the process of bringing together the hard and soft components of a piece of furniture to form the finished article.

Costs	Typical Applications	Suitability
• No tooling costs • Low to high unit costs (depending on complexity and textile material)	• Furniture • Marine and automotive interiors • Public transport seating	• One-off to high volume production

Quality	Related Processes	Speed
• Very high quality that depends on the skill of the upholsterer and type of material	• Steam bending • Weaving • Wood lamination	• Moderate to long cycle time depending on size and complexity of product

INTRODUCTION

A typical upholstered chair consists of a structural frame, foam padding and a textile cover. Sofas and lounge chairs may also have a sprung seat deck. Springs are mounted onto the frame or suspended within it. There are 3 main spring systems: 8-way hand-tied, drop-in and sinuous springs.

Hand-tied springs are the most expensive and considered superior. Using them increases cycle time considerably. A wooden box is fabricated, webbing is stretched across it and the springs are hand-tied to the webbing up to 8 times.

Drop-in and sinuous springs are machine-made. They are as their names suggest: drop-in springs are a prefabricated unit fixed into the framework, and sinuous springs are continuous lengths of steel wire bent into 's' shapes and fixed at either end.

The structural framework is generally fabricated in wood or metal. Its strength determines the durability of the product.

Nowadays, the padding is polyurethane (PUR) foam. It is either molded using the RIM process (page 64) or cut and glued together. Modern foams have reduced the need for sprung seat decks, and many contemporary sofas and lounge chairs no longer have springs. Instead, foam density is designed to provide maximum comfort.

The covering is fabric or leather and is permanently fixed to the padding and underlying framework.

TYPICAL APPLICATIONS

Upholstery is used extensively in furniture and interior design. It is used to make 'soft' furnishings for the automotive, marine, home, office and transport industries. The materials for each of these applications will vary according to the likely wear and environmental conditions.

Typically, upholstered products include lounge chairs, sofas, task and office chairs, car seats and interiors, boat seats and interiors, break out seating and padded walls.

RELATED PROCESSES

There are many processes used in the upholstery procedure. Wooden frames tend to be laminated (page 190), steam bent (page 198) or CNC machined (page 182). Metal frames are bent (page 98) or cut and welded (page 308). The cover is cut out on a CNC x- and y-axis cutter, or by hand, sewn together and glued and stapled onto the foam and framework.

The foam padding is either molded over the framework, or cut and glued onto it. The choice of RIM is not solely dependent on quantities because certain shapes cannot be produced feasibly in any other way and have to be molded.

Alternatives to upholstery include mesh stretched over the framework, wooden, plastic or metal slats, injection molding (page 50), weaving (page 332) and wood lamination.

Machine Stitching Process

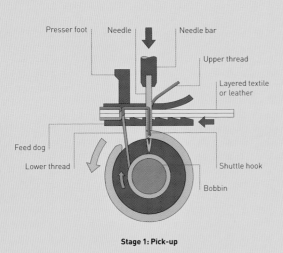

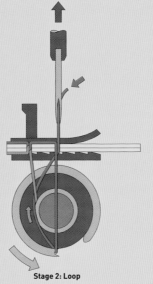

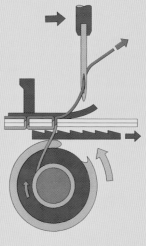

Stage 1: Pick-up   Stage 2: Loop   Stage 3: Finish

TECHNICAL DESCRIPTION

Machine stitching is a simple form of mechanical joining that requires a complex series of operations to execute. There are 3 main types of machine sewing: lockstitch, chain stitch and overlock. The diagram illustrates lockstitching, which is used to join the leather cover in the Boss Eye chair case study.

Lockstitching is a mechanized process. The needle and shuttle hook are synchronized by a series of gears and shafts, powered by an electric motor.

In stage 1, the upper thread is carried through the fabric by the needle, and the lower thread is wound on a bobbin.

The needle pierces the layers of material and stops momentarily. In stage 2, a spinning shuttle hook picks up the upper thread. The shuttle hook loops behind the lower thread, which is held under tension on the bobbin. In stage 3, as the shuttle continues to rotate, tension is applied to the upper thread, which pulls it tight, forming the next stitch. Meanwhile, the feed dog progresses forward, catches the fabric and pulls it into place for the next drop of the needle. The fabric is supported between the presser foot and feed dog. Industrial sewing machines can repeat this sequence over 5,000 times every minute.

Chain stitching, or loop stitching, is the process of making stitches from a single thread. The downfall of this method is that if the thread is broken at any point it readily comes apart.

Overlocking, or serging, is the process of finishing the edge of fabric with multiple threads. It protects the fabric from fraying by casting a net of interlocking stitches over the edge.

QUALITY

The look of the upholstery is largely dependent on the skill of the upholsterer. The comfort is determined by the quality of foam and springs, while the longevity of the product is affected by the rigidity and strength of the framework.

DESIGN OPPORTUNITIES

The density and hardness of foam can be chosen to suit the application. The fabric or leather cover is adhesive-bonded to the foam, so undercuts and overhangs can be upholstered. Shape limitation is determined by what can be cut and shaped in the textile; single axis curves

Right
With a highly skilled operator, an industrial sewing machine can produce over 5,000 accurately placed stitches per minute.

are easily covered with textile, whereas multiple axis curves require elastic material or 2 or more pieces of material stitched together.

Each product has to be upholstered individually and so manufacturers are often prepared to run single colours, or short runs of colour. The textiles available are as varied as the clothes you wear. Different weaves and materials have varying levels of durability, and the toughness of the covering is determined by the application of the product. For example, a domestic lounge chair may be used for only 3 or 4 hours on an average day, whereas break out areas in offices will see much heavier usage.

Fabrics have varying levels of elasticity and bias. Slightly elastic fabrics reduce the number of pieces required to cover a 3D shape. However, they can only be used on convex shapes because they will bridge concave profiles.

Upholstered covers can be finished in a number of different ways. Possible details include button tufting, beading and twin stitching.

DESIGN CONSIDERATIONS
Whatever the design, it must be possible to cut and stitch a fabric shape that will cover it. There are various techniques used to conceal the open end of the fabric cover. Conventionally, the fabric cover is pulled on and the edges stapled onto a single face, typically the bottom or back of the product. This is then covered with a separately upholstered panel.

Concave shapes can be upholstered, but the cover will need to be secured in place with a panel, pins, ties or adhesive to maintain the cover in place.

Fabrics come on rolls that are typically 1.37 m (4.5 ft) wide. Leather comes in various sizes: for example, cow hides range from 4 m² to 5 m² (43–54 ft²) and sheep skins can range from 0.75 m² to 1 m² (8–10 ft²).

The choice of covering material is an important factor in determining the unit price. High value materials, such as high quality leather from a top supplier like Elmo Leather in Sweden, can double the price of the product

COMPATIBLE MATERIALS
The area of application, such as automotive, marine, domestic, office, education, healthcare or public transport, determines the specification of upholstery material.

Contract grade materials suitable for high wear applications include polyamide (PA) nylon, thermosetting and thermoplastic polyester, synthetic leather polyurethane (TPU), polyvinyl chloride (PVC), polypropylene (PP) and other hardwearing fibres.

General upholstery materials for low wear applications include leather, flock, raffia, mohair, cotton and canvas.

Outdoor application materials have to withstand exposure to the elements and ultraviolet light. Examples include synthetic leather TPU and polycarbonate (PC) coated materials.

COSTS
There are no tooling costs in this process, but there may be tooling costs in the RIM foam process and for any wood laminating or steam bending.

Cycle time is quite long but depends on the size and complexity of the piece.

Batch production does not reduce the labour costs, but material costs may be reduced due to increased buying power. Labour costs tend to be high due to the level of skill required.

ENVIRONMENTAL IMPACTS
Upholstery is the culmination of many processes and as a consequence a typical sofa will include many different materials that have varying environmental implications.

The choice of covering material will affect the environmental impact of the product. For example, the William McDonough collection of fabrics produced by Designtex is biodegradable and manufactured in a closed loop industrial process. Many leather materials have to go through a series of potentially environmentally harmful tanning and dying processes before they can be used in upholstery. By contrast, Elmo Leather produce a chrome-free leather called Elmosoft. There are no hazardous substances used in the production process.

A lot more waste is produced when upholstering with leather, 1 reason why it is so much more expensive. The net shapes can be nested very efficiently on fabric, producing only 5% waste, whereas leather may have imperfections that cause up to 20% waste. Offcuts can sometimes be used in furniture with smaller pieces, or gloves.

→ Upholstering with cut foam

These images illustrate a technique for upholstering foam-padded chairs. An example of a chair produced in this way is the Sona chair, designed by Paul Brooks (**image 1**). Production began in 2005. This technique is suited to geometries that have a constant wall thickness; RIM is more cost effective for complex and undulating shapes such as the Eye chair.

This case study shows upholstering the Neo chair using the cut foam technique. The structure is laminated wood (**image 2**), onto which the foam is bonded (**images 3** and **4**). The covering material is stretched over the foam and stapled onto the plywood substrate. Each panel is made up in this way and then fitted together to conceal the inner workings of the chair.

1

2

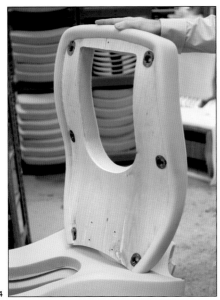

3

4

Featured Manufacturer

Boss Design
www.bossdesign.co.uk

→ Upholstering a RIM foam chair

The Eye chair was designed by Jackie Choi for Boss Design and production began in 2005. It is covered with fabric or leather (**image 1**).

The polyurethane foam is made by Interfoam using the RIM process. The foam has a metal structure for support and plastic or wooden panels to which the covering material can be fixed.

Upholstering this product is a time-consuming and highly skilled operation. First of all, the foam is coated in a thin film of adhesive (**image 2**) and covered in soft polyester or down lining, which helps smooth over imperfections and gives the surface a better feel (**image 3**).

Meanwhile, the leather is prepared by a pattern cutter (**image 4**), who carefully has to work out the least wasteful arrangement of patterns on the sheet. If possible, the patterns are cut from a single skin and stitched together (**image 5**). A tape is sewn on the inside for additional strength. Many different stitch patterns and details can be applied; this is twin stitch. The patterns are carefully designed so that the seams line up with the corners of the chair.

Once the leather patterns have been assembled, the cover is pulled over the foam inside out and sprayed with adhesive (**image 6**). The adhesive is not sticky at this point because it is triggered by steam.

The cover is removed and fitted onto the foam (**image 7**). It is stapled in place using the plastic panels molded into the product for this reason (**image 8**). Another panel, upholstered separately, snap fits onto the plastic panel to conceal the trimmed edges (**image 9**).

The leather cover is now securely in place. However, there are undercuts to which the cover must be bonded if it is to retain its shape. This is done with steam and a soft cloth (**images 10** and **11**). This process requires a great deal of patience; the heat from the steam softens and activates the adhesive, so that as the cover is gently pushed into the desired shape, the adhesive bonds to the foam overhang.

The final chair is inspected prior to packing and shipping (**image 12**).

1

2

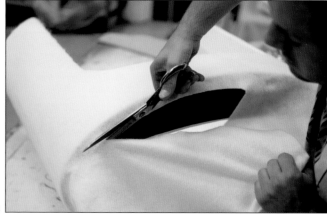
3

4

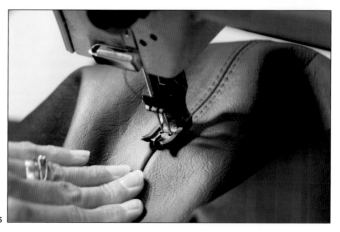
5

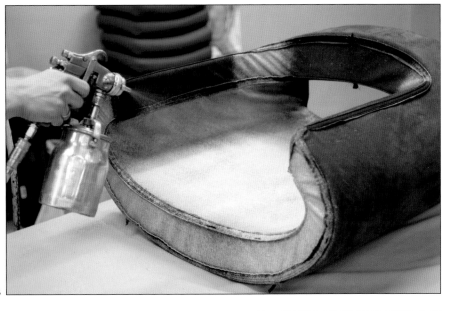

6

7

8

9

10

11

12

Featured Manufacturer

Boss Design

www.bossdesign.co.uk

Joining Technology
Timber Frame Structures

Large-scale timber frame construction uses a variety of fixing methods. Softwood and engineering timber structures can span large gaps. Amongst other applications, they are used for house building because construction is fast and efficient.

Costs	Typical Applications	Suitability
• No tooling costs • Moderate unit costs, depending on complexity	• Temporary structures • Theatre and film sets • Timber frame housing	• One-off to medium volume production

Quality	Related Processes	Speed
• Lightweight and long lasting	• Joinery	• Moderate cycle time (5–30 minutes per frame)

INTRODUCTION

Timber frame structures are used in factory-made low-rise buildings, interior structures, roofs, large enclosures and freestanding structures. There are many standard products, including roof and attic trusses, wall panels and floor cassettes. Bespoke structures are manufactured for each application, which may have different functional and decorative requirements.

Engineering timbers utilize the strength and stability of laminated wood and include plywood, oriented strand board (OSB), laminated strand lumber (LSL), parallel strand lumber (PSL) and composite I-beams. These materials are used in a variety of combinations to produce lightweight structures, engineered to precise requirements. A variety of softwoods are used, and each piece is strength graded. Structural designers select timber strong enough for each application.

Buildings and other large-scale structures are simulated in CAD software. They are tested for load-bearing strength and location-specific factors such as wind speed.

APPLICATIONS

As technology in this area progresses, so does the range of applications. Approximately three-quarters of the world's housing construction is timber. Timber frame is the most popular form of construction in countries with cold climates because it is such a fast and efficient method of building.

Some typical products include factory-constructed houses, theatre and film sets, temporary structures (such as pavilions

Fixing Methods

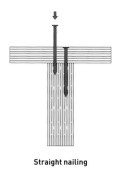

Straight nailing

Skew nailing

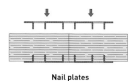

Nail plates

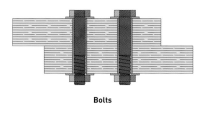

Bolts

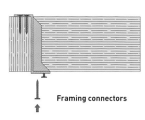

Framing connectors

and exhibitions), permanent and large-scale structures (including airports, government offices and residential housing), and interior structures (such as stairwells, warehouse conversions and wide spanning floors and ceilings).

RELATED PROCESSES

Joinery (page 324) and timber frame structures overlap. Joinery tends to use adhesives and wooden fixtures, whereas timber frames use metal fixtures for speed and ease of manufacture.

There are 2 main housing alternatives in the UK construction industry, which are masonry and steel framework. The choice of material will determine the method of construction. Timber frame is steadily becoming more popular in the UK due to its beneficial environmental impact and the speed of production.

QUALITY

Wood is a natural composite made up of lignin and cellulose. This has many advantages, but means that designs need to accommodate dimensional changes as a result of variation in moisture content. In timber frame construction, the strength and lightness

of wood is harnessed and dimensional stability achieved by laminating layers with strong adhesives to form engineering timbers (page 190).

Like other natural materials, wood has unique characteristics associated with visual patterns (growth rings), smell, touch, sound and warmth. The finish can be exposed to make use of its sensual qualities, or concealed, depending on the requirements of the application.

DESIGN OPPORTUNITIES

There are many reasons for building with timber: it has a low environmental impact, it is lightweight and strong, it is inexpensive and each piece is unique. It can be shaped using similar processes to plastic and metal, but will always retain natural qualities.

Timber frame construction is not limited to a range of standard products. Many structural designs have been refined to give optimum performance, but this does not limit the designer with regard to size or shape. Manufacturers have to accommodate a range of products, so there is little cost difference between making 10 or 100 identical products using this process.

TECHNICAL DESCRIPTION

Various fixing methods are used in timber frame construction to accommodate different joint types and access to them. They are generally semi-permanent, and so can be removed if necessary.

Nailing is used for tee and overlap joint configurations. It is always preferable to nail through the thinner material and into the thicker material. This provides obvious mechanical advantages. Generally, two-thirds of the nail shank should penetrate the lower material.

The strength of a nailed joint is determined by the angle at which the nail intersects the grain because wood is anisotropic. The nail shank parts the grain as it is forced into the wood. The grain contracts around it, forming a tight grip. Nailing into end grain is the weakest, and nailing across the grain is strongest.

Nail plates are effective for tee and butt joints. They are steel plates that have been stamped to form a rack of short nails. Even though these do not penetrate deep into the material they are strong due to the number of nails. They eliminate the need for overlapping joints.

Bolts tend to require more joint preparation. They are not self-tapping, like screws and nails, so require a pre-hole slightly larger than the shank. However, they tend to be stronger than other fixing methods because they grip the joint from both sides.

An alternative method for tee joints in load-bearing applications is framing connectors. These are bent metal fixings nailed in place to provide support in the critical areas and reduce stress on the joint. They are similar to a housing joint (see Joinery).

Metal fixtures come in a variety of finishes. They tend to be made in carbon steel, which is galvanized for improved longevity. For exterior applications, especially near the coast, stainless steel fixings should be used to avoid corrosion, which can cause staining and ultimately joint failure.

DESIGN CONSIDERATIONS

Designed properly, wooden structures can achieve large spans. They are capable of bearing significant loads, even in cantilever configurations. Care must be taken to ensure correct support to reduce stress on load-bearing joints. For example, there are guidelines for drilling holes, which weaken structures.

Wood is anisotropic and is stronger along the length of its grain than across. It can only be used for load-bearing applications in low-rise construction, although it is capable of carrying significant load and has been used in the UK for up to 7 storeys.

The strength of timber panels relies on the combination of the frame and the skin. Therefore, panels tend to be constructed flat and then fixed together to form 3D structures.

The size of mechanical fixture is calculated on CAD software to ensure adequate support. Generally, nails are used for tee and butt joints, while bolts are used for overlapping joints.

COMPATIBLE MATERIALS

All timber products including softwoods, hardwoods, veneers and engineering timbers can be used.

COSTS

There are no tooling costs: assembly can be achieved with hand tools, but the industry is evolving more sophisticated CNC assembly operations. Cycle time is moderate. Each frame takes 3 to 30 minutes to construct. A typical timber frame house can be constructed on site at a rate of a floor per day, but large, complex or curved structures may take longer than this.

Labour costs are typically high due to the level of skill and number of operators required for large construction.

ENVIRONMENTAL IMPACTS

One of the main reasons for specifying timber frame construction is the environmental advantage that it has

over competing methods. Timber is a renewable material. It has lower levels of 'embodied energy' than alternatives – in other words, it uses less energy to grow, extract, manufacture, transport, install, use, maintain and dispose of than other construction methods. In fact, timber frame structures have up to 50% less embodied energy than steel and concrete equivalents. Timber is biodegradable, and it can be reused or recycled.

→ Constructing a timber frame building

The load bearing structure of this 4-storey timber frame house (**image 1**) is the woodwork. It is being clad with a masonry skin, which does not come into contact with the timber framework; there is a cavity between them.

The structure of the building is manufactured as flat panels off-site in a factory. The pine struts for the roof and attic trusses are cut to length in preparation (**image 2**). They are assembled according to the requirements of the drawing, and nail plates are placed on either side of the joint (**image 3**). Each joint is placed in a hydraulic press, which force the nail plates into the timber (**images 4** and **5**). Once the joints are assembled they are pressed simultaneously. Each piece has been strength tested to verify its structural integrity and thus its location in the framework (**image 6**).

The wall panels are nailed together using pneumatic guns (**images 7** and **8**). Once again, all of the wood is cut to length and packaged as a bundle for assembly. The assembly process is computer-controlled and adjusts the bed and clamps to fit each part. The operator and machine work together nailing the joints.

The strength of these panels is achieved with a combination of timber frame and an OSB skin, which is nailed onto 1 side (**image 9**). The skin contributes rigidity to the framework and maximizes its strength without adding too much weight.

Floor cassettes are assembled using lightweight I-beams, which are a composite of LVL and OSB (**image 10**). They can span distances up to 7.5 m (24.6 ft). They are fully assembled in the factory and then dismantled into units for transportation.

The units are brought together on site and assembled very quickly (**image 11**). Each floor takes approximately 1 day to construct. Framing connectors are used to tie the I-beams together (**image 12**). They are also used to tie timber beams into masonry and steel framework.

8

9

10

11

12

Featured Manufacturer

Howarth Timber Engineering
www.howarth-te.com

Finishing Technology

Finishing Technology
Spray Painting

Spray painting is a fast and efficient means of applying adhesive, primer, paint, lacquer, oil, sealant, varnish and enamel. The surface determines the applicable finish and how long the process takes.

Costs	Typical Applications	Suitability
• No tooling costs, but may require jigs • Low to high unit costs, depending on size and paint	• Aerospace • Automotive and transportation • Consumer electronics and appliances	• One-off to mass production

Quality	Competing Processes	Speed
• Variable because it depends on the skill of the operator	• Hydro transfer printing • Powder coating • Vacuum metalizing	• Variable cycle time, depending on size and drying or curing time

INTRODUCTION

Spray coating is the application of liquid borne materials onto a surface. The sprayed material generally has 1 or more of the following functions: filler, primer, colour, decoration and protection.

High gloss, intense and colourful finishes are produced by a combination of meticulous surface preparation, basecoat and topcoat. The role of the base coat is to supply a monotone backdrop for the high gloss topcoat. The topcoat is clear and contains platelets or flakes of colour. As it is applied to the basecoat, the platelets or flakes are

propelled onto the surface of the base coat. This produces a topcoat that is multilayered: it is rich with colour near the basecoat and almost clear on top. This promotes a glossy, rich and intense colour finish.

The majority of spray painting is manually operated. However, there are sufficiently high volumes in the automotive, consumer electronics and appliance industries to justify robotic spraying systems.

TYPICAL APPLICATIONS

Spray painting is used in a vast range of applications including prototyping, repairs, low volume and mass production. It is used in the automotive industry for painting metalwork, and in the consumer electronics industry to colour plastic injection moldings.

RELATED PROCESSES

Powder coating is a dry coating technique (page 356) with a finish similar to 2-pack thermosetting paints (see technical description, page 353). It produces a uniform and glossy coating. Some techniques are electrostatic and so attract the coating particles to the surface of the workpiece. Combined with collection and recycling of the dry powder, up to 95% material utilization can be achieved with powder coating.

Hydro transfer printing (page 408) can produce effects that were previously spray painted with airbrushes and masking. It is a dipping process and reproduces patterns and print much more rapidly than spray painting.

Vacuum metalizing (page 372) is essentially spray painting with pure

aluminium. The aluminium is locked onto the surface of the workpiece by a basecoat and topcoat that are sprayed. It produces a highly reflective finish that can be tinted or coloured.

QUALITY

The quality of surface finish in this process depends on the skill of the operator. Spray coatings are built up in thin layers, typically between 5 and 100 microns (0.0002–0.004 in.) thick, with the exception of high-build systems such as combined filler-primer. It is essential that the surface is prepared to a high finish.

Above
These swatches are a range of Kandy colours over different basecoats. Each basecoat produces a new range of slightly different colours. Colours can also be matched to Pantone or RAL reference charts.

The level of sheen on the coating is categorized as matt or egg shell, semi-gloss, satin or silk and gloss.

The surface finish may not be critical if the role of the coating is to protect the surface. Protective paints form a barrier between the workpiece and atmosphere, to prevent rusting, for example.

Spray Coating Process

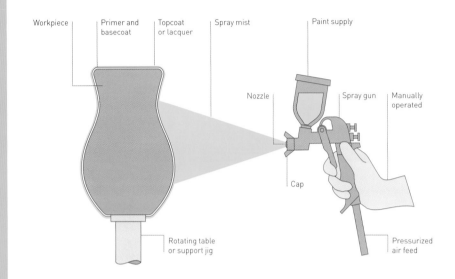

Workpiece — Primer and basecoat — Topcoat or lacquer — Spray mist — Paint supply

Nozzle — Spray gun — Manually operated

Cap

Rotating table or support jig

Pressurized air feed

TECHNICAL DESCRIPTION
Spray guns are gravity fed, suction fed or pressure fed. The diagram illustrates a gravity-fed gun. They all use a jet of compressed air to atomize the paint into a fine mist. The trigger controls the valve through which the pressurized air flows. The atomized paint is blown out of the nozzle and makes a cone shape. Pressurized air is blown out of the cap onto the cone, which forces it into an elliptical shape. This gives the operator more control because the shape can be adjusted to suit the application and required film thickness. The coating is applied onto the surface in overlapping strokes.

Suction feed guns have the paint pot on the underneath instead of on top. The paint is drawn into the jet of air by suction. Pressure feed systems are supplied by a paint pot that is connected to the spray gun via a flexible pipe. The paint is pressurized to force it into the spray gun.

However, paint is not integral to the surface and so is prone to peeling and flaking. This is often the result of pinholes or porosity in the coating, which allows the surface of the workpiece to corrode underneath.

Some aesthetic problems include 'pebbling', 'orange peeling', runs and sags. Pebbling is caused by a coating that is too dry; the thinner evaporates before it has had a chance to settle and smooth. Water borne paints are prone to orange peeling. As the name suggests, it is when the surface resembles orange peel. It is caused by the coating not flowing on the surface, or improper paint thinning. Runs and sags occur when there is too much wet coating material.

DESIGN OPPORTUNITIES

There is an almost unlimited range of colours and finishes for paints. Standard colour ranges include RAL and Pantone.

Colour is supplied by pigments, which are solid particles of coloured material. They can be replaced or enhanced by platelets of metallic, pearlescent, dichroic, thermochromic or photoluminescent materials.

Entire surfaces can be painted, including different materials, to produce a seamless finish. Otherwise, masks and templates can be used to create different colours, patterns, logos and text. Masks and templates can be made from tape, paper or cardboard for example.

There are many different techniques used to apply decorative effects. For example, overlaying colours and tones using different spraying techniques will produce graduation, shading and mottling. Marbling is achieved by draping materials over a wet topcoat, which drags and smears it over the basecoat. Covering it with a clear topcoat seals in the pattern with dramatic effect (see image, page 350).

A flock finish can be achieved by spraying adhesive followed by fibres of cotton, paper, silk or similar material.

There is even a varnish employed to produce a cracked topcoat. Crackle varnish is painted over a colour. Onto this is applied a contrasting colour, typically gold on top of black. The crackle varnish allows the topcoat to crack but not peel as it dries and contracts. The effect is similar to antique gold leaf.

DESIGN CONSIDERATIONS

The opportunities for paint effects and quality are heavily dependent on the skill of the operator. With the correct preparation and basecoat almost any combination of materials can be coated.

An important consideration is that spraying applies coatings in line-of-sight, so deep undercuts and recesses are more difficult to coat evenly. Electrostatic techniques draw more spray to the surface, but there is always overspray.

There is no size restriction. If the part will not fit in a spray booth or cannot be moved then it can be sprayed on site. For example, very large aeroplanes are

Conventional spray guns operate at approximately 3.45 bar (50 psi). This depends on the type and viscosity of paint.

Painted surfaces are nearly always made up of more than 1 layer. If necessary the surface is prepared with filler and primer. The primer may provide a surface onto which paint will adhere such as an acidic etch primer used to make a good bond with metallic surfaces. Some topcoats require a basecoat to provide a coloured backdrop. Opaque topcoats can be applied directly onto the primer.

Paints are made up of pigment, binder, thinner and additives. The role of the binder is to bond the pigment to the surface being coated. It determines the durability, finish, speed of drying and resistance to abrasion. These mixtures are dissolved or dispersed in either water (water borne) or a solvent (solvent borne).

The ingredients of varnish, lacquer and paint are essentially the same. Lacquers and varnishes can be pigmented, tinted or clear.

Another type of paint, 2-pack, has emerged. They are so called because they are made up of 2 parts, the resin and the catalyst or hardener. They are thermosetting and bond to the surface in a 1-way reaction. They are also hazardous to use because they contain isocyanates.

Water-borne paints are made up of pigment and a binder of acrylic emulsion, vinyl emulsion or polyurethane, dissolved or dispersed in water. They are applied to the surface and dry as the water evaporates and the binder adheres to the workpiece. They are flexible and so are suitable for painting wood and other materials that are likely to expand and contract in their lifetime.

Solvent-borne paints (also known as alkyd and oil-based) are made up of pigment and a binder of alkyd resin dissolved in thinners. Solvent-based paints are slow drying and off-gas polluting and harmful volatile organic compounds (VOCs).

Additives provide specific qualities such as quick drying, antifouling, anti-mould or antimicrobial properties.

There are 2 different types of enamel painting. The first is similar to the above, except that it is a mixture of paint with a high level of lacquer, which produces a high gloss finish. The second is applying a coating of glass, which is bonded onto the surface of the workpiece at high temperatures (above the melting point of the glass). This limits the finish to materials with a higher melting point ceramics and metal, for instance.

sprayed in booths that are modified aircraft hangers.

COMPATIBLE MATERIALS
Almost all materials can be coated with paint, varnish and lacquer. Some surfaces have to be coated with an intermediate layer, which is compatible with both the workpiece and topcoat.

Enamel paints that contain glass are only suitable for high-melting-point materials such as ceramics and metal.

COSTS
Tooling is not required. However, jigs or framework may be necessary to support the workpiece during spraying.

Cycle time is rapid but depends on the size, complexity, number of coats and drying time. A small consumer electronic product may be complete within 2 hours; in contrast, a car may take several days. Automated spraying techniques are very rapid. The parts are sprayed as the product is assembled to avoid masking.

All surface coatings are left for at least 12 hours to harden fully. Water borne paints and varnishes will dry in 2–4 hours. Solvent borne products will take roughly 4–6 hours.

Labour costs are typically high because these tend to be manual processes, and the skill of the operator will determine the quality of the finish.

ENVIRONMENTAL IMPACTS
Solvent borne paints contain volatile organic compounds (VOC). Since the 1980s automotive manufacturers have been making the switch to water-based paint systems to reduce their VOC emissions. It was not until the late 1990s that water-based systems were capable of producing a finish that equalled the previous solvent-based technologies.

Water-based paints are less toxic and easily cleaned with water.

Spraying is usually carried out in a booth or cabinet to allow the paints to be recycled and disposed of safely.

One system uses running water, which catches the overspray and transfers it to a central well where it settles and separates ready for treatment.

→ Spray painting a Pioneer 300 light aircraft

This is the Pioneer 300 (**image 1**). The kit components are manufactured by Alpi Aviation, Italy (**image 2**). The fuselage is assembled prior to spray painting. The joints between components are filled and rubbed down to make a smooth surface for the primer. Masking is used to protect the glass hood and areas of the bodywork (**image 3**).

A 2-pack polyurethane paint is used to prime the surface. All paint sprayed in this way has to be filtered to make sure that there are no lumps, dust or other contamination that might block the spray nozzle (**image 4**). The first coat is sprayed on in overlapping strokes (**image 5**). Each stroke is overlapped by about 50% to ensure a uniform coating.

The whole process is carried out in a spray booth that cleans and circulates the air continuously. It takes only a few minutes for the giant fans to remove and replace all the air in the booth.

In total 3 coats are applied and left to dry (**image 6**). The length of the drying tie depends on the cure system. Once the primer is sufficiently dry, the fuselage is rubbed down with abrasive paper (**image 7**). Painting several coats and then abrading the surface produces a much smoother finish. More coats of primer are required if the finish is not good enough at this stage because the topcoat will show up every imperfection.

In preparation for the topcoat the top layer of masking tape is removed (**image 8**). Underneath is a second layer of masking, which marks where the topcoat will be sprayed up to. The edges are staggered in this way to avoid a build up of paint; otherwise the primer would produce a visible white line along the masked edge.

The topcoat is applied (**image 9**) to each component. The 'racing' stripe is masked and the second topcoat colour is applied (**images 10** and **11**). The finished fuselage is assembled (**image 12**), ready for the wings.

1

2

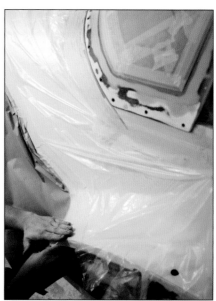

3

4

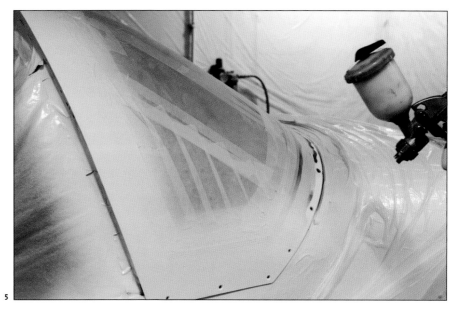

5

6

7

8

9

10

11

12

Featured Manufacturer

Hydrographics
www.hydro-graphics.co.uk

Finishing Technology

Powder Coating

This dry finishing process is used to coat a range of metalwork by either spray or a fluidized bed. The powder adheres to the workpiece electrostatically and is cured in an oven to produce a glossy protective coating.

Costs	Typical Applications	Suitability
• No tooling costs • Low unit cost	• Automotive • Construction • White goods	• One-off to mass production

Quality	Related Processes	Speed
• High quality, gloss and uniform	• Dip molding (as a coating) • Galvanizing • Spray painting	• Rapid application depending on size of part and level of automation • Curing takes 30 minutes or more

INTRODUCTION

Powder coating is primarily used to protect metalwork from corrosion and damage: the polymer forms a durable skin on the surface of the metal. Because it is a polymer there are many associated benefits such as a range of vivid colours and protective qualities that are engineered to suit the application. Furthermore, it is a dry process that has low environmental impacts.

Powder coating was developed in the 1960s for in-line coating aluminium extrusions (page 360) such as window frames. Architects and designers enjoyed the possibilities of powder coating and so demand increased for powder coating on other metals, including steel. Currently, aluminium and steel are the most widely powder coated materials, but many other materials are being explored that could benefit from a powder coated finish, including plastics and wood composites.

The powder can be applied as an electrostatic spray or in a fluidized bed.

The most common method of application is electrostatic spraying because it is more versatile. Fluidized bed is limited by the size of the tank and colour batches, and is therefore most suitable for mass production applications such as wire rack shelves in fridges and automotive parts. It is more commonly used for thermoplastic powder coatings.

TYPICAL APPLICATIONS

Powder coating is suitable for both functional and decorative applications. Functional applications include products for abrasive, outdoor and high-temperature environments such as automotive, construction and agricultural parts. Powder coating is also used in white goods that demand good temperature and chemical resistance. Other than that, it is used a great deal for parts of indoor and outdoor furniture, domestic and office products.

RELATED PROCESSES

Powder coating is often used in conjunction with other finishing processes for particularly demanding or outdoor applications. For example, steelwork must be galvanized (page 368) prior to powder coating for a durable and long-term finish. Without galvanizing, the steel would rust beneath the powder coating, causing it to blister and flake.

Alternatives to powder coating include wet spray painting methods (page 350), especially 2-part thermosetting paints which are extremely durable. These techniques are more suitable if a variety of materials are being coated, or if the materials cannot withstand the high temperatures in the powder coating baking process.

QUALITY

The polymer coating is baked onto the workpiece in an oven, which produces a glossy, durable and tough finish. By its very nature, powder coating produces

Electrostatic Spraying Process

Fluidized Bed Powder Coating Process

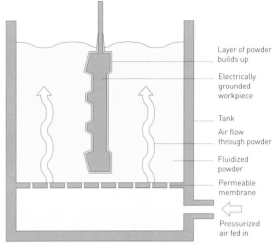

Charged spray nozzle (-)

Spray guns

Voltage

Powder feed

Plume of electrostatically charged powder

Electrically grounded metal workpiece

Layer of powder builds up

Layer of powder builds up

Electrically grounded workpiece

Tank

Air flow through powder

Fluidized powder

Permeable membrane

Pressurized air fed in

a uniform finish. This is because the airborne or fluidized powder is drawn to and adheres to the surface of the workpiece electrostatically. There is greater electrical potential difference where the coating is thinnest, which encourages the powder to build up in a uniform manner.

Thermoset coatings are the most commonly used and have similar characteristics to 2-part epoxy, or polyester paints. The baking process forms cross-links in the polymer that increase hardness and resistance to acids and chemicals. In contrast, thermoplastic coatings do not cross-link when heated. Instead, the dry powder coating melts and flows over the surface as it warms up. Cooling re-solidifies the polymer to produce a smooth and even coating.

DESIGN OPPORTUNITIES

These processes are used to provide a layer of protection, often with the addition of colour. There are many associated benefits, especially for application in children's play areas or public furniture, for example. The range of colours is unlimited and even metallic, speckled, textured and wood grain finishes are possible.

As well as pigments, many other additives can be mixed into the powder. Additives are used to improve the

TECHNICAL DESCRIPTIONS

ELECTROSTATIC SPRAYING

Powder coating materials are made by blending together the ingredients; resin, pigment, filler and binder. The mixture is then ground into a fine powder so that each particle contains the necessary ingredients for application.

The most common powder coating technique is electrostatic spraying. In this process the powder is suction fed or pressure fed to the spray gun. Air pressure forces it through the spray nozzle, which applies a high-voltage negative charge to each particle. This negative charge creates electrical potential difference between the particle and electrically grounded workpiece. The electrostatic force draws the plume of powder towards the workpiece, causing it to wrap around and lightly coat the reverse side. The negatively charged polymer powder adheres to the workpiece with static energy. All parts of the workpiece must be exposed to the stream of powder to ensure a uniform coating.

The whole process takes place in a spray booth. The parts are generally fed into and out of the booth on a conveyor belt, which provides the electrical grounding. After spraying, the workpiece is baked in an oven at approximately 200°C (392°F) for 30 minutes or so.

FLUIDIZED BED POWDER COATING

The fluidized bed technique is used to apply plastic coatings to parts both with and without the addition of electrostatic energy.

The fluidized powder is made up of the same ingredients as those used in electrostatic spraying. In fluidized bed coating, the powder is contained in a tank. Air is pumped through it to create a powder–air mixture that is dynamic, similar to a liquid. The part is dipped into the fluidized powder, which adheres to its surface. For thermosets, the powder has to be cured after dipping in the same way as electrostatic spraying.

It is frequently used to apply thermoplastic coatings. The workpiece is pre-heated to just above the melting temperature of the thermoplastic and dipped into the fluidized powder, which adheres to the surface of the workpiece on contact. In this way, thick coatings can be applied in a single dip or with multiple dips. Thicker coatings provide a greater level of protection as well as smoothing over rough surfaces and joints.

This technique, without the addition of electrostatic energy, improves the coating of wire racks and other parts that cause problems during electrostatic spraying due to the 'Faraday cage' effect.

→ Electrostatic spraying a gate

In this case study, Medway Galvanising are powder coating some previously galvanized steelwork. Preparation is key to the success of the finish. First of all, the parts have been galvanized (page 368) to protect the base metal, which is then sanded to produce an even higher quality finish (**image 1**).

The cleaning process consists of a series of 10 baths that progressively clean, degrease and prepare the surface for powder coating. A bath of zinc phosphate (**image 2**) provides an intermediate binding layer between the clean galvanized surface and the thermosetting powder.

The metalwork, in this case part of a steel gate, is loaded onto an electrically grounded conveyor belt that delivers it into the powder-coating booth (**image 3**). This process is manually operated, due to the level of versatility that is required in the factory. The operator is protected as he sprays a plume of electrostatically charged powder over the metalwork (**image 4**).

The powder coated finish is very delicate at this point, so handling is kept to a minimum. The part is loaded onto a conveyor that takes it through a camelback oven (**image 5**). The temperature inside the oven is 200°C (392°F). The part is heated as it is conveyed up and over the heating compartment, hence the name. At this temperature the thermosetting polymer chemically reacts, forming cross-links between the molecular strands to produce a very durable and tough coating. It is a gradual process and lasts 30 minutes or so.

The cured part has a rich-looking red plastic coating (**image 6**). It is still very warm and is left to air dry for a short period while the thermosetting plastic fully hardens.

1

coating's UV stability, chemical resistance, temperature resistance and durability, for example. Antimicrobial additives can also be incorporated, which inhibit the growth of mould, mildew and bacteria on the surface of the coating.

The choice of thermoplastic and thermosetting powders provides designers and manufactures with a large scope of opportunity. The coating can be engineered with additives, binders and pigments to suit a specific application if the volumes justify it. Otherwise, there is a vast range of polymer characteristics that can be utilized in standard powder-coated finishes.

There is very little risk of powder coatings running into sags in either technique, even when a heavy build up is required. The thickest coatings can be achieved with fluidized bed techniques, which can be built up by dipping a hot workpiece several times in thermoplastic powder. Electrostatic spraying can also produce thick coatings, but it is more time consuming and less practical. Thin films can be very difficult to achieve, especially with electrostatic methods and thermosetting powders.

DESIGN CONSIDERATIONS

Part design will affect the choice of technique. Each technique is generally suited to different materials: thermoplastic powders are coated using the fluidized bed technique and thermosets are electrostatically sprayed. But this is not always the case.

Electrostatic spraying is a versatile and widely used process. Most parts can be coated in this way. Fluidized bed coating offers improved coverage on complex and intricate geometries, especially if they will cause the 'Faraday cage' effect in electrostatic spraying.

Surface finish is determined by the quality of the finish prior to coating, unless a very heavy coating is produced in a fluidized bed. Castings, for example, will produce a textured finish if they are not prepared effectively. Preparation consists of a number of cleaning and etching baths, which ready the surface for powder coating. These stages are essential to produce a sufficient bond between workpiece and coating to ensure its longevity.

COMPATIBLE MATERIALS

Many metals can be coated in this way. However, the majority of powder coating is used to protect and colour aluminium and steelwork. Technologies are emerging that make it possible to powder coat certain plastics and composite wood panels, although they are new and relatively specialist. It is also possible to powder coat glass by the fluidized bed method.

The coating materials include thermoplastics and thermosets. Typical thermosets include epoxy, polyester, acrylic and hybrids of these polymers. They are characterized by good resistance to chemicals and abrasion, and their

2

3

4

5

6

hardness and durability. Certain grades
offer added protection against UV.

Thermoplastic powders include
polyethylene (PE), polypropylene (PP),
polyamide (PA), polyvinyl chloride (PVC),
fluoropolymers and many more. These
coatings make up a small percentage
of the powder coating market. They
can be built up in thick layers using the
fluidized bed method to provide superior
performance characteristics.

COSTS

There are no tooling costs, although
equipment costs are relatively high.

One coat is usually sufficient, so cycle
time is very rapid. The baking process to
cure the resin adds approximately half
an hour to the cycle.

This is a simple process: powder is
easy to spray and the potential electrical
difference actively encourages a uniform
coating to develop. Therefore, labour
costs can be quite low.

ENVIRONMENTAL IMPACTS

Powder coating is an efficient use of
materials and produces less waste than
wet spray painting methods. This is
partly due to electrostatically charging
the particles, but also because it is
possible to collect and filter powder
overspray. Production lines powder
coating with continuous colour can
achieve over 95% powder utilization.

Everything required for the coating
is contained in each particle: resin,
pigment, filler and binder. Therefore it is
not necessary to suspend the powder in a
solvent or water, which can be harmful to
the operator and environment.

Featured Manufacturer

Medway Galvanising Company
www.medgalv.co.uk

Finishing Technology
Anodizing

The surface of aluminium, magnesium and titanium can be anodized to form a protective oxide layer. It is naturally light grey, but can be electrolytically coloured, or dyed, with a range of vivid colours including red, green, blue, gold, bronze and black.

Costs	Typical Applications	Suitability
• Tooling is not usually necessary • Low unit costs, but increasing with film thickness	• Architectural • Automotive • Consumer electronics	• One-off to high volumes

Quality	Related Processes	Speed
• High quality, lightweight and very hard	• Powder coating • Spray painting	• Moderate cycle time (approximately 6 hours)

INTRODUCTION

Anodizing refers to a group of processes that are used to treat the surface of metals. The workpiece is made the anode and submerged in an electrolytic solution. The process builds up the naturally occurring oxide layer on the surface of the metal. The film is hard, protective and self-healing; aluminium oxide is inert and among the hardest materials known to man.

There are 3 main methods, which are natural anodizing, hard anodizing and chromic acid anodizing. Most architectural, automotive and general anodizing is carried out in sulphuric acid using the natural or hard anodizing methods. Chromic acid anodizing is a more specialized process.

Natural anodizing produces finishes 5–35 microns (0.00019– 0.0014 in.) thick and grey in colour. The finish can be coloured to range of vivid shades, such as the Bang & Olufsen BeoLab 4000 speakers (see image, opposite).

Hard anodizing produces films up to 50 microns (0.0020 in.) thick. It is used for more demanding applications because the thicker film improves wear and temperature resistance.

TYPICAL APPLICATIONS

Anodizing is used to protect and enhance metal for both indoor and outdoor use. Indeed, most aluminium in the automotive, construction, leisure and consumer electronics industries is treated this way.

Well known examples include the Maglite , Apple iPod and G5 Powermac. Other products include karabiners and general climbing equipment, televisions, telephones, appliances, control panels, picture frames, cosmetic packaging, shop fronts and structural products.

RELATED PROCESSES

Painting (page 350) and powder coating (page 356) add a layer of material to the surface of the metal. The advantages of these processes include building up thick layers, a wide colour range and how easily they can be repaired.

Aluminium, magnesium and titanium are relatively expensive metals and are selected for their superior strength to weight ratio. Anodizing improves the material's natural resistance to weathering without significantly increasing weight, and is therefore the most popular surface technology.

Chemically colouring stainless steel is a similar process: the naturally occurring passive film on the surface of the metal is enhanced.

QUALITY

Anodizing is unmatched for surface treating aluminium, magnesium and titanium. It is light, very hard, self-healing and resistant to weathering. The anodic film is integral to the underlying metal and so will not flake or peel like some coating processes. It has the same melting point as the base metal and

Anodizing Process

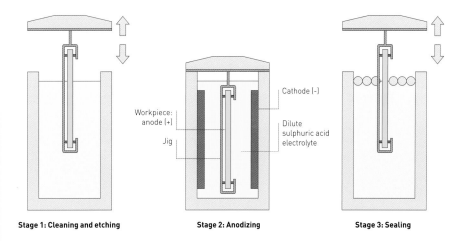

Stage 1: Cleaning and etching

Stage 2: Anodizing

Workpiece: anode (+)

Jig

Cathode (-)

Dilute sulphuric acid electrolyte

Stage 3: Sealing

TECHNICAL DESCRIPTION

Anodizing has 3 main stages: cleaning and etching, anodizing and sealing.

In stage 1, the cleaning and etching baths prepare the workpiece for anodizing. The surface is either etched or brightened in chemical baths. Etching produces a matt or dull finish and minimizes preparatory operations. Brightening (chemical polishing) produces a very high gloss surface suitable for decorative applications. Alkali and acid baths are used in succession to neutralize each other.

In stage 2, the anodizing takes place in an electrolytic solution, which is generally dilute sulphuric acid. A current is passed between the workpiece (anode) and electrode (cathode). This causes oxygen to gather on the surface of the part, which reacts with the base metal to form a porous oxide layer (aluminium produces aluminium oxide). The length of time in the bath, the temperature and the current determine the rate of film growth. It takes approximately

15 minutes to produce 5 microns (0.00019 in.) of anodic film.

Anodizing adds an anodic film to the surface of the metal. In doing so, a small amount of base material is consumed (roughly half the thickness of the anodic film). This will affect the roughness of the surface. Only thin coatings will retain a high gloss finish and these are therefore not suitable for high-wear applications.

Colour is applied in 3 main ways. The Anolok™ system, recommended for outdoor applications, is an electrolytic colouring process. A range of colours are created by depositing cobalt metal salts in the porous anodized surface prior to sealing. The

colour is produced by light interference. An alternative technique uses tin instead of cobalt. This is popular, but the colour is less resistant to UV light than the Anolok™ system. The third technique is known as 'dip and dye'. In this, the colour is created by simply dying the anodic film prior to sealing. This process can produce the widest range of colours, but it is the least UV stable and consistent, so is generally only used for decorative and indoor applications.

In stage 3, the surface of the porous film is sealed in a hot water bath. Sealing the film produces the hardwearing and weather-resistant characteristics that are associated with anodizing.

certain colour systems are guaranteed for up to 30 years.

The appearance of the finished piece is determined by the quality of the mill finish prior to anodizing. It is important that the material is from the same batch if it is used together because different batches will have varying colour effects. It is not possible to colour match to specific swatches such as Pantone.

DESIGN OPPORTUNITIES

Anodizing has many benefits including hardness, ease of maintenance, colour stability, durability, heat resistance and corrosion resistance, and it is non-toxic.

Anodic films have very high dielectric strength, so electrical components can be mounted onto them.

Anodizing can be applied to all types of finishes and textures, including satin, brushed, embossed and mirror polished.

The anodic film can be selectively removed by laser engraving (page 248) or photo etching (page 292). This is used to create patterns, text or logos by contrasting the colour of the anodized surface with that of the base metal.

DESIGN CONSIDERATIONS

Anodizing adds a film thickness to the dimensions of the product, but the

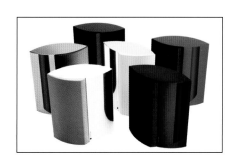

Above
The vivid colours of these Bang & Olufsen BeoLab 4000 speakers are produced by anodizing aluminium.

→ Anodizing automotive trims

Heywood Metal Finishers use the Anolok™ colouring system. This technique is used to anodize parts for exterior and demanding applications such as in the construction and automotive industries. It is distinguished from other processes by the colouring phase: cobalt metal salts are deposited in the porous film during colouring. The colours (grey, bronze, gold and black) are produced by light interference and so are more durable and less prone to fading than other methods.

In this case, aluminium extrusions are being anodized for automotive application (**image 1**). The lengths are loaded onto an adjustable jig, which is set up to accommodate different lengths of metal (**image 2**).

The parts are mounted at a slight incline to allow the chemicals to drain off during dipping (**image 3**). There are 10 baths in the anodizing cycle and the whole process can take up to 6 hours (**images 4** and **5**).

At the other end of the plant some aluminium parts are being anodized. They are dipped into the sulphuric acid electrolyte for between 15 and 60 minutes (**image 6**). Longer dip times produce thicker films, which are generally more

hardwearing. After this, they are sealed in hot water with additives at a temperature of 98°C (208.4°F).

The anodic film is non-toxic and can be handled immediately after sealing. The parts are removed from the jig and packed (**image 7**).

etching and cleaning process usually compensates for this as it removes about the same amount of material. However, this will depend on the anodizing system and hardness of the material because softer material will etch more rapidly. Therefore, it is essential that the anodizing company is consulted for parts that have critical dimensions.

The colouring process will determine the UV stability of the part. The Anolok™ colouring system, which applies cobalt metal salts electrolytically, is guaranteed for up to 30 years in architectural applications, but the colour range is limited to grey, bronze, gold and black.

Variation in the chemical composition of the metal affects the colour. This is a concern both for welded joints and for fabrications containing metal from different batches.

This is a dipping process and so liquids have to be able to drain from the parts. The maximum size of part that can be anodized is limited by the capacity of the baths, and fabrications can be up to 7 x 2 x 0.5 m (23 x 6.6 x 1.6 ft).

COMPATIBLE MATERIALS

Aluminium, magnesium and titanium can be anodized.

COSTS

There are no tooling costs, but jigs may have to be made up to accommodate the parts in the anodizing baths. Cycle time is approximately 6 hours.

This process is generally automated, so labour costs are minimal.

4

5

6

ENVIRONMENTAL IMPACTS

Waste from anodizing production
is non-hazardous. Companies are
monitored closely to ensure they do not
allow contamination into wastewater
or landfill. Although acidic chemicals are
used in the anodizing process there are
no hazardous by-products.

The baths are continuously filtered
and recycled. Dissolved aluminium
is filtered from the rinse tanks as
aluminium hydroxide, which can be
safely recycled or disposed of.

Anodized surfaces are non-toxic.

7

Featured Manufacturer

Heywood Metal Finishers
www.hmfltd.co.uk

Finishing Technology

Electroplating

This is an electrolytic process used to apply a thin film of metal to another metal surface. A strong metallurgical bond is formed between the base material and coating. Electroplating produces a functional and durable finish.

Costs	Typical Applications	Suitability
• No tooling costs, although jigs may be required to support parts • High unit costs, determined by materials	• Consumer electronics • Furniture and automotive • Jewelry and silversmithing	• One-off to high volume production

Quality	Related Processes	Speed
• Coating material is chosen for specific qualities such as brightness or resistance to corrosion	• Galvanizing • Spray painting • Vacuum metalizing	• Moderate cycle time, depending on type of material and thickness of coating

INTRODUCTION

Electroplating is used to produce functional and decorative finishes on metals. Thin layers of metal, from less than 1 micron (0.000039 in.) up to 25 microns (0.0098 in) thick, are deposited on the surface of the workpiece in an electrochemical process.

Electroplated metals benefit from a combination of the properties of the 2 materials. For example, silver-plated brass combines the strength and reduced cost of brass with the long lasting lustre of silver.

It is possible to plate certain plastics in an electrochemical process. However, this is not strictly electroplating. It is a slightly different process because the finish can only be achieved by coating the plastic with an electroless intermediate layer. This provides a stronger base for the desired material to be plated onto, but it is difficult to produce a long lasting finish because there is no metallurgical bond between the coating and substrate.

TYPICAL APPLICATIONS

This process is used a great deal by jewellers and silversmiths, who can form parts in less expensive materials that have suitable mechanical properties, then coat them with silver or gold to give a bright, inert and tarnish-free surface finish. Examples include rings, watches and bracelets. Tableware includes beakers, goblets, plates and trays.

Trophies, medals, awards and other pieces that will not come into prolonged human contact can be plated with rhodium or nickel for an even longer lasting and brighter finish.

Examples of electroplated plastic include automotive parts (gear sticks, door furniture and buttons), bathroom fittings, cosmetic packaging and trims on mobile phones, cameras and MP3 players.

Electroplating has many important functional roles, such as improved levels of hygiene, ease of joining (such as silver and gold soldering) and improved thermal and electrical conductivity. Gold is used in critical applications to improve conductivity and ensure a tarnish free surface finish.

RELATED PROCESSES

Electroplating is the most reliable, repeatable and controllable method of metal plating.

There are many other techniques used to coat materials with metal, including spray painting (page 350) with conductive paints, galvanizing (page 368) and vacuum metalizing (page 372).

Spray painting technology has been improving steadily. There are now paints with high metallic content. Finishes can be polished and buffed like solid metal. These processes rely on the polymer carrier within which the metal platelets are suspended. Like vacuum metalizing, these surface coatings do not form a metallurgical bond with the workpiece.

Electroplating Process

- Connected to power source (-)
- Connected to power source (+)
- Wire jig
- Electrically charged workpiece (-)
- Electroplated metal coating
- Electroplating tank
- Metal anodes
- Dissolved metal ions
- Electrolytic solution

QUALITY

Electroplated films are made up of pure metal or alloys. An integral layer is formed between the workpiece and metal coating because each metal ion forms a strong metallurgical bond with its neighbour.

Metals that will not plate to each other can be joined with intermediate layers that are compatible with both the coating and base material. For example, brass will affect the strength and corrosion resistance of a gold electroplating. Therefore, a nickel intermediate layer is used as a barrier and provides a strong inter metallic bond and protection to the brass.

The quality of surface finish is largely dependent on the surface finish of the workpiece prior to electroplating. The metal coating is too thin to cover scratches and other imperfections.

DESIGN OPPORTUNITIES

The main benefit of this process is that it can produce the look, feel and benefits of 1 metal on the surface of another, allowing parts to be formed in materials that are less expensive or have suitable properties for the application. Electroplating then provides them with a metal skin that possesses all the desirable aesthetic qualities.

The most common electroplating materials include tin, chrome, copper, nickel, silver, gold and rhodium. Each of these materials has their own particular properties and benefits.

Rhodium is a member of the platinum family and is very expensive. It has a long-lasting lustre that will not tarnish easily in normal atmospheric conditions, and

TECHNICAL DESCRIPTION

Electroplating is made up of 3 main stages, which are cleaning, electroplating and polishing. The parts are mounted onto a jig to support them through the electroplating process.

The cleaning stage consists of degreasing in a hot caustic solution. Then parts are immersed in a dilute cyanide solution, to remove surface oxidization, which is neutralized in sulphuric acid.

Electroplating occurs in an electrolytic solution of the plating metal held in suspension in ionic form. When the workpiece is submerged and connected to a DC current, a thin film of electroplating forms on its surface. The rate of deposition depends on the temperature and chemical content of the electrolyte.

As the thickness of electroplating builds up on the surface of the workpiece the ionic content of the electrolyte is replenished by dissolution of the metal anodes. The anodes are suspended in the electrolyte in a perforated container.

The thickness of electroplating depends on the application of the product and material. For example, nickel used as an intermediate levelling layer may be up 10 microns (0.00039 in.), while electroplated gold needs to be only 1 micron (0.000039 in.) thick for decorative applications.

After electroplating the parts are finely polished (buffed) in a process known as 'colouring over'.

is hard, highly reflective and resistant to most chemicals and acids. It is used for decorative applications that demand superior surface finish such as medals and trophies.

Gold is a unique precious metal with a vibrant yellow glow that will not oxidize and tarnish. The alloy content of the electrolyte bath will affect whether it has a rose or green tint. The purity of gold is measured in carat (ct), or karat (kt) in North America: 24 ct is pure gold;

18 ct is 75% gold by weight; 14 ct is 58.3% gold; 10 ct is 41.1% gold and 9 ct is 37.5% gold. Many national standards, including the US standard, allow 0.3% negative tolerance: the British standard does not.

Silver is less expensive than the previous metals. It is bright and highly reflective, but the surface will oxidize more readily and so has to be frequently polished or 'coloured over' to maintain its brightness. Silver's tendency to oxidize is used to emphasize details by 'blackening' it in a chemical solution and then polishing raised details back to bright silver. This technique is often used in jewelry to emphasize relief patterns.

Nickel and copper are often used as intermediate layers. They help to produce a very bright finish because they provide a certain amount of levelling. If they are built up in sufficient quantities they will cover small imperfections and produce a smooth layer to electroplate onto. As intermediate layers they provide a barrier between the electroplated metal and workpiece. This is especially useful for metals that either contaminate each other, or are not compatible.

Copper is an inexpensive material for electroplating, but tarnishes very quickly and so is rarely used as a topcoat.

DESIGN CONSIDERATIONS

Parts have to be connected with a DC current to be electroplated. This can be done in 2 ways: they are either 'loose wired' or mounted onto a rigid jig. Loose wiring is an effective way to hold parts for low volume electroplating. Jigs have a fixed point of contact, which is visible after electroplating. This is minimized by contacting the workpiece between 2 small points, or on an inconspicuous part of the product.

It is possible to mask areas with special waxes or paint so they are not electroplated. This will increase unit cost.

Chrome plating was used extensively in the automotive and furniture industries. It is still used a great deal but is steadily being replaced, due to the heavy metal content of the process. At present there is nothing to compete with the brightness and durability of chrome electroplate onto a nickel basecoat. Similar levels of reflectivity can be achieved with other metal combinations, but they may not be as durable.

COMPATIBLE MATERIALS

Most metals can be electroplated. However, metals combine with different levels of purity and efficiency.

1

2

The plastic that is most commonly electroplated is acrylonitrile butadiene styrene (ABS). It is able to withstand the 60°C (140°F) processing temperature, and it is possible to etch into the surface to form a relatively strong bond between it and the electroless plated metal.

COSTS

There are no tooling costs, but jigs may be required to hold the parts in the tank.

Cycle time depends on the rate of deposition, temperature and electroplated metal. It is approximately 25 microns (0.00098 in.) per hour for silver and up to 250 microns (0.0098 in.) per hour for nickel. The rate of deposition will affect the quality of the electroplating: slower processes tend to produce more precise coating thickness.

Labour costs are moderate to high depending on the application. For example, silverware and jewelry has to be finished to a very high level because appearance and durability are critical.

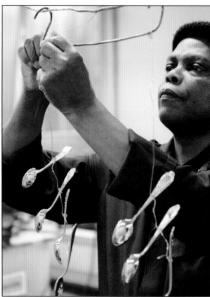

3

ENVIRONMENTAL IMPACTS

Many hazardous chemicals are used in all of the electroplating processes. These are carefully controlled with extraction and filtration to ensure minimal environmental impact.

Coating thickness is measured in microns, and these processes apply only the necessary amount of material.

Gold and silver are inert and suitable for all types of products including medical implants, beakers, bowls and jewelry. Nickel should not be used in contact with the skin because it is irritating and can be poisonous.

→ Silver electroplating nickel-silver cutlery

The quality of the finish is largely dependent on the smoothness of the finish prior to electroplating. These nickel-silver spoons are polished to a high gloss in preparation for silver electroplating (**images 1** and **2**).

They are being coated entirely with a thin layer of silver and so they are connected to the DC current with a loose fitting wire (**image 3**). The parts are agitated in the electroplating bath so they receive an even coating. If they were jigged on a frame there would be a small area left unplated.

To prepare the metal for electroplating it is immersed in a series of cleaning solutions, including a dilute cyanide solution, in which the surfaces of the spoons can be seen fizzing (**image 4**). All traces of contamination, such as polishing compound and grease, are removed.

In this case 25 microns (0.00098 in.) of silver is being electroplated onto the surface (**image 5**). This takes approximately an hour in the electroplating bath. The whole process is computer-controlled to ensure maximum precision and quality of surface finish.

The electroplated parts are cleaned and dried (**image 6**). At this point the finish is not a high gloss. A brightening agent is sometimes added to the electroplating bath to produce a more reflective finish. The surface is improved and finished with a very fine iron powder polishing compound, known as 'rouge'. This is applied in a buffing process known as 'colouring over' to produce the highly reflective finish (**image 7**).

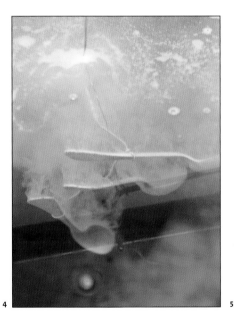

4

5

6

7

Featured Manufacturer

BJS Company
www.bjsco.com

Finishing Technology
Galvanizing

In this process steel and iron are hot dipped in molten zinc that alloys to its surface metallurgically and provides electrochemical protection against the elements. It produces a bright, distinctive pattern on the surface, which over time becomes dull and grey.

Costs	Typical Applications	Suitability
• Low cost (roughly doubles material cost)	• Architectural and bridges • Automotive • Furniture	• One-off to mass production

Quality	Competing Processes	Speed
• Excellent protection • Appearance affected by quality of steel	• Electroplating • Spray painting • Vacuum metalizing	• Rapid cycle time (typically complete within 10 minutes)

INTRODUCTION

Zinc and iron combine to produce a very effective alloy that increases the life of steelwork and ironwork substantially. Unprotected metalwork continuously corrodes and its structure is undermined, whereas galvanized metalwork is protected from the elements and so retains its structural integrity. Waterloo Station, for example, has recently had its Victorian steel roof re-galvanized for the first time (see image, opposite).

During galvanizing, zinc bonds with iron metallurgically to produce a layer of zinc-iron alloy coated by a layer of pure zinc. The coating is therefore integral to the base material; the intermediate alloy layer is very hard and can sometimes exceed the strength of the base material.

Galvanizing is carried out in 2 ways: hot dip or centrifugal galvanizing. They are essentially the same, except that in centrifugal galvanizing the parts are dipped in baskets and then spun after submerging in molten zinc. This removes excess zinc and produces a more uniform, even coating. This is particularly useful for threaded fasteners and other small parts that require accurate coating.

Galvanizing is resilient to aggressive handling and over the last 150 years has proven to provide a long-lasting, tough and low-maintenance coating. Galvanized steel can be recycled and so has an almost indefinite lifespan.

TYPICAL APPLICATIONS

Typical applications include architectural steelwork such as stairwells, walls, floors and bridges; agricultural hardware, automotive chassis and furniture.

RELATED PROCESSES

Other techniques used to coat materials with metal include electroplating (page 364), vacuum metalizing (page 372) and spray painting (page 350) with conductive paints. Unlike these processes, galvanizing is limited to coating steel and iron with zinc.

QUALITY

When the steelwork is removed from the galvanizing bath the zinc has a bright, clean finish. Over time and with exposure to atmospheric conditions this becomes dull and grey. It is tough and protects the base material against corrosion from oxygen, water and carbon dioxide. Levels of corrosion can vary from 0.1 microns (0.0000039 in.) per year for indoor uses to 4–8 microns (0.00015–0.00031 in.) per year for outdoor applications near the coast. A typical coating is between 50 and 150 microns (0.0020–0.0059 in.) thick depending on the application technique. The thickness of the coating is normally determined by the thickness of the base metal. The exceptions are centrifugal galvanized coatings, which produce a slightly thinner coating, or thicker coatings produced by roughening the surface of the part or adding silicon to the steel during production.

By its very nature the zinc coating protects the steel, even if it is penetrated.

Hot Dip Galvanizing Process

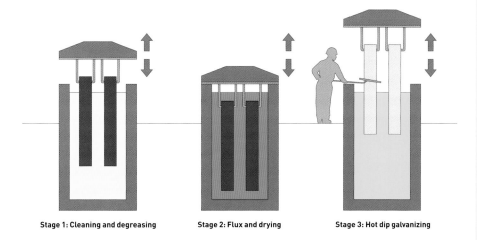

Stage 1: Cleaning and degreasing   Stage 2: Flux and drying   Stage 3: Hot dip galvanizing

The zinc will react with atmospheric elements more readily than iron and forms a deposit over the exposed area that protects the base material from further corrosion.

The look of the zinc coating is also affected by the quality of the steel. The final effect can appear bright and shiny to dull and grey; the latter is caused by steel with high silicon content.

DESIGN OPPORTUNITIES
This is a versatile process that can be used to protect small items such as nuts and bolts as small as 8mm (0.315 in.) diameter, to very large structures up to 12 m x 3 m (40 x 10 ft). Structures larger than this can be fabricated post-galvanizing. Complex and intricate shapes can be galvanized in a single operation, including hollow and open-sided vessels.

DESIGN CONSIDERATIONS
Galvanizing coats the entire surface of a part with zinc. High temperature tape, grease or paint can be used to mask areas. Certain hollow geometries can be galvanized only on the outside, but this requires special coating techniques.

The galvanizing bath is maintained at 450°C (840°F), so all parts must be able to withstand that temperature. Another important consideration is potentially explosive design elements that include sealed tubes and blind corners. All welding slag, grease and paint have to be removed pre-treatment.

COMPATIBLE MATERIALS
Because galvanizing relies on a metallurgical bond, only steel and iron can be coated in this way.

COSTS
The costs of this process are low, especially in the long term. No specific tooling is required. Cycle time is rapid.

Labour costs are moderate; the quality of finish is affected by the skill of the operators, amongst other factors.

ENVIRONMENTAL IMPACTS
This process can increase the life of steelwork to between 40 and 100 years. It is widely accepted that half of all new steel produced is used to replace corroded steel. In some countries this costs up to 4% of GDP. Galvanizing dramatically increases the longevity of steel fabrications, reducing their environmental impact.

The process uses zinc efficiently to protect the surface of steelwork. After each dip the unused zinc drains back into the galvanizing bath for reuse. Zinc can be recycled indefinitely without loss of any physical or chemical properties.

TECHNICAL DESCRIPTION
There are typically 6 baths in the hot dip galvanizing process. The first 4 are in the cleaning and degreasing stage of the process. The parts are dipped in hot caustic acid for degreasing. Next, the parts are dipped in 2 progressive hydrochloric acid pickling baths to remove all mill scale and rust. Lastly, they are washed at 80°C (176°F) in preparation for the zinc flux.

During stage 2 the metalwork is dipped in a flux of hot zinc ammonium chloride to condition the clean surface for galvanizing and ensure a good flow of zinc over the internal and external surfaces of the metalwork.

Finally, the metal work is immersed in a bath of molten zinc, which is maintained at 450°C (840°F). The zinc bonds with the iron metallurgically to create a zinc-iron alloy that is inherent to the surface of the metalwork. Strata of zinc-iron alloy layers of varying concentration are built up during the process, and the outer layer is typically pure zinc. The dipping process can last up to 10 minutes, depending on the depth of zinc coating required. The parts are gradually withdrawn from the zinc bath to allow excess zinc to drain off.

Above
The roof of Waterloo Station in London was galvanized nearly 100 years ago. It has not needed re-galvanizing until recently and will probably last another century before any further treatment.

→ Hot dip galvanizing steelwork

First of all the parts are cleaned and checked for any air or solution traps that may cause problems during the galvanizing process. The metalwork is then rigged onto beams at a 30° angle to facilitate draining (**images 1** and **2**). Drainage holes may have to be designed into the product to ensure that solution can drain off during the process.

The metalwork is moved through to the galvanizing plant where it is dipped in progressive cleaning and pickling baths (**image 3**). The preparation, cleaning and base material content determines the quality of the galvanized part, so this stage of the process is essential to ensure consistent levels of quality. The penultimate bath (**image 4**) contains a flux that conditions the surface in preparation for the hot dip galvanizing. The metalwork is removed from the flux steaming in preparation for galvanizing (**image 5**);

its temperature has by now been raised to 80°C (176°F).

The parts are then dipped into a bath of molten zinc at 450°C (840°F) (**image 6**). The zinc spits as it comes into contact with the cooler metal. During galvanizing the surface of the molten zinc is continuously skimmed to remove any contamination and flakes of metal that might affect the quality of the galvanized finish (**image 7**).

The parts are removed from the molten zinc slowly to allow excess zinc to drain back into the bath (**image 8**). The use of zinc is very efficient; a ratio of roughly 1:15 of zinc to metalwork is usual. The parts are removed from the bath and air dried or quenched, according to the customer requirements (**image 9**) and they are loaded for delivery (**image 10**).

1

2

3

4

5

6

7

8

9

10

Featured Manufacturer

Medway Galvanising Company
www.medgalv.co.uk

Finishing Technology
Vacuum Metalizing

Also known as physical vapour deposition (PVD) and sputtering, the vacuum metalizing process is used to coat a wide range of materials in metal to create the look and feel of anodized aluminium, chrome, gold, silver and other metals.

Costs	Typical Applications	Suitability
• No tooling costs, but jigs are required • Moderate unit costs	• Consumer products • Reflective coatings • RF, EMI and heat shielding	• One-off to mass production

Quality	Related Processes	Speed
• High quality, high gloss and protective finish with similar characteristics to spray painted coatings	• Electroplating • Galvanizing • Spray painting	• Moderate cycle time (6 hours including spray painting)

INTRODUCTION

This process combines very high vacuum and an electrical discharge that vaporizes almost pure metal (most commonly aluminium) in a vacuum deposition chamber. The plume of vaporized metal condenses onto surfaces, coating them with a high-gloss film of metal.

It is a means of coating many different materials, including plastic, glass and metal, with metal. There are no tooling costs, and the process is controllable and repeatable, making it suitable for coating everything from single prototypes to mass produced items. Prototypes and

models in suitable materials can be coated to give the look and feel of a metal part. In contrast, mass produced metal parts can be coated for added value.

Coating thickness depends on the application. Cosmetic finishes are typically less than 6 microns (0.00024 in.) thick, with a metal film of less than 1 micron (0.000039 in.). For functional coatings, thicknesses of 10 to 30 microns (0.00039–0.0012 in.) are produced using plasma vaporizing techniques, which can build up film thickness indefinitely.

TYPICAL APPLICATIONS

Vacuum metalizing is used equally for decorative and functional applications. Decorative uses include jewelry, sculptures, trophies, prototypes, kitchen utensils and architectural ironmongery.

Coatings can be functional, providing electromagnetic interference (EMI) or radio frequency (RF) shielding, improved wear resistance, heat deflection, light reflection, an electrically conductive surface or a vapour barrier. Some typical products are torch and automotive light reflectors, machine parts, metalized plastic films and consumer electronics.

RELATED PROCESSES

Other processes used to coat materials with metal include electroplating (page 364), galvanizing (page 368), and spray painting (page 350). Spray painting with conductive paints is also suitable for RF and EMI shielding. These processes are closely related; spray painting is used to apply a base coat pre-metalizing and seal in the delicate metal film with a topcoat. Spray painting and vacuum metalizing can coat the widest range of materials.

Vacuum Metalizing Process

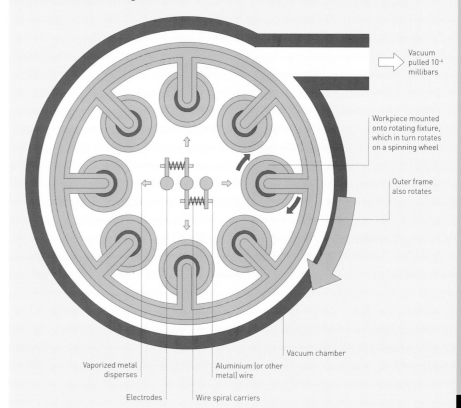

Vacuum pulled 10⁻⁴ millibars

Workpiece mounted onto rotating fixture, which in turn rotates on a spinning wheel

Outer frame also rotates

Vacuum chamber

Vaporized metal disperses

Aluminium (or other metal) wire

Electrodes

Wire spiral carriers

VACUUM METALIZING

373°

TECHNICAL DESCRIPTION

The parts are first cleaned and coated with a base coat by spray painting. The base coat has 2 main functions: improving surface finish and encouraging the metal vapour to adhere to the workpiece.

The workpieces are mounted onto rotating holding fixtures (custom made for each part), which are in turn rotated on spinning wheels. The assembly is suspended within a frame, which also rotates. All in all, the parts are being rotated around 3 parallel axes simultaneously. This is to ensure an even coating with line-of-sight geometry.

Before the vacuum metalizing can take place, a vacuum has to be generated within the metalizing chamber. To reach 10⁻⁴ millibars (0.0000145 psi) takes approximately 30 minutes, depending on the materials being coated. The metalizing process can operate at a lesser degree of vacuum, but the quality of the finish will be inferior.

When the correct pressure is reached, an electrical discharge is passed through the wire of aluminium (or other metal) by the electrodes. The combination of the electric current and high vacuum cause the almost pure metal to vaporize in an instant. It bursts into a plume of metal vapour, which condenses on the relatively cool surface of the workpiece. The condensing metal adheres to the base coat on the parts in a thin, uniform layer.

The vacuum metalized film is protected by the application of a topcoat. This is water-clear, but can be coloured to mimic various metallic materials. The topcoat is then cured in a warm oven for 30 minutes or so. The end result is a metallic layer encapsulated between 2 coats of lacquer, which is durable and highly reflective.

→ Vacuum metalizing brass hinges with aluminium

The process starts with the application of a base coat lacquer (**image 1**). This is essential for a high-quality finish; not only does it promote good adhesion between the workpiece and metallic coating, it also gives a smoother finish. Once applied, the workpieces, which are mounted onto their jigs, are loaded into a warm oven to accelerate the curing of the base coat (**image 2**).

After 30 minutes or so, the jigs are loaded, with the parts, onto the rotating holding fixtures (**image 3**). Loading the parts by hand means that they are individually checked, which limits waste. The parts are secured to the holding fixtures by clips that will not affect the quality of the coating. Each design will require a different method for connecting it to the holding fixtures.

The wire spiral holders that connect the positive and negative electrodes are loaded with 95% pure aluminium wire

(**image 4**). The whole assembly is loaded into the vacuum chamber (**image 5**). It takes about 30 minutes or so to pull a sufficient vacuum. An electrical discharge is passed through the wire, causing it to heat up and vaporize (**image 6**). It glows white-hot and a film of aluminium begins to form on the workpiece. The vacuum metalizing process takes only a few minutes. The chamber is brought back up to atmospheric pressure and the door opened.

Everything that goes into the vacuum chamber emerges with a thin coating of vaporized metal. The unprotected metal film can be easily rubbed off at this stage. Spraying a lacquer topcoat onto the workpiece secures the thin metallic film and bonds it to the base coat. The parts before and after they have been vacuum metalized appear markedly different (**image 7**).

1

2

QUALITY

Vacuum metalizing is used to increase the surface quality by improving reflectivity, wear and corrosion resistance. It also improves colouring capability; the topcoat can be impregnated with a wide range of metallic colours. The quality of the finish is determined by the quality of the surface prior to coating.

The coating is applied on line-of-sight geometry. This means that the parts have to be rotated during metalizing to encourage an even coating, and deep undercuts and recesses can escape the coating process. The vacuum can be replaced with argon gas to encourage a more vigorous coating, but this will be more expensive to carry out because it is a specialist process.

DESIGN OPPORTUNITIES

Vacuum metalizing is an inexpensive and versatile metal coating technique. There is no tooling cost, which makes a

smoother transition from prototyping to production, and also means that designers can explore the look and feel of their objects in metal at an early stage.

The coating produced by vacuum metalizing is fine and uniform. Passing the workpiece through the process more than once increases the film thickness.

Vivid colours can be used to replicate anodized aluminium, bright chrome, silver, gold, copper or gunmetal, among others. The advantage of this is that relatively inexpensive materials can be formed and then vacuum metalized to give the look and feel of metal.

DESIGN CONSIDERATIONS

The quality of the coating is affected by the surface quality of the workpiece. In other words, the metal finish will only be as smooth as the uncoated finish. If the desired effect is distressed, this must be achieved prior to metal coating.

The maximum size of the part that can be metalized is determined not only

by the vacuum chamber, but also by the geometry of the part. Flat parts up to 1.2 m x 1 m (3.94 x 3.3 ft) can be coated, where as 3D parts are limited to 1.2 m x 0.5 m (3.94 x 1.6 ft) because they have to rotate as they are metalized.

COMPATIBLE MATERIALS

Many materials are suitable, including metals, rigid and flexible plastics, resins, composites, ceramics and glass. Natural fibres are not suitable; it is very difficult to apply vacuum if moisture is present.

Aluminium is the most commonly used metal for coating. Other metals that can be used include silver and copper.

COSTS

There are no tooling costs, but jigs often have to be made to support the workpiece in the vacuum chamber.

Cycle time is moderate (up to 6 hours).

It is quite a labour intensive process; the parts have to be sprayed, loaded, unloaded and sprayed again. This means the labour costs can be quite high, but depend on the complexity and quantity of parts.

ENVIRONMENTAL IMPACTS

This process creates very little waste. Spraying the base coat and topcoat has impacts equivalent to spray painting.

Metalizing extends the life of products by increasing their resistance to corrosion and wear. Very little material is needed for the metal coating because it is generally a very thin film, but this depends on the application.

Featured Manufacturer

VMC Limited
www.vmclimited.co.uk

Finishing Technology

Grinding, Sanding and Polishing

Surfaces are eroded by abrasive particles in these mechanical processes. The surface finish ranges from coarse to mirror, and can be a uniform texture or patterned depending on the technique used and type and size of abrasive particle.

Costs	Typical Applications	Suitability
• No tooling costs for many applications • Non-standard profiles may require tooling • Unit costs dependent on surface finish	• Automotive, architectural and aerospace • Cookware, kitchens, sanitary and medical • Glass lenses, storage jars and containers	• One-off to mass production
Quality	**Related Processes**	**Speed**
• High quality finish that can be accurate to within fractions of a micron (0.000039 in.)	• Abrasive blasting • Electropolishing	• Rapid to long cycle time, depending on size and type of finish

INTRODUCTION

Grinding, sanding and polishing cover a wide range of processes used to finish metal, wood, plastic, ceramic and glass products. The size and type of abrasive particle, combined with the method of finishing, will determine the range of surface effects that can be achieved. These terms are used to describe different techniques for surface cutting.

Grinding is used to produce surface finishes on hard materials. The process fulfils a range of functions, including deburring metals, preparing surfaces for further processes, cutting into or right through materials, and precision finishes. There are many different grinding techniques, including wheel, belt and platen grinding, honing and barrel finishing. Many different abrasive materials are used, including metal, mineral, diamond and maize seed.

Sanding is used to describe the process of eroding surfaces with

Mechanical Grinding, Sanding and Polishing Techniques

Wheel cutting

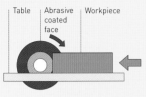

Abrasive coating | Spinning disk
Workpiece | Magnetic table

Surface cutting

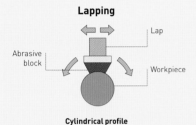

Table | Abrasive coated face | Workpiece

Edge cutting

Lapping

Lap
Abrasive block
Workpiece

Cylindrical profile

Spinning head | Abrasive pad
Workpiece
Table

Flat profile

Belt sanding

Table | Workpiece | Spinning abrasive belt | Rotating platen

Rotary

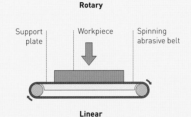

Support plate | Workpiece | Spinning abrasive belt

Linear

Honing

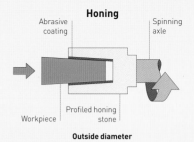

Abrasive coating | Spinning axle
Workpiece | Profiled honing stone

Outside diameter

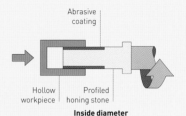

Abrasive coating
Hollow workpiece | Profiled honing stone

Inside diameter

TECHNICAL DESCRIPTION

These are common techniques used for cutting surfaces in industrial applications. Each is capable of applying a range of finishes, from super bright to very coarse, depending on the type of abrasive material. They all rely on lubrication, which reduces the build up of heat and wear on the cutting tool.

To achieve a highly reflective and super-bright finish, the material will pass through a series of stages of surface cutting, which will use gradually finer grits of abrasive. The role of each abrasive is to reduce the depth of surface undulation. In mirror polishing this is measured in terms of Ra (roughness average); a mirror polish is less than Ra 0.05 microns (0.0000019 in.).

Wheel cutting is carried out at high speed. Either the outside edge or the face of the wheel is coated with abrasive particles, generally made from metal and so providing a hard surface to grind or polish. They are equally suitable for coarse grinding and diamond polishing. The set up of the wheel and table will determine the precision of the finish.

Belt sanding is used in both woodworking and metalworking. There are many different types of machine, including free standing and portable types. Like wheel cutting machines, they operate at high speed and are designed to apply the desired finish very quickly.

They are not suitable for applying a super-bright finish. The diagrams illustrate how they are set up to cut different geometries of product. The rotary method is suitable for circular and irregular tube and rod profiles; as the belt spins, the platen rotates to produce an even finish around the outside diameter.

Honing is suitable for grinding and polishing internal and outside diameters of rotationally symmetrical parts. It is a precise method of surface cutting; an example application is finishing the inside diameter of cylinder blocks in engines. The tooling for this can be made up specially for each job. The abrasive surface wears away gradually as it is used, and therefore has to be replaced to retain accuracy.

Lapping is typically used to produce a very fine and super-bright finish on hard metallic and glass surfaces. The abrasive is not coated onto the pad or block, but is integral to it. Blocks are rigid abrasive blocks, and pads are flexible rubbers impregnated with abrasives of a defined grit size. The surface finish is therefore very controllable. Lapping can be carried out at different speeds depending on the material being polished; mirror finishes on flat and cylindrical surfaces can take many hours to complete.

abrasive-coated substrates. The abrasive particles consist of sand, garnet, aluminium oxide or silicone carbide; each has its benefits and limitations. The grade of paper (grit) is determined by the size of particle and ranges from 40 to 2,400. Lower numbers indicate fewer particles for the same area and so produce a courser finish.

These processes are known as polishing when they are used to produce a bright and lustrous finish on hard surfaces. Polishing is characterized by the use of compounds in the form of pastes, waxes and liquids in which the

Case Study

→ Wheel grinding

Wheel grinding is used to produce a very flat surface that is suitable for rapid prototyping onto using the direct metal laser sintering (DMLS) process (page 232). First the surface is milled to provide an even surface for grinding (**images 1** and **2**). This speeds up the process. Then the metal plate is placed onto the magnetic grinding table and carefully ground for up to 45 minutes to produce the desired Ra value (**images 3** and **4**).

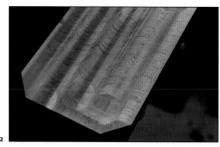

Featured Manufacturer

CRDM
www.crdm.co.uk

---

abrasive particles are suspended. Cloths are commonly used to apply the abrasive paste, either by hand or spun at high speed on a wheel.

Mixing water with abrasive-coated substrates, such as wet and dry paper, produces a similar effect. The water suspends abrasive particles that form during sanding, creating a paste that contributes to a very fine surface finish.

TYPICAL APPLICATIONS

These processes are used in all manufacturing industries both for surface preparation and as precise finishing operations. Applications are widespread and cover both industrial and DIY projects.

As well as finishing, grinding can cut into or right through fibre reinforced composites, metals and glass. This is especially useful for brittle materials.

RELATED PROCESSES

Electropolishing (page 384) is used to produce bright, lustrous and burr-free finishes on metal. It is not as accurate as these techniques and cannot produce such a bright finish. The advantage of electropolishing is that part geometry is not a significant consideration and will not affect cost or cycle time.

Abrasive blasting (page 388) is commonly used to smooth and finish metal castings. It is a versatile and rapid process, but produces a limited range of surface finishes, all of which are matt.

Polishing and grinding of embossed patterns in metal will emphasize the depth of texture.

QUALITY

It is possible to grind and polish surfaces accurate to within fractions of a micron. Precision operations are considerably more expensive and time consuming, but are sometimes the only viable method of manufacture. For example,

→ Honing glass

In this case honing is being used by Dixon Glass to produce an airtight seal between a glass container and a stopper.

The profiled metal honing stone is machined specially for this application. It is loaded into the chuck of a lathe and coated with a mineral based cutting compound (**image 1**). As the part is ground, the coarseness of abrasive compound is reduced to produce a finer finish (**image 2**). Abrasive honing increases the size of hole very gradually until the glass stopper fits perfectly into it (**image 3**). The glass stopper is finished in the same way.

The finished products (**image 4**) are used to store scientific specimens for many decades, and honed glass is the only material capable of this.

1

2

3

4

glass specimen jars (see case study) are ground with a honing tool to form a hermetic seal capable of maintaining the contents for more than 100 years.

Mechanical polishing produces hygienic surfaces, suitable for catering and medical applications.

Polished metal finishes reflect over 95% of the light that hits their surface. Finely polished stainless steel is 1 of the most reflective metals and can be used as a mirror.

Depth of surface texture at this scale is measured as a value of Ra (roughness average). Approximate Ra values for metal finishes are as follows: a ground finish obtained with 80–100 grit will be Ra 2.5 microns (0.000098 in.); a dull buffed finish from 180–220 grit will be Ra 1.25 microns (0.000049 in.); dull polish from 240 grit will be Ra 0.6 microns (0.000024 in.); and a bright polish produced with a polishing compound will be Ra 0.05 microns (0.0000019 in.).

DESIGN OPPORTUNITIES

Polished surfaces are more hygienic and easier to clean, which makes them very suitable for high traffic areas and human contact. In contrast, satin and finely textured finishes are prone to fingerprints and other marks.

Featured Manufacturer

Dixon Glass
www.dixonglass.co.uk

→ Rotary sanding

This is a typical application for the rotary belt sanding method (**image 1**). It produces an even satin finish on the surface of stainless steel (**image 2**). It is equally suitable for non-uniform profiles.

1

2

Featured Manufacturer

Pipecraft
www.pipecraft.co.uk

Repeating patterns can be polished into surfaces, like those found on café tables. These patterns reduce the visual effects of wear and dirt.

It is often more cost effective to polish stock material, such as sheet or tube, prior to fabrication. The joints can then be polished manually afterwards.

DESIGN CONSIDERATIONS

The shape of the part will determine the effectiveness of these processes. For example, precision grinding and polishing operations are limited to flat, cylindrical and conical shapes. This is because the operations are reciprocal and rely on either rotating or moving back and forth.

In contrast, cosmetic grinding and polishing operations can be applied to most shapes because they can be carried out manually if necessary.

The hardness of the material will affect the surface finish. Stainless steel is very hard and so can be polished to a very fine surface finish; aluminium is a softer metal and so cannot be polished to such a bright lustre.

COMPATIBLE MATERIALS

Any material can be ground, sanded or polished, but they may not produce desirable results. The hardness of the material will affect how well it finishes, as well as how long it will take.

COSTS

Much grinding, sanding and polishing can be carried out with standard tooling. Consumables are a consideration for the unit price. Specific tooling can be very costly, but depends on the size.

Cycle time is largely dependent on the size, complexity and smoothness of surface finish. Brightly polished surfaces can take many hours to achieve.

Labour costs are also dependent on the size, complexity and smoothness of surface finish required. A dull polish will add about 10% of the raw material cost, a measured finish about 25% and a reflective finish about 60%.

ENVIRONMENTAL IMPACTS

Even though these are reductive processes, there is very little waste produced in operation.

Case Study

→ Vibratory finishing

This is a method used for finishing high volumes. It is especially suitable for deburring metal, but is also used to cut back painted surfaces and a range of other applications. A deep drawn (page 88) metal part is placed into the vibrating barrel, which is filled with smooth, hard pellets (**image 1**). It works in much the same way as pebbles eroding each other on the beach to produce smooth, rounded stones. Many products can be in the barrel together.

Crushed maize seed (**image 2**), which is used to produce a very fine finish on hard surfaces, is the next stage in the abrasive process.

Case Study

→ Manual polishing

This is the most expensive and time-consuming method of polishing. It is used to produce an extremely high surface finish on the surface of parts that cannot be polished by mechanical means. This polishing method can be used to work every surface of this colander (**images 1–3**). The size of spinning mop can be adjusted to fit smaller profiles such as spoons. The density of the mop is adjusted for different grades of cutting compound.

The final stage is a buffing process, which uses the finest polishing compound to produce a highly reflective finish. It is a labour intensive process, but is nonetheless used to finish a high volume of products (**image 4**).

Featured Manufacturer

Alessi
www.alessi.com

→ Diamond wheel polishing

Diamond particles are used to finish a range of materials that are too hard for other polishing compounds. They are also used for high speed finishing of plastics. This acrylic box has been CNC machined. The lid and base are paired up and then the 4 sides are cut on a circular saw (**image 1**). This produces an accurate finish, but it has an undesirable texture. The workpiece is placed on the cutting table and clamped in place.

The diamond cutting wheel is spinning at high speed and produces a super fine finish in a matter of seconds (**images 2** and **3**). This is a mass-production method suitable for finishing a high volume of products (**image 4**).

1

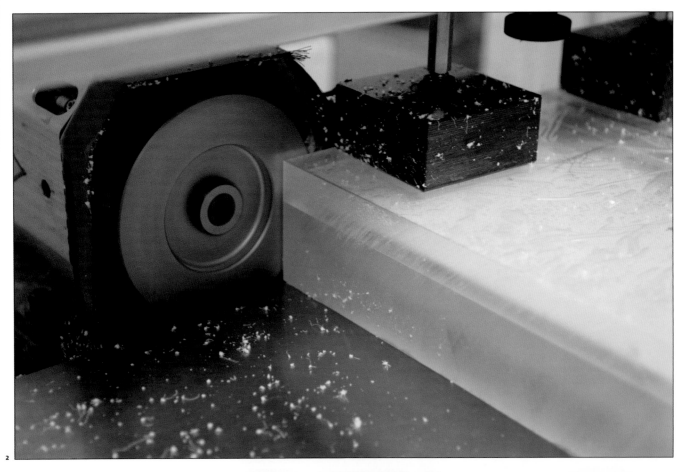

2

3

4

Featured Manufacturer

Zone Creations
www.zone-creations.co.uk

→ Lapping

Lapping is used to produce a range of finishes, including reciprocal patterns, dull and super bright finishes. To polish a sheet of stainless steel to a bright finish, the polishing compound is applied with a roller each time the lapping pad passes over the surface (**image 1**). Pressure is applied to maintain a very accurate finish. Afterwards a lime substitute is spread over the surface of the stainless to remove any remaining moisture before packing (**image 2**).

When lapping to produce a patterned finish, the edges of the sheet are deburred prior to polishing (**image 3**) because a protective plastic film is applied by the machine directly after polishing. The circular pads that make the pattern (**image 4**) are impregnated with abrasive particles. The pattern (**image 5**) is a typical example of the finish that can be achieved by this method.

1

2

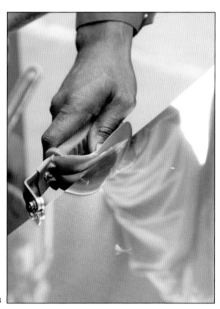
3

4

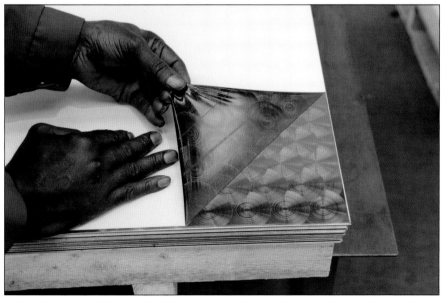
5

Featured Manufacturer

Professional Polishing
www.professionalpolishing.co.uk

Finishing Technology

Electropolishing

This is the reverse of electroplating; material is removed from the surface of the workpiece by electrochemical action to leave metal parts with a bright and clean finish.

Costs	Typical Applications	Suitability
• No tooling costs, but jigs are required • Low unit cost (roughly 5% material cost)	• Architectural and construction • Food processing and storage • Pharmaceutical and hospitals	• One-off to mass production

Quality	Related Processes	Speed
• High quality, bright, lustrous and hygienic surface finish	• Electroplating • Grinding, sanding and polishing	• Moderate cycle time (5–30 minutes)

INTRODUCTION

Electropolishing produces a bright lustre on the surface of metals. It is an electrochemical process; surface removal takes place in an electrolytic solution, which can be very accurate. As well as polishing, this process cleans, degreases, deburrs, passivates and improves the corrosion resistance of metal surfaces. It is becoming more widely used because the environmental impacts are less harmful than those of other processes.

Electropolished stainless steel has similar visual characteristics to chrome-plated metals. Electropolishing is a

Electropolishing Process

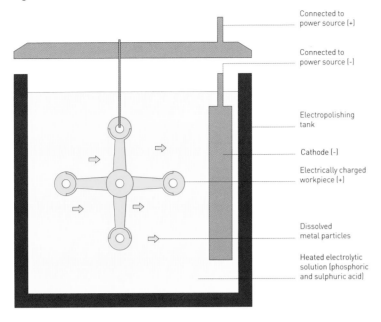

Connected to power source (+)

Connected to power source (-)

Electropolishing tank

Cathode (-)

Electrically charged workpiece (+)

Dissolved metal particles

Heated electrolytic solution (phosphoric and sulphuric acid)

Microscopic detail

Before   **After**

TECHNICAL DESCRIPTION
The electropolishing process is made up of 3 elements: the pre-cleaning (where necessary), the polishing (see diagram) and the final cleaning to remove chemical contaminants.

The process takes place in a bath of electropolishing solution, which is made up of phosphoric and sulphuric acid. The bath is maintained between 50°C and 90°C (122–194°F), depending on the rate of reaction; the warmer the solution, the more rapid the reaction. The workpiece is suspended on an electrically charged jig and becomes the anode. The cathode is also placed in the electropolishing solution and is generally made of the same material as the workpiece. For electroplating, conversely, the workpiece becomes the cathode.

When an electric current is passed between the cathode and the workpiece, the electropolishing solution dissolves metal particles from the surface of the workpiece. Surface dissolution takes place more rapidly on the peaks because that is where the power density is greatest. Low points are dissolved more slowly, and thus the surface of the material is gradually made smoother.

After electropolishing the parts are neutralized, rinsed and cleaned.

simpler process than electroplating chrome (page 364) and uses less water and chemicals, so is more cost effective. As a result of the increase in use of electropolishing, it is steadily becoming less expensive.

TYPICAL APPLICATIONS

Electropolishing has become a widely accepted finishing process for metals. In architectural and construction metalwork, it is used for both aesthetic and functional reasons (resistance to corrosion and reduced stress concentration). In the pharmaceutical and food industries it is used for mainly functional reasons (hygiene and resistance to corrosion). Aesthetic applications make up roughly 90% of its total use.

RELATED PROCESSES

Electropolishing has been steadily replacing more ecologically harmful processes such as chrome plating (see electroplating, page 364). This is

driven by industry's aim to reduce the consumption of chrome and other heavy metals (very toxic chromic acid is used for chrome plating). Electropolishing does produce chemical waste that must be cleaned and pH neutralized, but it is less harmful than these other processes.

Similar to mechanical polishing (page 376), electropolishing is a reductive process. However, it is possible to electropolish complex shapes that would be impractical by mechanical means.

QUALITY

The quality of finish that can be achieved with electropolishing enhances the surface of metalwork both aesthetically and functionally. Aesthetically, it increases the brightness and reflective index by smoothing (polishing) the surface of the metal. On a microscopic level, the high points are dissolved more rapidly than the troughs. The result is reduced surface roughness and reduced surface area. However, the end result is largely dependent on the

surface finish prior to electropolishing because material removal is usually no greater than 50 microns (0.002 in.), so a rough finish will be smoothed but not completely removed. Even so, this process is frequently used for deburring as well as polishing because burrs are like high points and so are dissolved very rapidly.

Functionally, electropolishing produces a finish that is clean, hygienic, less prone to stress concentration and more resistant to corrosion. The electrochemical action removes dirt, grease and other contamination.

Case Study

→ Electropolishing architectural steelwork

This case study demonstrates a recent development in electropolishing technology. The equipment has been modified to reduce effluents and improve the environmental credentials of the process.

Throughout the process test pieces are checked to measure the rate of electropolishing accurately (**image 1**). Electropolishing is generally used to remove up to 40 microns or so: this sample shows a reduction of 30 microns (0.0012 in.). The sample indicates the amount of time required in the electrolyte to remove sufficient materials.

These are investment cast stainless steel architectural spiders (see page 133), which are designed to hold panes of glass in buildings (**image 2**). They are jigged with electrical contact (**image 3**). This particular process requires only a small number of baths to electropolish effectively, neutralize and rinse the parts. They are mounted onto a rotating frame which suspends them in each tank (**image 4**).

They are dipped in the electrolytic solution for 10 minutes or so. As they are raised up, the green acidic mixture drains back into the tank (**image 5**). The polishing process is complete, and the rest of the process neutralizes and rinses.

The spiders are hung to drain after they have been rinsed (**image 6**), and are then washed a final time to remove any remaining contamination. The parts are then heated to dry (**image 7**), after which they are safe to handle and so are packed and shipped.

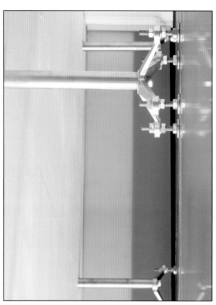

It produces a more hygienic surface because potential microscopic bacterial and dirt traps are opened up to allow more thorough sterilization. The process dissolves iron more readily than other metallic elements, and this phenomenon means that electropolished stainless steels have a chromium-rich layer on the surface. This layer protects the steel from corrosion because it reacts with oxygen to form chromium oxide, which passivates the surface and makes it less reactive to atmospheric elements. As well as protecting the steel, the chromium-rich layer can be so bright that it also gives the illusion of chrome plating.

DESIGN OPPORTUNITIES

The entire process takes place submerged in baths of solution, which has many advantages over mechanical operations. The process is equally effective on small technical and large structural parts. Size and complexity have no bearing on cycle time. It does not apply any mechanical stress to parts and so can be used to deburr and finish delicate parts that are not suitable for mechanical polishing techniques.

This process is also useful in the preparation of parts that will be used in applications that require contact between surfaces. This method of polishing deburrs and so reduces friction between mated surfaces such as threaded assemblies.

DESIGN CONSIDERATIONS

The level of finish is determined by the quality of the metal surface prior to electropolishing, so preparation is key. Marks from blunt tools on the forming or cutting machine, deep scratches and other imperfections will not disappear with electropolishing. Instead, they may be emphasized by the enhanced surface finish. Metalwork that requires preparation, such as a casting, is abrasive blasted (page 388) to produce a uniform surface finish that will respond well.

COMPATIBLE MATERIALS

Most metals can be electropolished, but this process is most widely used for stainless steels (particularly austenitic grades). Generally, an electropolishing plant will be set up for a specific material because different materials cannot be polished together or even in the same electrolytic solution. Other metals that can be treated in this way include aluminium and copper.

COSTS

There are no tooling costs, but jigs are required for electrical contact.

Cycle time is rapid, but depends on the amount of material removal and cleanliness of the part.

This process tends to be either fully or partially automated, so labour costs are relatively low. It adds only 5% to the cost of the base material, as opposed to metal plating, which can add up to 20%.

3

4

5

6

ENVIRONMENTAL IMPACTS

The environmental impacts of this process are threefold. Firstly, it is replacing the use of chrome-plated steel, which is a harmful material due to the chemicals used in its production. Electropolishing uses less harmful chemicals, needs less water and is simpler to operate. Secondly, it increases the longevity of stainless steel because it increases the material's natural protection against corrosion. And thirdly, this process removes material, rather than coating with an additional material, so reduces material consumption and eliminates delamination and other associated problems.

Approximately 25% of the chemical solution is replaced annually. The consumption of chemicals represents only a small proportion of the overall costs. They tend to be far less hazardous to the operator, and the resultant residue is more easily treated than with other metal finishing processes.

7

Featured Manufacturer

Firma-Chrome Ltd
www.firma-chrome.co.uk

Finishing Technology
Abrasive Blasting

This is a generic term used to describe the process of surface removal by fine particles of sand, metal, plastic or other abrasive materials, which are blown at high pressure against the surface of the workpiece and produce a finely textured finish.

Costs	Typical Applications	Suitability
• No tooling costs, but masks may be required • Low to high unit cost	• Architectural glass • Decorative glassware • Shop fronts	• High volume removal of surface material • Low to medium volume fine finishing

Quality	Competing Processes	Speed
• Very fine details can be produced by skilled operators	• Chemical paint stripping • Photo etching • Polishing	• Rapid cycle time, but slowed down by masking, layering and carving

INTRODUCTION

Abrasive blasting encompasses sand blasting, dry etching, plastic media blasting (PMB), shot blasting and bead blasting. All of these have 2 main functions. Firstly, they are used to prepare a surface for secondary operations by removing contaminants and applying a fine texture that will aid surface addition processes. Secondly, they can be used to apply a decorative texture to the surface of a part such as affecting the transparency of glass or applying a pattern. Other functions include deburring, cutting and drilling.

Glass carving is used in architectural glass and shop fronts. It is the process of sculpting the glass with 3D relief patterns, with and without the use of masking (see image, page 484, above right).

TYPICAL APPLICATIONS

Typical surface removal operations include the deburring and preparation of metal surfaces for further finishing operations and removing damaged or weak material from a surface such as rotten wood from sound wood or paint and rust from metal.

Typical decorative applications include glassware, signage, awards and medals and art installations.

RELATED PROCESSES

CNC engraving (page 396), photo etching (page 392) and laser cutting (page 248) are all used to produce similar effects on glass and a range of other materials. Textured plastic films laminated onto glass sheet produce similar effects to abrasive blasting.

Abrasive Blasting Process

Blasting gun

Blasting nozzle

Jet of abrasive particles

Workpiece

Exposed areas will be etched by the abrasives

Protective wax resist or mask of resist material

Abrasive material fed into gun at high pressure

QUALITY

With manual operations the success of this process is largely dependent on the skill of the operator. Very fine details can be reproduced accurately. Material removal is permanent, so care must be taken to etch the surface of the workpiece in a controlled fashion.

DESIGN OPPORTUNITIES

This is a fast and effective way of removing surface material. The various grades and types of blasting media mean that a range of textures and effects can be achieved, from a very fine texture (similar to that achieved with acid etching), to a more rugged sand-blast look. The depth of texture will affect the level of transparency in clear materials by increasing light refraction. This quality can be used to enhance the visual depth of an image, which will also be affected by layers of colour in the workpiece.

DESIGN CONSIDERATIONS

Like spray painting (page 350), this process is limited by line-of-site geometry. This also means that erosion will not occur under the edges of the mask, but rather perpendicular to it. The abrasive material must be matched to the workpiece and it is advisable to test the abrasives, pressure and distance from workpiece prior to full production. Textures can act as dirt traps, emphasizing fingerprints on glass for example; to reduce visual degradation they are often encased in a laminate or a clear, protective coating.

TECHNICAL DESCRIPTION

This is a simple process that can be used to attain extremely sophisticated results on many different materials. Abrasive blasting can be either pressure fed, or siphon (suction) fed. Both can be used inside a sealed booth (glove box) or a walk-in booth. Alternatively, blasting apparatus can be used on-site or outdoors if the workpiece is very large or impractical to move. With all operations, the steady stream of compressed air generally operates at pressures between 5.51 bar and 8.62 bar (80–125 psi), except some specialized systems such as PMB, which operate at 2.76 bar (40psi).

Pressure fed systems are the most effective. The abrasive material is forced through the blasting gun from a pressurized tank, attached by a flexible pipe. Siphon fed systems are more versatile because they suck the blast material through a feed pipe, which is simply inserted into a container of blast material. Both processes are controlled by 5 main variables: air pressure, nozzle diameter, distance from workpiece, abrasive flow rate and type of abrasive.

The choice of abrasive material is crucial and will be affected by many factors, which include the material being etched, the depth, grade and speed of cut and economics. The abrasives are graded in much the same way as abrasive papers. The choice of materials includes sand, glass bead, metal grit, plastic media and ground walnut and coconut shells.

Sand is the most commonly used because it is the most readily available and versatile. Metal grit abrasives include steel and aluminium oxide. Steel grit is the most aggressive and is used to create a rough texture on the surface of the workpiece. Aluminium oxide is gentler and less durable, resulting in a softer texture. Glass beads are the least abrasive and reserved for more delicate materials. Plastic media are the most expensive blast materials and require specialist equipment. They are much softer than other blast materials and the shapes, from round to angular, can be selected to suit the application. Certain natural materials, such as walnut shells, will not affect glass or plated material. This can be useful if, for example, the operator only wants to erode the metal part of a product made up of metal and glass.

Wax resists are used for decorative applications. The abrasive materials cannot penetrate 'slippery' materials and bounce off, leaving the masked area untouched. Wax resists are produced from artwork that has been generated in 2-tone and are applied to the surface of the workpiece using an adhesive backing.

→ Decorative sand blasting

The glass vessel being decorated here was designed by Peter Furlonger in 2005. It was shaped by glassblowing (see page 152). For this project, sand blasting has been used to decorate the surface of the workpiece with a multilayered artwork. The same principle can be used to apply decorative patterns to surfaces, except the wax resist is replaced with masking.

Firstly, the artwork or pattern is translated onto a wax resist, either by hand or by printing (**image 1**) and applied to the workpiece (**image 2**). Negative etching is used to describe the process of etching the image and leaving the background intact. This case study is an example of positive etching because the image is left intact and the background is textured.

Once the wax resist has been applied, the inside surface is masked to avoid damaging it with the abrasive (**image 3**).

The product is held in a glove box with protective gloves. The operator (artist) carefully etches away the unprotected areas of the workpiece with virgin aluminium oxide, ranging from 180 grit to 220 grit (**image 4**). Once the first stage of etching is complete, specific areas of the wax resist are removed, and other delicate areas have their protection reinforced (**image 5**). This will enable the artist to create a multilayered etched image. This technique is particularly effective on this product because the coloured surface has been built up in layers, in that the top layer has been fired with a blowtorch to produce metallic effect, the layer underneath is non-metallic and the base layer is clear glass.

The product is returned to the booth for the second stage of the etching process (**image 6**). Finally the wax resist and all other masking is removed to ready the workpiece for cleaning and polishing (**image 7**).

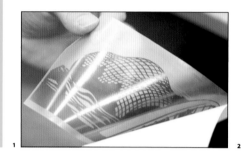

COMPATIBLE MATERIALS

This process is most effective for etching metal and glass surfaces, but can be used to prepare and finish most materials such as wood and some polymers.

COSTS

There are no tooling costs. However, masks may be required, the cost of which is affected by the complexity and size of the area to be abrasive blasted.

Cycle time is good, but is slowed down by complex masking and multilayering. Automation of simple tasks will greatly improve cycle time.

For manual operation labour costs are relatively high.

ENVIRONMENTAL IMPACTS

The dust that is generated during abrasive blasting can be hazardous. Booths and cabinets can be used to contain the dust that is created, otherwise breathing apparatus will be required, much the same as those used for spray painting (page 350) and other hazardous finishing processes.

Working in a sealed booth means that the blast materials can be reclaimed easily for re-use. Attaching a vacuum system to the blast head makes it possible to reclaim spent material for field stripping applications.

Featured Manufacturer

The National Glass Centre
www.nationalglasscentre.com

4

5

6

7

Preparation · Colour · Appearance · Protection · Information

Finishing Technology
Photo Etching

This is surface removal by chemical cutting. It has a similar appearance to abrasive blasting. The surface of the metal is masked with a resist film and unprotected areas are chemically dissolved in a uniform manner.

Costs	Typical Applications	Suitability
• Low cost tooling • Moderate to high unit costs	• Jewelry • Signage • Trophies and nameplates	• Prototype to mass production

Quality	Competing Processes	Speed
• Prolonged exposure to chemical attack will cause undercutting	• Abrasive blasting • CNC machining and engraving • Laser cutting and engraving	• Moderate cycle time, typically 50–100 microns (0.002–0.0039 in.) per pass (5 minutes)

INTRODUCTION

Photo etching, also referred to as acid etching and wet etching, is the process of surface removal by chemical dissolution.

This process is precise and low cost. Phototooling is printed acetate, which is inexpensive to replace. The accuracy of the etching is determined by the resist film, which protects areas of the sheet that will remain unchanged.

Surface removal is slow, typically 50–100 microns (0.0020–0.0039 in.) in a 5 minute pass. Depths of 150 microns (0.0059 in.) are suitable for decorative applications. Patterns, text and logos can be filled with colour.

TYPICAL APPLICATIONS

Applications include signage, control panels, nameplates, plaques and trophies. Photo etching is also employed by jewellers and silversmiths for decorative effect.

Photo Etching Process

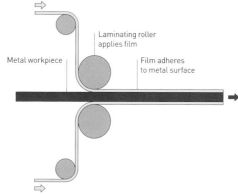

Stage 1: Applying photosensitive resist

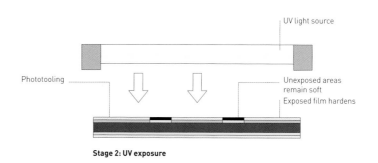

Stage 2: UV exposure

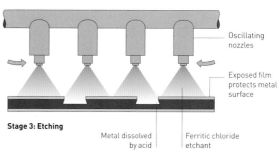

Stage 3: Etching

RELATED PROCESSES

Laser cutting (page 248), CNC engraving (page 396) and abrasive blasting (page 388) are used to produce similar effects in a wide range of materials. However, lasers and CNC engraving heat up the workpiece and can cause distortion in thin materials.

QUALITY

A major advantage of photochemical machining is that it is less likely to cause distortion because there is no heat, pressure or tool contact; the final shape is free from manufacturing stresses.

The chemical process does not affect the ductility, hardness or grain of the metal structure.

DESIGN OPPORTUNITIES

This process is suitable for prototyping and high volume production. Tooling costs are minimal; the negatives can be produced directly from CAD drawings, graphic software or artwork and last for many thousands of cycles. Small changes are inexpensive and adjustments can be made to the design, so this process is suitable for experimentation and trials during the design process.

Lines, dots and areas of etching can be filled with colour if the etch is more than 150 microns (0.0059 in.) deep. Very small details can be coloured independently such as patterns, logos, text and halftone images (visible dot pattern). It is possible to etch multiple layers with repeated masking and processing; layers significantly increase cycle time.

Thin sheet materials, up to 1 mm (0.04 in.), can be cut as well as etched (see photochemical machining, page 244).

TECHNICAL DESCRIPTION

The material is meticulously prepared; it is essential that the metal workpiece is clean and grease free to ensure good adhesion between the film and the metal surface. In stage 1, the photosensitive polymer film is applied by dip coating (page 68) or, as here, hot roll laminating. The coating is applied to both sides of the workpiece because every surface will be exposed to the chemical etching process.

Phototooling (acetate negatives) are printed in advance from CAD or graphic software files or artwork. In stage 2, the negatives are applied to either side of the workpiece and the workpiece, resist and negative are exposed to UV light. Both sides of the part are exposed to ensure that the resist on the reverse is fully hardened and protective. The soft, unexposed photosensitive resist film is chemically developed away. This exposes the areas of the metal to be etched.

In stage 3, the metal sheet passes under a series of oscillating nozzles that apply the chemical etch. The oscillation ensures that plenty of oxygen is mixed with the acid to accelerate the process.

Finally, the protective polymer film is removed from the metalwork in a caustic soda mix to reveal the finished etching.

→ Photo etching a stainless steel plaque

This case study illustrates the entire etching sequence where graphics are applied to only 1 side of the metalwork. The part is a building plaque for the British Embassy in Beirut. The final product is a combination of photo etching and colour fill (**image 1**).

The process begins with a printed acetate negative (**image 2**). Each negative can be used for a single etching process or multiple etchings; there is very little wear and tear, so they can last a very long time. In preparation, the metal workpiece is carefully cleaned in a series of baths. The first bath contains 10% hydrochloric acid and degreases the metal, which is then washed in mild detergent and water (**image 3**). The surface of the metal is dried with pressurized air.

The process takes place in a dark room to protect the photosensitive film. The polymer film is applied to both sides of the metal sheet by hot roll laminating (**image 4**). The

negative is secured to 1 side of the workpiece and placed in the light booth, where it is exposed to ultraviolet light on both sides (**image 5**). Unexposed polymer film is washed off in a developing process (**image 6**). Close inspection often reveals minor blemishes in the protective film, which can be touched up with a liquid chemical resist that dries onto the film instantly (**image 7**). This is a time-consuming process but it ensures a high quality finish.

The workpiece now passes through the etching process, which takes up to 20 minutes (**image 8**). The depth of the etch is checked, 150 microns (0.0059 in.) is sufficient for filling with colour (**image 9**). The remaining polymer resist and chemical etchant are cleaned off with a caustic soda mix and then finally with pure water (**image 10**). It is desirable to trim the edges at this stage because it

is likely that the chemical etchant will have attacked them during processing.

Filling the etched areas with colour is a 2-stage process and usually takes no longer than 30 minutes. The colour is mixed, using cellulose-based paints, and squeegeed over the workpiece surface (**image 11**). Multiple colours can be applied to very intricate patterns, but this takes considerably longer. Each colour will require 20 minutes drying time. Finally, excess paint is polished off and the plaque is finished (**image 12**).

1

2

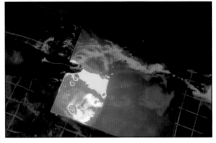

3

4

Chemical cutting processes, including photo chemical machining and etching, are generally limited to sheet materials. It is possible to etch very thick materials, or 3D surfaces using a paste. The stencil is applied in the same way, but instead of passing the workpiece under a series of oscillating nozzles, a chemical paste is applied, which has a similar effect.

DESIGN CONSIDERATIONS

The intricacy of detail is limited only by the quality of the photosensitive film. Very small details, down to 0.15 mm (0.006 in.) in diameter, can be reproduced. Therefore, it is essential that

the film and phototooling are free from dust and other contamination because this is large enough to be visible on the finished workpiece.

Lines and large areas of surface material can be removed without causing any stress to the workpiece.

COMPATIBLE MATERIALS

Most metals can be photo etched, including stainless steel, mild steel, aluminium, copper, brass, nickel, tin and silver. Aluminium is the most rapid to etch and stainless steel takes the longest.

Glass, mirror, porcelain and ceramic are also suitable for photo etching,

although different types of photo resist and etching chemical are required.

COSTS

Tooling costs are minimal. The only tooling required is a negative that can be printed directly from data, graphics software or artwork.

Cycle time is moderate. Processing multiple parts on the same sheet reduces cycle time considerably.

Labour costs are moderate.

5

6

7

8

9

10

11

ENVIRONMENTAL IMPACTS

In operation, metal that is removed
from the workpiece is dissolved in the
chemical etchant. However, offcuts and
other waste can be recycled. There are
very few rejects because it is a slow and
controllable process.

The chemical used to etch the metal
is one-third ferric chloride. Caustic soda
is used to remove spent protective film.
Both of these chemicals are harmful and
operators must wear protective clothing.

12

Featured Manufacturer

Mercury Engraving
www.mengr.com

Finishing Technology
CNC Engraving

A precise method for engraving 2D and 3D surfaces, CNC engraving is a high quality and repeatable process. Filling engravings with different colours and the use of clear materials are effective ways to enhance design details.

Costs	Typical Applications	Suitability
• No tooling costs • Moderate unit costs	• Control panels • Toolmaking and diemaking • Trophies, nameplates and signage	• One-off to batch production

Quality	Competing Processes	Speed
• Very high	• Laser cutting • Photo etching • Screen printing	• Moderate, depending on the size and complexity of engraving

INTRODUCTION

CNC and laser technologies (page 248) are the 2 main methods used for engraving. These processes have replaced engraving by hand with chisels or a pantograph; both of these methods are still practised, but the labour costs are too high for them to compete. Another important factor is that CNC and laser technology can be used to engrave a wider range of materials, including stainless steel and titanium.

The CNC engraving process is carried out on a milling or routing machine. These machines will operate on a

CNC Engraving Process

Movement in x, y and z axes

Track and bellows

Workpiece

Adhesive film to secure

Table

Chuck

Engraving tool

Tungsten (or other) cutting tip

TECHNICAL DESCRIPTION
The cutting speed is determined by the material and engraving tool. Tungsten is the most commonly used material for cutting tips. It can be re-sharpened and even re-shaped several times. It is not uncommon to cut a fresh tool for each job, depending on the requirements of the design. Harder materials, such as granite, will require diamond-coated cutting bits.

This is a 3-axis CNC machine, with all movement controlled by the track and bellows. The operating programme engraves the design either in straight lines, or following the profile of the design and creating a centrifugal pattern. The choice of cutting path depends on the shape of the engraving.

Cutting speeds are generally 1 mm to 50 mm (0.04–1.97 in.) per second. Harder materials and deeper engravings require slower cutting.

minimum of 3 axes: x, y and z. Machines operating on 3 axes are suitable for engraving flatwork; 5-axis machines are able to produce more complex engravings and accommodate 3D surfaces, but are more expensive to run.

TYPICAL APPLICATIONS

Almost every industry uses engraving in some form. Some products that stand out are trophies, nameplates and signage. Other products include control panels, measuring instruments, tool surfaces for plastic molding and metal casting and jewelry.

RELATED PROCESSES

Laser technologies are suitable for engraving very fine details down to 0.1 mm (0.004 in.). However, equipment costs are high and so it is less commonly used than CNC milling and routing, which are more readily available.

Photo etching (page 392) is suitable for shallow engraving of metals. Screen printing (page 400) and cut vinyl are inexpensive alternatives to engraving.

QUALITY

This is a high quality, repeatable process precise to 0.01 mm (0.0004 in.). A compromise has to be made between the size of the cutting head and the

speed of cut. Smaller cutting heads, which are often 0.3 mm (0.0012 in.) or less, reproduce fine details and internal radii more accurately. In contrast, larger cutting heads will complete the engraving in less time, making them more economical.

DESIGN OPPORTUNITIES

In clear materials engravings can be applied to the reverse of the part, which greatly improves the visual qualities of the engraving. In a single pass the engraving is deep enough to fill in with colour. Other than the obvious benefit of colour matching to, for example, a company logo, filling in with colour visually eliminates any evidence of the cutting operation.

Any thickness of material can be engraved down to approximately 1 mm (0.04 in.). Even with 3-axis machines, the cutting depth can be varied across the engraving by stepping up and down. Fixing points and other markings can be made during the engraving process to reduce time and increase accuracy.

DESIGN CONSIDERATIONS

Very fine and intricate details are reproducible, but large engravings with very fine details will take much longer to process. The size of the workpiece is

determined by the CNC bed size. Some factories are equipped with beds large enough to machine full size models of cars. However, standard beds are often no larger than 2 m² (21.53 ft²), which is adequate for the majority of applications and less expensive to run.

The CNC process operates from CAD data. Illustrator files are sufficient for 2-axis engraving. Older versions of certain programmes are more stable than newer editions, so it is sensible to work with the manufacturer's preferred version of any software to ensure maximum compatibility.

Typefaces are often engraved, especially for signage, nameplates and trophies. They should always be supplied to the manufacturer as outlines, or vectors, otherwise they may be replaced by another font in the transition.

COMPATIBLE MATERIALS

Almost any material can be engraved in this way, including plastic, foams,

→ CNC engraving a trophy

CNC engraving can be as complex as the designer chooses. Here, a relatively simple design is engraved onto a pre-cut 10 mm (0.4 in.) poly methyl methacrylate (PMMA) acrylic blank (**image 1**). Multiple parts can be cut simultaneously to reduce set-up and programming time. The 3 sheets are secured on the work bed using an adhesive film.

Before the engraving process begins, the cutting tip is zeroed to the top right-hand corner of the workpiece (**image 2**). This synchronizes the CAD data with the CNC engraving machine and ensures accurate tolerances on the part, which will not be trimmed post-engraving.

It takes 45 minutes to engrave the 3 trophies on 1 side with the straight-line method of cutting (**image 3**). A cutting tip of 0.3 mm (0.012 in.) is used because the internal radii are very tight. A balance has to be struck between larger cutters, which will remove material more rapidly, and definition of detail. Fine lines visible in the water clear plastic can be reduced with polishing or slower cutting speeds. Cellulose-based paint is applied to the engraving to emphasize the design (**image 4**). The engraving is cut just deep enough to contain the paint effectively, at 0.2 mm (0.008 in.). Multiple colours can be applied to very intricate, interweaving patterns, but this takes considerably longer. Each colour will require 20 minutes drying time. This trophy uses a single colour and so can be cleaned up after 20 minutes. The use of clear plastic demonstrates the precision of the CNC engraving process (**image 5**).

wood, metal, stone, glass, ceramic and composites. Even so, it is not common to find manufacturers that machine all these materials. There are many reasons for this, including their cutting tools, cutting speeds, and the fact that dust from certain materials can become volatile when combined.

COSTS

There are no tooling costs. Cycle time is moderate, but depends on the size and complexity of the engraving. Simple engravings with large internal radii can be cut very quickly, while intricate designs will take considerably longer due to the reduced size of the cutting tool.

Labour costs are minimal. The operation can usually run without any intervention from an operator.

ENVIRONMENTAL IMPACTS

All material that is removed is waste and is not normally recycled.

1

2

3

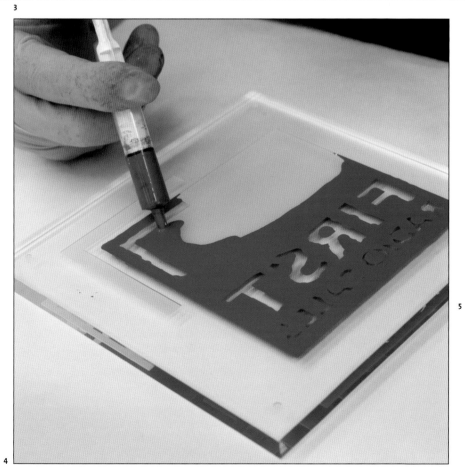

4

5

Featured Manufacturer

Mercury Engraving
www.mengr.com

Finishing Technology
Screen Printing

Traditionally known as silkscreen printing, this is a wet printing process used to apply graphics to flat and cylindrical surfaces. It is inexpensive and can be used on a variety of materials including textiles, paper, glass, ceramic, plastic and metal.

Costs	Typical Applications	Suitability
• Low tooling costs • Low unit costs, but dependent on number of colours	• Clothing • Consumer electronics • Packaging	• One-off to mass production
Quality	**Related Processes**	**Speed**
• High quality and sharp definition of detail	• Foil blocking • Hydro transfer printing • Pad printing	• Manual systems (1–5 cycles per minute) • Mechanized production (1–30 cycles per minute)

INTRODUCTION

This versatile printing process is used to apply accurate and registered coatings to a range of substrates. It is not just used to apply ink; any material the right consistency can be printed. For example, solder paste is screen printed onto circuit boards in reflow soldering (page 312), in-mold decoration films (page 50) are screen printed and even butter is screen printed onto bread in the mass production of sandwiches.

All of the screen printing techniques use the same simple process. It is suitable for manually operated or mechanized

The Screen Printing Process

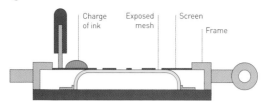

Stage 1: Load

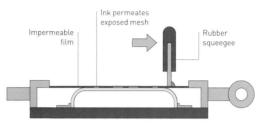

Stage 2: Screen print

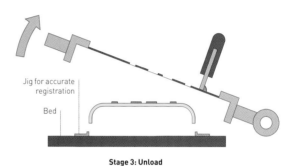

Stage 3: Unload

systems and the quality is similar. Most types of ink are suitable, which means this method can print graphics onto almost any surface.

TYPICAL APPLICATIONS

Screen printing is used across many industries because it can print onto many different substrates and is inexpensive. Typical products include wallpaper, posters, flyers, bank notes, clothes, signage, artwork and packaging.

Ink can be screen printed directly onto a product's surface, or onto an adhesive label that is bonded to the surface. Graphics are screen printed onto films used for in-mold decoration in the production of consumer electronics and similar products.

Scratch-off inks, which are used for direct mail, mobile top-up cards and security applications, are applied by screen printing.

Printed circuit boards (PCB), radio frequency identification (RFID) chips and other electronic applications are typically made by covering the surface with copper and then selectively removing it to produce the circuitry, but it is now possible to screen print circuitry with conductive inks. Flexible materials can be printed in this way, and there are even transparent conductive inks.

RELATED PROCESSES

Developments in varnishes and inks cured by ultraviolet light mean this process can produce decorative effects that are similar to, or even richer than, solid colours in foil blocking (page 412). Varnish can be applied over the entire surface to enrich and protect the colours,

TECHNICAL DESCRIPTION

This is a wet printing process. A charge of ink is deposited onto the screen, and a rubber squeegee is used to spread the ink evenly across the screen. Those areas protected by the impermeable film (stencil) are not printed.

The screens are made up of a frame, over which a light mesh is stretched. The mesh is typically made up of nylon, polyester or stainless steel.

Each colour requires a separate screen. A full-scale positive image of each colour is printed onto a separate sheet of acetate. The areas to be printed with ink are black and the areas not to be printed are clear. The full-scale positive is mounted and registered on the screen, which is coated with photosensitive emulsion. The emulsion is exposed to ultraviolet light, which causes it to harden and form an impermeable film. The areas that were not exposed, under the black parts of the acetate, are washed away to produce the stencil.

There are 4 main types of ink: water-based, solvent-based, polyvinyl chloride- (PVC-) based plastisol, and UV curing.

Water and solvent-based inks are air-dried or heated to accelerate the process. PVC-based plastisol inks are used mainly to print textiles. They have varying levels of flexibility, determined by the quantity of plastisol, which can cope with stretching fabrics. They polymerize, or harden, when heat is applied. UV inks contain chemical initiators, which cause polymerization when exposed to UV light. These inks have superior colour and clarity, but are also the most expensive.

or in selected areas, when it is known as 'spot varnishing'.

Like foil blocking, screen printing is generally limited to flat and cylindrical parts. It is not possible to print undulated or concave surfaces. This is where pad (page 404) and hydro transfer (page 408) printing take over.

QUALITY

Screen printing produces graphics with clean edges. The inks have a paint-like consistency and so will not run or bleed in most cases.

The definition of detail and thickness of printed ink is determined by the size of mesh used in the screen. Heavier gauges will deposit more ink, but have lower resolution of detail. These tend to be used in the textile industry, which requires copious amounts of ink, whereas light meshes are used to print on paper and other less absorbent materials.

DESIGN OPPORTUNITIES

There is a vast range of colours, including Pantone and RAL ranges. Equally, there are numerous types of ink such as clear varnish, metallic, pearlescent, fluorescent, thermochromatic and foam.

In a process known as 'window printing', the reverse of a clear panel is printed, so the ink is protected beneath the panel and has a high gloss finish. This is used on mobile phone screen covers and televisions, for example.

In a similar way, multicoloured designs can be applied to a range of products by screen printing a reverse of the image onto the back of a clear label. The label is bonded to the product to give the impression of direct printing with no label. This technique, sometimes referred to as 'transfer printing', can be used to apply screen printed graphics to any shape that will accept an adhesive label.

DESIGN CONSIDERATIONS

Each colour is applied with a different screen, which will require registration. Each colour also has to dry, or cure,

between applications, but this can be accomplished in a matter of seconds with UV curing systems.

The type of ink is generally determined by the application and the material being printed.

COMPATIBLE MATERIALS

Almost any material can be screen printed, including paper, plastic, metal, ceramic and glass.

COSTS

Tooling costs are low, but depend on the number of colours because each colour will require a separate screen.

Mechanized production methods are the most rapid, and can print up to 30 parts per minute.

The labour costs can be high for manual techniques, especially for complex and multicolour work. Mechanized systems can run for long periods without intervention.

ENVIRONMENTAL IMPACTS

Inks for printing on light-coloured surfaces tend to be less environmentally harmful. PVC-, formaldehyde- and solvent-based inks contain harmful chemicals, but they can be reclaimed and recycled to avoid water contamination.

→ Window printing a glass screen

In this case study a pane of glass is printed on both sides with contrasting colours. The white print on the face of the glass is informative and the black printing on the back conceals the component behind it.

The white details have already been printed and have dried and cured sufficiently for the black coat to be applied (**image 1**). The colours are applied in batches because a different screen has to be used for each. The part is loaded face down underneath the screen (**image 2**). A small amount of black ink is applied to the screen (**image 3**).

For a high-quality print on glass, a solvent-based 2-pack ink is used. This has excellent adhesion to a wide range of 'difficult' surfaces, including glass, plastics, metals and ceramics. It is also resistant to abrasion, wear and many solvents.

A rubber squeegee is used to spread the ink across the surface of the screen (**image 4**). Pressure is applied during spreading, to ensure that the ink penetrates the permeable areas of mesh to build up a dense layer of colour with clean edges.

The part is removed while the ink is still wet (**image 5**). Because it is solvent-based, this ink will dry and cure at room temperature in about 1 hour, but for optimum results, the parts are stacked in racks (**image 6**) and the drying process is accelerated in an oven at 200°C (392°F).

3

4

5

Screen printing is an efficient use of ink; applying ink directly to the product's surface reduces material consumption.

Screens are recycled by dissolving the impermeable film away from the mesh so that it can be reused.

6

Featured Manufacturer

Instrument Glasses
www.instrument-glasses.co.uk

Finishing Technology
Pad Printing

Pad printing, also known as tampo printing, is a wet printing process used to apply ink to 3D and delicate surfaces. It can be used to apply logos, graphics and other solid colour details to almost any material.

Costs	Typical Applications	Suitability
• Low tooling costs • Low unit costs	• Automotive interiors • Consumer electronics • Sports equipment	• Batch to mass production

Quality	Competing Processes	Speed
• High quality and sharp edges, even on undulating surfaces	• Hydro transfer printing • Screen printing • Spray painting	• Printing takes 2 to 5 seconds • Oven curing takes 20 to 60 minutes

INTRODUCTION

This process produces repeatable print on surfaces that are flat, concave, convex or even both. This is possible because the ink is applied to the product by a silicone pad. When the silicone is brought into contact with the surface, it wraps around it while stretching very little, so graphics can be applied to a range of surfaces without loss of shape and quality.

This process lays down a much thinner film of ink than screen printing (page 400). It is typically less than 1 micron (0.000039 in.), so for backlit and critical applications more than 1 layer may have to be applied.

TYPICAL APPLICATIONS

Pad printing is used to decorate a range of products where appearance or information is important on undulating or delicate surfaces. Keypads in handheld devices, remote controls and some mobile phones are pad printed.

It is used a great deal in the production of consumer electronics where in-mold decoration (page 50) is not suitable. Typical examples include the application of logos, instructions and images.

Sports products can be complex shapes. For example, golf balls are round and have dimples covering their surface. It is possible to pad print graphics that cover these undulations with high definition of detail and clean edges. Other balls, including footballs, baseballs and basketballs, are also printed this way.

RELATED PROCESSES

Pad printing and hydro transfer printing (page 408) are the only printing methods suitable for applying ink directly to undulating surfaces. Spray painting (page 350) can be used to similar effect, but is a different process and requires masking and finishing.

The difference between pad and hydro transfer printing is that hydro transfer is used to cover entire surfaces. Pad printing tends to be much more accurate, rapid and used for the application of graphical details.

QUALITY

The quality of print is determined by the definition of detail on the cliché plate. The smooth silicone will transfer all of the ink it picks up onto the surface of the part. The definition of detail can be very fine, down to 0.1 mm (0.004 in.) lines spaced 0.1 mm (0.004 in.) apart.

TECHNICAL DESCRIPTION

A positive image of 1 colour of the design is engraved onto the cliché. In stage 1, it is flooded with ink and wiped clean with a squeegee to ensure a good covering of ink in the engraved pattern, but none on the plate itself. This is because the silicone pad will pick up all the ink it comes into contact with.

The engraved design in the cliché is very shallow. Therefore, the layer of ink is very thin and begins to dry almost immediately. In stage 2, the silicone pad comes down and presses onto the ink and cliché. The ink adheres to the surface of the silicone as it is drying out.

In stage 3, the silicone pad moves over to the product. Meanwhile, the squeegee tracks back across the cliché, which is about to be flooded with ink again.

In stage 4, the silicone pad is compressed onto the workpiece. It wraps around the profile of the workpiece, ensuring adequate pressure between the surfaces to transfer the ink. The silicone has very low surface energy, so the ink comes off very easily.

In stage 5, the part is finished, and the silicone pad tracks back to the cliché, where a fresh charge of ink has been flooded and wiped clean. The next part is loaded in, and the process starts all over again.

Because it is possible to pad print fresh ink on top of wet ink, machines often work in tandem. As soon as 1 colour has been printed, the part is transferred to a second machine and printed with subsequent colours.

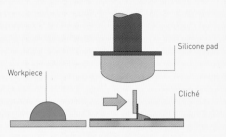

Stage 1: Preparation

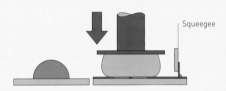

Stage 2: Pick-up

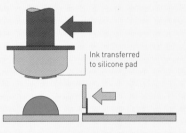

Stage 3: Transfer

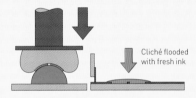

Stage 4: Print

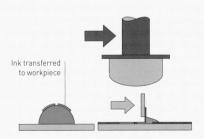

Stage 5: Finish

→ Pad printing a backlit keypad

This is a compression-molded (page 44) rubber keypad for a handheld electronic device. It is backlit, so the printing quality must be very high because any imperfections will be emphasized.

In this case, pad printing is being used to apply the negative graphics. The buttons have already been printed white and yellow (**image 1**). These are the colours that will be seen when it is illuminated.

The keypad is loaded onto the pad printing machine (**image 2**). To prevent the buttons from moving during printing, a small acrylic jig is placed over the part.

The cliché is flooded with ink and wiped by the squeegee (**image 3**). Next, the silicone pad is brought into contact with the cliché and pressed down lightly to pick up the ink (**image 4**).

As the silicone pad transfers to above the workpiece, the squeegee floods the cliché with fresh ink (**image 5**). The ink on the silicone pad is a thin film (**image 6**).

The printing process is very rapid. The silicone pad aligns with the part and is pressed onto it (**images 7** and **8**). Pressure is applied and the ink is transferred onto the surface of the workpiece.

On the finished part (**image 9**), the black masks the light so that coloured numbers are displayed. The parts are placed onto racks and in an oven to cure the ink fully (**image 10**).

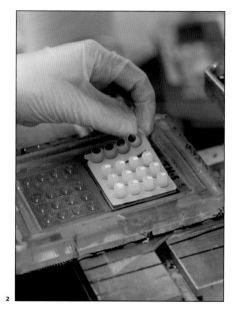

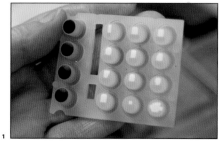

1

2

3

4

DESIGN OPPORTUNITIES

As in the other printing methods, the ink is printed with linear or rotary application. Rotary techniques make it possible to print continuously onto undulating surfaces; they can also print cylindrical products, such as cosmetic packaging, all the way round.

Like screen printing, pad printing can be used to apply conductive inks. Therefore, it is possible to print circuitry on curved, concave and convex shapes.

DESIGN CONSIDERATIONS

This process is capable of printing flat, convex and concave shapes. However, there is a limit to how far around a profile the silicone pad will form. For example, it is not possible to print more than half the way round a cylindrical part. For this, rotary pad printing is used.

It is limited to graphical details no larger than 100 mm x 100 mm (3.94 in.). This is the maximum size that can be picked up by the pad from the cliché.

The inks used in pad printing are typically limited to solvent-based types because water-based inks will not pick up on the silicone pad.

COMPATIBLE MATERIALS

Almost all materials can be printed in this way. The only materials that will not accept print are those with lower surface energy than the silicone pad such as polytetrafluoroethylene (PTFE). This is because the ink has to transfer, and such materials are just as non-stick as silicone.

Some plastic materials will require surface pre-treatment to ensure high print quality.

COSTS

Tooling costs are low. The cliché is typically the most expensive but is limited to 100 mm x 100 mm (3.94 in.). It is cut by laser (page 248), photochemical (page 244) or CNC (page 182) machining.

Cycle time is rapid. Inks can be laid down wet-on-wet, which is an advantage for multiple-colour printing.

Labour costs are low because most of the process is mechanized.

ENVIRONMENTAL IMPACTS

This process is limited to solvent-based inks and associated thinners that may contain harmful chemicals.

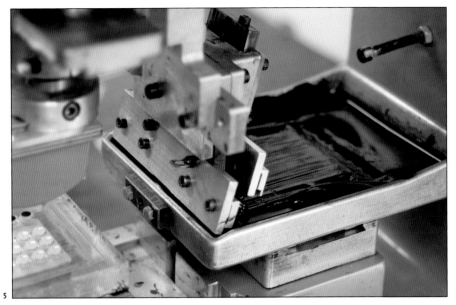

5

6

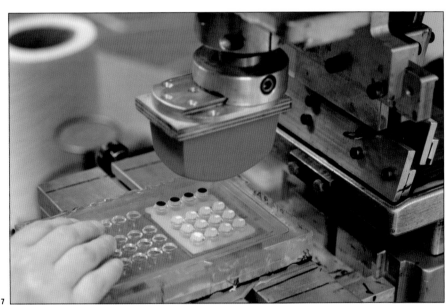

7

8

Featured Manufacturer

Rubbertech2000
www.rubbertech2000.co.uk

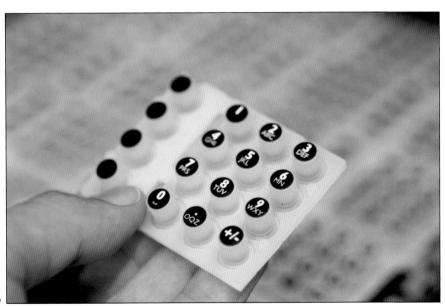

9

10

Hydro Transfer Printing

Hydro transfer printing is used to apply decorative finishes to 3D surfaces. A range of vivid graphics can be digitally printed onto the transfer film, which is wrapped around the product with water pressure.

Costs	Typical Applications	Suitability
• No tooling costs, but small products require jigs • Low to moderate unit costs	• Automotive • Consumer electronics • Military	• Low volume to mass production

Quality	Related Processes	Speed
• High-definition images • Wrap with very little stretch	• In-mold decoration • Pad printing • Spray painting	• Good cycle time (10–20 cycles per hour)

INTRODUCTION

Hydro transfer printing is known by many different names, including immersion coating, cubic printing and aqua graphics. They are all basically the same process, but different companies supply the various printing technologies.

This is a relatively new process, but has already been applied across a wide range of products and industries. It has 2 main functions: imitation and decoration. The entire surface of a product can be coated with a print of wood grain, marble, snake skin or carbon fibre (see image, left), for example. It is hyper-realistic and can transform the appearance of a flat or 3D product.

Alternatively, geometric patterns, flags, photographs or a company's own graphics can be applied for decorative effect. The transfer films are printed digitally , so images can be block colour, multicoloured and continuous tone with no effect on cost.

TYPICAL APPLICATIONS

This process is used for applications where surface appearance and cost are critical. For example, car interior trims can be injection molded plastic (page 50), but decorated to look like walnut. The automotive industry has used the process a great deal, partly because it integrates as part of the spray painting process (page 350). Other products include alloy wheels, door linings, gear sticks and steering wheels.

Hydro transfer printing is a cost-effective method for applying decorative finishes to mobile phone covers, computer mice, sunglasses and sports equipment. It allows for much shorter

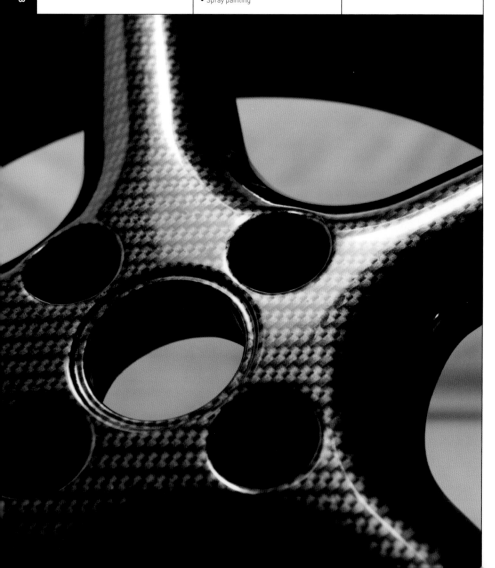

Hydro Transfer Printing Process

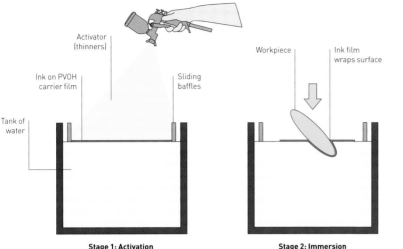

Activator (thinners)

Ink on PVOH carrier film

Sliding baffles

Tank of water

Stage 1: Activation

Workpiece

Ink film wraps surface

Stage 2: Immersion

Printed surface

Stage 3: Finish

production runs that in-mold decoration (page 50), and does not require specially designed tooling.

As demonstrated in the case study, these processes lend themselves to applying camouflage and other concealing finishes to weapons. The finish is as durable and resistant to ultraviolet light as a spray painted coating and so is suitable for application on rifle stocks, barrels and scopes.

RELATED PROCESSES
Hydro transfer printing is an addition to conventional spray painting. The immersion process makes up one-third of production; the other two-thirds are spraying before and after immersion.

Spray painting can produce similar affects to hydro transfer without the immersion process, using masking and airbrushing. These techniques are still used on large and impractical surfaces, but they are time consuming and labour intensive. The hydro transfer process is more cost effective for applying many decorative surface finishes.

Pad printing (page 404) is the only other printing process capable of applying graphics to undulating surfaces, but it is limited to small areas, whereas hydro transfer is used to coat entire products. Pad printing is a more precise printing technique.

TECHNICAL DESCRIPTION
The diagram illustrates the immersion cycle, which is only one-third of the whole process. Prior to applying the print the surface is prepared, typically by coating it with an opaque basecoat (primer). At this stage it is possible to smooth any blemishes and improve the surface finish.

The immersion process is carried out in a tank of warm water at 30°C to 40°C (86–104°F). The transfer film is made up of a polyvinyl alcohol (PVOH) backing and ink surface. In stage 1, the PVOH side is laid on to the surface of the warm water and triggered with a spray activator. Each stage of the process has to be timed accurately because once activated the film becomes gelatinous and delicate. If left for too long, it will disperse across the water. Sliding baffles are used to stop it moving about.

In stage 2, the part is immersed and the pressure of the water forces the ink to follow the contours of the surface. The inks are transferred to the surface of the part as it is submerged. The whole process takes only 3–4 minutes.

After printing, the inks are locked in with a transparent topcoat (lacquer). This is applied by spraying and can be matt or gloss depending on the application.

QUALITY
The quality of print in this process is equal to digital printing and the colour range is limitless. The ink is less than 1 micron (0.000039 in.) thick and is made durable by sandwiching between a basecoat (primer) and topcoat (lacquer). The basecoat ensures that the ink will bond to the substrate, while the topcoat seals it in.

DESIGN OPPORTUNITIES
With this process it is possible to make products look as if they are formed from materials that would be too expensive. There is a wide range of standard prints

Above
Hydro transfer printing is used to apply a vast range of hyper-realistic finishes, including wood grain, figured, grained and camouflage, onto flat or 3D surfaces

Case Study

→ Hydro transfer printing a rifle stock

These are plastic injection molded rifle stocks that are hydro transfer printed (**image 1**). Each is different, mainly because it is impractical to make them all exactly the same. However, making them individual also gives the consumer greater choice.

Some parts can be printed in a single dip, others have to be masked and printed in several cycles. These parts are masked in 2 halves (**image 2**) because there is a re-entrant angle on the trigger guard which it is not possible to print well in a single dip.

The ink is supplied on a PVOH backing film on rolls typically 1 m (3.3 ft) wide (**image 3**). The film is cut to length and floated on top of the warm water bath (**image 4**). It is allowed to rest for 45 seconds, in which time the PVOH backing film will begin to dissolve. The baffles are brought in to surround the film and stop any lateral movement, and the film is sprayed with

an activator (**image 5**). This prepares the ink for transferring onto the surface of the workpiece.

After only a few seconds the part is carefully dipped into the tank (**image 6**). It is immersed in the water at an angle that ensures no air bubbles form on its surface. The ink is gelatinous, but remains intact as it wraps the 3D shape.

The surface of the water is cleared before the product emerges (**image 7**). At this point the ink has adhered but is not protected. It is rinsed to remove any residue (**image 8**) and then sprayed with a hardwearing and resistant topcoat. The finished item (**image 9**) shows how well the film conforms to the shape of the part because it is not distorted and has even filled small recesses, channels and other design details.

1

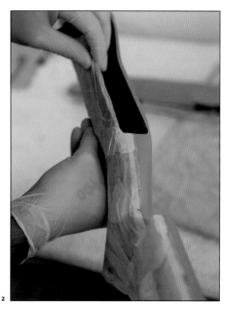

2

to choose from; it is also possible to print your own design.

Complex shapes that may not be feasible in the desired material can be coated to look like it. For example, it is unlikely that an injection molded part with snap fits and other fixing points is practical or even possible to replicate in solid walnut.

An experienced printer can coat all shapes, angles, curves, protrusions and recesses. Shapes that are too complex for a single dip can be masked and coated twice. The ink will not stick to itself, so it is possible to make a clean join line. Even so, it is best to design the part so that the join line is underneath or out of sight.

All of the print and colour is applied in a single operation, so there is no need to register different colours.

DESIGN CONSIDERATIONS

This process is only suitable for printing patterns onto surfaces. It is possible to align the product with the pattern, but

due to the nature of the process it is not feasible to position graphics precisely. This means it is not practical for graphics that require precise application such as numbers on control panels.

Deep recesses, holes and re-entrant angles require an outlet for air at the top, otherwise a bubble will form, and the ink will not make contact with the surface. Immersing the part at the optimum angle will often overcome problems with re-entrant angles and shallow recesses.

Some surfaces are suitable for printing onto directly, but others may need preparation with a basecoat. The basecoat is also important for determining the colour and quality of the print. The inks are very thin and almost transparent. The basecoat provides the optimum colour to view the inks against.

Flat surfaces, gentle curves and bends along a single axis are the simplest shapes to print. It is also possible to print 5 sides of a cube and bends on many axes. However, the more complex and

undulating the shape, the more difficult it will be to print. Cones and sharp edges are probably the most difficult, and the pattern will not reproduce as well as on flat surfaces.

The size of part that can be printed is limited to the immersion tank and width of film. This is typically 1 m² (11 ft²).

COMPATIBLE MATERIALS

Almost any hard material can be coated. If it can be spray painted, then it is also feasible to apply hydro transfer graphics. The most commonly used substrates are injection molded plastics and metals.

COSTS

There are no tooling costs. However, small parts will need to be mounted on specially designed jigs so that many parts can be printed simultaneously.

Cycle time depends on the size and complexity of the part, but is typically no more than 10 minutes. Masking adds labour and time to the process.

3

4

5

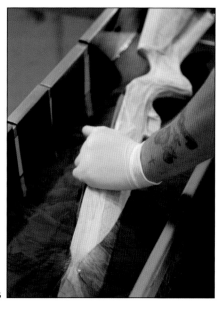

6

7

Labour costs are moderate, because the majority of applications are manually dipped. If there are sufficient volumes to justify mechanized dipping then it is likely that in-mold decoration (page 50) will be used instead.

ENVIRONMENTAL IMPACTS

A variety of sprays, thinners and chemicals are used in the process. It is similar to spray painting, but is a more efficient use of materials, and there is very little waste. All contamination can be filtered from the water and disposed of safely.

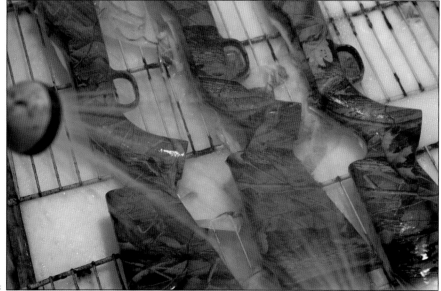

8

9

Featured Company

Hydrographics
www.hydro-graphics.co.uk

Finishing Technology

Foil Blocking and Embossing

These are dry processes used to apply decorative finishes to a range of substrates. A profiled metal tool is pressed onto the surface and leaves behind either a reverse image in foil, or a relief pattern.

Costs	Typical Applications	Suitability
• Very low tooling costs • Low unit costs	• Consumer electronics • Packaging • Stationery and printed matter	• Very low volumes to mass production

Quality	Related Processes	Speed
• High quality repeatable finish with fine edge detail	• Pad printing • Screen printing • Spot varnish	• Rapid cycle time (approximately 1,000 cycles per hour)

INTRODUCTION

Foil blocking is known by many names, including foil stamping, hot stamping and gold foil blocking. It is a pressing operation, which lends itself to use with embossing.

The foil or relief pattern (see images, opposite) is impressed onto the surface of a material with precisely machined metal tooling. This is a rapid and repeatable process used a great deal in the packaging and printing industries and suitable for both small and high volume production runs.

It is possible to combine foil blocking and embossing into a single operation, but for the highest quality finish they should be used independently. This is because the tooling is very slightly different. Foil blocking is carried out on the face of the material with a square edge tool, whereas emboss tooling has a small radius and the best results are achieved when pressing from the opposite side into a matched tool.

TYPICAL APPLICATIONS

This process is used a great deal in the printing industry to decorate book covers, packaging, invitations, flyers, posters, CD cases and corporate stationery.

Foil can be printed onto a range of materials, including paper, wood, plastic and leather. It is used to print logos and text directly onto stationery and cosmetics packaging. Holographic foils are used for security on bankcards, driving licences, concert tickets and gift vouchers.

Foil blocking is used to print films for in-mold decoration (page 50), mainly for consumer electronic applications.

Rotary (continuous) foil blocking techniques are even used to apply imitation wood finish to plastic architectural trim. In this area it overlaps with pad printing (page 404). Foil blocking is generally not suitable for surfaces with undulations; pad printing uses semi-rigid silicone pads and so overcomes this problem.

RELATED PROCESSES

Foil blocking and embossing are gratifying and inexpensive processes. Other printing methods are limited to flat colour, except spot varnishing.

Spot varnishing is a coating process that can be used to decorate printed finishes. It is used for similar reasons to foil blocking, but it is not opaque like foils; it works by enriching colours beneath it on the substrate. It is commonly used to enhance logos, headers and other design details on printed surfaces. Spot varnishing is generally applied by screen printing (page 400) or digital printing. The varnish is cured instantly in UV light.

Clear foils have been developed to compete with spot varnishing. Using the same colour for the foil and substrate, such as a rich red on matt red, can achieve similar effects and give the impression of a spot varnish.

QUALITY

If the process is set up correctly and on a reliable machine, the metal tool will make precise and repeatable impressions over long production runs.

Different colours tend not to be overlapped when printing with other methods; a benefit of foil is that it is opaque and so registration is not usually so critical. Even so, the metal tooling can be set up to precise requirements.

Foil blocking forms a slight impression in the surface of the material due to the pressure applied. This can be an aesthetic advantage and helps to protect the foil from abrasion. The depth of impression depends on the material's hardness; thin materials will emboss all the way through and so have a raised surface on the reverse, which may not be desirable.

DESIGN OPPORTUNITIES

There are many different foils, including matt, gloss, metallic, holographic, patterned and clear. The colour range is vast, including Pantone and RAL colour charts. The foils can be used to add value, such as gold leaf, emphasize a design detail, or apply print, including type.

Different colours of foil can be applied directly onto each another, so multicolour designs are often laid down in solid colours on top of each other, to avoid registration problems.

Foil will bond well to most substrates. The thickness of the material is not a consideration for foil blocking as long as it can be fed through the machine.

It is possible to foil block and emboss simultaneously, even though it is not preferable; the benefit is a reduction in cycle time and cost.

Top left
This magnesium tool is for embossing thin sheet materials.

Middle left
This 6 point type has been photochemically machined onto the surface of a magnesium foil blocking tool. Laser engraved copper is best for very high volumes and very fine details.

Left
Foil blocking is not limited to flat parts. Cylindrical cosmetic packaging is often decorated in this way for added value.

Top right
Embossing is used for subtle and high quality effects on printed matter. It is known as 'blind embossing' when it is carried out on non-printed materials.

Above
Foil blocking can reproduce lines as thin as 0.25 mm (0.01 in.) in most colours.

DESIGN CONSIDERATIONS

Very complex and intricate designs can be block foiled and embossed. However, different types of foil may or may not lend themselves to fine details. There is a large range of silver and gold foils, some of which can be used for type down to 1.5 mm high (6 point). Other colours, such as black, white or red, may not be suitable for such fine details. Trials should be carried out with the printer to establish whether small designs are feasible.

Bold typefaces are more difficult to print because the holes in the letters are smaller, and half-toned graphics (tints) can be difficult due to the size of the dots.

Only flat and cylindrical surfaces can be foil blocked. Flat is the most common and least expensive. Cylindrical parts require more specialized tooling and so are more expensive. Therefore, this technique is generally limited to large volume production.

The maximum size of tooling is limited by the impression platen, which is typically up to A1, or 594 x 841 mm (23.38 x 33.11 in.).

COMPATIBLE MATERIALS

Most materials can be foil blocked, including leather, textile, wood, paper, card and plastic.

Foil Blocking and Embossing Processes

Foil blocking

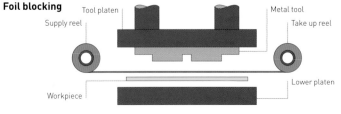

Stage 1: Load

Embossing

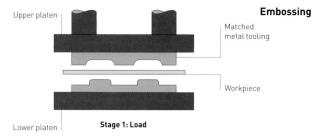

Stage 1: Load

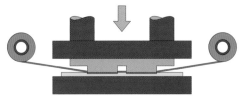

Stage 2: Foil block

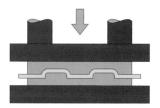

Stage 2: Emboss

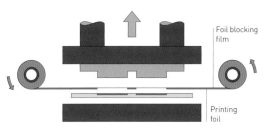

Stage 3: Unload

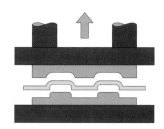

Stage 3: Unload

The thickness depends on the density and resilience of materials. Embossing is generally limited to paper and card up to 2 mm (0.08 in.) or 500 gsm, plastic up to 1 mm (0.04 in.) and leather. Any thickness can be debossed (impressed on 1 side).

COSTS

Tooling costs are very low for both foil blocking and embossing. Rotary tools and matched tooling are more expensive.

Cycle time is rapid, and up to 1,000 parts per hour can be processed.

Labour costs in production are not high, but changeover time is variable, depending on the complexity of the design because the tools have to be registered with existing print.

ENVIRONMENTAL IMPACTS

Environmental impacts are very low. Recycling used foil is impractical, so some designs, such as borders, waste all of the foil within the design area. Embossing produces no waste.

TECHNICAL DESCRIPTION

Foil blocking and embossing are essentially the same process. The difference is that in foil blocking, a layer of foil is placed between the tool and substrate during stamping.

The tooling is metal and is machined by laser cutting (page 248), photochemical machining (page 244) or CNC engraving (page 396). The raised areas of the tool apply the image, and can be used to apply either a positive or negative impression of the design.

In each process, heat and pressure are applied. The metal tools are typically 100–200ºC (212-392ºF). They can be linear or rotary in orientation: rotary processes are very rapid and can be used for continuous production of sheet materials and cylindrical parts.

Metallic foils are very thin deposits of aluminium, applied by vacuum metalizing (page 372). Non-metallic colours, prints, patterns and clear foils are thin films of plastic. Both types are supported by a plastic backing film. This also maintains the integrity of the film once the foil has been printed. On the surface of the foil is a thin film of adhesive that bonds the delicate foil to the substrate.

The combination of heat and pressure bond it to the substrate on contact. It is embedded into the surface, and the depth of impression is determined by the hardness of the materials and the pressure applied to the tool. The tool has square edges, which help to provide a clean edge on the cut out.

Embossing is carried out between matched tooling, whereas debossing only requires an impression tool. Emboss tools have a small radius at the perimeter of the design for aesthetic and functional reasons; a sharp edge would stress the fibre of the workpiece. In a foil block-emboss combination, the foil cut off is not as clean as in separate operations.

Case Study

→ # Foil blocking paper

There are many different types of foil, which are supplied on rolls (**image 1**). Rolls are supplied 640 mm (25 in.) wide and 122, 153 or 305 m (400, 500 or 1,000 ft.) long. The rolls are cut to length and loaded onto the press (**image 2**). The tooling is photochemically machined magnesium (**image 3**). It is mounted onto the press behind the foil. Each piece of paper is picked up by a series of vacuum nozzles, which feed it onto the impression bed (**image 4**). The pressing takes less than a second (**image 5**), and the paper is unloaded.

After an impression, the foil (**image 6**) is wound onto a take up reel. It is now considered scrap. On the finished foil blocking, the gold foil catches the light to produce a shimmering and eye catching design detail (**image 7**).

1

2

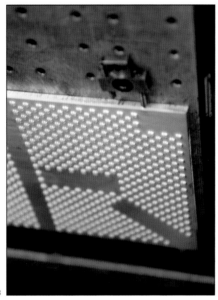

3

4

5

6

7

Featured Manufacturer

Impressions
www.impressionsfoiling.co.uk

Part

5

Materials

Introduction

This section is dedicated to new and emerging technologies, as well as commonplace materials that make up everyday products. Manufacturing processes are affected by a material's properties and so the choice of material will determine what can be achieved and also, crucially, how much it will cost. Sometimes there are subtle differences between material family members such as specific grades of plastic, or metal alloys. In such cases, the typical applications demonstrate individual material use and the additional functionality that it brings to the product.

Materials continue to be a source of inspiration, and emerging material developments provide us with a rare insight into the well-guarded laboratories where they were created. Advanced materials, such as electroluminescent polymers (see image, opposite), amorphous metals (page 453) and shape memory alloys (page 452), are providing solutions to problems with life-changing effects, whilst providing designers with new areas of opportunity.

Materials have the capacity to affect design language. For example, high-performance composite materials (pages 419 and 438) are used to make lightweight structures that would not otherwise be feasible. The opportunity for designers is no longer just to produce gravity-defying structures, but to use these materials to create new product typologies. An example of this is the lace Crochet table (see image, above right). To make this table, crocheted cotton has been soaked in epoxy and cured over a mold. It takes the conventional lace table cover and turns it into the table itself.

A material fulfils both functional and emotional roles. The function of a material is to deliver the expected level of performance. But its influence is more far reaching because materials have sensual qualities too. For example, wood

Cosmos

Designer/client:	Naoto Fukasawa/Swarovski Crystal Palace Project
Date:	2006
Material:	Electroluminescent polymer and Strass® Swarovski® crystal
Manufacture:	Various

has a distinctive smell, is warm to the touch and dents on impact. In contrast, glass is hard, cold and brittle. These properties affect our conception of them and consequently how they are applied. Material selection is therefore integral to the design process.

The longevity of a product is affected by many external factors such as trends and economic value. However, material performance plays a vital role in the long-term relationship that develops between a user and their product. Certain materials encourage users to bond emotionally with the product more than others. For example, a wooden kitchen table that needs polishing and looking after promotes interaction and so encourages a bond to form and develop. Over time, natural materials age and wear according to use and location and so become unique to that application. Designers have picked up on this phenomenon and developed products and materials that actively encourage

Crochet table

Designer/client:	Marcel Wanders/Moooi
Date:	2001
Material:	Cotton and epoxy resin (EP)
Manufacture:	Crocheted cotton is saturated in EP and then formed over a single sided mold

emotional interaction. The Material Memories collection (page 442) is an example of a material whose surface wears away to reveal new patterns of material beneath and in doing so promotes a long-term bond between the product and its owner.

The mechanical properties of each group of materials – plastic, metal, wood, ceramics and glass – vary as a result of their different molecular make up.

PLASTICS
Plastics are divided into 2 main groups: thermoplastic and thermosetting. Both are made up of long chains of repeating

Recycled stationery

Made by:	Remarkable
Material:	Recycled materials including car tyres, car parts, CD cases, plastic boxes, polystyrene packaging, drinks cups and juice cartons
Manufacture:	Various

Self-healing plastic microcapsule

Made by:	Beckman Institute, USA
Notes:	This scanning electron microscope image shows the fracture plane of a self-healing epoxy with a ruptured urea-formaldehyde microcapsule in the centre.

units, known as polymers. The polymeric structure of thermoplastics means that they become plastic and then fluid when heated and so can be molded in a range of processes. Their properties can be fine tuned by adjusting the polymer structure very slightly. Therefore, there are many different types and new ones emerge all the time. Thermoplastics can be melted and reprocessed, but their strength will be slightly reduced each time. Even so, plastics do not break down rapidly in landfill and are energy efficient to recycle, so new products are constantly emerging that make use of these properties (see image, above left). Thermosetting plastics, on the other hand, form polymer chains when 2 parts react together, or when 1 part is catalyzed. They differ from thermoplastics because during polymerization they form permanent cross-links, so cannot be heated, melted and formed; they are shaped in a mold by pouring (see vacuum casting, page 40), injecting (page 50), or vacuum drawing

(see composite laminating, page 206). The rate of polymerization can be adjusted to suit the process and application. This is critical because the reactions are exothermic, so large wall thicknesses may build up too much heat during curing and affect the strength of the part.

Self-healing plastic sample

Made by:	Beckman Institute, USA
Notes:	Image of a self-healing fracture specimen after testing. This shows 1 half of the specimen after it has been fractured into 2 pieces.

The fact that thermosetting plastics are formed by mixing 2 parts means that they can be developed to self-heal if a crack forms. Autonomic Healing Research at the Beckman Institute in the USA recently developed a structural plastic that has self-healing properties (see images, opposite, above right and below). The breakthrough was made possible by the development of microcapsules of dicyclopentadiene (DCPD) that acts as a healing agent with a wall thickness that would rupture when the material began to crack, but not before. The microcapsules release the healing agent, which is catalyzed by chemicals also encapsulated in the material. The liquid material is drawn into the crack by capillary action and polymerizes to form a strong bond with the parent material. Up to 75% of material toughness is recovered by the self-healing process.

Generally, the properties of families of materials are similar. There are obviously exceptions to the rule, and among the most intriguing is polyethylene (PE), which is a member of the thermoplastic polyolefin family (page 430). Low-density polyethylene (LDPE) is a commodity plastic used in a wide range of packaging applications. Ultra high-density polyethylene (UHDPE), on the other hand, is an exceptionally strong fibre used in applications ranging from medical devices to bullet-proof armour.

WOOD

Wood is also made up of polymers: lignin, cellulose and hemicellulose. Lignin is a polymer composed of repeating and cross-linked phenylpropane units; cellulose is made up of repeating glucose molecules; and hemicellulose is a complex branched polymer that forms cross-links between cellulose and lignin. Some biopolymers, such as cellulose and starch (page 446), can be molded and shaped in similar ways to thermoplastics. Unlike synthetic plastics, they will break down naturally and over a shorter time.

Droog Design Table by Insects

Designer/owner:	Front (Sofia Lagerkvist, Charlotte von der Lancken, Anna Lindgren and Katja Sävström)/Droog Design
Date:	2003
Material:	Wood
Manufacture:	Surface decoration created by wood eating insects

As a natural material, wood is prone to attack by insects, animals and disease. No product demonstrates this more cleverly than Table by Insects (see image, above). In this case wood eating insects were let loose on the table after it was constructed. The patterns formed by the insects are intriguing and will be different each time a product is made in the same way.

Wood has anisotropic properties as a result of its grain: both tensile and compression strength are greater along the grain than across it. Steam bending (page 198) works by heating and softening the lignin into its plastic

Glass bottles

Made by:	Beatson Clark
Material:	Soda lime glass
Manufacture:	Glassblowing (machine blow and blow)
Notes:	Clear glass cannot tolerate colour contamination, but brown or green glass can use a high percentage of mixed cullet (recycled glass).

state so that the wood can be shaped over a former. This process utilizes wood's inherent strengths because the grain is formed into the direction of bend. The length of the grain also affects the strength of wood. Processes that reduce grain length, such as CNC machining (page 182), therefore reduce tensile and compressive strength.

METALS
Metals are made up of metallic elements, which are combined in different amounts to form metals with specific properties. The temperature at which the bonds between the elements break down and the metal melts varies according to its ingredients. Low melting point metals, such as zinc (420°C/788°F), become liquid at a temperature that is below the point of degradation of some thermosetting plastics. They can therefore be shaped in a plastic mold (page 144), which has obvious financial advantages. Metals with a higher melting point can be cast in molds made from steel. However, steel and refractory metals cannot be cast in this way and so are either investment cast (page 130) in refractory ceramic shells, or formed in their solid or plastic state. Unlike polymer chains, the strength of bond between metallic elements does not decrease over time or with repeated recycling. This is of great importance because the process of mining metals is more expensive and energy intensive than the synthesis of plastics.

When formed in their plastic state, metals have improved grain alignment compared with liquid state formed parts. Forging (page 114) is used to produce bulk shapes in a range of metals. The hammering or pressing action forces the metals into shape; as a result the grain follows the contours of the part. Liquid state forming produces turbulence in the hot metal, and so the resulting part has inferior mechanical properties.

CERAMICS AND GLASS
Ceramics are made up of non-metallic substances and are formed at high temperatures. Therefore, they have good high temperature properties, but this also means that a lot of energy is required during production. Ceramics are either crystalline or non-crystalline. Crystalline ceramics, known simply as ceramics, can be shaped by hand (page 168), pressing (page 176), slip casting (page 172) or sintering, and are then fired at high temperature in a kiln. Non-crystalline ceramics, known as glass, are formed in their hot, molten state by hand (page 160), blowing (page 152), press molding, press bending or kiln forming.

Violin

Made by:	Frederick Phelps
Material:	Spruce
Manufacture:	Hand carved from solid wood
Notes:	Knot-free and very high quality spruce is still unsurpassed as a material for making violins.

These materials are relatively high cost compared to the other groups, so nowadays are generally only used when no others are suitable. For example, clear or tinted glass is used in reusable packaging applications because it is inert and can be cleaned, sterilized, reused or recycled. Refractory ceramics used for investment casting have very high temperature resistance, but are brittle enough to be broken away afterwards without damaging the metal part within.

COMPOSITES

Composites combine the properties of different material groups and so blur the boundaries between them. The most readily combined materials are plastics with fibre reinforcement. The type of plastic determines the method of forming. Fibre reinforcement is selected to suit the performance demands of the application and has a similar effect to grain in wood, producing materials with anisotropic properties. Processes such as 3D thermal laminating (page 228), filament winding (page 222) and composite laminating (page 206) take advantage of this property because the fibres can be laid down according to the direction of stress. Even injection molded (page 50) fibre reinforced plastic parts have anisotropic properties because the fibres align with the direction of flow. Finite Element Analysis (FEA) software is employed to maximize the efficiency of fibre reinforcement whilst minimizing molding issues.

TRADITIONAL AND NEW MATERIALS

Composite materials create new opportunities for designers because

they have the potential to replace the conventional materials for any given application. Plastics are making the biggest impact by replacing metals in critical applications. However, the properties of some materials are irreplaceable. High quality acoustic musical instruments, such as violas, violins and cellos, are still hand crafted by highly skilled makers. Every aspect of the product from material selection to tuning is overseen by the maker (see image, above), and a signature on the finished piece represents their dedication. Traditionally, spruce (page 470) is used for the soundboard of a

violin; maple (page 475) is used for the back; and the neck, bridge and tuning buttons are carved from rosewood or ebony (page 478). The finest violins are produced from wood that was felled 10 or more years previously. The tree is felled in the winter months, when it is said to be 'sleeping', so that there is as little oil and moisture in the wood as possible. These instruments are worth a lot of money, but the harmony that has been struck between the materials and manufacturing remains unchallenged by modern methods.

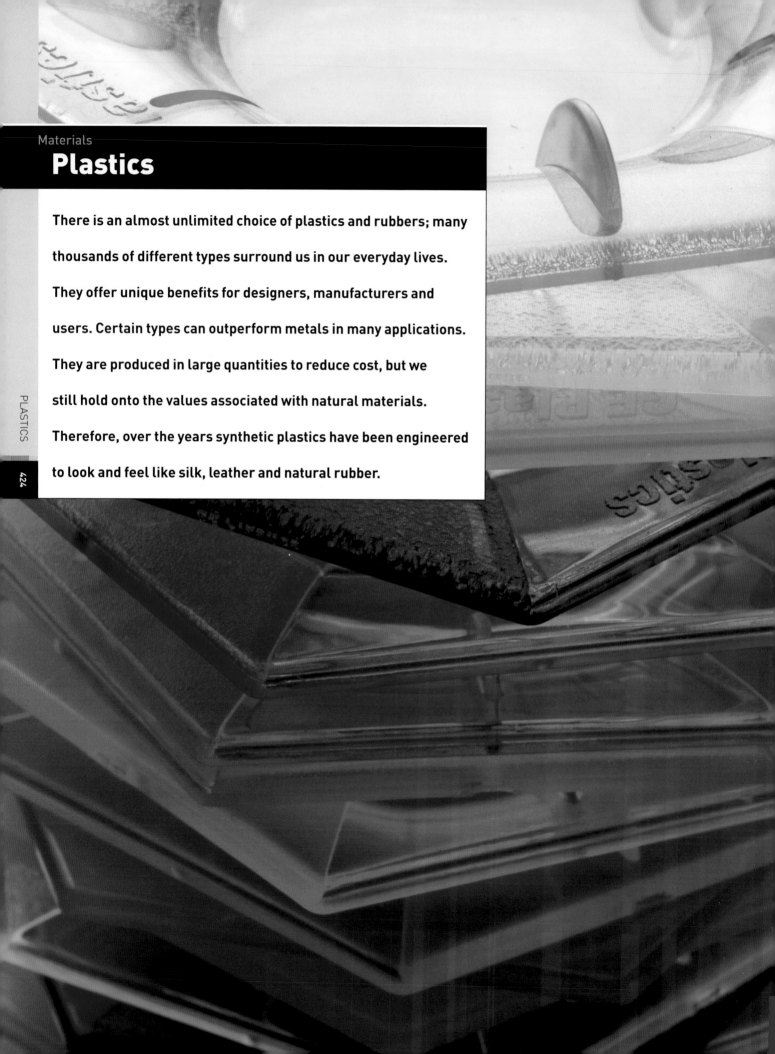

Plastics

There is an almost unlimited choice of plastics and rubbers; many thousands of different types surround us in our everyday lives. They offer unique benefits for designers, manufacturers and users. Certain types can outperform metals in many applications. They are produced in large quantities to reduce cost, but we still hold onto the values associated with natural materials. Therefore, over the years synthetic plastics have been engineered to look and feel like silk, leather and natural rubber.

TYPES OF PLASTIC

Plastics are made up of polymers: long chains of repeating units (monomers) which occur naturally. Synthetic plastics are produced by the petrochemical industry. Some natural polymers are refined and molded without the addition of petrochemicals. These are known as bioplastics, and they include starch-based, cellulose-based and natural rubber materials.

Bioplastics require less energy to produce as raw materials and are fully biodegradable at the end of their useful life. There are also grades of synthetic plastics available that have been modified to break down within a specified time period. This is made possible by the addition of bioactive compounds, which do not affect the physical qualities of the plastic but enable it to be digested by microorganisms in the ground within 1 to 5 years. Products made with this material include some clear plastic bottles, food trays and refuse bags.

Synthetic plastics are divided into 2 groups: thermoplastics and thermosetting plastics (thermosets). The distinction is not always clear-cut because some materials can be both thermoplastic and thermosetting depending on their polymeric structure, for instance, polyester and polyurethane. Materials can also be compounded to produce composites of both groups. The defining difference between these 2 groups is that thermosetting plastics form permanent cross-links between the polymer chains. This creates a more durable molecular structure that is generally more resistant to heat and

chemicals. The formation of cross-links means that the curing process is 1-way; they cannot be remolded.

Thermoplastics can be remolded many times. Off cuts and scrap (regrind) can be reprocessed with virgin material without a significant effect on the properties of the material. Regrind is typically limited to less than 15%, but some materials may contain a much higher percentage.

Thermoplastics are further categorized by molecular structure and weight. The structure or crystallinity differs according to the type of material and method of manufacture. For example, if polyethylene terephthalate (PET) is cooled quickly, the polymer chains form a random and amorphous structure. However, if it is allowed to cool slowly, the polymer chains form an orderly crystalline structure. Amorphous plastics tend to be transparent and have greater resistance to impact. In contrast, crystalline materials tend to have better resistance to chemicals.

Polymers are blended together to form copolymers and terpolymers such as styrene acrylonitrile (SAN) and acrylonitrile butadiene styrene (ABS).

All 3 groups of plastic – bioplastics, thermoplastics and thermosets – contain elastomeric (rubber) materials.

Foam swatches

Made by:	Beacons Products
Material:	Various
Manufacture:	Molded and cut
Notes:	Most plastics can be foamed with open or closed-cell structures. The density, colour and hardness of foam can be specified to suit the requirements of the application.

Elastomers are characterized by their ability to stretch and return to their original shape. Thermoplastic elastomers (TPE) can be melt processed in exactly the same way as rigid thermoplastics. They are frequently over-molded and multi-shot injection molded (page 50) onto rigid thermoplastic substrates to harness the qualities of both.

Like synthetic plastics, synthetic rubbers are produced by the petro-chemical industry. Natural rubber (NR), on the other hand, is produced from a sap tapped from the Para rubber tree (*Hevea brasiliensis*) and is used to make products such as lorry and aeroplane tyres.

Most types of plastic can be foamed. This is carried out with a blowing agent. There are many different types of foam, ranging from flexible to rigid, with either open or closed cells (see image, above). Foams are used for a spectrum of applications such as upholstery, safety, model making and as a core material in composite laminating (page 206).

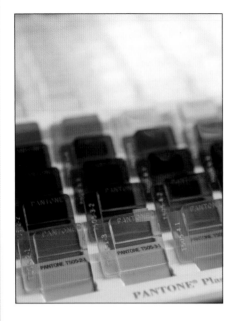

Pantone swatches

Made by:	Pantone
Material:	Polycarbonate (PC)
Manufacture:	Injection molded
Notes:	International reference used to select, specify and match colours.

E-glass fibre reinforcement

Made by:	Various
Material:	Woven glass fibre
Manufacture:	Extruded and woven
Notes:	E-glass, or electrical grade glass, is borosilicate and was first made for electrical insulation. The type of weave used is determined by the application.

ADDITIVES, FILLERS AND REINFORCEMENT

Additives and fillers are used to enhance the properties of plastics. They are used to improve colour, specific mechanical properties, electrical conductivity, mold flow and antimicrobial properties and to increase the material's resistance to fire, UV light and chemical attack.

Some plastics, such as polycarbonate (PC), polystyrene (PS) and PET, are water clear and so take colour very well; they can be tinted or opaque. The colours are typically made up of a carrier resin coated in pigment, which is added to the raw material during processing.

Colour is generally determined by reference to either Pantone (see image, above left) or RAL. These are recognized as international standards and ensure that the colour that is specified by the designer in 1 country is exactly the same as the colour that is injection molded by a factory in another country. Other colour effects that can be achieved in

plastics include metallic, pearlescent, thermochromatic, photochromatic and photoluminescent.

Fillers to improve mechanical properties include talc, minerals, fibres and textiles. Thermoplastics can be injection molded (page 50), extruded and compression molded (page 44) with all of these fillers. Fibre reinforcement may be less than 1 mm (0.004 in.) long in thermoplastic molding. Long and continuous strand fibre reinforcement is incorporated into thermosets by compression molding, composite laminating and 3D thermal laminating (page 228). The types of fibre used include glass (see image, above right), aramid, carbon and more recently hemp and jute. Fibre reinforced composite materials have superior strength for their weight, several times greater than metal, and different grades are suitable for everything from boat hulls to lightweight stacking furniture (see image, opposite).

Carbon fibre is manufactured by oxidizing, stretching and heating a polymer, such as polyacrylonitrile (PAN), to over 1500°C (2732°F) in a controlled atmosphere. The process of carbonization takes places over several stages of heating and so requires a great deal of energy. Eventually, long ribbons of almost pure graphite are formed, and the width of ribbon will affect the strength of the carbon fibre. There are different grades (strengths) of carbon fibre, which depend on the manufacturer.

Carbon fibre has very high strength to weight and as a result of its own success is becoming more difficult to purchase because production cannot keep up with demand. It is a relatively expensive material, but is becoming more widespread in consumer products such as sports equipment. This is partly due to the added value that carbon fibre brings to an application, but also as a result of improved manufacturing techniques, including recent breakthroughs in recycling methods. Traditional composite laminating requires a great deal of highly skilled labour and is therefore very expensive. New techniques are now being used, such as resin transfer molding and resin infusion (page 206), which are more rapid and mechanized and so reduce labour costs considerably.

NOTES ON MANUFACTURING

Plastics with different molecular structures lend themselves to different methods of manufacture. Thermoplastics are shaped by heating them until they are soft or liquid enough to be formed, so they are generally supplied in granulated form. Processes such as thermoforming (page 30) require sheet materials, which are extruded from granules. This increases the material costs due to the extra processing that is required. Thermoplastics are also available as drawn fibres and blown film.

Different amounts of crystallinity also affect processing. For example, amorphous materials do not have a sharp melting point like crystalline materials. Therefore, they are more suited to processes like thermoforming, because the material stays soft and formable over a wider temperature range. In contrast, crystalline materials will flow more easily in the mold due to their sharp melting point and are therefore more suitable for thin wall sections and complex features.

Due to their different properties, not all thermoplastics are compatible. Care must be taken when using processes such as multi-shot injection molding and plastics welding to ensure a strong inter-material bond. Material incompatibility is sometimes used as an advantage, for example, in release agents and lubricants.

Thermosets are cured in the mold. Some are engineered to cure at room temperature such as polyester, vinyl ester, epoxy (EP) and polyurethane (PUR); these are supplied in liquid form and mixed with a catalyst or hardener. They can be poured or injected into a mold. Alternatively, powdered, liquid or solid thermosets are heated in a mold to trigger the formation of cross-links.

The production of bioplastics crosses over with both thermoplastics and thermosets; they can be molded with thermoplastic technology, but the process may not be repeatable. Cellulose acetate (CA) and polylactic acid (PLA) are processed using similar techniques to thermoplastics. For example, Potatopak food packaging (see image, page 446 top right) is produced by compression molding powdered starch.

MATERIAL DEVELOPMENTS

The plastics industry is constantly evolving. New blends, compounds and additives are being developed to create opportunities, improve performance and reduce cost. Many developments will affect the design industry in the coming year: below is a brief outline of some of the most important.

Stacking stool

Designer:	Rob Thompson
Date:	2001
Material:	Polyester and glass fibre
Manufacture:	Composite laminating

L'Oreal shop, Paris

Client:	L'Oreal
Date:	Completed 2004
Material:	Bencore Starlight polycarbonate (PC)
Manufacture:	CNC machining

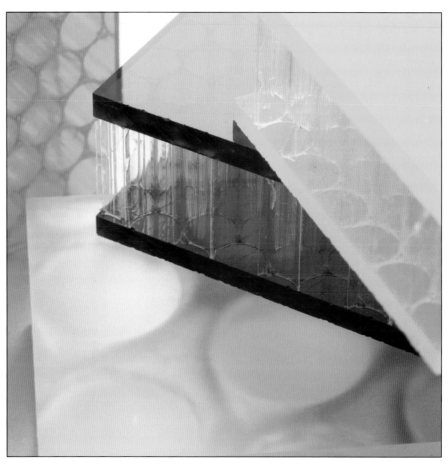

There are many different types of lightweight composite plastic panels. Core materials include aluminium honeycomb, DuPont™ Nomex® honeycomb and corrugated and rigid foam. Bencore, an Italian manufacturer, produce a range of plastic panels used in applications such as furniture, trade fair stands, shop interiors and exteriors (see images, above). The polycarbonate (PC) core, whose commercial name is Birdwing®, is thermoformed into 19 mm (0.75 in.) diameter 3D cells. This core is laminated between clear or coloured PC sheets (see images, above). These panels are also available in styrene acrylonitrile (SAN), high impact polystyrene (HIPS) or polyethylene terephthalate modified with glycol (PETG).

Rapid prototyping (page 232) has revolutionized the design process. Not only are models accurate to within a few microns, but also this process is capable of producing plastic parts that cannot be made in 1 piece in any other way. Plastics

Bencore Starlight and Lightben

Made by:	Bencore
Material:	Polycarbonate (PC)
Manufacture:	High pressure laminated
Notes:	Available in polycarbonate (PC), styrene acrylonitrile (SAN), high impact polystyrene (HIPS) and polyethylene terphthalate modified with glycol (PETG).

available include epoxy resin, unfilled nylon, and glass or carbon filled nylon. Traditionally these materials have been engineered to mimic production grade plastics such as acylonitrile butadeine styrene (ABS), polypropylene (PP) and poly methyl methacrylate (PMMA) acrylic. Recent developments have made it suitable for direct manufacturing low volume parts and one offs as well as pre-production models.

Conductive polymers combine the benefits of plastics technology with the conductivity of metals. Developments include printed circuitry, light emitting plastics and electroactive polymers (EAP).

Printed circuitry is used in Plastic Logic's flexible active-matrix display (see image, opposite above left). This technology has the potential to revolutionize the way we read books, reports and even the morning paper. The rigid glass backpane required for conventional amorphous silicon-based displays has been replaced by a plastic, which means the display can be flexible, thinner and more lightweight.

Light emitting polymers, or polymer light emitting diodes (PLED), are based on the principles of electroluminescence. Sandwiched between 2 electrodes, the polymer lights up when an electrical current is passed through it. It is now possible to print a matrix of plastics that emit different colours of light to produce very thin flat panel displays.

EAPs are similar to electroactive ceramics and shape memory metal alloys: they respond to electrical stimulation with a change in shape or size. They are in the very early stages of

Flexible active matrix display

Made by:	Plastic Logic
Material:	Plastic display using E Ink® Imaging Film
Manufacture:	Active-matrix LCD backplane fabricated onto plastic substrate
Notes:	Displays made from plastic are more flexible and lighter than those made of glass.

development, and at present there are very few commercial applications.

ENVIRONMENTAL IMPACTS

There have been many developments that reduce the environmental impacts of plastics. For example, many disposable products are produced in bioplastics or a thermoplastic with bioactive additives.

Thermoplastics are very efficient to recycle. There is minimal degradation of quality and some products are even made from 100% recycled materials. Smile Plastics in the UK manufacture a range of plastic sheets by made up of 100% recycled plastic material (see image, above right). They are produced from packaging, footwear and even old mobile phones. Patagonia, an American based company, produce a range of fleeces from 100% recycled plastics drink bottles (PET). The role of the designer is to ensure that plastics can be recycled with minimal contamination. Therefore, parts should be designed in a single material if possible and with recycling and disassembly in mind.

This book cannot avoid mentioning the negative environmental impacts of certain plastics. Not only do they take thousands of years to biodegrade, some are also harmful in production. For example, polyester and epoxy

Recycled plastic sheets

Made by:	Smile Plastics
Material:	100% recycled thermoplastic
Manufacture:	Compression molded
Notes:	Many thermoplastics can be preprocessed in this way, including polyethylene (PE), PET, PVC and polycarbonate (PC).

contain styrene; phenolic resin, urea and melamine contain formaldehyde; polyurethane resin (PUR) contains diisocyanates; and polyvinyl chloride (PVC) contains dioxins. These chemicals are known pollutants and are toxic. Some are referred to as volatile organic compounds (VOC) and are known carcinogens. During production care must be taken to avoid inhaling them, and some plastics will continue to off-gas during their lifetime and so both reduce indoor air quality and pollute the atmosphere.

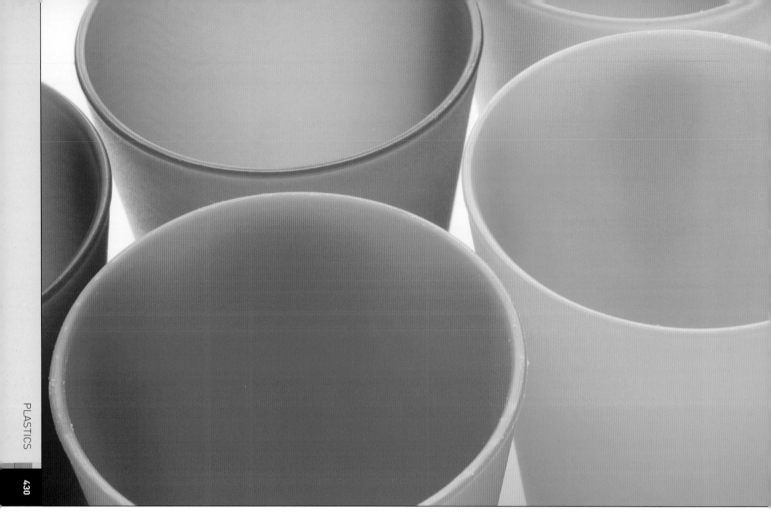

Polyolefins

- Polyethylene (PE)
- Polypropylene (PP)
- Ethylene vinyl acetate (EVA)
- Ionomer resin

PP and PE are the building blocks of the plastic industry and make up more than half of total global production.

Qualities: These plastics have a low coefficient of friction and are noted for their resistance to water absorption and attack by many acids and alkalis. They are non-toxic and are available as transparent, tinted or opaque. They are generally not suitable for applications above 100°C (212°F).

PE has good resistance to punching and tearing, even at low temperatures. Its properties are partly determined by its molecular weight, and it is classified as high-density (HDPE), ultra high-density (UHDPE), low-density (LDPE), ultra low-density (ULDPE) and linear low-density (LLDPE).

POLYOLEFINS
IKEA Kalas mug

Designer/client: Monika Mulder/IKEA
Date: 2005
Material: Polypropylene (PP)
Manufacture: Injection molded

UHDPE is an exceptional material with very high impact resistance. It was developed in the 1970s by DSM, who manufacture it under the trade name Dyneema®. As a drawn fibre or sheet it is up to 40% stronger than para-aramid (DuPont™ Kevlar®, see polyamides page 438) and 15 times stronger than steel.

PP has exceptional fatigue resistance, so is ideal for integral hinges and snap fits. It has a waxy feel, due to low surface energy, which means it is difficult to coat with adhesives and paints. Even so, it is easily welded and mechanically joined.

EVA can be semi-rigid, or very flexible, depending on the vinyl acetate content.

POLYOLEFINS
Plastic cork

Made by: Various
Material: Ethylene vinyl acetate (EVA)
Manufacture: Co-extrusion
Notes: The plastic is tasteless, long lasting and can be extracted with a conventional corkscrew.

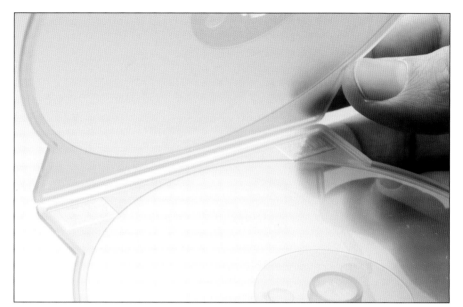

Clamshell CD packaging

Designer/client:	Unknown
Date:	Unknown
Material:	Polypropylene (PP)
Manufacture:	Injection molding

Délices de Cartier

Made by/client:	Alcan Packaging Beauty/Cartier
Date:	2006
Material:	DuPont™ Surlyn® ionomer resin cap on a glass bottle
Manufacture:	Cap is injection molded

Thin walled bottles

Made by:	Various
Material:	High density polyethylene (HDPE)
Manufacture:	Extrusion blow molding (EBM)
Notes:	Polypropylene (PP), polyethylene (PE), polyethylene terephthalate (PET) and polyvinyl chloride (PVC) are all suitable for EBM.

As a semi-rigid polymer it has similar characteristics to LDPE. Increasing the vinyl acetate content increases its flexibility and produces similar characteristics to rubber.

A typical ionomer is ethylene methacrylic acid (E/MAA). Ionomers are thermoplastics with ionic cross-links.

They have superior abrasion resistance and tear strength, high levels of transparency and good compatibility and adhesion to many metals and polymers. Their exceptional impact resistance makes them virtually shatterproof, even at low temperatures.

Ionomers can be shaped by conventional thermal forming processes. As they cool and solidify the metal ions (sodium or zinc) form cross-links in the thermoplastic structure, giving them similar characteristics to thermosetting plastics at low temperatures. On heating, the ionic cross-links break down so the material can be recycled and reprocessed.

Applications: The whole family of materials are used to make textiles for consumer and industrial applications. PE is used for thin walled plastic packaging. Examples include shopping bags, milk bottles, medical and cosmetic packaging. It is non-toxic and so is used for toys and chopping boards.

UHDPE is used for high performance applications such as bullet proof garments, ropes and parachute strings.

PP is a versatile material that lends itself to a wide range of applications. It is used for food packaging, drinking cups, caps and enclosures. Other uses include furniture, lighting and water pipes.

STYRENES

MacBook Pro packaging

Made by:	Unknown
Material:	Expanded polystyrene (EPS)
Manufacture:	EPS molding
Notes:	EPS is lightweight, protective and insulating.

STYRENES

Attila can crusher

Designer/client:	Julian Brown/Rexite
Date:	1996
Material:	Glass filled acrylonitrile butadiene styrene (ABS)
Manufacture:	Injection molding

PLASTICS

432

EVA is puncture and tear resistant and is often used for medical packaging such as blood transfusion bags. In recent years, EVA has seen an explosion in consumer and packaging applications such as luxurious carrier bags and flexible net packaging used for glass bottles and fruit. As foam it is used for trainer soles and imitation cork.

Ionomers are used in a range of applications, from blown films and coatings to injection molded and blow molded parts. It is possible to mold thick wall sections without sink marks or voids. It is water-clear and so can feel and look like glass and is therefore utilized in high-end packaging such as cosmetics. Other applications include golf balls covers and the soles of football boots.

Costs: PE, PP and EVA are low cost. Ionomers are medium cost.

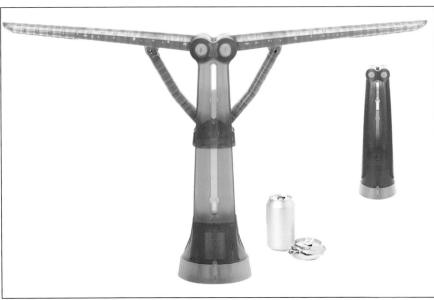

Styrenes

• Polystyrene (PS)
• Acrylonitrile butadiene styrene (ABS)
• Styrene acrylonitrile (SAN)
• Styrene butadiene styrene (SBS)
• Styrene ethylene butylene styrene (SEBS)
This is a broad range of thermoplastics that are easy to process and have good visual properties.

Qualities: PS is categorized as general purpose (GPPS), expanded (EPS) and high impact (HIPS). GPPS is naturally transparent and brittle. HIPS is produced by blending PS with polybutadiene; it has improved impact resistance and toughness. EPS is molded by expanding PS beads with gas and steam, which inflates them into foam that is 2% material and 98% atmosphere.

ABS has similar properties to HIPS with superior chemical and temperature resistance. It has a high gloss surface (like GPPS) and is produced in vivid colours.

SAN is resistant to impact and scratching and has excellent light transmission qualities (90%).

SBS and SEBS are copolymerized with synthetic rubber (page 445) and so have a large range of flexibility, with a shore hardness range of 0A–50D. They remain flexible even at low temperatures.

These materials are generally limited to applications below 80°C (176°F).

Kokon Furniture 'Double Chair'

Designer/owner:	Jan Konings and Jurgen Bey/ Droog Design
Date:	1997
Material:	Polyvinyl chloride (PVC) covering and used furniture
Manufacture:	PVC covering stretched over used furniture to create new objects

Water soluble pouch

Made by:	Various
Material:	Polyvinyl alcohol (PVOH)
Manufacture:	Plastic welded film
Notes:	The film dissolves in cold water to deliver a measured dose of detergent to laundry.

Applications: GPPS is used for disposable food packaging, cutlery and drinking cups, CD 'jewel' cases, lighting diffusers and model making kits. HIPS is used for vending cups, product housings and toys. EPS is used in 3 main ways: either molded into the desired shape (packaging, helmets and toys), molded into pellets (loose fill), or as sheet material that can be cut (model making, thermal insulation and packaging). EPS is the cheapest polymer foam.

ABS is used for a range of applications including housings for power tools, telephones, computers and medical equipment, ear defenders and other safety equipment, children's toys and automotive parts.

SAN has high thermal stability and so is often used for transparent kitchen appliance mixing bowls, water purifiers and other products that require a very high surface finish and durability combined with the convenience of dishwashing. It is also used in the automotive and medical industries.

SBS and SEBS are used extensively in multishot injection molding (page 50) and co-extrusion applications with thermoplastics including PE and PP (see polyolefins, page 430), ABS, PC (see polycarbonate, page 435) and PA (see polyamide, page 438). Products include babies' teats and dummies, ice cube trays, soft touch grips and children's toys.

Costs: Low.

Vinyls

- Polyvinyl chloride (PVC)
- Polyvinyl alcohol (PVOH)

PVC has a glossy appearance and its shore hardness can be adjusted to suit the application. PVOH is a water soluble barrier film.

Qualities: PVC is a long lasting material and more than half of global production is for use in the construction industry. It has a glossy surface finish and is available in a range of vivid colours. It is either plasticized or not. Unplasticized PVC (uPVC) is rigid; where as plasticized PVC is flexible with a shore hardness range of 60–95A. It is naturally flame retardant and has good resistance to UV light, but PVC is not suitable for exposure to prolonged periods at temperatures above 60°C (140°F).

The downside of PVC is that it contains chlorine and dioxins. This has led to many campaigns against its use, especially in food, medical and toy applications. However, the production of PVC is increasing as it replaces more wood and metal products in the building and construction industries.

Hydrolyzed polyvinyl acetate (PVA) makes PVOH, a water-soluble film. PVA is used in adhesives and coating because it is not suitable for molding. PVOH is produced as fibres and films. It dissolves easily in water and the rate of dissolution is adjusted by the structure of the polymer (crystallinity) and temperature; PVOH will absorb more water as it warms up. This means that PVOH can be used in water at room temperature without biodegrading; when the water is warmed to its trigger temperature, the polymer chains start breaking down rapidly.

Applications: PVC is used to make 'vinyl' records. Now it is more widely used for extruded window frames, doorframes and guttering. Other applications include identity and credit cards, medical packaging, tubing, hoses, electrical tape and electrically insulating products.

Vinyl is used to coat materials and fabrics to protect them from chemicals, stains and abrasion. Examples include coated wallpaper and upholstery fabrics.

PVOH is used in hospital laundry bags, water-soluble packaging, fishing bait rigs and, when mixed with sodium tetraborate, children's 'slime'.

Costs: Low.

Acrylic and composites

• Poly methyl methacrylate (PMMA)
Acrylic is used for its combination of clarity, impact resistance, surface hardness and gloss.

ACRYLIC AND COMPOSITES
Lymm Water Tower kitchen

Designer/client:	Ellis Williams Architects/ Russell and Jannette Harris
Date:	Completed 2005
Material:	DuPont™ Corian® (acrylic and aluminium trihydrate)
Manufacture:	Thermoformed, CNC machined and adhesive bonded

Qualities: Acrylic sheet materials are available as extruded or cast. There are small differences between the 2 that will affect how they are machined and formed. Sheets can be up to 60 mm (2.36 in.) thick, but become very expensive over 10 mm (0.4 in.). Sheet materials can be thermoformed (page 30), drape formed and machined (page 182). Acrylic can be machined to tight tolerances because it is a hard material and it can be polished to a high gloss finish (especially by diamond polishing, page 376). Sheet materials can be cut with conventional wood cutting equipment, as long as the tool does not

heat the material above 80°C (176°F), because it will start to soften. It is an ideal material for laser cutting, scoring and engraving (page 248) because the heat of the laser produces a gloss finish.

Edge glow (see image, opposite) is caused by light picked up on the surface of the sheet and transmitted out through the edges. Cut edges and scored lines glow in the right lighting conditions. This effect is exploited in signage and decorative applications.

It is possible to incorporate materials and objects in cast acrylic sheet or blocks. Examples included barbed wire, razor blades, flowers, grass and broken glass.

DuPont™ Corian® (see image, opposite) is a blend of PMMA and natural mineral (aluminium tri-hydrate). This forms a hard, durable sheet material. It is relatively expensive, but has a luxurious quality unmatched in thermoplastics. Sheets are 3680 x 760 mm (145 x 30 in.), 6 mm to 12.3 mm (0.024–0.48 in.) thick and available in a wide range of colours.

Applications: Acrylic is used in point of sale displays, furniture, signage, light diffusers, control panels, screens, lenses and architectural cladding.

Costs: Moderate.

ACRYLIC AND COMPOSITES
Discs showing edge glow

Made by:	Various
Material:	Poly methyl methacrylate (PMMA) acrylic
Manufacture:	Cast or extruded
Notes:	Tinted acrylic materials produce edge glow from ambient light.

Polycarbonate

• Polycarbonate (PC)
PC has excellent clarity and superior mechanical properties. Rich and luminous colours make this an ideal material for electronic products such as mobile phones and Apple Mac computers.

Qualities: PC is a suitable and safe alternative to glass for certain products such as beakers and spectacle lenses. It has an amorphous structure, which contributes to it being the toughest clear plastic. However, this also means that it is prone to degradation by UV light and certain chemicals.

PC is blended with other polymers, such as ABS (see styrenes, page 432), to increase rigidity and impact resistance, especially for thin walled parts. Blending and molding materials together harnesses the desirable properties of each; PC-ABS is less expensive than PC and improves the properties of ABS, with

POLYCARBONATE
Bluetooth Penelope*Phone

Designer/client: Nicolas Roope and Kam Young/
Hulger Ltd
Date: 2005
Material: Polycarbonate (PC)
Manufacture: Injection molded

POLYCARBONATE
Miss K table lamp

Designer/client: Philippe Starck/Flos
Date: 2003
Material: Polycarbonate (PC) internal and
external diffuser, and poly methyl
methacrylate (PMMA) acrylic frame
Manufacture: Injection molded and vacuum
metalized with aluminium

Thermoplastic polyurethane

- Thermoplastic polyurethane (TPU)
There is a range of flexibilities from shore
hardness 55A to 80D. TPU is resistant to
abrasion, tearing and puncturing.

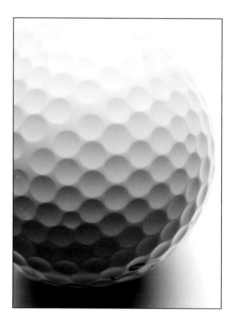

THERMOPLASTIC POLYURETHANE
Golf ball

Made by: Various
Material: Thermoplastic polyurethane (TPU)
Manufacture: Injection molded
Notes: Golf balls are made up of 2 or
more layers: the core is typically
synthetic rubber and the outer
layer is TPU.

better surface finish and processing, for example. PC is a relatively compatible material, which can be blended, multishot injection molded (page 50) and coextruded with a number of different materials.

Applications: Applications include lighting diffusers (indoor, outdoor and automotive), police riot shields and safety shields, beakers, beer glasses and babies bottles, motorcycle helmets and CDs and DVDs.

Costs: Moderate.

Qualities: The properties of TPU are similar to thermoset polyurethane (page 443). They are tough, durable, resistant to fuels, oils and greases and resistant to flexural fatigue across a broad temperature range, from -45°C to 75°C (-49–167°F). They exhibit low levels of creep and high levels of resilience and so are suitable for load bearing applications.

TPU is available as a resin, sheet, film and tube. It is naturally opaque but can be produced in clear.

Applications: TPU is breathable and so is used for many clothing and sportswear applications such as shoes

THERMOPLASTIC POLYESTERS
Sei water bottle

Company:	Sei Global
Date:	2006
Material:	Polyethylene terephthalate (PET)
Manufacture:	Injection blow molded

THERMOPLASTIC POLYURETHANE
Feu D'Issey

Designer/client:	Curiosity Inc./Beaute Prestige International
Date:	2001
Material:	The red casing is thermoplastic polyurethane (TPU)
Manufacture:	Injection molded

and weatherproof outdoor clothing. It is a tactile material and is used to make synthetic leather.

It is suitable for abrasive applications such as conveyor belts, automotive interior linings, instrument panels, wheels and over-molded gear sticks and levers. As well as molded products, TPU is applied as a coating.

Costs: Moderate.

Thermoplastic polyesters

- Polyethylene terephthalate (PET)
- Polyethylene terephthalate modified with glycol (PETG)
- Polybutylene terephthalate (PBT)
- Polycyclohexylene dimethylene terephthalate (PCT)
- Liquid crystal polymer (LCP)
- Thermoplastic polyester elastomer (TPC-ET)

There is a range of thermoplastic polyester engineering materials. All have high dimensional stability and are resistant to chemicals.

Qualities: PET is the most widely used thermoplastic polyester (for thermoset polyesters, see page 442). It can be amorphous or semi-crystalline depending on how it is processed and cooled (rapid cooling does not allow the formation of a crystalline structure). It is possible to modify PET with glycol (PETG):

this reduces the brittleness and premature ageing of PET. PETG has improved thermoforming properties and so is ideal for lighting diffusers and medical packaging, for example.

PBT is electrically insulating and flame retardant. It can be glass filled, which improves its heat stability and temperature range up to 200°C (392°F).

PCT has all the mechanical properties of PET and PBT with greater resistance to high temperatures and water absorption.

LCP is an aromatic polyester. It is non-flammable, has exceptionally low mold shrinkage and very high dimensional stability. LCP has rod-like molecules, which align during melt processing. They produce superior physical properties over a wide temperature range with very little thermal expansion or shrinkage, especially in the direction of flow; the molecular alignment produces anisotropic properties.

TPC-ET is an engineering elastomer, it has excellent fatigue resistance and

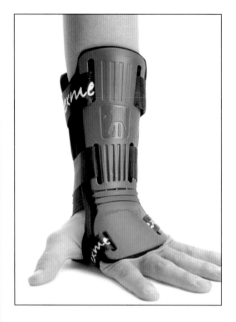

THERMOPLASTIC POLYESTERS
Flexometer® wrist guard

Designer/client:	Dr Marc-Hervé Binet/Skimeter
Date:	2003
Material:	DuPont™ Hytrel® thermoplastic polyester (TPC-ET)
Manufacture:	Injection molding

THERMOPLASTIC POLYESTERS
Chevrolet HHR headlamp

Designer/client:	DuPont™ Automotive/Chevrolet
Date:	2006
Material:	Headlamp bezel and trim ring are DuPont™ Crastin® polybutylene terephthalate (PBT)
Manufacture:	Injection moulded

Polyamides

• Polyamide (PA)
• Aromatic polyamide (Kevlar® and Nomex®)
PA, commonly known as nylon, is a versatile material used in a wide range of applications. Aromatic polyamides are stronger than steel for their weight.

its flexibility ranges from 30–80D. It has high impact strength, tear strength and impact resistance.

Applications: PET is used for drinks bottles because it is transparent in its amorphous state, is easily blow molded and can withstand internal gas pressure from carbonated drinks. Post-consumer PET bottles are reprocessed into new products such as textiles and fleeces.

PET is used as an industrial film. An example is Madico's Lumisty view control film. The textured surface on the film refracts light and controls the viewing angle either horizontally or vertically. It is used as a coating on glass and plastic for security, decoration, to reduce ultraviolet light and to prevent glass disintegrating if it breaks. It is a high performance material: Northsails 3DL and 3Dr technology (page 228) laminates aramid and carbon fibres between PET film.

PBT and PCT are used for oven handles and knobs, electrical connector blocks, switches and light bulb housing. LCP is used for lighting connectors, computer chip carriers, microwave oven parts and cookware, mobile phone parts and aerospace optical components.

TPC-ET is used in keypads, tubing, seals, sports goods, medical devices and soft touch handles.

Costs: Costs vary according to type: PET is low to moderate cost, TPC-ET is moderate cost, PBT and PCT are moderate to high cost and LCP is high cost.

Qualities: There are many different grades of nylon, which are categorized as follows: nylon 6; nylon 6,6; nylon 6,12; amorphous nylon (transparent) and high temperature nylon. These grades can be modified in a number of different ways to produce reinforced, flame retardant, tough and super-tough versions.

Nylon is self-lubricating, has a low coefficient of friction, and has good resistance to abrasion and chemicals.

There are 2 main types of aromatic polyamides, including para-aramids known by the DuPont™ trademark name Kevlar®, and meta-aramids, which are also known by the DuPont™ trademark name Nomex®. They are only available as spun fibre or sheet material because there is no other practical way to form them. They have superior resistance to temperatures up to 500°C (932°F) and are inherently flame retardant.

Para-aramid fibres have exceptional strength to weight, far superior to steel. Alongside UHDPE (see polyolefins, page

ACETAL
Flo-Torq® IV propeller hub

Designer/client: Mercury Marine
Date: 2006
Material: DuPont™ Delrin® acetal
Manufacture: Over-molded by injection molding onto titanium rods

POLYAMIDES
Makita Impact Driver

Designer/client: Makita
Date: 2006
Material: Polyamide (PA) nylon body and thermoplastic elastomer (TPE) pistol grip
Manufacture: Multishot injection molded

430), they are as close as scientists have come to mimicking spider web.

Meta-aramids are used mainly for their high heat performance, electrical insulation and flame resistance. They do not have the same level of resistance to cutting and abrasion as para-aramids and so they are often reinforced or laminated into papers.

Applications: Nylon is used in bushes and bearings, electrical equipment housings, furniture parts (mechanisms and feet) and sports equipment. As a drawn fibre, it exhibits similar properties to silk and so is widely used as a substitute in garments, apparel and carpets. Following its conception in the 1930s, nylon rapidly became a material in its own right. Stockings and tights made out of nylon are very successful because they are dramatically cheaper than silk versions. Cordura® is the trade name for nylon fabrics from Invista; they are tough, durable and resistant to abrasion.

Powdered nylon is fused into 3D forms in rapid prototyping (page 232).

Para-aramids are used as a single material for ropes and sheets, and commonly as fibre reinforcement in composite materials. Meta-aramids are found in applications that demand very high performance such as power generation, aircraft, aerospace, racing cars and fire fighting. They are often used as fabrics such as in flame resistant undergarments worn by racing drivers.

Costs: Nylon is moderate cost and aromatic polyamides are high cost.

Acetal

• Polyoxymethylene (POM)
POM, known as acetal, is an engineering polymer that bridges the gap between plastics and metals.

Qualities: A high level of crystallinity means acetal is dimensionally stable at peak temperatures and can operate close to its melting point. It is resistant to many chemicals, fuels and oils, and its low coefficient of friction reduces wear.

It is tough, even at low temperatures, with low warpage, and has high fatigue endurance, so is suitable for spring and snap fits. It is ideally suited to injection molding (page 50) and machining (page 182), but can be rotation molded (page 36), blown (page 22) and extruded.

Applications: Acetal is used if the less expensive thermoplastic polyester (page 437), PE (see polyolefins, page 430) and polyamides (opposite) are not suitable. It is important in the appliance, hardware and automotive, industries where it has replaced metal parts. Uses include speaker grilles, sports goods, seatbelt systems and automotive fuel systems.

Costs: Moderate to high.

Polyketone

- Polyetheretherketone (PEEK)
PEEK is used for its hardness, dimensional stability and wide working temperature range of -20°C to 250°C (-4–482°F). It is suitable for molding and machining and is available as sheet, plate, rod and tube.

Qualities: PEEK is an opaque grey material. It is non-flammable, electrically insulating and hard, with a low friction coefficient. Mechanical properties are maintained at temperatures as high as 250°C (482°F), and it is resistant to many acids and chemicals.

PEEK reinforced with carbon fibre has dramatically improved mechanical properties with little increase in weight, and is used in the aerospace industry.

Applications: Aerospace, medical equipment and electronic products.

Costs: Very high.

Fluropolymers

- Polytetrafluoroethylene (PTFE)
- Ethylene tetrafluoroethylene (ETFE)
- Fluorinated ethylene propylene (FEP)
Commonly known by the DuPont™ trademark name Teflon®, fluoropolymers are well suited to use in extreme environments: they are inert and stable across a wide temperature range.

Qualities: These are classified as a family containing fluorine atoms, which give them superior resistance to virtually all chemicals and an extremely low friction coefficient. The low friction coefficient acts like a lubricant and stops other materials sticking to them.

They are non-flammable, resistant to stress cracking and fatigue, ultraviolet light and virtually all chemicals. They operate effectively at a wide temperature range from -250°C to 250°C (-418–482°F).

They can be formed by a host of melt processes including injection molding (page 50), rotation molding (page 36) and extrusion, but they are very expensive and so often limited to coatings and linings. They can be transparent or opaque, so coatings can appear invisible.

Applications: These are high cost materials, but they are effective as a thin coating. Examples include non-stick cookware and self-cleaning coatings on fabric and glass. Fluoropolymers are also used for automotive, aerospace and laboratory applications.

Costs: Very high.

Thermoplastic rubber compounds

- Melt-processible rubber (MPR)
- Thermoplastic vulcanizate (TPV)
- Engineering thermoplastic vulcanizate (ETPV)
These combine the performance of rubber with the processing advantages of thermoplastics.

Qualities: These materials contain elements of rubber but do not cross-link during shaping, so can be recycled many times. They have similar mechanical and physical properties to rubber but can be molded on conventional thermoplastic equipment. This means much more rapid processing times. They typically range from shore hardness 20 to 90A.

These materials are resistant to abrasion, chemicals, fuels and oils. They have high impact strength, are flexible even at low temperatures and resistance to flexural fatigue. MPR can withstand constant use at 125°C (257°F) and ETPV up to 150°C (302°F).

TPV and ETPV are made up of a thermoplastic matrix embedded with particles of EPDM (see polyolefins, page 430). The thermoplastic matrix of TPV is PP. It is known as ETPV when the PP is replaced by a higher performance thermoplastic.

Applications: The range of applications include keypads, soft touch handles and grips, hot water tubing, seals, gaskets, sportswear and goods, medical devices and automotive parts.

Costs: Moderate to high.

Formaldehyde condensation resins

- Phenol formaldehyde resin (PF)
- Urea formaldehyde (UF)
- Melamine formaldehyde (MF)
These thermosetting materials have a hard and glossy finish that is resistant to scratching. They are resistant to most chemicals, have good dielectric properties and operate across a wide temperature range.

Qualities: PF, under the trade name Bakelite, revolutionized the plastics and design industry. It was invented around 1900 and became the first commercially molded synthetic polymer in the 1920s. PF is naturally dark brown or black and so does not colour easily.

UF and MF are sub-categorized as amino resins. They are naturally white or clear and so are available in a range of vivid colours. In the 1930s, they began to replace PF for use in domestic products like tableware.

Formaldehyde resins are resistant to moisture absorption and so do not stain easily. They are tasteless and so are

FORMALDEHYDE CONDENSATION RESINS
Aramith billiard balls

Designer/client: Saluc
Date: N/a
Material: Phenolic resin
Manufacture: Molded, ground and polished

FORMALDEHYDE CONDENSATION RESINS
Le Creuset casserole

Designer/client: Le Creuset
Date: Unknown
Material: Phenolic resin handle
Manufacture: Compression molded

suitable for food and drink applications, and are also resistant to burning.

PF is suitable for operating temperatures up to 120°C (248°F). MF forms more cross-links than the other 2, and so has superior mechanical and chemical resistance properties.

Applications: Examples include electrical housing, plugs and sockets, kitchen equipment, tableware, ashtrays, knife and oven handles, light fittings and snooker, pool, bowling and croquet balls. Demand has recently increased for under-the-bonnet applications in battery-powered cars because these materials provide the electrical insulation and stability that is required at a reasonable price.

They are used in adhesives for laminating plywood and they are impregnated into papers and fibres. For example, DuPont™ Nomex® (see polyamides, page 438) is impregnated with PF, laminated wood is bonded

together with UF and Formica is a laminate of paper or cotton and MF.

Costs: PF and UF are low cost. MF is more expensive.

FORMALDEHYDE CONDENSATION RESINS
Mebel clam ashtray

Designer/client: Alan Fletcher/Mebel
Date: 1970
Material: Melamine
Manufacture: Compression molded

Polyesters and composites

• Polyester
Polyesters are the least expensive composite laminating resins.

POLYESTERS
Material memories

Designer:	Rob Thompson
Date:	2002
Material:	Polyester resin and natural fibres
Manufacture:	Composite laminating

Qualities: These differ from thermoplastic polyesters (page 437) because they are unsaturated, with double bonds between molecules.

Styrene is added to assist polymerization. It makes the material more liquid, so it can be cast or laminated at room temperature. Polymerization is catalyzed with methyl ethyl ketone peroxide (MEKP). MEKP is not part of the polymerization process; it activates it and causes a rapid exothermic reaction.

Polyesters exhibit high levels of shrinkage, in the region of 5–10%. This causes fibre reinforcement and other fillers to 'print through' and create a

visible texture on the surface. Gel coats are used to minimize print through.

They are naturally translucent and can be coloured to form an opaque material.

Applications: A range of composite applications such as automotive body panels, lorry cabs, cherry picker cradles and boat hulls. They are prone to water absorption, so have to be covered with a gel coat or vinyl ester for increased lifespan in outdoor and marine settings.

Costs: Low. These are the least expensive of the 3 main laminating resins: polyesters, vinyl esters and epoxies.

Vinyl esters and composites

• Vinyl ester
Vinyl esters bridge the gap between polyester and epoxy resins.

Qualities: These materials are similar to polyesters. They are also catalyzed with methyl ethyl ketone peroxide (MEKP). However, the reaction does not produce as much heat, so thicker sections can be cured. This leads to faster cycle times.

They are tougher and more resilient than polyesters, and they are less prone to water absorption and more resistant to chemicals.

As with polyesters, styrene is off-gassed by the material.

Applications: Vinyl esters have become popular laminating materials because they are less expensive than epoxy, but have superior properties to polyester. They are used in mold making, racing cars, aeroplane fuselage, wings and interiors.

Costs: Low to moderate.

Epoxies and composites

• Polyepoxide resin (EP)
Epoxies are high performance resins. They form very strong bonds with other materials and so are applied as coatings and adhesives as well as laminated and molded products.

Qualities: Epoxies are naturally clear and rigid. They can be pigmented and filled with metallic flake. It is also possible to fill them with objects and materials during casting.

Epoxies are not catalyzed. The reaction is facilitated by a hardener, which makes up 1 half of the resin. They shrink very

EPOXIES AND COMPOSITES
Lux table

Designer:	Penny Forlano, Vujj
Date:	2006
Material:	Carbon fibre reinforced epoxy resin (EP)
Manufacture:	Composite laminating

EPOXIES AND COMPOSITES
Carbon chair

Designer/ made by:	Bertjan Pot and Marcel Wanders/ Moooi
Date:	2004
Material:	Carbon fibre reinforced epoxy resin (EP)
Manufacture:	Carbon fibre saturated in EP is hand-coiled over a mold.

little during polymerization, typically less than 0.5%. Epoxies are electrically insulating and have high resistance to chemicals. They are suitable for high temperature applications, up to 200°C (392°F), and also have very good resistance to fatigue.

Applications: Epoxies are used in many different formats such as liquid for casting and laminating, paint, coating and encapsulation, structural adhesives and impregnated in carbon fibre weaves. Laminated products include furniture, boat hulls, canoes, automotive panels and interiors.

Epoxies are suitable for outdoor and marine applications because they are not prone to water absorption.

Costs: Moderate.

Polyurethanes

- Polyurethane resin (PUR)
PUR is a versatile material available in a range of densities and hardnesses, used as a solid cast material, foam, adhesive and liquid coating.

Qualities: PUR is naturally tan or clear, and available in an almost unlimited colour range (see image, above right). Thermoplastic polyurethanes (see page 436) are used in molding and coating applications such as trainer soles, sports wear and golf balls. PUR prototyping materials are designed to mimic

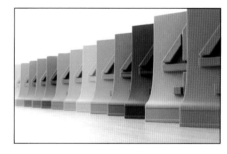

POLYURETHANES
Maverick Television awards

Designer/client:	Maverick Television/Channel 4
Date:	2005
Material:	Polyurethane resin (PUR)
Manufacture:	Vacuum casting

injection molded (page 50) materials like PP (see polyolefins, page 430), ABS (see styrenes, page 432) and nylon (see polyamides, page 438). Snap fits, live hinges and other details otherwise limited to injection molding can be made by vacuum casting (page 40) and reaction injection molding (page 64).

POLYURETHANES
Open-cell foam balls

Made by:	Various
Material:	Foamed polyurethane resin (PUR)
Manufacture:	Reaction injection molded
Notes:	Most plastic can be foamed with open or closed-cell structures.

PUR materials are resilient and durable, and resistant to flexural fatigue, abrasion and tearing. They operate at temperatures up to 120°C (248°F), but they are affected by ultraviolet light and will become hard and brittle with prolonged exposure.

Applications: More than half of all PUR is foamed. There are 3 main types: spray foam, flexible foam and rigid foam.

Spray foam is used for filling holes, as insulation or adhesive and to fill cavities in products to make them more rigid.

Flexible foams are used to provide cushioning in gym mats, carpet underlining, most upholstered foam furniture, other seating, mattresses, car interiors, packaging, shoe soles and linings in clothes and bags.

Rigid foam is used as an insulating liner in fridges and freezers, in sports and protective equipment and as a core material in composite lamination (page 206).

PUR is used for prototyping and low volume production. Examples include car bumpers, dashboards and interiors, toys, sports equipment and housings for consumer electronics.

PUR drawn fibre is resistant to cutting and abrasion and is used in textiles, shoe uppers and sports equipment.

As an adhesive and coating PUR is used to protect and bond wood with other materials. It is quick drying and forms a watertight and rigid seal.

Costs: Low to moderate.

Silicones

• Silicone resins
These are low strength but highly versatile materials. They are used as adhesives, gels, rubbers and rigid plastics. They have excellent electrical resistance and high heat stability, and are chemically inert.

Qualities: This family of synthetic polymers are derived from the element silicon (14 in the periodic table), but their properties should not be confused.

Silicone resins are typically resistant to water, chemicals and heat. They have good insulating and lubricating properties. These materials are non-toxic

SILICONES
Animal rubber bands

Designer/client:	D-Bros/H Concept
Date:	2003
Material:	Silicone
Manufacture:	Extruded and cut

and have low levels of volatile organic compounds (VOCs), which makes them suitable for food preparation and packaging and medical applications.

Silicones have a wide operating temperature range from -110°C to 315°C (-166–599°F). They are naturally electrically insulating, but can be modified with a carbon additive to make them conductive.

They are shaped by injection molding (page 50), compression molding (page 44), casting (page 40) and extrusion. They are used by model makers because certain grades cure at room temperature. These are known as room temperature vulcanizing (RTV) rubbers. They can be coloured with a range of plain, pearlescent, metallic, thermochromatic and fluorescent pigments.

Applications: These are relatively expensive materials, but their high performance properties mean they are suitable for a range of applications.

SILICONES
Skin Light

Designer:	PD Design Studio
Date:	2006
Material:	Silicone
Manufacture:	Cast

SILICONES
Flexible ice cube tray

Designer/client:	Unknown
Date:	Unknown
Material:	Silicone
Manufacture:	Compression molded

Synthetic rubbers

- Isoprene rubber (IR)
- Chloroprene rubber (CR)
- Ethylene propylene rubbers (EPM and EPDM)
- Butyl rubber (IIR)
- Butadiene rubber (BR)
- Acrylonitrile butadiene rubber (ABR)
- Styrene butadiene rubber (SBR)

SYNTHETIC RUBBERS
Divisuma 18 portable calculator

Designer/client:	Mario Bellini/Olivetti SPA
Date:	1973
Material:	Synthetic rubber, acrylonitirile butadiene styrene (ABS) and melamine resin
Manufacture:	Compression molding and injection molding

Typical uses include O-rings, gaskets, weather strips, seals, medical and surgical equipment, and food processing (for example, molds for chocolate).

Certain grades are considered safe for children's toys, baby bottle teats and medical implants. However, it is uncertain whether silicone causes tissue disease and so it is not recommended for more than 25 days in the body.

As a liquid it is used as a lubricant and release agent in RIM (page 64), composite laminating (page 206) and compression molding (page 44).

Costs: High cost.

There are several different types of synthetic rubber. They are used in place of natural rubbers. They have shape memory and so will return to their original shape after deformation.

Qualities: They are resilient, resistant to abrasion, have good low temperature properties, good resistance to chemicals and are typically electrically insulating. IR is the synthetic version of natural rubber (page 447) and consequently has similar characteristics and applications.

CR was the first synthetic rubber, invented by DuPont™ in the 1930s, and is also known by their trademark name of Neoprene®. it has very good resistance to oils and atmospheric corrosion, making it a suitable substitute for natural rubber.

EPM and EPDM have good resistance to atmospheric corrosion and heat. They range from 30A to 95A shore hardness.

IIR is impermeable to gas and resistant to atmospheric corrosion. It is utilized in tyre inner tubes.

ABR has superior resistance to oil and fuel, and SBR is the cheapest and most common form of synthetic rubber.

SBR is a widely used low cost rubber with properties similar to natural rubber.

Applications: Synthetic rubbers are used in place of natural rubbers in a range of applications. Typical products include fuel hoses, gaskets, washers, O-rings, diaphragms, tank linings and seals.

Neoprene® is well known for use in wetsuits, footwear and other sportswear.

Costs: Low to high cost.

Potatopak plate

Made by:	Potatopak
Material:	Potato starch
Manufacture:	Compression molded
Notes:	100% biodegradable plates and packaging.

Cellulose-based

- Cellulose acetate (CA)
- Cellulose acetate propionate (CAP)
These are thermoplastic self-shining materials that maintain a glossy finish. They are used in products that have a lot of user contact because they feel 'natural'.

Irwin Marples chisels

Made by:	Irwin Industrial Tools
Date:	The first acetate chisel handles in the 1950s, still in production today
Material:	Cellulose acetate propionate (CAP) handle and tool steel blade)
Manufacture:	Cast handle with forged and tempered blade

Envirofill

Made by:	Pro-Pac Packaging
Material:	Potato or wheat starch
Manufacture:	Extrusion
Notes:	A reusable alternative to polystyrene (PS) and 100% biodegradable in water.

Qualities: Cellulose-based polymers are partly made up of cellulose, which is a naturally occurring polymer and can be derived from wood pulp or cotton. It is similar to starch, which is also a polysaccharide (glucose polymer). The difference is that cellulose is structural (found in cell walls), while starch is a form of storage.

These polymers are tough, impact resistant and transparent, and certain grades are biodegradable. The raw material has to be processed heavily and chemically modified to produce a commercially viable and stable material. Sheets of cellulose are produced by

mixing together cellulose, pigments and various chemicals, which are rolled into flat sheets and left to dry. This produces colour patterns that cannot be achieved with injection molding (page 50) such as marbling and tortoiseshell.

CAP is available in transparent grades and is stiffer and more impact resistant than CA. It is the more expensive of the 2.

Applications: Cellulose-based polymers are available as sheet material, drawn fibre or molded product. Sheet materials are used in the production of spectacle frames, hairclips, cutlery handles, toys and watchstraps, for example.

Cellulose is also used to make photographic film and is drawn into fibres for textiles. Edible films are made with cellulose; they can be flavoured and are used for packaging food.

Costs: Moderate.

Starch-based

- Polylactic acid (PLA)
- Rice, potato and corn starch
Starch is made up of carbon dioxide, water and other elements. Thermoplastic starch plastics will breakdown into those elements readily when exposed to microorganisms or water.

Qualities: PLA has similar characteristics to LDPE (see polyolefins, page 430). It is transparent and is available as a blown film, fibre or foam. The starch used in the production of PLAs can be derived from agricultural by-products of barley, maize and sugar beet crops, all of which are annually renewable. It has

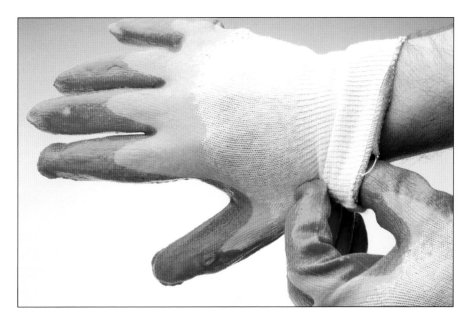

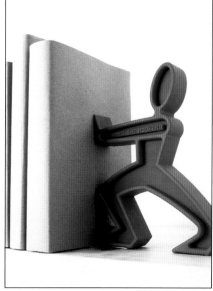

to be chemically refined into dextrose, which is fermented into lactic acid and then distilled to form polymerized PLA. Roughly 2.5 kg (5.51 lb) of maize are required to produce 1 kg (0.22 lb) of PLA.

These materials require about 50% less energy to produce than conventional thermoplastics. They can be thermoformed (page 30), injection molded (page 50) and extruded.

Potato and corn starch materials are not so chemically refined as PLA; instead, starch rich mulch or powder is placed into a mold and baked. It is possible to extrude and compression mold (page 44) starch-based plastics. Starch-based plastics typically contain 70% or more starch. The higher the starch content, the more rapidly the plastic will break down: high starch content plastics will dissolve in water within 15 minutes.

Applications: PLA is used in the production of compost bags, disposable cutlery, food and beverage packaging, women's hygiene products and pharmaceutical intermediates.

Rice, potato and corn starch are used in food serving and packaging (plates, trays, bowls and punnets), electronics packaging and loose fill packaging.

Costs: Low to moderate.

NATURAL RUBBER
Latex coated gloves

Made by:	Unknown
Material:	Natural rubber (NR) latex
Manufacture:	Dip coated (dip molding)
Notes:	Rubber makes a flexible, protective and non-slip coating.

Natural rubber

• Natural rubber (NR)
Also known as latex, NR is derived from naturally occurring sap that is drained from trees.

Qualities: NR is stabilized by vulcanizing it using sulphur. It is used in almost pure form (dip) or mixed with other materials. It has excellent flex, tear and abrasion resistance and is more elastic than most synthetic rubbers (page 445).

The majority of NR production is carried out in Malaysia, Indonesia and Thailand. It is drained from trees in a process known as tapping. The trees produce latex for up to 30 years, after which time they are replaced, so NR is available from renewable sources. In a factory the latex is coagulated and rolled into sheets ranging from 0.2 mm to 3 mm (0.0079–0.118 in.) thick.

NATURAL RUBBER
James the bookend

Designer:	Black+Blum
Date:	2002
Material:	Natural rubber (NR)
Manufacture:	Compression molded

Applications: Natural rubber is used as liquid coatings, foams and solid materials. The majority of NR is used in truck and aeroplane tyres. It is dip molded (page 68) to produce gloves, condoms, clothes and medical equipment. Gloves are coated with latex to improve grip, puncture, tear and abrasion resistance.

Costs: Moderate.

Metals

Metals are elements, many essential for sustaining life on earth. In the correct quantities they play a vital role in our diets and the existence of plants and animals. They are even useful in the fight against diseases: certain metals are known to have natural antimicrobial properties. However, these same properties mean that certain metals are toxic and polluting in sufficient quantities. Metals are used for structural and aesthetic applications from packaging to bridges. A range of finishes can be produced, some of which are self healing and so dramatically increase longevity.

Metals are extracted from their mineral ore, refined and processed to form combinations of metallic and other elements suitable for use in products. This is an energy intensive process, and so metals are typically more expensive than plastics that have equivalent strength characteristics.

In recent years plastics have replaced metals in many applications, due to the ease with which they can be produced and their physical and mechanical properties modified for specific applications. Even so, metals are still widely used in engineering, medical, large-scale, performance and high perceived value applications.

As a group of materials they are typically heavy and have high melting points. There are many different types of metal suitable for an unlimited range of applications. Some are non-toxic and sufficiently inert to implant into the human body; examples include titanium joint replacements and gold teeth. Others are poisonous, prone to corrosion or explosive.

TYPES OF METAL

Metals are divided into 2 main groups: ferrous and non-ferrous. The defining feature is that ferrous metals contain iron. Ferrous metals are produced and used in larger quantities than non-ferrous. This is mainly due to the versatility of steel; it is used in products, from domestic appliances to the automotive, construction and shipbuilding industries, for example.

Even though ferrous metals are limited to those containing iron, they make up more than half of all metal consumption. There is a large number of different iron alloys.

Most, but not all, ferrous metals are magnetic, due to their iron content. Non-magnetic ferrous metals include austenitic stainless steels. Non-ferrous metals are nearly always non-magnetic, but cobalt and nickel are exceptions to the rule. Even so, iron is by far the strongest naturally magnetic element.

Alloys are hybrids of different metallic elements combined to enhance properties and reduce cost. They can be made up with properties suitable for the demands of specific applications. For example, Arcos knives (see image, above) use a specially developed stainless steel. The alloying elements are chromium, molybdenum and vanadium. The proportions and combination of properties produce an exceptionally sharp and long lasting cutting edge that can be maintained by sharpening.

Molybdenum is part of a class of materials known as refractory metals. This group of metals has extremely high melting points: while steels melt at approximately 1350°C (2462°F), refractory metals have a melting point above 1750°C (3182°F). This makes them very difficult and expensive to manufacture. To utilize their superior properties, they are alloyed to form hybrid materials.

Arcos chef's knife

Made by:	Arcos
Material:	Molybdenum vanadium stainless steel
Manufacture:	Forged and polished

CORROSION, PATINA AND PROTECTIVE OXIDE FILMS

Certain ferrous metals are prone to surface corrosion. In the presence of oxygen and water the iron reacts to form a layer of iron oxide, commonly known as rust. This is a degradation process that gradually consumes metal that contains iron from the outside inwards. Protective coatings, such as galvanizing (page 368), painting (page 350) and powder coating (page 356) are used to prevent corrosion. Stainless steels are produced by alloying steel with chromium and other metallic elements. Chromium forms a protective oxide on the surface of steel and reduces subsequent corrosion of the base metal. Surprisingly, pure iron is more resistant to atmospheric corrosion than steel. This is due to its lower carbon content.

Cor-ten is high-strength low alloy steel. It is distinguished by a protective patina that develops on its surface in the atmosphere. This reduces the need for painting or other protective coatings.

Detail of Fulcrum, London, UK

Designer:	Richard Serra
Date:	Completed 1987
Material:	Cor-ten high strength low alloy steel
Manufacture:	N/a

BeoSound and BeoLab

Designer/client:	David Lewis/Bang & Olufsen
Date:	2003
Material:	Aluminium alloy
Manufacture:	Anodized

It is used in architecture and outdoor sculpture. Over time the surface develops a rust-like appearance, as in the case of the sculpture Fulcrum (see image, left).

Some non-ferrous metals, such as gold and platinum, are practically inert. Others react with the atmosphere and produce a surface oxide layer that forms a protective barrier against further corrosion. The oxide layer is also very durable, such as anodized (page 360) aluminium. The layer of oxide that is produced by anodizing is among the hardest materials known. The patina can be used for decorative effect such as coloured anodized aluminium (see image, above) or pre-weathered copper (see image, page 460). These self-protecting metals are virtually maintenance-free and so are often used in building façades, roofs and cladding.

Galvanic corrosion occurs when dissimilar metals are electrically connected and submerged in water (electrolyte). The electrolyte may be seawater or even a thin film of condensation. The reaction is similar to electrolytic processes such as electroplating (page 364) and electropolishing (page 384): 1 of the metals becomes the anode and the other the cathode. The metal that becomes the anode will erode more quickly than normal. Metals will only corrode each other in this way if there is sufficient potential difference between them, and the rate of corrosion is determined by the strength of the potential difference. For example, aluminium and steel must be insulated in application, otherwise the steel may corrode the aluminium.

Black-Light

Designer:	Charlie Davidson
Date:	2005
Material:	Aluminium foil
Manufacture:	Handmade

Aluminium extrusions

Made by:	Various
Material:	Aluminium alloy
Manufacture:	Extrusion
Notes:	Most non-ferrous metals can be extruded into profiles with internal features.

NOTES ON MANUFACTURING

There are 3 categories of forming process: liquid, plastic and solid state forming. As the names suggest, the processes are categorized by the relative temperature of forming: solid state forming is carried out with no additional heat, while in plastic state forming the metal is heated up to below its melting point and in liquid state forming it is heated to above its melting point. Plastic and solid state forming can be used to shape most metals. Casting processes are liquid state forming, and most use steel molds, so the metal being cast must have a melting temperature below that of steel. Steel is cast in ceramic molds such as those used for investment casting (page 130).

Certain metals are reactive and even explosive in certain conditions. For example, titanium is reactive with oxygen in its molten state. Therefore, liquid forming is carried out in a vacuum or inert atmosphere. This increases the cost of production.

Primarily, there are many standard products such as foil (see image, above), sheet, strip, rod, tube, angle and other extruded profiles (see image, above right). These can be further processed into coloured, perforated, woven, expanded, corrugated, embossed and textured materials.

Many metals are available as either cast or wrought. Wrought metal is cast into bars and then rolled (see forging, page 114) into profiles, sheets, strips and other profiles. The difference between the products is the grain structure. Cast products have a random grain structure, while forged and rolled metals have a linear grain alignment. This alignment has superior resistance to stress and fatigue. Products used in construction and high performance applications are typically formed in their plastic state by rolling or forging.

Grain structure can be further improved with heat treatment. This is the process of controlled heating and cooling. The rate of cooling affects grain structure and produces either hard and brittle or ductile and tough structures.

MATERIAL DEVELOPMENTS

Metal foam is produced by a similar technique to polymer foam; a foaming

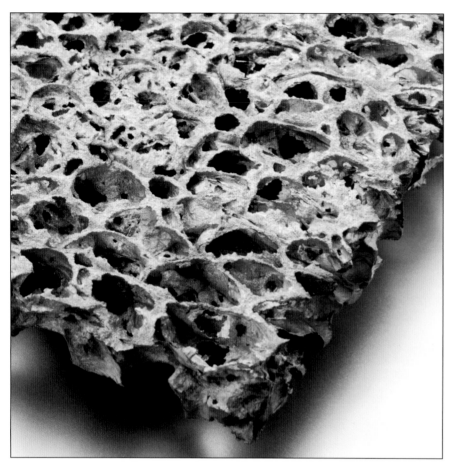

Foamed metal

Made by:	Various
Material:	Aluminium alloy
Manufacture:	Foamed (open-cell)
Notes:	Aluminium and copper are the most common metal foams.

agent is mixed with the metal and is triggered at a specific temperature in the mold to produce an open or closed cell structure (see image). There is a wide range of alloys, densities, porosities and shapes. They are lightweight, rigid, conductive and impact absorbing. The cells can be less than 1 mm (0.04 in.) across, or very large and giving a foam that is up to 95% air. Metal foam can be used as a core material, either skinned or foamed inside a hollow metal profile to produce lightweight panels and tubes.

Superplastic alloys include certain grades of aluminium and titanium. They exhibit extremely high levels of elasticity (more than 1000%) when heated up to 450°C to 500°C (840–932°F) for aluminium or up to 850°C to 900°C (1562–1652 °F) for titanium. They are superformed (page 92), which is a similar process to thermoforming plastics (page 30). Superforming requires less pressure than conventional techniques and can form deep profiles with tight radii from a single sheet.

Metal powders compressed into 3D bulk parts are porous. There are different processes used to shape metal powders and each produces a different level of porosity. For example, pressing and sintering produces a high level of porosity. This is removed by soaking the part in molten nickel (or other suitable metal), which solidifies to fill the voids. In contrast, direct metal laser sintering (rapid prototyping, page 232) produces parts that are 98% metal. The metal is sufficiently dense to make tooling for injection molding (page 50).

Powder metallurgy makes it possible to form materials that it is not practical to liquid state form such as refractory metals. It is also used to combine the properties of dissimilar materials. For example, flexible magnets are made up of iron powder in a polymer matrix.

Metal composites, also known as metal matrix composites, are made up of non-ferrous metal reinforced with particles of high performance ceramic such as silicon carbide or alumina (page 490) or carbon fibres. They exhibit higher levels of wear resistance, temperature resistance and stiffness. Typical carbon reinforced aluminium panels are lighter than aluminium with the stiffness of steel.

Shape memory alloys are novel materials that exhibit molecular rearrangement at specific temperatures. In other words, they can be deformed and then, when triggered at a specific temperature, they will return to their original shape. The temperature that triggers the rearrangement can be engineered to suit specific applications and can range from approximately -25°C to 100°C (-13°– 212°F). An example is Nitinol, which is a nickel titanium alloy.

This property is referred to as superelasticity when the temperature for rearrangement is set lower than room temperature. The result is a material

Vertu Ascent

Made by:	Vertu
Date:	2004
Material:	Liquidmetal®
Manufacture:	Various
Notes:	This is the first use of amorphous metal in a mobile phone. Other recent applications include tennis rackets and baseball bats.

Scrap metal

Made by:	N/a
Material:	Steel wire
Manufacture:	N/a
Notes:	All types of metals are cost effective to recycle.

that is constantly springing back to its original shape. The original shape is formed and set by heat-treating the raw material in the desired shape.

Amorphous metals combine the physical properties of metals with the processing advantages of plastics (see image, above left). This is made possible by the random arrangement of the atomic structure, which is similar to glass; the atoms are aligned randomly, as opposed to the typical crystalline structure of metal. Casting conventional metals produces inferior grain structure, but cast amorphous metals have exceptional properties such as high impact strength, strength to weight (twice that of titanium), hardness and resistance to corrosion and wear.

ENVIRONMENTAL IMPACTS
Metal products are typically long lasting and have higher perceived value than plastic equivalents. However, mining and extracting metals from their ores

is energy intensive and produces a lot of waste and hazardous by-products. Metals are extracted from their ores in a reduction process. Electrolysis or a chemical reducing agent (carbon or hydrogen) removes the oxygen atoms from the metal atoms. The concentration of metal in the ore varies, and low concentrations will produce more waste as a result. Inert metals do not always need to be extracted from ore because they do not react with oxygen in the first place. For example, gold can be found in its raw metallic state.

Much less energy is required to recycle post-consumer and industrial waste

metals into new metal products. The economic value of metal and efficiency of recycling means that nearly all industrial metal scrap is recycled (see image, above). Consumer waste has to be collected and separated, which increases the costs slightly. But it is still up to 90% more energy efficient to recycle than mine raw material. Unlike plastics, metals retain their strength when they are recycled. Therefore, they can be shaped and recycled many times without any loss of strength. Endless recycling reduces the environmental impact of the mining and extraction of the material.

Iron

- Wrought iron
- Cast iron

Iron is a low cost structural material whose use predates steel by many hundreds of years.

Qualities: Iron ore is an abundant material in the earth's crust and it is believed to make up a large percentage of the mantle and core. It is a heavy and soft material that is relatively easy to form hot or cold. It is classified as wrought or cast. Wrought iron contains less than 0.2% carbon and cast iron contains between 2–4% carbon. Iron with a carbon content between 0.2–2% is classified as steel.

Wrought iron has a very low carbon content and very few contaminants. It can have a visible grain, produced by forging or rolling (page 114) the iron and slag together in the production process.

It is soft, ductile and has been replaced by carbon steel for most applications due to its inferior strength.

Cast iron is made up of iron, carbon and small amounts of silicon. There are different types, including grey, white, ductile and malleable. They are differentiated by the formation of carbon in the iron matrix and alloying elements. Generally, cast iron has good dampening properties and machinability, is resistant to fatigue and corrosion and is difficult to weld due to the high carbon content.

Applications: Wrought iron has been largely replaced by steel. Traditional applications include architectural metalwork and fencing. It remains a useful material in the construction industry, for manufacturing tools, automotive crankshafts and suspension arms, and also for cooking equipment, due to its high thermal conductivity.

Costs: Low to moderate.

IRON
York railway station roof, UK

Architect:	William Peachey and Thomas Prosser
Date:	Completed 1876
Material:	Wrought iron I-beams
Manufacture:	Hot and cold rolled

IRON
Le Creuset giant grill and griddle

Made by:	Le Creuset
Material:	Cast iron
Manufacture:	Cast and enamelled

STEEL
VW Beetle

Designer/client:	Ferdinand Porsche/Volkswagen
Date:	1960s model
Material:	Low carbon steel
Manufacture:	Cold metal pressing (stamping and deep drawing), welding and painting

STEEL
Bird Kettle

Designer/client:	Michael Graves/Alessi
Date:	1985
Material:	Stainless steel
Manufacture:	Deep drawing, welding and polishing

Steel

- Carbon steel
- Low alloy steel
- Stainless steel
- Tool steel

These are the most common metals and can be found in many industrial and domestic applications. The specific properties of each type are determined by the carbon content and alloys.

Qualities: Carbon steels have a low, medium or high carbon content, ranging from approximately 0.2–2%. Higher carbon content produces a harder, less ductile and more brittle material. Mild steel (plain carbon steel) is a term that covers a range of carbon steels up to 0.25% carbon content. They are distinguished by ease of solid state forming and welding. Carbon steels are prone to oxidization and corrosion, so are protected with a coating in some form.

Low carbon steels are relatively ductile, malleable and easy to shape. In contrast, high carbon steels are hard and as a consequence they are both resistant to abrasion and more brittle. Medium carbon steels have levels of carbon and alloys that are ideal for hardening by heat treatment.

Low alloy steels are made up of iron, carbon and up to approximately 10% of other metals, such as nickel (page 462) and chromium. The additional alloys are used to improve certain properties of the steel such as resistance to corrosion, formability and toughness. Certain grades of these materials are also referred to as high strength low alloy steels (HSLA).

Stainless steels are a group of alloy steels that contain iron, less than 1% carbon, 10% chromium or more and other alloys. The high levels of chromium result in very good resistance to corrosion. There are 4 main types, which are austenitic, ferritic, martensitic and precipitation hardening. Austenitic grades are ductile, strong and non-magnetic; ferritic grades are less strong, magnetic and generally used for indoor

STEEL
Leatherman Wave

Made by:	Leatherman Tool Group
Date:	2004
Material:	Stainless steel
Manufacture:	Punching, grinding, polishing and heat treating

STEEL
Coloured stainless steel

Made by:	Rimex Metals
Material:	Stainless steel
Manufacture:	Chemical colouring process
Notes:	The range of colours includes red, green, blue, gold, bronze and black.

and decorative applications; martensitic are the hardest but least resistant to corrosion; and precipitation hardening grades are high strength and have moderate resistance to corrosion.

Tool steels are so called because they are used for cutting tools and dies. The carbon and alloy content make them hard, tough and resistant to abrasion even at high temperatures. Specific examples include high-speed steel (HSS) and mold steels.

Steels with a carbon content between approximately 0.3–0.7% are suitable for hardening by heat treatment. This is the process of heating up the steel and cooling it at different rates to form different microstructures. Normalized steel is heated to between 800°C and 900°C (1472–1652°F) and then slowly cooled, which allows the microstructure to develop into a strong formation. Quenched steel is cooled very rapidly in cold water and so is very hard and very brittle. Tempered steel is quenched and

then heated up to 200°C (392°F) for an hour before cooling, which allows the carbon particles to diffuse and develop the steel's toughness and ductility.

Applications: More than three-quarters of all steel production is carbon steel. Low carbon steel is used a great deal in construction, automotive metalwork and mill products such as sheet, strip, beams and sections. Medium carbon steel is used for crankshafts, chassis, axles, springs, forgings and pressure vessels. High carbon steel is used for springs, high strength wire and low cost cutting tools.

Low alloy steels are also used in construction. Interesting examples in the UK are Richard Serra's sculpture Fulcrum, in London (see image, page 450), Antony Gormley's sculpture The Angel of the North, at Gateshead and Heatherwick Studio's sculpture B of the Bang, in Manchester, all made in Cor-ten steel. The alloys in this particular grade of steel eliminate the need for protective

coatings. The material develops a protective oxidized layer that prevents further corrosion of the metal.

Stainless steels are used in a range of decorative and functional applications. They are expensive and so are only used if necessary, usually for decorative appeal. Examples include sinks, worktops, food preparation and cooking equipment, cutlery, architectural metalwork, furniture, lighting and jewelry.

Tool steels are typically used in tools such as screwdrivers, hammers, cutting tips and saw blades. They are also used to make dies for melt processing plastics and some metals.

Costs: Like all metals, the price of steel fluctuates according to fuel prices and global demand. Prices ranges according to the type of steel: carbon steels are the least expensive, but still generally more expensive than iron, followed by low alloy steels and stainless steels. Tool steels are the most expensive.

ALUMINIUM ALLOYS
Lightweight composite panel

Made by:	Cellbond Composites
Material:	Aluminium alloy honeycomb
Manufacture:	N/a
Notes:	Wood, metal, glass reinforced plastic and polycarbonate (PC) are all used as surface materials.

ALUMINIUM ALLOYS
Fluted tart tins

Made by:	Various
Material:	Aluminium alloy
Manufacture:	Pressed

ALUMINIUM ALLOYS
Coca-Cola Pocket Dr.

Made by:	Coca-Cola
Material:	Aluminium alloy
Manufacture:	Various

Aluminium alloys

• Aluminium alloys
This is a lightweight and conductive metal that is non-toxic and does not affect the taste of food or drink. It is used in a range of decorative and functional applications.

Qualities: Bauxite ore, from which aluminium is extracted, is an abundant material in the earth's crust. But extracting the aluminium is an energy intensive process; it takes roughly 3 kg (6.6 lb) of bauxite to produce 0.5 kg (1.1 lb) of aluminium. This is why it is so efficient to recycle aluminium rather than extract if from new bauxite.

Pure aluminium is quite soft and ductile. It is alloyed with small amounts of copper (page 460), manganese, silicon, magnesium (page 458) and zinc (page 459) to improve hardness and durability.

It has good strength to weight; the same strength can be achieved with roughly half the weight of aluminium as of steel (page 455). It is also a very good electrical and thermal conductor.

In the presence of oxygen the surface reacts to form a protective layer, which makes it almost maintenance free. The protective layer is enhanced and can be coloured with the anodizing process (page 360). The surface of aluminium can also be polished (page 376) to a bright and high quality surface finish. Aluminium is a highly reflective metal, and this quality is exploited by vacuum metalizing (page 372), which is the process of applying a very thin film of aluminium onto a high gloss surface.

Applications: It is used in a wide range of applications including packaging, drinks cans and cooking equipment. Housings and frameworks for consumer electronics and appliances are made in aluminium, and automotive applications include engine parts, bodywork and chassis. Aeroplanes, trains and ships use a great deal of aluminium. In the construction industry uses include window frames, trims and doors.

Costs: Moderate to high.

Joint Strike Fighter blisk

Made by:	Rolls-Royce plc
Date:	2006
Material:	Titanium alloy
Manufacture:	Blades linear friction welded onto central disk
Notes:	A blisk is a 1-piece bladed disk rotor design.

Magnesium alloys

• Magnesium alloys
These have better strength to weight than aluminium, but are more expensive.

Qualities: Magnesium is extracted by a very energy intensive electrolytic or oxide reduction process, which makes it more expensive than aluminium (page 457). It is often alloyed with aluminium, silicon and zinc to improve its performance in specific forming applications and reduce its susceptibility to stress cracking.

It is suitable for many aluminium forming and finishing processes such as die casting (page 124), superforming (page 92) and anodizing (page 360). It is less reflective and conductive than aluminium, and more prone to corrosion (especially in salt water).

Magnesium is explosive, especially in powdered form. It has a bright flame and is used for pyrotechnics and flares.

Applications: Many of the applications are similar to aluminium; magnesium is used when greater strength to weight is required. In the automotive industry it is used for the chassis, engine parts and bodywork of performance cars.

S9 Matta Titanium bicycle frame

Made by:	Bianchi
Date:	2006
Material:	Titanium alloy
Manufacture:	Extruded tube cut to length, welded and polished

Other examples include casings for electronics products such as mobile phones, MP3 players, camcorders and laptops, as well as sports equipment such as bicycle frames and tennis rackets.

Costs: Moderate to high.

Titanium alloys

• Titanium alloys
Titanium alloys are an expensive alternative to aluminium and magnesium, so are limited to applications that demand high strength to weight and superior corrosion resistance.

Qualities: Titanium is more energy intensive to produce than aluminium (page 457) and magnesium and so is more expensive. It has excellent resistant to corrosion, especially to salt water and certain chemicals.

Like aluminium, titanium is protected by naturally occurring oxide that

forms on its surface. Anodizing (page 360) thickens the layer and increases protection. The oxide is porous and can be dyed with a range of vivid colours.

Titanium is reactive with oxygen and so high temperature processing is typically carried out in a vacuum or inert atmosphere. Low temperature processes, such as spinning (page 78), stamping (page 82) and superforming (page 92), can be carried out in normal atmospheric conditions, so titanium can be processed with similar ease to aluminium.

Applications: A famous application is the metal cladding on the Guggenheim Museum in Bilbao, Spain, designed by Frank Gehry and completed in 1997.

Titanium is non-toxic and so is used for medical implants such as bone and joint replacement and strengthening or dental implants, for piercings and for jewelry. Like aluminium and magnesium, it has also been used for casing for mobile phones, cameras and laptops.

It is particularly suitable for springs, due to its low density and low modulus of elasticity. Springs in titanium can be less than half the size and weight of steel (page 455) equivalents, and it is therefore used for this type of application in performance motorbikes, cars and aerospace applications.

Metal uses only make up a small percentage of the total use of titanium. It is more commonly found as titanium dioxide, a white pigment used in the production of paints and paper.

Costs: High.

Zinc alloys

- Zinc alloys
Zinc alloys exhibit high resistance to corrosion and so a great deal of zinc production is for galvanizing steel.

Qualities: Zinc has low viscosity and a relatively low melting point, approximately 420°C (788°F). These qualities make it particularly suited to casting. It is suitable for forming small, bulk, sheet, complex and intricate shapes. Zinc castings are often electroplated (page 364) with another metal to combine the casting opportunities of zinc with the aesthetic properties of other metals, such as nickel (page 462), chrome or precious metals (page 462).

It is resistant to atmospheric corrosion and many acids and alkalis. However, it tarnishes very quickly and develops a rich patina on its surface. This is a visible layer that forms on the surface of the zinc and protects it from further corrosion. Galvanized (page 368) parts are very bright at first. Without treatment, the surface will become less bright over time. Building products are often treated in a process known as pre-weathering to avoid colour variation.

As a building material it is maintenance free and in normal conditions will last for 80–100 years before it needs to be replaced.

Applications: Most zinc is used in galvanizing or is alloyed with copper to produce brass (page 460). Zinc alloys (other than brass) are used for castings in

ZINC ALLOYS
Zinc sheet

Made by:	Rheinzink
Material:	Zinc alloy
Manufacture:	Cast and rolled
Notes:	The 3 finishes are bright, pre-weathered blue-grey and pre-weathered slate-grey.

many industries. Examples include door handles, bathroom furniture, automotive parts and jewelry.

In construction zinc is used for gutters, drain pipes, roofs, wall cladding and façades. Sheets of zinc are also used as worktops in kitchens and bars. Although it is less common now, worktops in hospitals and laboratories were once covered with zinc.

Costs: Moderate.

ZINC ALLOYS
Heavy-Weight tape dispenser

Designer:	Black+Blum
Date:	2004
Material:	Zinc alloy
Manufacture:	Die cast

Copper alloys

- Copper
- Brass
- Bronze

Copper alloys are ductile, have a low melting point and are easy to form. Copper develops a protective and decorative patina on its surface, which changes colour over time.

Qualities: Copper is an efficient thermal and electrical conductor. It is considered to be a hygienic material with antimicrobial properties. Many types of bacteria are neutralized when they come into contact with it, which is one reason it is used in hospital door handles.

Copper is a bright reddish pink when first produced. This does not last long. The surface quickly develops a layer of oxide, which is dark brown in colour. This will gradually become greenish in colour (verdigris) as it develops. With prolonged exposure, the film becomes very durable and protective. This means the copper is maintenance free and long lasting.

The rate of colour change depends on the oxygen, sulphur dioxide and moisture content of the atmosphere. Indoors, copper will change very slowly, whereas next to the sea, or an industrial area, copper will develop a green sulphate layer in as little as 5 years.

Alloyed with zinc (page 459) or tin (page 462), it has a lower viscosity and so is suitable for casting complex and intricate shapes.

Brass is an alloy of copper and 5–45% zinc. A greenish brown patina develops on its surface and becomes a dark brown over time. There are many different types of brass, which are categorized by the quantity of zinc. Higher levels of zinc produce harder and more brittle brasses.

Bronze is an alloy of copper and up to 40% tin. The patina develops much more slowly and is a brownish colour.

Applications: Many uses for copper are associated with its conductive properties. Examples include electrical cables,

(page 459)

(page 462)

COPPER ALLOYS
Villa ArenA, Amsterdam

Architect:	Virgile & Stone Associates Ltd. and Benthem Crouwel Architecten Leebo bouwsystemen
Date:	Completed 2001
Material:	Tecu® Patina copper alloy
Manufacture:	The process pre-weathers metal by mimicking natural oxidization

heating elements, and conductive bases for pots and pans.

It is also used for roofs and cladding (see images, above and opposite above), typically pre-weathered for decorative or restoration purposes. Other outdoor applications include decorative metalwork and sculptures: the Statue of Liberty in New York is made from copper and shows the green patina that develops with prolonged exposure.

Instruments, including trumpets, saxophones, bells and symbols, are produced in brass and bronze.

Costs: Low to moderate.

COPPER ALLOYS
Urban Entertainment Centre

Architect:	Will Alsop Architects
Date:	Completed 2005
Material	Tecu® Brass copper and zinc alloy
Manufacture:	Cold shaped and soft soldered
Notes:	This building is part of a major redevelopment of Almere, in the Netherlands.

COPPER ALLOYS
Copper coffee pot

Designer:	Unknown
Date:	Unknown
Material:	Copper alloy and tin
Manufacture:	Metal spinning and pressing

COPPER ALLOYS
Brass darts

Designer:	Unknown
Date:	Unknown
Material:	Brass
Manufacture:	Turned on a lathe

Nickel alloys

- Nickel alloys
Nickel is used mainly for electroforming, electroplating and as an alloy in stainless steel.

Qualities: Nickel is a bright metal and has very good resistance to oxidization and corrosion. The properties are used in low alloy and stainless steels (page 455).

It has a rapid rate of deposition in electrolytic solution and so is an efficient and useful material for electroplating (page 364) and electroforming (page 140).

Skin contact can result in the user's sweat dissolving nickel salts, which may result in allergic reaction, or more specifically, allergic contact dermatitis.

Nickel is the base alloy element in so-called 'superalloys' that are stable and can operate at over 600°C (1112°F). They also have very high resistance to corrosion and oxidization. These materials are typically shaped by investment casting (page 130).

Applications: Very large electroformed parts can be produced in nickel. It is used to make molds for the aerospace, marine and performance automotive industries.

It is used to electroplate trophies and plaques because it produces a bright and long lasting finish, but is generally not used on products that have prolonged and intimate contact with the user's skin. Short-term contact, such as with nickel plated coins and keypads, is not usually problematic.

Austenitic grades of stainless steel contain nickel. They have superior resistance to corrosion and oxidization and so are commonly found in outdoor and construction applications.

The properties of nickel-based superalloys are exploited in jet engine parts and other extreme applications.

Costs: Moderate to high.

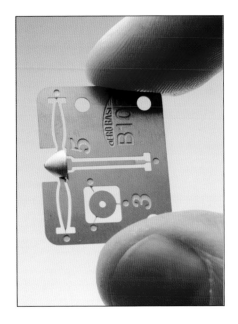

NICKEL ALLOYS
Scale model part

Made by:	Aero Base
Material:	Lead alloy nose cone and nickel-silver (alloy of nickel, copper and zinc) sheet
Manufacture:	Centrifugal cast nose cone and photochemical machined sheet

Lead and tin alloys

- Lead alloys
- Tin alloys
- Pewter
These are soft metals that are suitable for casting. They are sometimes referred to as 'white metals'.

Qualities: These alloys have low melting points: tin is below 250°C (482°F), lead is below 350°C (662°F), and alloys of these metals can be as low as 200°C (392°F). They have low viscosity when they are molten and so are efficient materials to cast. Low pressure casting techniques can produce high definition of detail.

Lead is toxic and a heavy metal, and its use in many applications is now being phased out due to its link with blood and brain disorders.

Tin is non-toxic and has excellent resistance to corrosion. It is alloyed with other metals to reduce their melting point and increase corrosion resistance.

Pewter is an alloy of tin with small amounts of copper (page 460) and lead. It is so soft it can be carved by hand.

Applications: So-called 'white metals' are used in many casting applications such as jewelry, architectural models and scale models. They are often plated with another metal or painted.

Lead is still used in construction and as radiation shielding. It is also useful as a weight, for example, in the keel of sailing boats.

Tin is used a great deal for alloying and plating other metals. Tin plated products include steel packaging, toys, cooking utensils and furniture.

Costs: Low.

Precious metals

- Silver
- Gold
- Platinum
These are rare and therefore expensive metals. Most have exceptional resistance to corrosion. They are very efficient thermal and electrical conductors, and are also non-toxic.

Qualities: Silver is bright and highly reflective, but the surface oxidizes quite readily and so has to be frequently polished or 'coloured over' to keep its brightness. Like the other precious metals, it is a very efficient conductor. Silver ions have antimicrobial properties. Electroforming (page 140) and electroplating (page 364) can produce products in almost pure silver.

Gold is a very soft, malleable and ductile material. It can be beaten into very thin sheets, known as gold leaf. It has a shiny and tarnish free surface finish.

Pure gold is yellow. Different colours are produced by varying the alloy

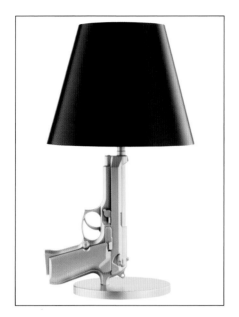

PRECIOUS METALS
Bedside Gun light

Designer/client:	Philippe Starck/Flos
Date:	2005
Material:	Gold and aluminium body
Manufacture:	Die cast body, electroplated with 18 ct gold

content. Red or pink gold contains copper (page 460); white gold contains platinum, silver or zinc (page 459); purple gold contains a precise measure of aluminium (page 457); blue and black are also possible. Alloying gold with other metallic elements will alter the mechanical properties as well as colour.

The purity of gold is measured in carat (ct): 24 ct is pure gold; 18 ct is 75% gold; 14 ct is 58.3% gold; and 10 ct is 41.1% gold by weight. Unlike many gold standards, the UK allows no negative tolerance.

Platinum is the most rare and precious of these 3 metals and so the most expensive. It is hard, durable and ductile and is resistant to corrosion by abrasion, oxygen and many chemicals. Platinum is a very good conductor and catalyst. It is a member of a group of precious metals including rhodium, palladium and iridium among others.

Applications: Silver is used for both industrial and decorative applications.

Decorative uses include jewelry, cutlery and tableware. Silver ions are used in paints, powder coats and varnish to inhibit the growth of bacteria or fungi.

Gold has a rich and lustrous appearance that is useful in jewelry, tableware and medals. Its non-toxic qualities are used in tooth repairs and capping. It is even used in food and drink such as gold flake in Goldschläger. It is plated onto cables for high quality sound systems and telecommunications due to its high conductivity and tarnish free surface finish.

Platinum is used in jewelry and medical applications because it is does not corrode or tarnish and is non-toxic. It is also a very effective catalyst and this quality is employed by catalytic converters: platinum catalyzes the conversion of polluting exhaust fumes into water, carbon dioxide and other less harmful substances. It plays an important role in fighting cancer because it is an active ingredient in

PRECIOUS METALS
Memento Globe necklace

Designer:	Rachel Galley
Date:	2006
Material:	Silver globe and 9 ct gold ball
Manufacture:	Centrifugal cast, soldered and polished

certain chemotherapy drugs. The role of platinum is to damage the DNA in cancerous cells to inhibit cell division.

Costs: High to very high. Silver is the least expensive, and platinum is considerably more expensive than gold.

Wood and natural fibres

Wood is a natural composite material made up of xylem tissue, which is a fibrous material consisting mostly of elongated, rigid walled cells that provides trees and shrubs with an upwards flow of water and mechanical support. Its strength and lightness have been exploited for millennia. Fibrous woody materials include bamboo, reed, rush and vines such as rattan. These grow fast and when treated are suitable for similar applications to wood. Demand for bamboo is increasing because it is proving to be an economic, environmental alternative to wood in many projects.

Machine stress-rated pine

Made by:	Various
Material:	Pine (EW) or spruce (ER)
Manufacture:	Sawn timber
Notes:	This is a construction softwood categorized as EW/ER (European whitewood/European redwood), which has been kiln dried (KD) to less than 20% moisture content.

The qualities of wood are the result of natural growth and the influence of the elements. Each species of tree produces timber with particular strengths, weaknesses and visual characteristics. Some grow fast, tall and straight; others are slow growing with interlocking grain. Over the years certain woods have become essential and irreplaceable in the construction of buildings, musical instruments, tools and furniture.

Wood is a sensual material that is warm to the touch. Some species are more aromatic than others, such as cedar (page 470), which has been traditionally used in coat hangers to deter moths, as shoetrees to counteract foot odour, and in pencils because it has a pleasant taste and is resistant to splintering. As a natural, edible and biodegradable material, wood is prone to disease, insect attack and decay. A famous example is Dutch elm disease, which wiped out much of Europe and North America's elm population. Many species of tree take several decades to mature ready for timber production. Elm was once a popular timber but may never recover from the impacts of the disease, which is still active.

TYPES OF WOOD

Wood is used as logs, lumber, veneer, panel products, engineering timber, pulp and paper. It is classified as either softwood or hardwood.

Softwoods are coniferous and typically evergreen trees, and include pine, spruce, fir and cedar. Hardwoods are typically deciduous and broad leaved trees. The terms softwood and hardwood are misleading. For example, balsa is very soft

and light but is classified as a hardwood, while certain softwoods are dense and hardwearing like some hardwoods.

Lumber is sawn timber. It comes in a range of profiles and sizes from only a few millimetres (under 0.25 in.) up to 75 mm x 250 mm (3 x 10 in.). Larger sections are produced as laminated engineering timber. Softwood is typically graded as either structural or aesthetic. Stress grading is carried out either visually or by non-destructive testing in machine stress rating (MSR). MSR is carried out by non-destructive bending, or ultrasound. There are many different grades such as truss rafter (TR) and construction (C) grades. The stress rating is indicated on a numerical scale. This grading is used by engineers to specify the correct bearing strength and elasticity of timber in architectural projects (see image, top).

Veneers (see image, above) are produced by cutting very thin strips of wood, between 1 mm and 5 mm

Rotary cut (peeled) veneer

Made by:	Various
Material:	Birch
Manufacture:	Rotary cut
Notes:	The grain stretches and compresses on either side of the veneer as it is bent. It will bend more easily in the direction that it was peeled from the log.

(0.04–0.2 in.) thick, from logs. They are either peeled continuously around the circumference of the log (rotary cut) or cut across the width of the log. Veneers are used in the construction of laminates and for surfacing other materials.

Panel products (see image, page 466 top left) are made up of wood veneers, particles or core materials that are bonded together with strong adhesives and high pressure. They are an efficient use of materials and include plywood, particleboard, oriented strand board (OSB) and composite constructions.

Engineering timbers utilize the strength and stability of laminated

Panel products

Made by:	Various
Material:	Various
Manufacture:	High pressure laminated
Notes:	From the top down: veneered MDF by Tin Tab; DuPont™ Nomex® Decore™; birch core with ash faces by Tin Tab; beech ply by Tin Tab; and cork rubber by Tin Tab.

Structural timber beam

Made by:	Trus Joist
Material:	Parallam® parallel strand lumber
Manufacture:	Laminated with adhesive
Notes:	PSL is made up of strips of veneer up to 2 m (6.6 ft) long, which are pressed with an adhesive to form billets up to 20 m (66 ft) long.

Embossed coasters

Designer:	Shunsuke Ishikawa
Date:	2006
Material:	White board
Manufacture:	Blind embossing
Notes:	A metal tool forms an impression on 1 side of the board with heat and pressure.

wood and include laminated strand lumber (LSL), parallel strand lumber (PSL) (see image, top right) and I-beams. The construction industry is the largest user of these products.

Wood pulp is used in the production of paper and pulp. Xylem contains both cellulose fibres and lignin. The fibres are structural and are bonded with lignin. Paper and board can be embossed effectively with heat and pressure because the lignin softens when heated, to allow the structure to be formed (see image, right). Recycling paper and pulp reduces the length of the fibres and so reduces their strength; it is therefore not an infinitely repeatable process.

WOOD GRAIN, STRUCTURE AND APPEARANCE

The strength and appearance of lumber are determined by many important factors. These include the type of tree, defects, the method of drying and how it was sawn.

Wood grain is produced by growth rings and rays. These are made up of cell structures that transport water and nutrients around the tree. Annual growth rings develop as a consequence of seasonal change and can be used to tell the age of the tree. Early in the growing season tree growth is rapid and the wood is typically lighter in colour. Darker rings indicate slower growth, usually from later in the growing season. Rings are intersected with rays, which are structures radiating from the centre of the tree to transport food and waste laterally. The combination of rings and rays produces patterns and flecks of colour on the surface of sawn timber.

As a tree matures the centre darkens, and this is known as heartwood; the lighter wood near the bark is sapwood. The depth of sapwood and colour contrast depend on the species of tree.

Grain pattern is further influenced by the angle of sawing. After felling, trees are sawn tangentially or radially (see image, opposite left). Tangential sawing, known as plainsawn, is the most efficient and economic method for cutting a log. Radial cutting, known as quartersawn, produces a more wear resistant surface finish with an even grain pattern.

The direction of grain affects the strength, working properties and

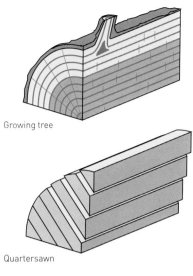

Growing tree

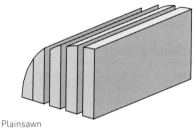

Quartersawn

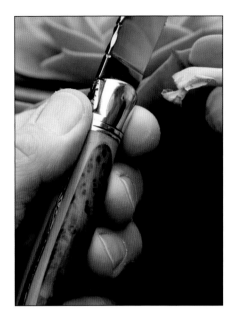

Plainsawn

Cross-section of tree grain, with rings and rays, and sawing techniques

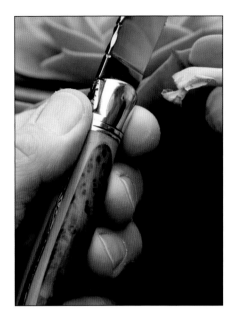

Laguiole knife handle

Made by:	Laguiole
Material:	Hardwood burl
Manufacture:	N/a
Notes:	Other types of figured grain include bird's-eye, curly and flame.

durability of wood. The surface of a plank is typically flat grain; in other words the plank is cut lengthways from the tree. Planing with the grain will produce a smooth finish, whereas planing against or across the grain may cause 'tear out'. End grain is produced by cutting across the width of the tree and is therefore at the end of sawn lumber. Fasteners such as nails and screws will easily pull out of end grain. Joint designs in joinery (page 324) have evolved to maximize the contact of flat-grain at an end grain junction. For example, dowel, biscuit and finger joints maximize flat grain contact and thus improve joint strength. However, end grain does have its benefits; it is frequently used in chopping boards and butcher's blocks because it does not splinter like flat grain and causes less wear on the knife.

Wood is a natural and variable material. Defects and figured grain patterns are not always predictable. Figured grain patterns are rare and

expensive and so are often only available as veneers. There are many types including burl (see image, above right), bird's-eye, curly and flame. Bird's-eye is rare, but it occurs more frequently in sugar maple than any other tree. It is unclear what causes this distortion of the annual growth rings.

Knots are the most common defect found in timber. They occur where branches and trunks come together and affect the surrounding grain pattern. They affect the aesthetics and structural integrity of wood and so are often used as an indicator of its quality. They are classified as either dead or alive. Alive knots are where branches were attached to the trunk when the tree was felled. They can be sound and tight and typically have little effect on strength. Dead knots are the result of branches and twigs that became detached from the trunk earlier in its life; the tree continued to grow and so encapsulated the knot. These knots can come loose and fall out

to leave knotholes. All knots produce sap and so have to be sealed with shellac or 'knotting' prior to painting.

Other defects, such as warping, twisting, bowing, checking and splitting, are caused by drying the wood. Reducing the moisture content of wood (up to a certain point) is important for many reasons. Drier wood is stronger, stiffer, lighter and less prone to decay than 'green' wood. Traditionally, wood was seasoned in a process known as air-drying at a rate of 1 year per 25.4 mm (1 in.) to give a final moisture content of 18–20%. Nowadays, modern kiln-drying techniques can reduce the moisture content of 25.4 mm (1 in.) thick lumber to under 20% in 10 days. However, kiln-drying procedures of 3 months or more are preferred for good quality timber even though this is more expensive.

The moisture content of wood continues to change even in application. This is inevitable, and so a combination of joinery, sealants and design is employed to reduce the effects of shrinkage and expansion.

Wood is prone to greater shrinkage across the grain. Quartersawn timber is cut in a pattern radiating from the centre of the log and so has more even grain pattern. It is therefore less liable to twist and warp as it dries and shrinks.

NOTES ON MANUFACTURING
Wood is available for manufacturing as a sheet material, solid lumber, and as chips, particles and shavings.

Sheet materials include veneers and panel products. They are versatile, strong and lightweight. They are also typically more stable than solid lumber because they are made up of thin plies bonded with strong adhesives. Thin sheet materials can be shaped by sawing, laser cutting (page 248), laminating (page 190), bending (page 198) and machining (page 182). Thick sheet materials are typically used flat and profiled by sawing or machining. Slight bends may be achieved with kerfing (page 190).

Treeplast® golf tee

Made by:	PE Design and Engineering
Material:	Treeplast® biocomposite of wood, corn and natural resins
Manufacture:	Injection molded

Environ® panels

Made by:	Phenix Biocomposites
Material:	Soy based or thermosetting resin reinforced with natural fibre
Manufacture:	High pressure laminated
Notes:	Panels are used for construction, interiors and furniture.

Most timber can be purchased as solid lumber. Limiting factors are price and availability such as in the case of bird's-eye maple. Planks more than 150 mm (5.91 in.) wide are typically stabilized by cutting the wood into strips and bonding them back together alternating opposing directions of growth. This means that as it shrinks and expands the wood works against itself and is less likely to buckle, bow or twist. Solid lumber is typically shaped by sawing, machining, steam bending or carving. Large radii bends can be achieved by kerfing and laminating.

Chips, particles and shaving are typically bonded into sheet materials or molded into products. Other uses include wood shavings as a protective packaging material and bedding for animals.

MATERIAL DEVELOPMENTS

Much of the development in wood has been in producing stronger, lighter and more reliable products such as engineering timbers.

Another interesting area of development is biocomposites. These are moldable materials made up of natural fibres (commonly wood) bonded together with either natural or thermosetting adhesives. In some cases they crossover with bio-fibre reinforced plastics and bioplastics (see page 425). Two examples are Environ® and Treeplast®. Phenix Biocomposites produce a range of panel products made up of natural fibres bonded with soy based or synthetic resins (see image, above left). Types of fibre reinforcement include agricultural by-products such as wheat straw, and recycled newspapers.

The panels are conventional sizes, 1220 mm x 2440 mm (4 x 8 ft), and between 12.7 mm and 21.4 mm (0.5–1 in.) thick. They can be machined with conventional woodworking equipment.

Treeplast® is made of wood (50–70%), crushed corn and natural resins (see image, above right). It can be processed with conventional injection molding (page 50) equipment. Shape is unlimited, except for a minimum 3 mm (0.118 in.) wall thickness and 3° draft angle. The basic material will biodegrade very rapidly in water and within 4–6 weeks in soil. Another grade is available that is water resistant and longer lasting.

Bendywood® is an exciting new material development (see image, opposite). It is made by steaming and compressing hardwoods along their length, and its advantage is that it can then be bent in a cold and dry state. Bend radius is approximately 10 times greater than the thickness of the material. So far it has been used to make handrails,

furniture and sculpture. The maximum size of the raw material is currently 100 mm x 120 mm x 2200 mm (3.94 in. x 4.72 in. x 7.22 ft). It is possible to bend thin sections by hand; thicker sections are bent on a ring roller (see page 98) or other suitable bending equipment.

ENVIRONMENTAL IMPACTS

Wood is an environmentally beneficial material. It is non-polluting, biodegradable, can be recycled and should be from renewable sources.

Deforestation is a problem in many developing countries. Systems are in place to minimize the use of illegally sourced timber such as origin and chain of custody certification. These systems verify the flow of wood from forest to factory and end use, and ensure that the timber comes from renewable sources.

Wood has relatively low 'embodied energy': harvesting, sawing, drying and transporting it does not take a great deal of energy. The cost and environmental impact of drying timber is often reduced by air-drying the first stage so that it requires less kiln-drying to reduce the moisture content to 18–20%.

Oily woods that are self-protective and virtually maintenance-free outdoors include cedar, larch, teak and iroko.

Bendywood®

Made by:	Candidus Prugger
Material:	Hardwood
Manufacture:	Formed by hand around a mandrel
Notes:	This treated hardwood can be bent in its cold and dry state.

However, they also tend to be difficult to bond with adhesives, impregnate with preservative or coat finish.

The dust produced by machining and sanding certain woods is irritating to the lungs and eyes. Examples include teak, wengé, iroko and walnut.

Softwoods

- Pine
- Douglas fir
- Larch
- Spruce
- Cedar

Softwoods have moderate strength, straight grain and a distinctive odour. They are predominantly evergreen, fast growing and useful in the DIY, construction, pulp and paper industries.

Qualities: These coniferous softwoods are grown all over the world, but some species are limited to certain latitudes and altitudes. The rate of growth determines the strength and economic value of the wood. Slower grown trees have fewer knots and more tightly packed grain, and so they are harder and more expensive.

Pine (*Pinus* species) includes Scots pine (also known as European redwood), jack pine, red pine and eastern white pine. There are over 100 different species. Even though many of the trees cannot be easily differentiated, the properties of the sawn timber are slightly different

and suitable for different applications. Common properties are a coarse and straight grain, off-white to dark brown colouring and a distinctive odour that is produced by the resin. Unless impregnated with preservative, these woods are generally not suitable for exterior applications. Their colour will darken over time.

Spruce (*Picea* species) has a long, straight grain and is an even, yellowish white colour. There are 2 main types, the Norway spruce and Sitka spruce. Sitka spruce trees are some of the tallest in the world and are commonly 40 m to 50 m (130–165 ft) or more high. The wood has a uniform, knot-free appearance and is stable after drying. This produces superior acoustic properties, utilized in soundboards in musical instruments.

Douglas fir (*Pseudotsuga* species) is grown across North America. The trees are very tall and can reach heights of 100 m (328 ft). The height and properties of the tree are affected by distance from

the coast: coastal Douglas firs grow faster and taller than inland varieties. The timber is off-white with yellow and red tones. It has good strength to weight and dimensional stability. It is stiff and so is used mainly in structural applications.

Cedar (*Cedrus* species) can be off-white, light yellow or light reddish brown. It is relatively soft and light, and so prone to scratching, abrasion and indentation. When cut, it produces oils that repel insects and protect it against decay. Of these woods it is the most resistant to decay and can be used untreated outdoors and submerged in water. Over time the surface will become silvery grey.

SOFTWOODS
Inn the Park, London

Designer:	Hopkins Architects
Date:	Completed 2004
Material:	Larch
Notes:	Inn the Park was shortlisted in the 2004 Wood Awards, UK.

SOFTWOODS
Colouring pencils

Made by:	Various
Material:	Cedar
Manufacture:	N/a
Notes:	Cedar is straight grained, aromatic and is resistant to splintering.

SOFTWOODS
Violin table (work in progress)

Made by:	Frederick Phelps
Material:	Spruce
Manufacture:	Hand carved from solid wood
Notes:	Knot-free and very high quality spruce is radially cut and air dried for 10 years or more, before carving into a violin or cello table (front).

Larch (*Larix* species) is a deciduous softwood grown in Europe, Asia and America. Like other coniferous woods, its strength depends on the climate. Trees grow more slowly in cold climates and so are more dense and hardwearing. Larch differentiates itself because the core turns into heartwood more rapidly than other softwoods. The benefit of this is that heartwood is naturally more resistant to decay, so larch is more resistant to salt water, fungus, temperature change, denting and abrasion. Natural oils produced by the sawn timber protect it from the elements and so it can be used untreated outdoors.

If left untreated, it will gradually turn from light golden brown to grey.

Applications: These woods are available as sawn timber and plywood.

Pine is widely used in furniture, flooring, construction, kitchen utensils and wall panelling.

Spruce's light colour and long fibres make it ideal for producing pulp and paper. It is also used for pallets, crates, flooring and structural applications, and the framework of small boats and light aircraft. Its acoustic properties are utilized in musical instruments including guitars, violins and pianos.

Douglas fir is mainly used in structural applications, such as timber frame construction (page 344), and the framework for light aircraft.

Cedar is the most resistant to decay of all the softwoods and so is used in outdoor decking, furniture, fencing, roof shingles, wall cladding, saunas (indoors and outdoors) and baths. Its distinctive smell and insect repelling qualities are used in the construction of wardrobes, clothes hangers and shoetrees. It is also used to make pencils because it is resistant to splintering. It also tends to be the most expensive of the softwoods.

Larch is used in flooring, house construction, fences, walkways, bridges and piles. A well-known use of larch is in the construction of Venice, the support structure of which has been submerged in water for hundreds of years.

Costs: Low to moderate.

POPLAR
Wooden packaging

Made by:	Various
Material:	Poplar
Manufacture:	Stapled or glued
Notes:	Poplar is relatively fast growing, inexpensive and can be formed into tight bends without splitting.

POPLAR
Matryoshka dolls

Made by:	Unknown
Material:	Aspen
Manufacture:	Turned on a lathe
Notes:	Other readily available woods suitable for turning include lime, birch and alder.

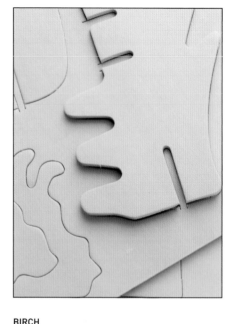

BIRCH
Die cut plywood Tyrannosaurus

Made by:	Muji
Material:	Birch plywood
Manufacture:	Die cut
Notes:	Plywood up to 15 mm (0.59 in.) thick is suitable for die cutting.

Poplar

- Poplar
- Aspen

This is a lightweight and relatively soft hardwood. It is a fast growing commodity timber used in pulp, paper, packaging and engineering timbers.

Qualities: Poplar (*Populus* species, some commonly called aspens) has a light yellowish green colour. It is a soft timber and so machines very easily. However, this also means that it dents and wears rapidly. It is low density, lightweight and has good flexibility. It is porous and resistant to splitting, so thin sheets can be bent into moderately tight curves, but it is not suitable for steam bending because it has relatively short fibres and these are prone to warping as they dry.

It is used a great deal as a core and composite material in engineering timbers. In this application, poplar is used in much the same way as softwoods (page 470), including pine, Douglas fir and spruce.

It can be stained to take on the appearance of other woods such as cherry (page 478) or maple (page 475).

Applications: Poplar is available as a sawn and engineering timber. As an alternative to softwoods it is used as plywood core, laminated strand lumber (LSL, see page 466), furniture construction, pallets and boxes. Due to its high level of flexibility and relative low cost it is the most widely used wood for matchsticks and packaging soft cheese.

Costs: Low to moderate.

Birch

- Birch

Birch is relatively durable and strong. It has a light colour with a uniform texture.

Qualities: There are many different types of birch (*Betula* species), which grow across North America, Asia and Europe. Birch is a relatively fast growing deciduous tree that can reach heights of 30 m (98 ft) or so.

It is quite heavy and strong. Certain species are quite hard, while others are soft and flexible, which makes them easy to machine and work. Birch is suitable for steam bending, but is prone to warping as it dries and so has to be clamped in place for the duration of the drying cycle.

Unless it has been coated or impregnated with preservative, it is not suitable for outdoor use.

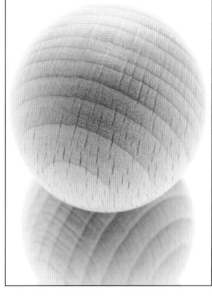

Applications: Solid birch and veneers are used to make disposable cutlery, toothpicks, toys and wooden puzzles. Birch is commonly used as plywood in furniture and interior doors.

Costs: Low to moderate.

BIRCH
Disposable wooden cutlery

Made by:	Various
Material:	Birch veneer
Manufacture:	Pressed and cut
Notes:	This cutlery is an alternative to disposable plastic cutlery with a lower environmental impact.

BEECH
Wooden spheres

Made by:	Unknown
Material:	Beech
Manufacture:	Turned on a lathe
Notes:	Beech is hard and uniform density, so a good finish can be achieved with machining.

Beech

• Beech
This is popular material used as both sawn timber and veneer. The sapwood is off-white and the heartwood is a reddish brown.

BEECH
Flight 1-Seat

Designer:	Artur Moustafa and Jonas Nordgren for Vujj
Date:	2006
Material:	Beech
Manufacture:	CNC machined

Qualities: Beeches (*Fagus* species) are hardwoods suited to northern temperate regions, which can grow to over 30 m (98 ft). The grain is short and tightly packed, resulting in a hard material that is resistant to denting and is relatively easy to work. The wood has an even density and so wears in a slow and uniform manner.

Beech is relatively porous and so prone to shrinking and expanding rapidly with changes in moisture content. Unless it has been impregnated with preservative, it is not suitable for outdoor use.

Due to its short grain, beech is liable to split under tension and is not very elastic.

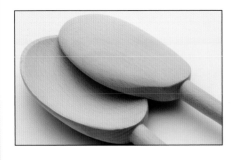

BEECH
Wooden utensils

Made by:	Various
Material:	Beech
Manufacture:	CNC machined
Notes:	Beech is non-toxic and safe for food use, but cannot be left in water for prolonged periods.

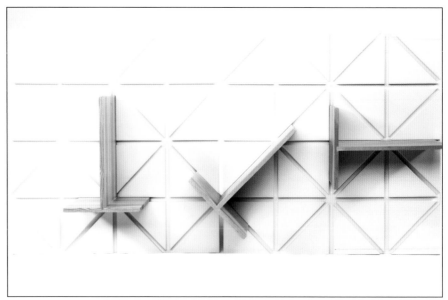

ASH
Alog shelving system

Designer:	Johannes Herbertsson and Karl Henrik Rennstam for Vujj
Date:	2006
Material:	Ash shelves and MDF (particle board) wall mount
Manufacture:	CNC machined

Hickory

- Hickory
- Pecan

These are North American trees used in the production of tool and axe handles. They are tough, hard and strong.

Therefore, it is not generally used for structural applications. It is suitable for steam bending (page 198), laminating (page 190) and machining (page 182).

Applications: Beech is 1 of the most widely used woods. It is available as sawn timber, plywood and veneer. Typical uses include woodworking tools, furniture, cooking utensils, doors, floors, beds, toys, disposable cutlery and clothes pegs.

Costs: Moderate.

Ash

- White ash
- Black ash

Ash is hard, elastic and resilient. Traditionally it has been used for tool handles that require a high degree of shock resistance.

Qualities: There are many trees in the ash (*Fraxinus*) family. The 2 main commercial timber producing types and the white ash (American) and black ash (European). These are deciduous trees that can reach 30 m (98 ft) in height.

The timber is off-white to light brown (tan). It has no distinctive odour or taste.

Unless impregnated with preservative, it is not suitable for outdoor use.

Ash has a long straight grain and long fibres. It is tough, resilient and has good strength to weight. The sawn timber is hard, durable and elastic with good resistance to shock. It is ideal for steam bending (page 198), turning and joinery (page 324). Machining (page 182) and polishing (page 376) will produce a high surface finish.

Applications: Ash is typically available as sawn timber. It has traditionally been used for tool handles and sports equipments such as baseball bats, hockey sticks, snooker cues and oars. Wooden car chassis were commonly made from ash. It is also commonly used in furniture, toys and flooring.

Costs: Moderate.

Qualities: These trees (*Carya* species) are native to North America. The properties vary according to the particular species. Generally, the sapwood is off-white with light brown streaks, and the heartwood is reddish brown; in some trees they contrast with dramatic effect.

Hickory is considered to be among the strongest and most durable hardwoods. It is tough and flexible and so is ideal for tool and axe handles. It is dense and so can be difficult to work by hand. The hardness produces a fine finish that is durable to abrasion and denting.

Applications: It is available as sawn timber and occasionally as a veneer. Typical applications include walking sticks, tool handles, axe handles, drum sticks, sports equipment, cabinets, furniture and flooring. In Europe it is largely replaced by ash, which is more readily available and so is less expensive.

Costs: Moderate to high.

HICKORY
Drumsticks

Made by:	Vic Firth
Material:	Hickory
Manufacture:	Turned on a lathe
Notes:	A dense wood is ideal, with little flex for more pronounced acoustic properties.

ELM
Windsor hall table

Made by:	Ercol
Date:	2005
Material:	Elm
Manufacture:	CNC machining

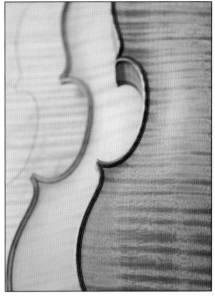

MAPLE
Violin backs

Made by:	Frederick Phelps
Material:	Hard maple or European sycamore
Manufacture:	Hand carved from solid wood
Notes:	The figured grain (flame) is essential for the acoustic properties of the violin because the grain is interrupted.

Elm

- Elm
Elm is a light brown timber with an interlocking grain. It is ideal for steam bending.

Qualities: There are many species of elm (*Ulmus*). Deciduous hardwoods, they grow across much of the northern hemisphere, but were almost completely wiped out in Europe and parts of North America by Dutch elm disease in the latter half of the 20th century. This has made elm much harder to purchase.

It is a medium brown timber that has a similar appearance to oak, with an interlocking grain that is resistant to splitting. It is supple when wet and so is very good for steam bending (page 198). Certain species are resistant to decay outdoors, especially when constantly submerged in water.

Applications: Limited availability as sawn timber. Applications include boat and barge hulls, furniture and beds.

Costs: Low to high.

Maple

- Maple
- Sycamore or field maple
These are deciduous hardwoods that have similar strength characteristics. They are off-white and darken gradually with age. They have no distinctive taste or smell.

Qualities: There are many different species of maple (*Acer*), which are categorized as either hard or soft. The timber has a uniform off-white colour with light brown heartwood. Hard maple is a heavy and strong timber with tightly packed grain; it is very hard and so is resistant to denting and abrasion.

Figured maples, such as bird's-eye and curly, are rare and high cost hard maples, typically only available as veneers. The European sycamore is a type of hard maple, often with interlocking grain. This reduces splitting and produces interesting patterns on sawn timber.

Soft maple has similar characteristics to hard maple and is a less expensive alternative in painted applications. But as the name suggests, it is a little softer.

Applications: These woods are available as sawn timber and veneer; some grades are suitable for steam bending. Typical applications include frameworks for upholstered furniture, cutting surfaces, chopping boards and butcher's blocks (particularly hard maple and sycamore), flooring, handles, buttons and knobs. Unless coated or treated, they are limited to indoor use. In North America, the sap of hard maple is refined into maple syrup.

Costs: Moderate.

OAK
Oak side table

Made by:	Unknown
Material:	Oak
Manufacture:	CNC machined
Notes:	Oak is widely used in furniture because it is dense and hardwearing.

OAK
Architectural cladding

Made by:	Various
Material:	Oak
Manufacture:	Sawn
Notes:	Oak gradually becomes silvery grey when used untreated outdoors. It is long lasting and durable against the elements.

WALNUT
Wing sideboard

Designer/client:	Michael Sodeau/Isokon Plus
Date:	1999
Material:	Walnut veneer
Manufacture:	Wood veneer laminating

Oak

- European oak
- American oak
- Asian oak

Oak is particularly hard, dense and resistant to denting. It is a heavy timber used in furniture construction, house building and floorboards.

Qualities: Oaks (*Quercus* species) are deciduous hardwoods and each type has slightly different physical and aesthetic characteristics. They are categorized by their country of origin, and each country may produce several different types such as the USA whose principle commercial oaks are known as white oak and red oak. These are relatively fast growing with a straight grain, and are very durable.

European and Asian oaks are slow growing. This produces a heavy and dense material with a tightly packed grain. Colour varies according to origin: European oak is light brown to light grey; American oak is light brown to dark reddish brown; and Asian oak is typically a lighter off-white colour.

Oak is hard, strong and stiff. The surface is resistant to denting and abrasion and so is particularly difficult to work by hand. Its hardness makes it prone to chipping and splitting.

Oak has a particularly high level of tannin. This is a preservative that has been exploited by the winemaking and leather tanning industries for centuries.

Oak is suitable for machining (page 182), joinery (page 324) and certain types can be steam bent (page 198). It can be used outdoors untreated, where over time it will gradually turn grey. Indoors, it will slowly become darker over time.

Applications: Oak is available as sawn timber and veneer. It is widely used in house building, boat building, furniture, doors, windows and floors. Wines and spirits are stored in oak casks or barrels.

Costs: Moderate to high.

Walnut

- Walnut

Certain species of walnut are prized as highly decorative timbers. They are also very hard and resistant to shock.

Qualities: Walnut (*Juglans* species) is a deciduous hardwood that grows across America, Europe and Asia. Certain species, including the black or American walnut and the Persian walnut, have hard and dense wood with a tight and densely packed grain. Walnut can be steam bent (page 198) and laminated

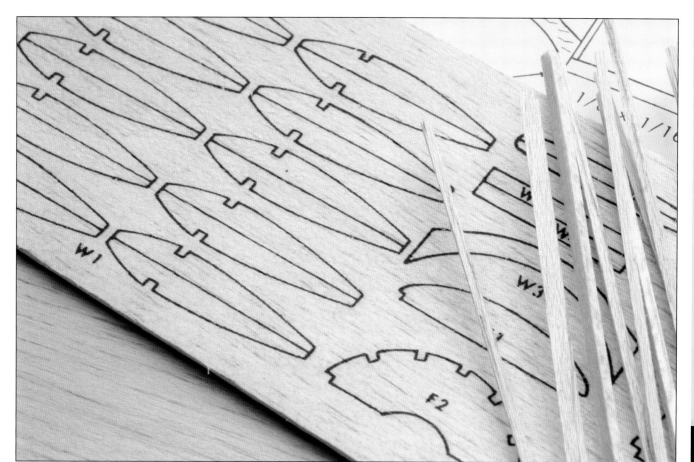

(page 190), because it does not split or splinter easily.

Due to its popularity as a decorative timber, it is mostly only available as a veneer. It has rich dark brown heartwood that contrasts with the lighter coloured sapwood. Steaming the wood makes a more uniform plank, but will dull the richness slightly. The grain is straight, but is commonly figured with burls and curls. Timber with such details is considerably more expensive.

It can be used untreated outdoors and in contact with the soil.

Applications: Walnut is available as sawn timber and veneer. Typical applications include floors, furniture, car interiors, tableware, knife handles, ornaments, musical instruments and gunstocks.

Costs: Moderate to high.

Balsa

• Balsa
Balsa wood is the lightest commercial grade timber. For its weight it is strong, resilient and shock absorbing.

Qualities: Balsa (*Ochroma pyramidale*) trees are fast growing and low density. They grow in humid environments such as central and southern America. Kiln drying reduces the water content to produce a low density and lightweight timber. The density varies from tree to tree; lower density balsa is less common and so is more expensive, even though heavier balsa is stronger.

Balsa contains less lignin than most other woods, contributing to its softness, and the cells are large and open, contributing towards its relative lightness. This means the tree is less rigid than most, and its stability depends on it

BALSA

Andreason BA4-B aeroplane model

Designer/client:	Walt Mooney/Peck-Polymers
Date:	N/a
Material:	Balsawood
Manufacture:	Die cut and inkjet printed

being pumped full of water; more than half of the living tree is made up of water.

The grain is coarse and open, so it is difficult to achieve a smooth surface finish. It is easily carved by hand and machine. Balsa is a light and uniform yellowish brown colour.

Applications: It is available as a sawn timber and veneer. Typical applications include architectural models, radio controlled aeroplanes, rockets and toys. It is also used as a core material in surfboards and boatbuilding.

Costs: Low to moderate.

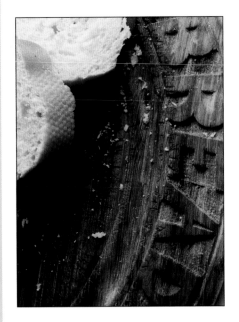

FRUITWOOD
Bread board

Made by:	Unknown
Material:	Cherry
Manufacture:	Hand carved
Notes:	Cherry is a hard, dense and decorative wood.

Apple (*Malus* species) is a slow growing and not very tall tree. This means that it is difficult to obtain large planks of it. The timber has a tightly packed grain and is heavy and hard. Like other fruitwoods, apple shrinks a great deal as it is dried, and this can cause warping and cracking.

Pear trees (*Pyrus* species) are similar to apple trees: they are slow growing and not very tall. The timber varies from light pinkish brown to yellowish brown. It has a tight grain and smooth, hard surface.

Applications: These woods are available as sawn timber and veneer. They are used for turning, machining (page 182) and as decorative veneers. Cherry is used in furniture, cabinets, floors and doors. Apple and pear are used for ornaments, tableware, wooden instruments and furniture. They have fragrant smoke, so are used to flavour food and tobacco.

Costs: High.

IROKO
Iroko work surface

Designer:	Retrouvius
Date:	Completed 2006
Material:	Reclaimed iroko
Manufacture:	Machining

Applications: Sawn timber and veneer are used for flooring, decking, outdoor furniture, worktops and general joinery. It has been used in laboratories due to its resistance to chemicals and water decay.

Costs: Moderate.

Fruitwood

- Cherry
- Apple
- Pear

This range of deciduous hardwood fruiting trees are used for their decorative appearance combined with moderate strength. Cherry is the most well known.

Qualities: Cherry (*Prunus* species) has distinctive and desirable heartwood of a rich reddish brown colour that gradually darkens with age. Sometimes it is speckled with darker patches. It is moderately slow growing and reaches heights of 30 m (98 ft) or so. There are many different species including American cherry (black cherry), wild cherry and Brazilian cherry. Once dried, it is stable, moderately dense and strong. It has a straight and uniform grain that can be polished to a high finish. The heartwood is resistant to decay. However, it is a high cost timber and so is often used as a veneer.

Iroko

- Iroko

Iroko is an African hardwood that is naturally resistant to chemicals and decay. It has similar properties to teak and so is often used as a cheaper alternative.

Qualities: Iroko (*Chlorophora* species) is an African hardwood. It is endangered and protected in some countries, but there is sufficient production from other sources for it to be competitively priced.

It has light yellow sapwood and brown heartwood. It is used for decorative appeal as well as durability and strength. Iroko contains natural oils that protect it from water decay and chemicals. Outdoors it gradually turns silvery grey.

Its interlocking grain produces good mechanical properties, but can cause tearing during machining (page 182). Hard calcium carbonate deposits in the wood fibres can blunt cutting tools.

Exotic hardwoods

- Mahogany
- Ziricote
- Keruing
- Rosewood
- Teak
- Ebony
- Wengé

This is a group of highly decorative timbers with lustrous colours. Certain species are very hard and durable. Some are endangered.

Qualities: These woods are highly priced and hard to come by, due to their places of origin, desirability and rate of growth. They are prized for their lustrous colours; ebony (*Diospyros ebenum*) is black or very dark red, ziricote (*Cordia dodecandra*) has a deep, contrasting swirling grain, wengé (*Millettia laurentii*) is a dark chocolate,

EXOTIC HARDWOODS
Reclaimed teak stool

Designer: Mandala
Date: 2006
Material: Reclaimed teak
Manufacture: Hand carved

rosewood (*Dalbergia* species) is a veined brown and mahogany (*Khaya* species) is a rich brown that darkens with age.

These materials, especially ebony and wengé, are very hard and resistant to denting. Teak (*Tectona* species), keruing (*Dipterocarpus* species) and mahogany contain natural oils that protect them against decay, so are popular for outdoor furniture, decking and boat building.

Care must be taken when specifying or purchasing exotic hardwoods. Certain endangered species (such as *Swietenia*, the original mahogany) are protected by international legislation, and permits are required for trading.

Applications: Limited availability as sawn timber and veneers. They are used in flooring, furniture, ornaments, musical instruments and for decorative inlay.

Costs: High to very high.

EXOTIC HARDWOODS
Bluetooth ziricote Pip*Phone

Designer/client: Nicolas Roope and Kam Young/
                  Hulger
Date: 2005
Material: Ziricote
Manufacture: CNC machined

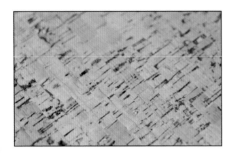

CORK
Cork sheet

Made by:	Various
Material:	Cork
Manufacture:	Compression molded
Notes:	Sheets and blocks are made up of scrap produced in the manufacture of cork stoppers.

CORK
Cork low table

Designer/client:	Jasper Morrison/Moooi
Date:	2002
Material:	Cork
Manufacture:	Turned from agglomerate cork

BAMBOO
Bamboo flooring

Made by:	Various
Material:	Bamboo
Manufacture:	Laminated and CNC machined
Notes:	Bamboo is fast growing, hardwearing and lightweight.

Cork

• Cork
Cork is light, buoyant and up to 85% air by volume. It is naturally resistant to decay.

Qualities: Cork is harvested from the bark of the cork oak (*Quercus suber*), which grows across the Mediterranean. Once a tree is old enough, the bark is peeled from the tree every 10 years or so. Production is sustainable because peeling the bark does not damage these trees permanently; most other trees will die if their bark is cut around the trunk.

Cork is naturally light tan, but can be coloured. Suberin, a waxy substance, is naturally present in the cork and protects it from water penetration and microbial attack. It is resistant to decay and difficult to burn, which makes it useful as an insulating material in construction.

Applications: Cork is available as compressed blocks and sheets mounted onto a backing material. Applications include fishing floats, bottle stoppers, seals, insulation, pin boards, furniture and flooring.

Costs: Low to moderate.

Bamboo

• Bamboo
This is a fast growing grass that is harvested and used as a hardwood-like material.

Qualities: There are many different species and genera of bamboo, which have different rates of growth. Certain species grow up to 1 m (3.3 ft) per day and reach heights of 30 m (98 ft) or more. China is the largest producer of bamboo and has utilized its properties in construction and furniture for many thousands of years.

Once harvested, bamboo continues to send out shoots. Its popularity is growing beyond the countries of production due to its rapid growth, sustainability and relatively low cost. It can be harvested every 4 years, compared with several decades for many hardwoods.

Bamboo has similar characteristics to hardwoods (pages 472–9). It is harder and lighter than many hardwoods and suitable for turning, machining (page 182) and working by hand. The stems are hollow with ringed joints, which produces decorative patterns when cut or split into planks and bonded together.

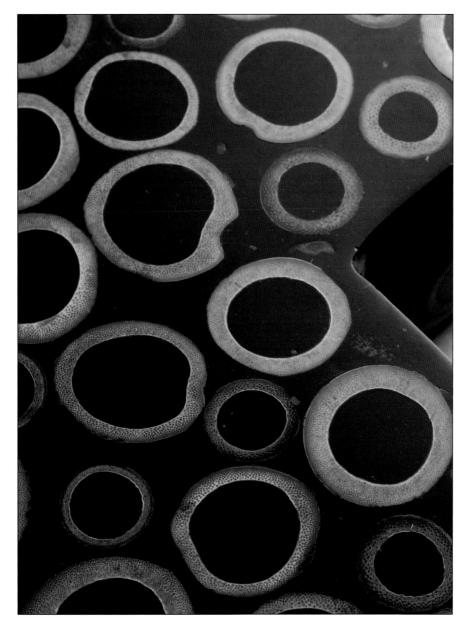

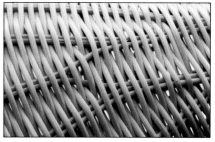

BAMBOO
Bamboo resin stool

Designer:	Mandala
Date:	2005
Material:	Bamboo in black resin
Manufacture:	Molded and machined

Applications: Bamboo is available round and cut to length, split into strands (wicker and cane) and compressed into planks. Typical applications include buildings, scaffolding, furniture, tableware, countertops and flooring.

Costs: Low to moderate.

Rattan, reed and rush

- Rattan
- Reed
- Rush

Collectively known as wicker (which also includes bamboo and willow), these materials are used in making baskets, tableware and furniture.

Qualities: Rattan is a diverse group of climbing palms that produce vines up to 200 m (656 ft) long and 70 mm (2.75 in.) in diameter. The largest producer is Indonesia, but they grow across Asia. The outer bark is removed and the core is dried to form material suitable for wicker furniture. The outer bark is often used

RATTAN, REED AND RUSH
Wicker lights

Designer:	Mandala
Date:	2006
Material:	Rattan
Manufacture:	Hand woven

to bind joints in the furniture. Once dry, the inner core has a similar hardness to many species of softwood (page 470). It is lightweight and flexible and so can be formed around tight bends. Once formed it holds its shape. Unlike bamboo, these materials have a solid core. The choice of material depends upon local availability. Reed and rush are grasslike plants that grow in wetlands and are used in seating and musical instruments. The flexible shoots of a willow, known as switches, are also suitable for weaving.

Applications: Available as lengths of fibrous material, which are typically used in the production of wicker baskets, tableware and furniture.

Costs: Low to moderate.

Ceramics and Glass

Ceramics and glass are classified as non-metallic materials that are made by firing. They are hard and brittle materials that have been used for millennia in the production of decorative and functional objects. In recent years the boundary between ceramics and glass has been blurred with the development of high performance materials including glass ceramic. This material is formed like glass, but develops a ceramic-like crystallized structure following heat treatment.

TYPES OF CERAMIC

There are 2 main groups of ceramics, which are clay-based and high performance ceramics. Clay-based ceramics are those associated with pottery and include earthenware, stoneware and porcelain. They are fine-grained materials made up of clay minerals (aluminium silicate), quartz and rock fragments. Historically, the quality of ingredients differed according to location. Nowadays, clay is purchased from manufacturers that produce raw materials in accordance with guidelines.

High performance ceramics have fewer impurities than clay-based ceramics and superior properties. They are resistant to high temperatures, most chemicals and corrosion, and outperform metals in many demanding applications. However, like clay-based ceramics they are hard and brittle, and their porous structure has high compression strength but low tensile strength, which typically falls between metal and plastic.

TYPES OF GLASS

There are several different glasses and the exact ingredients differ slightly according to location, processing plant and application. The most common glass, soda-lime glass, is widely used in design, architecture and jewelry. Other glasses are more expensive and so tend to be used only when necessary. For example, borosilicate glass is more resistant to high temperatures and thermal shock than soda-lime glass and so is used in kitchenware and laboratory products. Glass ceramic is even more resistant to chemicals, high temperature and thermal shock, and is used in glass cooker

Random Light

Designer/ made by:	Bertjan Pot/ Moooi
Date:	2001
Material:	Glass fibre reinforced epoxy resin (EP)
Manufacture:	Resin soaked yarn is draped around an inflatable mold

Chair #1

Designer:	Ansel Thompson
Date:	2001
Material:	Fibre optics, aluminium honeycomb and glass fibre reinforced epoxy resin (EP)
Manufacture:	Composite laminating

tops (see image, page 493) as well as industrial applications.

Glass fibres are used to produce textiles, insulation, structural reinforcement and fibre optics. The type of glass depends on the application, for example, textiles are generally a borosilicate or similar glass, whereas insulation (glass wool) is filaments of soda-lime glass. Glass fibres and textiles are used for structural reinforcement in GRP (glass reinforced plastic). The length of fibre used is determined by the application and manufacturing process. Glass fibre used to reinforce injection molded thermoplastic can be as short as 1 mm (0.04 in.). These materials have greater toughness, strength and resilience, and are used in the production of power tools, furniture and automotive parts, for example. Continuous and long strand fibre reinforcement produces maximum strength to weight properties. Processes used to mold continuous and long strand fibre reinforcement include composite laminating (page 206), compression molding (page 44), filament winding (page 222) and 3D thermal laminating (page 228). Typical products include racing cars, monocoque boat hulls, canoes, furniture (see image, above) and light shades (see image, top).

Ilac Centre, Dublin

Designer/client:	Christopher Tipping and Fusion Glass/British Land Ltd and Chartered Land Ltd
Date:	Completed 2006
Material:	Soda-lime glass
Manufacture:	Sand blasting, deep carving, gilding and colour rub

Fibre optics carry light along their length as a result of internal reflection. Modulated light is used to transfer information across vast distances in the communication industries via cables made up of bundles of fine glass fibres. Fractures and imperfections will cause light to escape along the length of the fibre. In recent years this quality has been used by designers to illuminate interiors and products such as furniture (see image, below, page 483) in novel ways.

NOTES ON MANUFACTURING CERAMICS

Ceramics and glass are processed in different ways. Clay-based ceramics are plastic and formable in their wet state. Processes used to shape them include throwing (page 168), slip casting (page 172) and press molding (page 176). They are fired to fuse the ingredients and lock the shape in place. Earthenware is fired at 1000°C to 1250°C (1832–2282°F), stoneware is fired at 1200°C to 1300°C

Prada Store, Tokyo

Architect/Client:	Herzog & de Meuron/Prada
Date:	Completed 2003
Material:	Laminated soda-lime glass
Manufacture:	Kiln formed

(2192–2372°F); and porcelain is fired at over 1300°C (2372°F).

High performance ceramics are typically formed in a powdered state. Processes used to shape them include powder injection molding (see metal injection molding, page 136), ceramic injection molding (CIM) and press molding followed by sintering. The particles are sintered or fired at around 2500°C (4532°F), which makes them much more expensive to fabricate than clay-based ceramics.

High performance ceramics can be machined (page 182), although this can be a lengthy process, laser cut (page 248)

and water jet cut (page 272). Tungsten carbide is sufficiently conductive to be cut and shaped using die sink EDM and wire EDM (page 254).

NOTES ON MANUFACTURING GLASS

Glass is shaped in a hot and molten state. The process begins in a furnace, where the raw materials are all crushed into uniform sized particles and mixed together. The raw materials are mixed with cullet (crushed recycled glass) at 1500°C (2732°F) and fuse together to form a homogenous, molten mass. At this stage they are either coloured with additives, or a decolourant is added to make clear glass. The material is conditioned for 8–12 hours, which is long enough for all the bubbles to rise to the surface and dissipate, and cooled to 1150°C (2102°F). The glass is now ready to be formed by blowing (page 152), pressing (page 152) or molding. Lampworking (page 160) is the process of local heating and forming of glass,

5

Stella Polare

Designer/client:	Georg Baldele/Swarovski Crystal Palace Project
Date:	2002–2005
Material:	Strass® Swarovski® crystal
Manufacture:	Diamond ground, polished, rouged and buffed

6

Pure Magic Candlestick

Designer:	Antonio Arevalo, Fusion Design
Date:	2005
Material:	Crystal glass
Manufacture:	Sub-surface laser engraving, CNC machining and polishing

7

Kaipo light

Designer/ made by:	Edward van Vliet/Moooi
Date:	2001
Material:	Silvered crystal glass
Manufacture:	Glassblowing and silvering

so it does not require the same level of conditioning beforehand.

Sheet materials (see image, opposite) can be shaped by kiln forming, which is also known as slumping and sag bending, or press bending. Car windshields are formed by precision press bending. Textured and glass sheet is produced by rolling hot glass between profiled rollers; wire reinforced glass is produced in a similar continuous process.

Once formed, glass is annealed by heating and slow cooling to remove internal stresses in the molecular structure. Borosilicate glass is heated up to around 570°C (1058°F), 'soaked' at this temperature, and then very gradually cooled down. Each type of glass is annealed at a different temperature. Thicker sections take longer to anneal because the temperature throughout the glass structure has to be equalized. Therefore, some pieces, particularly artwork and sculpture, have to be annealed for several days, or even

months. In such cases the temperature may be reduced by 1°C to 2°C (2.7–3.4°F) each day to minimize potential stress.

Glass sheets are laminated together for safety, security and decorative effect. Laminated safety glass is produced by bonding together sheets of glass with a polyvinyl butyral (PVB) film. The process uses heat and pressure to remove any air bubbles, so the sheet looks solid. This product is primarily used in automotive and construction applications. It is also possible to laminate decorative wire mesh, fabric, wood veneer, and coloured and printed PVB into glass sheets.

Glass can be modified by machining, abrasive blasting (page 388), deep carving, laser engraving (page 248) and water jet cutting (page 272). Deep carving is an abrasive blasting process. Instead of simply frosting the surface, a skilled operator can use the high-pressure jet of abrasive particles to 'carve' 3D shapes and patterns into the surface of the glass (see image, opposite). Lead

and crystal glass are relatively soft and so can be cut more easily. Cutting facets into these glasses enhances their sparkle and lustre. Chandeliers, ornaments and decorative tableware are commonly made in this way. Machining can produce shapes accurate to within 15 microns (0.00059 in.). Hard glasses, such as borosilicate, take much longer to cut and polish and so are less commonly worked in this way. Layering coloured and clear glass can enhance the effects of cutting.

Sub-surface laser engraving is the process of creating 2D and 3D designs within glass objects (see image, above centre). It is possible to engrave either a 2D illustration or a 3D CAD file or scan below the surface of glass. Careful consideration should be given to 3D designs because the engraving is see-through and it can easily become cluttered and overcrowded. Lead alkali glass (crystal glass) is the ideal medium for this process because it has excellent optical properties.

8

9

10

Sponge and sponge vase

Designer/ made by:	Marcel Wanders/Moooi
Date:	1997
Material:	Porcelain
Manufacture:	To make this form a natural sponge (top) is dipped in liquid ceramic slip and fired (above).

machined. When it breaks, the fragments tend to be small and blunt as a result of the modified structure, so it is used in safety applications such as door glazing, car windscreens and dinnerware.

High performance glass is made at elevated temperatures and so is more difficult and expensive to fabricate.

MATERIAL DEVELOPMENTS

Developments in ceramics and glass have been concentrated in the high performance materials and coatings. Materials are being combined in different ways to produce new composite materials such as light transmitting concrete and lightweight foamed panels.

Litracon® (see image, above left) is a light transmitting concrete invented by Hungarian architect Áron Losonczi in 2001. It is made up of glass fibres embedded in a fine concrete matrix. The glass fibres are in parallel alignment and so transmit light through the concrete without distorting the image. The fibres

Glass mirrors are produced by coating glass with a fine layer of silver, followed by copper (or other metal) and a protective lacquer. Twin-walled blown glass (see image, page 485, right) can be silvered on the inside and sealed without the need for a protective layer. Protecting or sealing is essential because otherwise silver will tarnish and corrode. The blown glass technique has both decorative and functional uses. Silver prevents radiation of heat, a quality used in the production of glass vacuum (thermos) flasks.

Tempered glass, also known as safety or toughened glass, is produced by heat treatment. Soda-lime glass is heated to

Litracon® light transmitting concrete

Made by:	Litracon Bt, Hungary
Material:	Glass fibres (4%) in a fine concrete matrix
Manufacture:	Cast

approximately 650°C (1202°F) and then quenched in blasts of cool air. This forces the surface of the glass to cool more rapidly than the core and thus forms a structure under compression. As a result, tempered glass is several times stronger than annealed glass. However, once tempered, it cannot be modified or

11

Aerogel and Peter Tsou, JPL scientist

Made by:	JPL (Jet Propulsion Laboratory)
Material:	Aerogel (NASA/JPL)
Manufacture:	N/a
Notes:	Silica with a porous, dendritic-like structure

99.8% atmosphere, so it is extremely lightweight. Like ordinary glass, it is fragile and brittle, but it is a much more efficient insulator because the internal structure has a vast surface area. It is considerably more expensive than ordinary glass and is not yet available in transparent grades. There are not many commercial applications outside aerospace. However, in 2007 Dunlop Sports released a range of tennis rackets with an aerogel core. Its extreme lightness is used to produce a racket with high strength to weight and stiffness.

A vast range of coatings have been developed to improve the characteristics of glass. These include dichroic, anti-reflective, thermally insulating, sound reflecting, self-cleaning and view control.

Pilkington developed self-cleaning glass in 2001. Pilkington Activ™ is produced by coating flat glass with a very thin, transparent layer of titanium oxide. This produces a surface that is both hydrophilic and photocatalytic. In combination, these qualities help to maintain a surface that is free of dirt and watermarks: UV radiation helps to break down surface dirt, which is subsequently washed away by rain.

ENVIRONMENTAL IMPACTS

Glass production is a hot and energy intensive process. Cullet mixed with the raw ingredients reduces the amount of energy required in production. Glass can be recycled indefinitely without any degradation to its structure. However, the use of cullet may have an impact on the colour of the glass. Clear glass cannot tolerate impurities and colour contamination, whereas brown and

are very thin and make up approximately 4% of the material, so they do not affect its structural integrity.

Ceramic and glass foams harness the high temperature properties of these materials in the form of lightweight structural materials. Ceramic foams are typically formed by saturating polymer foam (or similar porous material) with ceramic slurry. The firing process hardens the ceramic and simultaneously burns out the foam. A range of ceramics can be formed in this way. Two applications include high temperature filtration and thermal insulation. In a project for Droog Design and Rosenthal, Marcel Wanders

developed a similar technique for the production of domestic ceramics (see images, opposite).

Glass foams are formed by adding a foaming agent to the raw material. Bubbles form in the molten glass and are locked in place as it cools. Each bubble is sealed into the glass (closed cell), so it remains watertight and gastight. Foams can be formed from crushed (recycled) glass by mixing the ingredients in a mold. As the crushed glass is heated the gassing agent produces bubbles that are encapsulated in the molten structure.

Ceramic fibres and textiles can withstand temperatures up to 1450°C (2642°F). The material determines the temperature resistance. They are used for insulation (as mineral wool), fire protection and other high temperature applications. Products include textiles, non-wovens, foil and rope.

Aerogel (see image) is made up of an open cell silica structure. It is very low density and can be as much as

12

Blue Carpet

Designer/client:	Heatherwick Studio/Newcastle City Council
Date:	Completed 2002
Material:	Crushed glass in white resin
Manufacture:	Cast

green glass may contain a higher percentage of mixed cullet.

Alternatively, cullet can be used in its crushed state. Applications include glass foam for insulation and aggregate in concrete. Heatherwick Studio found a new and interesting use for Bristol Cream bottles in Newcastle upon Tyne, UK. Crushed blue glass is mixed with white resin to form hardwearing and decorative paving slabs, benches and kerbs (see image, above). Glass is non-toxic and safe to dispose of, although this is not desirable for such a valuable reusable material. It is long lasting and can be sterilized and refilled many times.

Pottery ceramics

- Earthenware
- Stoneware
- Porcelain

Clay combined with other minerals is plastic and can be formed by hand. When fired at high temperatures it becomes hard, brittle and in some cases vitreous.

Qualities: Earthenware is clay that has been fired at 1000°C to 1250°C (1832–2048 °F). Clay is made up of aluminium silicate, quartz and other fine rock particles, and the ingredients will vary slightly according to the geographical location. The type of clay determines the quality of the earthenware. Terracotta, for example, is a type of clay used in the production of earthenware. When fired, it is a characteristic reddish brown colour.

Earthenware tends to be porous and brittle, and will chip more easily than stoneware or porcelain. It has to be glazed to be made watertight. Unglazed pots that are left outdoors frequently

crack in freezing conditions due to moisture absorption.

Stoneware is fired at 1200°C to 1300°C (2192–2372°F). This produces a vitreous or semi-vitreous ceramic that is less porous and stronger than earthenware. Even though it is less porous, it is not completely watertight unless it has been glazed. The ingredients are not as refined as those of porcelain, so the raw material is often visibly speckled.

Porcelain is made up of a specific type of clay, kaolin (china clay), mixed with petuntse, quartz and other minerals. It is typically more refined than earthenware and stoneware. When fired at over

POTTERY CERAMICS
Is That Plastic? butter dish

Designer:	Helen Johannessen
Date:	2004
Material:	Earthenware
Manufacture:	Slip cast and glazed

1300°C (2372°F) these minerals combine to form a vitreous and translucent ceramic. It is dense and watertight, but is often glazed for decorative purposes.

Applications: Earthenware is typically used in the production of wall and floor tiles, toilets, sinks, garden pots and tableware. Stoneware is used for these applications as well as cookware. Porcelain is more expensive, so typical applications are more likely to include dinnerware, vases, cups and saucers. Porcelain is also used for dental implants and certain industrial applications.

Costs: Pottery ceramics are moderately expensive materials: earthenware is the least expensive and porcelain is the most expensive.

POTTERY CERAMICS
Ramekin

Material:	Stoneware
Manufacture:	Press molded
Made by:	Unknown
Notes:	Stoneware is more durable than earthenware and can be used as cookware.

POTTERY CERAMICS
Egg vase

Designer/ made by:	Marcel Wanders/Moooi
Date:	1997
Material:	Porcelain
Manufacture:	Slip cast

High performance ceramics

- Alumina
- Silicon nitride
- Tungsten carbide
- Zirconia
- Silicon carbide
- Boron carbide

A wide range of almost pure non-metallic materials for demanding or extreme applications.

Qualities: These are the most common of the high performance ceramics. They are hard and durable and have very good resistance to temperatures up to over 2000°C (3632°F), wear and corrosion. However, they tend to be brittle and have poor resistance to impact and shock.

Alumina (aluminium oxide) ceramic has high chemical stability. Silicon nitride ceramic is very hard and has good resistance to shock and heat. Tungsten carbide is exceptionally hard, so is used in cutting tools and as an abrasive powder for blasting (page 388), grinding and polishing (page 376). The hardness, abrasion resistance and fine surface

HIGH PERFORMANCE CERAMICS
Jamie Oliver Flavour Shaker

Designer:	Jamie Oliver and William Levene
Date:	2006
Material:	Alumina ceramic ball in a plastic flask
Manufacture:	Pressed, sintered, ground and polished ceramic

finish that can be achieved with zirconia is utilized in cutting blades; unlike metal, it requires no lubrication. Silicon carbide ceramic has a very high melting point, above 2500°C (4532°F). Boron carbide is 1 of the hardest materials known and is used in nuclear applications and armour. Some high performance ceramics are

non-toxic, do not absorb odours and are safe for food contact, but not all.

Applications: Applications include medical and dental implants, armour, cutting tools and blades and wear resistant nozzles, such as for water jet cutting (page 272). Due to their high cost and brittle nature they are often applied as high performance coatings, to provide protection against corrosion, chemicals and wear, without needing lubrication. They are used by the aerospace and high performance automotive industries.

Costs: Very high.

Soda-lime glass

- Soda-lime glass

Also known as 'commercial glass', this is the least expensive and most common glass. Applications range from blown glass packaging to windowpanes.

Qualities: Soda-lime glass is made up of silica sand (up to 75%), soda ash, lime (calcium oxide) and other additives. The exact ingredients depend on the processing and application.

It is a 'soft' glass that is relatively easy to mold and fabricate. It softens at around 400°C to 500°C (752– 932°F) and so is economical for mass production. However, this also means that soda-lime glass is prone to shatter at high temperatures or in response to sudden changes in temperature.

Soda-lime glass finishes with a smooth and non-porous surface. It is inert and tasteless and suitable for

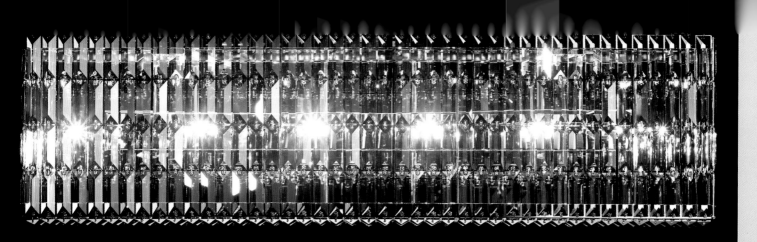

SODA-LIME GLASS
Condiment jar

Made by:	Various
Material:	Soda-lime glass
Manufacture:	Machine press and blow

LEAD ALKALI GLASS
Glitterbox

Designer/client:	Georg Baldele/Swarovski® Crystal Palace Project
Date:	2006
Material:	Swarovski® crystal
Manufacture:	Diamond ground, polished, rouged and buffed

Lead alkali glass

- Lead glass
- Crystal glass

Due to the lead content these glasses have a higher refractive index than other types. Increased refraction produces a clearer and more lustrous glass.

packaging liquids, food and many chemicals. However, borosilicate glass (page 492) is more resistant to many acids and alkalis and so is more suitable for packaging and storing certain products.

Applications: Applications are widespread and include windowpanes (float glass, page 494), automotive windows, mirrors, tableware, dinnerware, light bulbs, light shades, packaging, storage jars and laboratory glassware.

Costs: Low.

Qualities: Like soda-lime glass, lead alkali glass is silica based, but the lime is replaced by lead and the soda replaced with potash. If it has less than 25% lead it is known as crystal glass and when there is more than 25% lead it is known as lead glass. Over prolonged periods the lead content can leach, so this glass is not suitable for storing liquids and foods.

The lead oxide acts as a flux to reduce the softening temperature of the glass. As a consequence it is even 'softer' than soda-lime glass. It is relatively easy to form by glass blowing (page 152), press molding, cutting, CNC machining (page 182) and finish by polishing (page 376).

Cutting enhances the sparkle of the glass and as such is used in the production of decorative tableware, lighting (see image), ornaments and jewelry.

Lead content makes it suitable for certain radiation shielding applications. Such glasses typically have more than 50% lead content and are used in hospitals, airports and laboratories.

Like soda-lime glass, lead alkali glass is not suitable for use in high temperature applications or those that will experience rapid temperature change.

Applications: Typical applications include cut glass, packaging, tableware, vases, candlesticks and ornaments. For their superior optical properties, lead and crystal glass are used in jewelry, awards, trophies, prisms and lenses for telescopes and cameras. They are also used as clear radiation shielding in hospitals, laboratories and airports.

Costs: Moderate to high.

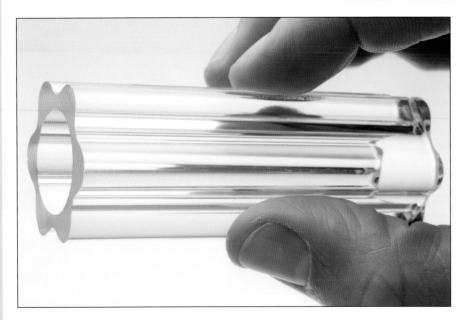

Borosilicate glass

• Borosilicate glass
This is also known under the trade names Duran, Simax and Pyrex (although the Pyrex consumer brand no longer uses borosilicate glass in North America). It is used primarily for its resistance to high temperatures and thermal shock.

Qualities: Borosilicate glass is so called because it contains up to 15% boric oxide and small amounts of other alkalis. It is 'harder' and more durable than soda-lime and lead alkali glass (pages 491). Borosilicate glass is more likely to survive being dropped or hit. It has low levels of thermal expansion, and is resistant to thermal shock, so it is useful for laboratory equipment that is repeatedly heated and cooled.

Its softening point is relatively high at 800°C to 850°C (1472–1562°F). This makes it more difficult to mold and fabricate, but means that it can be used for high temperature applications and can tolerate up to 500°C (932°F) for short periods.

It is more resistant to acids than soda-lime glass and has moderate resistance to alkalis. As a result, it is used to store chemicals and is suitable for long periods of storage. Museums use borosilicate storage jars for their precious collections of specimens. They have a precision

ground airtight seal, produced by honing (page 376), to preserve the contents.

Extruded glass profiles are typically borosilicate glass, because soda-lime is 'softer' and prone to breaking during processing. There is a range of diameters in each profile: tube can be up to 415 mm (16.34 in.) and complex profiles can be up to 120 mm (4.72 in.), but certain profiles (such as triangles) are limited to only 20 mm (0.79 in.).

Applications: Typical applications include ovenware, glass tea and coffee pots, kettles, scientific glassware (test tubes and distillation equipment) and pharmaceutical products. As well as industrial and performance applications, borosilicate has superior lampworking (page 160) characteristics, so is used in the production of jewelry, beads, paperweights, sculpture and ornaments.

Costs: Moderate to high.

High performance glasses

- Glass ceramic
- Aluminosilicate glass
- Quartz glass

These glasses have high working temperatures; they are relatively difficult to fabricate, but have superior resistance to heat and thermal shock.

Qualities: These are high performance and high cost materials. Glass ceramics are so called because they are shaped like glass in a molten state but heat-treated to give a high level of crystallinity, similar to ceramics. The resulting material is harder, more durable and resistant to rapid temperature change. It has very low levels of thermal expansion and can operate at temperatures ranging from -200°C to 700°C (-328–1292°F).

Aluminosilicate glass contains higher levels of aluminium oxide than other lower cost glasses. It is similar to borosilicate glass, but has improved resistance to chemicals,

high temperatures and thermal shock. It can operate at temperatures of up to 750°C (1382°F).

Quartz glass, also known as fused quartz and silica glass, is made up of almost pure silica (silicon dioxide). It is manufactured by heating up quartz to 2000°C (3632°F), which causes it to fuse together. It has exceptional resistance to high temperatures, thermal shock and most chemicals. It can be heated up to over 1000°C (1832°F) and rapidly cooled without any structural degradation.

Applications: Glass ceramics are used in stove and fireplace doors, light covers, cooker tops, oven windows, and ovenware that can be placed in a preheated oven straight from the freezer. Other high performance glasses are used in light covers mainly for industrial applications, but also halogen bulbs, cookware and even jewelry.

Costs: High to very high.

HIGH PERFORMANCE GLASSES
Glass ceramic cooker top

Made by:	Küppersbusch
Material:	Glass ceramic
Manufacture:	Formed as glass and then crystallized with heat treatment
Notes:	Glass ceramic can withstand thermal shock, is very durable and can withstand moderate impacts.

→ Pilkington float glass

Float glass is soda-lime glass with slightly modified ingredients to make it suitable for mass production. Alastair Pilkington developed the process, which was unveiled to the public in 1959 and has since become the standard method for mass producing flat glass. It is now widely used in the construction and automotive industries.

Flat glass sheets are available in a range of thicknesses from 0.4 mm to 25 mm (0.016–0.98 in.). It is produced in a 4 stage process. First of all silica sand, lime, dolomite and soda are mixed with cullet (recycled glass) in a furnace (**image 1**). The mix is heated by burning a combination of natural gas and pre-heated air at 1600°C (2912°F) (**image 2**). The hot molten glass leaves the furnace at approximately 1000°C (1832°F) and is floated on a bath of molten tin in a controlled atmosphere of hydrogen and nitrogen which prevents the tin from oxidizing (**image 3**). The glass is annealed and cooled (**image 4**). Finally, the sheet is scored and snapped into pre-determined sizes.

Modifications, such as coatings, are applied during production for specific functional improvements. Examples include self-cleaning Pilkington Activ™ produced by a very thin coating of titanium oxide; low emissivity Pilkington K Glass™ that reflects heat back into buildings and so reduces heat loss; and Pilkington Optifloat™, which is ideal for façades and furniture due to a higher level of clarity than 'clear' glass. This material is utilized in the Mall of Millennia, Florida, designed by JPRA Architects (**image 5**).

1

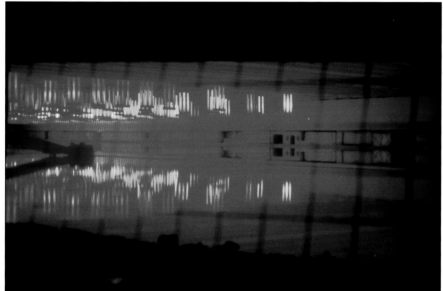

Featured Company

Pilkington Group Limited
www.pilkington.com

Glossary and Abbreviations

3DL
Three-dimensional thermal laminating (page 228).

3Dr
Three-dimensional rotary laminating (page 228).

ABR
Acrylonitrile butadiene rubber.

ABS
Acrylonitrile butadiene styrene.

A-class Finish
The automotive industry uses this term to describe a high gloss surface finish. Processes such as composite laminating (page 206), metal stamping (page 82), panel beating (page 72), polishing (page 376), spray painting (page 350) and superforming (page 92), are all capable of producing parts with an A-class finish.

A-side
Parts produced over a single-sided mold have an A- side and a B-side. The A-side does not come into contact with the tool and so has a higher surface finish than the B-side. Processes that use a single-sided mold include composite laminating (page 206), superforming (page 92) and thermoforming (page 30).

Acetal
Common name for polyoxymethylene (POM) (page 439).

Acetate
Alternative name for cellulose acetate (CA) (see cellulose-based, page 446).

Acrylic
Common name for poly methyl methacrylate (PMMA) (page 434).

Air-dried Timber
Lumber seasoned without the use of a kiln to 20% or less moisture content (see also kiln-dried timber, page 498, and seasoned timber, page 500).

Amorphous
Unorganized and non-crystalline molecular structure materials, which melt across a wider temperature range than crystalline materials.

Biocomposite
Natural materials, such as wood fibres or wheat straw, molded and bonded together with a natural or synthetic resin. Examples include Environ® and Treeplast® (see wood, page 468). They sometimes crossover with biofibre reinforced plastics.

Biofibre Reinforced Plastics
These are plastics reinforced with natural fibres such as cotton, hemp and jute; they are similar to biocomposites (page 468). Developments are concentrated in composite laminating (page 206) for automotive applications.

Bioplastic
Natural polymers that are made without petrochemicals: for example , cellulose-based (page 446), starch-based (page 446) and natural rubber materials (page 447). Some are completely biodegradable and require less energy to produce than synthetic polymers.

Biopolymers
Another name for bioplastics.

BR
Butadiene rubber.

CA
Cellulose acetate.

CAD
Computer-aided design is a general term used to cover computer programmes that assist with engineering, product design, graphic design and architecture. Some of the most popular 3D packages include Alias, Auto CAD, Maya, Professional Engineer (commonly known as Pro E), Rhino and Solid Works. Many products are now designed and engineered in this way. Some notable examples include aluminium forging (page 457), Chair #1 (page 483), the Eye chair (page 342) and Pedalite (page 53).

CAE
Computer-aided engineering is a general term used to cover the use of computer programmes in the design, simulation, analysis, production and optimization of products and assemblies. Examples include FEA and Moldflow (page 56).

CAM
Computer aided manufacturing.

CAP
Cellulose acetate propionate.

Checks
Splits that form in the end of a length of timber as it is seasoned or kiln dried.

CNC
Machining equipment that is operated by a computer is known as computer numeric control (CNC). This includes milling machines, lathes and routers used to manufacture all types of materials. The number of operational axes determines the types of geometries that can be achieved: 2-axis, 3-axis and 5-axis (page 183) are the most common.

COE
Coefficient of expansion.

Commodity Thermoplastic
A common name for the least expensive thermoplastics that make up the majority of total plastic production: for example, polypropylene (PP) and certain grades of polyethylenes (PE) (see polyolefins, page 430).

Copolymer
A polymer made up of long chains of 2 repeating monomers: for example, styrene acrylonitrile (SAN) (see styrenes, page 432).

CR
Chloroprene rubber.

CRP
Carbon reinforced plastic.

Crystalline
Highly organized molecular structure that has a sharper melting point than comparable amorphous materials.

DfS
Design for sustainability.

Direct Manufacturing
Manufacturing products directly from CAD data: for example, rapid prototyping (page 232).

DMC
Dough molding compound (page 218).

DMLS
Direct metal laser sintering (page 232).

Durometer Hardness
Another name for Shore hardness (see page 500).

Dyneema®
DSM trade name for ultra high-density polyethylene (UHDPE) (see polyolefins, page 430).

EBM
Extrusion blow molding (page 22).

EBW
Electron beam welding (page 288).

EDM
Electrical discharge machining (page 254).

EL
Electroluminescent.

Elastomer
A natural or synthetic material that exhibits elastic properties, and has the ability to deform under load and return to its original shape once the load is removed. Examples include rubber (page 425), thermoplastic elastomers (page 425) and thermoplastic polyurethane (page 436).

EMI
Electromagnetic interference.

Engineering Thermoplastic
High-performance thermoplastics materials for demanding applications. These tend to be more expensive and less widely applied than commodity thermoplastics (page 435), and include thermoplastics such as acetal (page 439), polyamides (page 438), polycarbonate (PC) (page 428) and thermoplastic polyesters (page 437).

Engineering Timber
These high-strength and dimensionally stable lumbers typically for architectural applications are produced by laminating wood with strong adhesives. Examples include laminated strand lumber (LSL) and parallel strand lumber (PSL) (see wood and natural fibres, page 466).

EP
Epoxy resin.

EPDM
Ethylene propylene diene monomer.

EPM
Ethylene propylene monomer.

EPS
Expanded polystyrene.

ETFE
Ethylene tetrafluoroethylene.

EVA
Ethylene vinyl acetate.

FEA
Finite element analysis is a computer simulation technique that uses FEM (page 16) to analyse designs, improve molding efficiency and predict part performance post-molding.

FEM
Finite element method is a numerical system that assists with engineering calculations by dividing an object in smaller parts, known as finite elements. The properties of these parts are mathematically formulated to obtain the properties of the entire object.

FEP
Fluorinated ethylene propylene.

Ferrous
Metals that contain iron: for example, steel (page 455) (see also non-ferrous, page 499).

Figured Grain
Wood grain with a distinctive pattern such as bird's-eye, fiddleback, flame and curly (see wood and natural fibres, page 475).

FRP
Fibre reinforced plastic is molded plastic reinforced with lengths of fibre, which can be carbon, aramid, glass or natural material such as cotton, hemp or jute. They are formed by a range of processes including composite laminating (page 206), DMC and SMC molding (page 218) and injection molding (page 50) (see also biofibre reinforced plastics, page 496, and GRP, page 498).

FSW
Friction stir welding (page 294).

GPPS
General-purpose polystyrene (same as PS).

Green Timber
Lumber with a moisture content above fibre saturation point, which is typically around 25%. Up to this percentage the structure of the wood will not be affected, because the water being removed is from the cells cavities as opposed to the cell walls. Green timber is used for steam bending (page 198).

GRP
Glass reinforced plastic is molded plastic reinforced with lengths of glass fibre. Injection molded (page 50) GRP has very short fibre length up to 3 mm (0.118 in.), whereas composite laminating (page 206) and DMC and SMC molding (page 218) have long or continuous lengths of glass fibre reinforcement (see also FRP, page 498).

Hardwood
Wood from deciduous and broad-leaved trees such as ash (page 474), beech (page 473), birch (page 472) and oak (page 476).

HAZ
Heat-affected zone.

HDPE
High-density polyethylene.

Heartwood
The central section of a mature tree trunk, which is usually darker in colour. It is often harder and denser than the surrounding sapwood (page 473) and no longer transports water, because it becomes saturated with resin (page 500) and tannin.

HIPS
High impact polystyrene.

HSLA
High strength low alloy steel.

IBM
Injection blow molding (page 22).

IIR
Polyisobutylene; but is commonly known as butyl rubber, because it is very similar in composition to natural rubber.

IR
Isoprene rubber.

Iridescent
A bright colour that changes according to the viewing angle.

ISBM
Injection stretch blow molding (page 22)

JIT
Just in time is a strategy employed by businesses to reduce stock and thus improve efficiency. Many factories are now set up to supply JIT, which is the process of supplying products in low volumes as and when they are needed. Some manufacturing techniques are not suited to short production runs and so stock has to be kept somewhere in the supply chain, which is usually with the manufacturer.

Kevlar®
DuPont™ trade name for aramid fibre (see polyamides, page 438).

Kiln-dried Timber
Lumber dried in a kiln. The moisture content of small cross-section lumber can be reduced to 12%. The moisture content has to be less than 14% for impregnated preservative to be effective (see also air-dried timber, page 467, and seasoned timber, page 500).

LBW
Laser beam welding (page 288).

LCA
Life cycle analysis.

LCP
Liquid crystal polymer.

LDPE
Low-density polyethylene.

LFW
Linear friction welding (page 294).

LLDPE
Linear low-density polyethylene.

LSL
Laminated strand lumber.

Lumber
Sawn wood for use in building and furniture making, for example.

Machining
Removing precise amounts of material with a cutting action: for example, by electrical discharge machinery (page 254), laser cutting (page 248) or milling.

MDF
Medium-density fibreboard is made up of particles of softwood bonded together with UF adhesive. The mix of wood particles and adhesive are formed into panels by heating and pressing in steel molds, which produces a smooth

surface finish. Fire retardant MDF grades are typically red, and moisture resistant grades are green.

MF
Melamine formaldehyde.

MIG
Metal inert gas.

MIM
Metal injection molding (page 136).

MMA
Manual metal arc welding (page 282).

Monomer
A small, simple compound with low molecular weight, which can be joined with other identical compounds to form long chains, known as polymers. Two similar monomers joined together in a long chain is known as a copolymer, and 3 similar monomers makes a terpolymer: for example, acrylonitirile butadiene styrene (ABS) (see styrenes, page 432).

Nd:YAG
Neodymium: yttrium aluminum garnet (Nd:YAG) is a crystal used in solid state lasers. This is one of the most common types of laser and is used for laser beam welding (page 288), laser cutting (page 248) and drilling a range of materials.

Neoprene®
The DuPont™ tradename for chloroprene rubber (CR).

Nomex®
DuPont™ trade name for aramid sheet (see polyamides, page 438).

Non-ferrous
Metals that do not contain iron: for example, aluminium alloys (page 457) and copper alloys (page 460) (see also ferrous, page 497).

NR
Natural rubber.

Nylon
Common name for polyamide (PA) (page 438). Nylon is an acronym of New York and London, where it is believed to have been discovered simultaneously.

OFW
Orbital friction welding.

OSB
Orientated strand board.

Overmolding
All molding and casting processes can be used to overmold. It is the process of molding over an insert, which becomes integral to the part on cooling.

PA
Polyamide.

PAP
Paper (abbreviation used for recycling purposes).

Patina
A surface layer that develops over time (such as verdigris copper, page 460) or a surface pattern that develops with frequent use (such as a smooth-worn wooden handle).

Pattern
An original design or prototype that is reproduced to form a mold: for example, in composite laminating (page 206). This mold can then be used to produce many identical parts.

PBT
Polybutylene terephthalate.

PC
Polycarbonate.

PCB
Printed circuit board.

PCT
Polycyclohexylene dimethylene terephthalate.

PE
Polyethylene.

Pearlescent
A lustrous translucence with a shimmering quality affected by the viewing angle.

PET
Polyethylene terephthalate.

PETG
Polyethylene terephthalate modified with glycol.

PF
Phenol formaldehyde resin.

Photochromic
A compound that changes colour with light intensity: for example, glasses that become sunglasses in bright light. It is added to materials as a dye.

PIM
Powder injection molding.

PLA
Polylactic acid.

Plastic State
The point at which a heated material becomes viscous and can be molded or formed.

PMB
Plastic media blasting.

PMMA
Poly methyl methacrylate.

Polymer
A natural or synthetic compound made up of long chains of repeating identical monomers. Examples include cellulose (page 446), starch (page 446), polypropylene (page 430), polystyrene (page 432) and polycarbonate (page 435). (See also copolymer, page 497, and terpolymer, page 501).

POM
Polyoxymethylene.

PP
Polypropylene.

PS
Polystyrene.

PSL
Parallel strand lumber.

PTFE
Polytetrafluoroethylene.

PU
Polyurethane [thermoplastic, almost exclusively injection molding].

PUR
Polyurethane resin [thermosetting, resin, ie most occurences].

PVA
Polyvinyl acetate.

PVC
Polyvinyl chloride.

PVOH
Polyvinyl alcohol.

PW
Plasma welding (page 282).

RAL
Reichsausschuss für Lieferbedingungen is a German colour chart system used mainly in paint and pigment colour specification.

Rays
Vessels radiating from the centre of a tree's stem and branches for transporting food and waste laterally. They are responsible for the characteristic flecks of colour in hardwoods (page 472) such as beech and oak.

Resin
A natural or synthetic, semi-solid or solid substance, produced by polymerization or extraction from plants, and used in plastics, varnishes and paints.

RF
Radio frequency.

RFW
Rotary friction welding (page 294).

RIM
Reaction injection molding (page 64).

RRIM
Reinforced reaction injection molding.

RTM
Resin transfer molding.

RTV
Room temperature vulcanizing. occurs when certain rubbers, such as silicone, are chemically cured at room temperature, as opposed to heat curing.

SAN
Styrene acrylonitrile.

Sapwood
Young and light-coloured wood that forms between the bark and heartwood (page 471) of a tree.

SAW
Submerged arc welding (page 282).

SBR
Styrene butadiene rubber.

SBS
Styrene butadiene styrene.

Seasoned Timber
Traditionally used to describe air-dried timber with a moisture content of less than 20%. Compared to green timber (page 467), dried timber has improved dimensional stability, toughness, bending stiffness, impact resistance, electrical resistance, resistance to decay and adhesive bond strength. The required level of moisture depends on the application (see also air-dried timber, page 496, and kiln-dried timber, page 498).

SEBS
Styrene ethylene butylene styrene.

Semi-crystalline
A molecular structure that contains both crystalline (page 425) and amorphous (page 425) patterns: for example, polyethylene terephthalate (PET) (see thermoplastic polyesters, page 437).

Shape Memory
The ability of a material to be heavily manipulated and then return to its original shape, sometimes with the application of heat. Materials with shape memory include synthetic rubbers (page 445) and certain metal alloys (page 452).

Shore Hardness
The hardness of a plastic, rubber or elastomer is measured by the depth of indentation by a shaped metal foot on a measuring instrument known as a Durometer. The depth of indentation is measured on a scale of 0 to 100; higher numbers indicate harder materials. These tests are generally used to denote the flexibility of a material. The 2 most popular are Shore A and Shore D. Indenter feet with different profiles are used in each case, and so each method is suitable for different material harnesses. For example, soft materials are measured on the Shore A scale, and hard materials are measured on the Shore D scale. There is not a strong correlation between different scales. Shore hardness is also known as durometer hardness.

SLA
Stereolithography (page 232).

SLS
Selective laser sintering (page 232).

SMC
Sheet molding compound (page 218).

Softwood
Wood from coniferous and typical evergreen trees, such as cedar, fir, pine and spruce (see softwoods, page 470).

SRIM
Structural reaction injection molding.

Sub-surface Laser Engraving
The process of laser-marking clear materials below the surface: 2D and 3D designs are reproduced as point clouds, which are sets of closely packed laser markings that outline the shape. Sub-surface laser engraving is also known as vitrography.

Tampo Printing
Another name for pad printing (page 404).

Teflon®
DuPont™ trade name for polytetrafluoroethylene (PTFE) and ethylene tetrafluoroethylene (ETFE) (see fluropolymers, page 440).

Terpolymer
A polymer made up of long chains of 3 repeating monomers, such as acrylonitirile butadiene styrene (ABS) (see styrenes, page 432).

Thermochromatic
A compound that changes colour as its temperature goes up or down. Thermochromatic pigments are available for inks, plastics and coatings.

Thermoplastic
A polymer that becomes soft and pliable when heated. In its plastic state it can be shaped and re-shaped by a range of molding processes, such as blow molding (page 22) and injection molding (page 50). Examples include polyamide (PA) nylon (page 438), polycarbonate (PC) (page 428) and polypropylene (PP) (see polyolefins, page 430) (see also commodity thermoplastic, page 497, and engineering thermoplastic, page 497).

Thermoplastic Elastomer
A general name used to describe thermoplastic materials that exhibit similar elastic properties to rubber.

Thermosetting Plastic
A material formed by heating, catalyzing or mixing 2 parts to trigger a 1-way polymeric reaction. Unlike most thermoplastics, thermosetting plastics form cross-links between the polymer chains, which cannot be undone, and so this material cannot be reshaped or remolded once cured. Thermosetting plastics tend to have superior resistance to fatigue and chemical attack than thermoplastic (page 435).

TIG
Tungsten inert gas.

TPC-ET
Thermoplastic polyester elastomer.

TPE
Thermoplastic elastomer.

TPU
Thermoplastic polyurethane.

UF
Urea formaldehyde.

UHDPE
Ultra high-density polyethylene.

ULDPE
Ultra low-density polyethylene.

uPVC
Unplasticized polyvinyl chloride.

UV
Ultraviolet.

Verdigris
Another name for the green patina that forms on the surface of copper (page 460). It comes from the French *vert de Grice*.

Vulcanize
The process of curing natural rubber with sulphur, heat and pressure in a 1-way reaction to form a thermosetting material (see thermoset, page 445).

Xylem
The fibrous material that makes up the stem and branches of trees and shrubs. It consists mostly of a series of elongated, rigid walled cells and provides trees and shrubs with an upward flow of water and mechanical support (see also heartwood, page 498, and sapwood, page 500).

Featured Companies

MANUFACTURERS

Beatson Clark
The Glass Works
Greasbrough Road
Rotherham S60 1TZ
United Kingdom
www.beatsonclark.co.uk
Processes: Machine blow and blow glassblowing

BJS Company
65 Bideford Avenue
Perivale Greenford
Middlesex UB6 7PP
United Kingdom
www.bjsco.com
Processes: Electroforming; Electroplating

Blagg & Johnson
Newark Business Park
Brunel Drive, Newark
Nottinghamshire NG24 2EG
United Kingdom
www.blaggs.co.uk
Processes: Arc welding; Laser cutting; Press braking; Roll forming

Branson Ultrasonics Corporation
Corporate Headquarters
41 Eagle Road, PO Box 1961
Danbury, CT 06813-1961
USA
www.branson-plasticsjoin.com
*Processes: Friction welding; Hot plate welding; Laser beam welding; Staking; Vibration welding; Ultrasonic welding
Products: see Processes*

Branson Ultrasonics (UK)
686 Stirling Road,
Slough Trading Estate
Slough
Berkshire SL1 4ST
United Kingdom
www.branson-plasticsjoin.com
*Processes: Friction welding; Hot plate welding; Laser beam welding; Staking; Vibration welding; Ultrasonic welding
Products: see Processes*

Chiltern Casting Company
Unit F
11 Cradock Road
Luton LU4 0JF
United Kingdom
www.chilterncastingcompany.co.uk
Processes: Abrasive blasting; Arc welding; Sand casting

CMA Moldform
Unit B6 The Seedbed Centre
100 Avenue Road
Birmingham B7 4NT
United Kingdom
www.cmamoldform.co.uk
*Processes: Centrifugal casting; Reaction injection molding; Vacuum casting
Products: scale models*

Cove Industries
Industries House
18 Invincible Road
Farnborough
Hampshire GU14 7QU
United Kingdom
www.cove-industries.co.uk
Processes: Arc welding; Dip molding; Press braking; Punching and blanking

Coventry Prototype Panels
Wheler Road
Seven Stars Industrial Estate
Coventry
West Midlands CV3 4LB
United Kingdom
www.covproto.com
Processes: Arc welding; Grinding, sanding and polishing; Panel beating

CRDM
Queen Alexandra Road
High Wycombe
Buckinghamshire HP11 2JZ
United Kingdom
www.crdm.co.uk
Processes: CNC machining; Electrical discharge machining; Rapid prototyping; Vacuum casting

Crompton Technology Group
Thorpe Park, Thorpe Way
Banbury
Oxfordshire OX16 4SU
United Kingdom
www.ctgltd.co.uk
*Processes: Composite laminating; Filament winding; CNC machining
Products: propshafts and driveline products*

Cromwell Plastics
53–54 New Street, Quarry Bank
Dudley
West Midlands DY5 2AZ
United Kingdom
www.cromwell-plastics.co.uk
Processes: Compression molding; DMC and SMC molding

Cullen Packaging
10 Dawsholm Avenue
Dawsholm Industrial Estate
Glasgow G20 0TS
United Kingdom
www.cullen.co.uk
Processes: Die cutting; Paper pulp molding

Deangroup International
Brinell Drive
Northbank Industrial Park, Irlam
Manchester M44 5BL
United Kingdon
www.deangroup-int.com
Processes: Investment casting

Dixon Glass
127–129 Avenue Road
Beckenham
Kent BR3 4RX
United Kingdom
www.dixonglass.co.uk
Processes: Lampworking
Products: extruded glass profiles;
laboratory glassware

Elmill Group
139A Engineer Road
West Wilts Trading Estate
Westbury, Wiltshire BA13 4JW
United Kingdom
www.elmill.co.uk
Processes: Swaging

ENL
Units 6–8, Victoria Trading Estate,
Kiln Road, Portsmouth
Hampshire PO3 5LP
United Kingdom
www.enl.co.uk
Processes: Injection molding; CNC
machining

Firma-Chrome
Soho Works, Saxon Road
Sheffield S8 0XZ
United Kingdom
www.firma-chrome.co.uk
Processes: Electroplating; Electropolishing

Heywood Metal Finishers
Field Mills
Red Doles Lane, Leeds Road
Huddersfield, HD2 1YG
United Kingdom
www.hmfltd.co.uk
Processes: anodizing

Hydrographics
Unit 4 Brockett Industrial Estate
York YO23 2PT
United Kingdom
www.hydro-graphics.co.uk
Processes: Hydro transfer printing; Spray
painting

Hymid Multi-Shot
Unit 10–12 Brixham Enterprise Estate
Rea Barn Road
Brixham
Devon TQ5 9DF
United Kingdom
www.hymid.co.uk
Processes: CNC machining; Electrical
discharge machining; Injection molding

Impressions Foil Blocking
31 Shannon Way
Thames Industrial Estate
Canvey Island
Essex SS8 0PD
United Kingdom
www.impressionsfoiling.co.uk
Processes: Foil blocking and embossing

Instrument Glasses
236–258 Alma Road
Ponders End
Enfield EN3 7BB
United Kingdom
www.instrument-glasses.co.uk
Processes: Glass scoring; Grinding, sanding
and polishing; Screen printing; Water jet
cutting

Kaysersberg Plastics
Madleaze Industrail Estate
Bristol Road
Gloucester GL1 5SG
United Kingdom
www.kayplast.com
Processes: Thermoforming
Products: packaging materials; packaging
systems; pallets

Kaysersberg Plastics
BP No. 27
68240 Kaysersberg
France
www.kayplast.com
Processes: as above
Products: as above

Lola Cars International
Glebe Road, St Peters Road
Huntington
Cambridgeshire PE29 7DS
United Kingdom
www.lolacars.com
Processes: Composite laminating

Marlows Timber Engineering Ltd
Howarth House, Hollow Road
Bury St Edmunds
Suffolk IP32 7QW
United Kingdom
www.marlows.com
Processes: Timber frame structures

Medway Galvanising
Castle Road
Eurolink Industrial Centre
Sittingbourne
Kent ME10 3RN
United Kingdom
www.medgalv.co.uk
Processes: Galvanizing; Powder coating

Mercury Engraving
Unit A5 Bounds Green Industrial Estate
Ringway
London N11 2UD
United Kingdom
www.mengr.com
Processes: CNC engraving; Photo etching;
Photochemical machining

Metal Injection Mouldings
Davenport Lane, Altrincham
Cheshire WA14 5DS
United Kingdom
www.metalinjection.co.uk
Processes: Investment casting; Metal
injection molding; Rapid prototyping

National Glass Centre
Liberty Way
Sunderland SR6 0GL
United Kingdom
www.nationalglasscentre.com
Processes: Glassblowing
Products: blown glassware

PFS Design & Packaging
Unit 403 Henley Park
Pirbright Road, Normandy
Surrey GU3 2HD
United Kingdom
www.pfs-design.co.uk
Processes: Die cutting; Foil blocking and
embossing; Screen printing; Ultrasonic
welding

Pipecraft
Units 6–7 Wayside
Commerce Way
Lancing
West Sussex BN15 8SW
United Kingdom
www.pipecraft.co.uk
Processes: Swaging; Tube and section
bending

Polimoon
Ruseløkkveien 6
PO Box 1943 Vika
N-0125 Oslo
Norway
www.polimoon.com
Processes: Blow molding; Injection
molding; Rotation molding;
Thermoforming

Polimoon Packaging
Ellough, Beccles
Suffolk NR 34 7TB
United Kingdom
www.polimoon.com
Processes: as above

Professional Polishing Services
18B Parkrose Industrial Estate
Middlemore Road, Smethwick
West Midlands B66 2DZ
United Kingdom
www.professionalpolishing.co.uk
Processes: Grinding, sanding and
polishing

Radcor
Hingham Road Industrial Estate
Great Ellingham
Norfolk NR17 1JE
United Kingdom
www.radcor.co.uk
Processes: Composite laminating

RS Bookbinders
61–63 Cudworth Street
London E1 5QU
United Kingdom
www.rsbookbinders.co.uk
Processes: Bookbinding

Rubbertech2000
Whimsey Trading Estate
Cinderford
Gloucestershire GL14 3JA
United Kingdom
www.rubbertech2000.co.uk
Processes: Compression molding; Pad
printing; Screen printing

S&B Evans & Sons
The City Garden Pottery
7a Ezra Street
London E2 7RH
United Kingdom
www.sandbevansandsons.com
Processes: Clay throwing
Products: garden pots

Superform Aluminium
Cosgrove Close, Blackpole
Worcester WR3 8UA
United Kingdom
www.superform-aluminium.com
Processes: Arc welding; Superforming

Superform USA
6825 Jurupa Avenue
Riverside, CA 92517-5375
USA
www.superformusa.com
Processes: as above

W. H. Tildesley
Clifford Works
Bow Street, Willenhall
West Midlands WV13 2AN
United Kingdom
www.whtildesley.com
Processes: Forging

VMC
Trafalgar Works
Station Road
Chertsey KT16 8BE
United Kingdom
www.vmclimited.co.uk
Processes: Spray painting; Vacuum metalizing

Windmill Furniture
Terrace Mews
London W4 1QU
United Kingdom
www.windmillfurniture.com
Processes: Joinery; Wood laminating

Yoshida Technoworks
11–12 Bunka2chrome
Sumida-ku, Tokyo
Japan
www.yoshida-tw.co.jp
Processes: Injection molding

Zone Creations
Unit 1 Chelsea Fields
278 Western Road
London SW19 2QA
United Kingdom
www.zone-creations.co.uk
Processes: CNC machining; Grinding, sanding and polishing; Laser cutting

MATERIALS AND FINISHES

B&M Finishers
201 South 31st Street
Kenilworth, NJ 07033
USA
www.bmfinishers.com
Products: coloured and pattern embossed stainless steel

Beacons Products
Unit 10 EFI Industrial Estate
Brecon Road
Merthyr Tydfil CF47 8RB
United Kingdom
www.beaconsproducts.co.uk
Products: expanded foam sheet and products

Bencore
Via S. Colombano 9
54100 Massa Z.I. (MS)
Italy
www.bencore.it
Products: structural plastic panels

Cambridgeshire Coatings
PO Box 34, St Neots
Huntingdon PE19 6ZG
United Kingdom
www.rage-extreme.com
Products: Outrageous® paints

Candidus Prugger
Via Johann Kravogl 10
39042 Bressanone (BZ)
Italy
www.bendywood.com
Products: Bendywood®; wooden carvings and architectural decorations

Designtex
200 Varick Street, 8th floor
New York, NY 10014
USA
www.dtex.com
Products: textiles

Distrupol
119 Guildford Street
Chertsey
Surrey KT16 9AL
United Kingdom
www.distrupol.com
Products: engineering thermoplastics

Dixon Glass
127–129 Avenue Road
Beckenham
Kent BR3 4RX
United Kingdom
www.dixonglass.co.uk
Products: extruded glass profiles; laboratory glassware

DSM Dyneema®
Mauritslaan 49
Urmond, PO Box 1163
6160 BD Geleen
The Netherlands
www.dyneema.com
Products: Dyneema®

DuPont™ Corian®
McD Marketing
10 Quarry Court
Pitstone Green Business Park
Pitstone LU7 9GW
United Kingdom
www.corian.co.uk
Products: as above

DuPont™ De Nemours
DuPont™ Surfaces
Chestnut Run Plaza 721-Maple Run
PO Box 80721
Wilmington, DE 19880-0721
USA
www.corian.co.uk
Products: DuPont™ Corian®

Elmo Leather
SE-512 81 Svenljunga
Sweden
www.elmoleather.com
Products: leather

Elmo Leather
505 Thornall Street, Suite 303
Edison, NJ 08837
USA
www.elmoleather.com
Products: as above

Fusion Glass Designs
365 Clapham Road
London SW9 9BT
United Kingdom
www.fusionglass.co.uk
*Products: structural architectural
decorative glass (abrasive blasted, kiln
formed and laminated)*

KM Europa Metal
TECU® Technical Consulting Center
Klosterstrasse 29
49074 Osnabrück
Germany
www.tecu.com
Products: copper sheet

KME
TECU® Technical Consulting Office
Knightsbridge Park, Wainwright Road
Worcester WR4 9FA
United Kingdom
www.kme-uk.com
Products: as above

Litracon
Tanya 832
H-6640 Csongrád
Hungary
www.litracon.hu
Products: Litracon building blocks

Nitinol Devises and Components
47533 Westinghouse Drive
Fremont, CA 94539
USA
www.nitinol.com
Products: shape memory alloys

Pantone
590 Commerce Boulevard
Carlstadt, NJ 07072-3098
USA
www.pantone.com
Products: colour solutions

PE Design and Engineering
PO Box 3051
nl-2601 DB Delft
The Netherlands
www.treeplast.com
Products: Treeplast®

Phenix Biocomposite
PO Box 609
Mankato, MN 56002-0609
USA
www.phenixbiocomposites.com
Products: biocomposite sheet materials

Pilkington Group
Prescot Road,
St Helens
Merseyside WA10 3TT
United Kingdom
www.pilkington.com
Products: glass and glazing products

Rheinzink
Bahnhofstrasse 90
45711 Datteln
Germany
www.rheinzink.com
Products: zinc sheet

Smile Plastics
Mansion House, Ford
Shrewsbury SY5 9LZ
United Kingdom
www.smile-plastics.co.uk
Products: recycled plastic sheet

Tin Tab
Unit 5–7 North Industrial Estate
New Road
Newhaven BN9 0HE
United Kingdom
www.multiplywood.com
*Products: specialist plywood and
engineering timber products*

Trus Joist
Weyerhaeuser
PO Box 9777
Federal Way WA 98063-9777
USA
www.trusjoist.com
*Products: architectural timber products;
engineering timber*

US Chemical & Plastics
PO Box 709
Massillon, OH 44648
USA
www.uschem.com
Products: Outrageous® paints

PRODUCTS

Aero Base
2–26 Takajomachi
Wakayama 640-8135
Japan
www.aerobase.jp
Products: scale models

Alessi
Via Privata Alessi 6
28882 Crusinallo di Omegna (VB)
Italy
www.alessi.com
*Products: children's objects; clocks and
watches; desktop accessories; glassware;
kitchen accessories; kitchenware;
tableware; telephones*

Bang & Olufsen
Peter Bangs Vej 15
7600 Struer
Denmark
www.bang-olufsen.com
*Products: loud speakers; music players;
telephones; televisions*

Bianchi
Via delle Battaglie 5
24047 Treviglio (BG)
Italy
www.bianchi.com
Products: bicycles

Biomega
Skoubogade 1, 1. MF
1158 København K
Denmark
www.biomega.dk
Products: bicycles

BJS Royal Silversmiths
65 Bideford Avenue
Perivale, Greenford
Middlesex UB6 7PP
United Kingdom
www.royalsilversmiths.com
Products: clocks; jewellery; tableware

Boss Design
Boss Drive off New Road, Dudley
West Midlands DY2 8SZ
United Kingdom
www.boss-design.co.uk
Products: office and contract furniture

Chevrolet
PO Box 33170
Detroit, MI 48232-5170
USA
www.chevrolet.com
Products: cars; commercial vehicles; SUVs; trucks; vans

Coca-Cola
PO Box 1734
Atlanta, GA 30301
USA
www.coca-cola.com
Products: refreshments

Crowcon Detection Instruments
2 Blacklands Way
Abingdon Business Park
Abingdon
Oxfordshire OX14 1DY
United Kingdom
www.crowcon.com
Products: control systems; fixed gas detectors; gas sampling systems; personal and portable gas monitors

Crowcon Detection Instruments
21 Kenton Lands Road
Erlanger, KY 41018-1845
USA
www.crowcon.com
Products: as above

Dixon Glass
127–129 Avenue Road
Beckenham
Kent BR3 4RX
United Kingdom
www.dixonglass.co.uk
Products: extruded glass profiles; laboratory glassware

Dunlop
Regent House
1–3 Queensway, Redhill
Surrey RH1 1QT
www.dunlopsports.com
Products: sports equipment

Ercol Furniture
Summerleys Road
Princes Risborough
Buckinghamshire HP27 9PX
United Kingdom
www.ercol.com
Products: furniture

Flos
Via Angelo Faini 2
25073 Bovezzo (BS)
Italy
www.flos.it
Products: indoor, outdoor and architectural lighting

Frederick Phelps
34 Conway Road
London N14 7BA
United Kingdom
www.phelpsviolins.com
Products: musical instruments

Hartley Greens & Co. (Leeds Pottery)
Anchor Road, Longton
Stoke-on-Trent ST3 5ER
United Kingdom
www.hartleygreens.com
Products: tableware

H Concept
2–13–2 Kuramae Taito-ku
Tokyo 111-0051
Japan
www.h-concept.jp
Products: desktop accessories; glassware; tableware

Irwin Industrial Tool
92 Grant Street
Wilmington, OH 45177-0829
USA
www.irwin.com
Products: hand tools; worksite products

Irwin UK
Parkway Works, Kettlebridge Road
Sheffield S9 3BL
United Kingdom
www.irwin.co.uk
Products: as above

Isokon Plus
Windmill Furniture
Terrace Mews
London W4 1QU
United Kingdom
www.isokonplus.com
Products: furniture

Küppersbusch
4920 West Cypress Street, Suite 106
Tampa, FL 33607
USA
www.kuppersbuschusa.com
Products: built-in ovens; coffee machines;
dishwashers; electric cooker tops; gas
cooker tops

Laguiole
4 Impasse des Avenues BP 45
42-601 Montbrison Cedex
France
www.laguiole.com
Products: corkscrews; cutlery; knives

Le Creuset
902 rue Olivier Deguise
02230 Fresnoy-le-Grand
France
www.lecreuset.com
Products: cookware

Leatherman Tool Group
PO Box 20595
Portland, OR 97294-0595
USA
www.leatherman.com
Products: tools; knives and accessories

Lloyd Loom of Spalding
Wardentree Lane, Pinchbeck
Spalding
Lincolnshire PE11 3SY
United Kingdom
www.lloydloom.com
Products: furniture

London Glassblowing
7 The Leather Market
Weston Street
London SE1 3ER
United Kingdom
www.londonglassblowing.co.uk
Products: glassware

Luceplan
Via E.T. Moneta 46
20161 Milano
Italy
www.luceplan.com
Products: indoor, outdoor and
architectural lighting

Magis
Via Magnadola 15
31045 Motta di Livenza (TV)
Italy
www.magisdesign.com
Products: furniture

Makita
14930 Northam Street
La Mirada, CA 90638
USA
www.makita.com
Products: cordless power tools

Mandala
408–410 St John Street
London EC1V 4NJ
United Kingdom
www.mandalafurniture.com
Products: furniture; lighting

Mathmos
22–24 Old Street
London EC1V 9AP
United Kingdom
www.mathmos.co.uk
Products: lighting

Mebel
Via Sulbiate 26
20040 Bellusco (Mi)
Italy
www.mebel.it
Products: melamine products

Mercury Marine
N-7480 County Road "UU"
Fond du Lac, WI 54935
USA
www.mercurymarine.com
Products: marine propulsion engines

Moooi
Minervum 7003
4817 ZL Breda
The Netherlands
www.moooi.com
Products: furniture; lighting; tableware

North Sails Nevada
2379 Heybourne Road
Minden, NV 89423
USA
www.northsails.com
Products: sails

Panasonic
1006 Kadoma
Kadoma City
Osaka 571-8501
Japan
www.panasonic.net
Products: audio; camcorders; cameras;
DVD and VCR; mobile phones; office
products; telephones; televisions

Peck-Polymers
A2Z Corp.
1530 W Tufts Avenue Unit B
Englewood, CO 80110
USA
www.peck-polymers.com
Products: scale models

Pedalite
12–50 Kingsgate Road
Kingston Upon Thames
Surrey KT2 5AA
United Kingdom
www.pedalite.com
Products: bicycle pedal lights

Pioneer Aviation
The Byre, Hardwick
Abergavenny
Monmouthshire NP7 9AB
United Kingdom
www.pioneeraviation.co.uk
Products: self-build aeroplane kits

Plastic Logic
34 Cambridge Science Park
Milton Road
Cambridge CB4 0FX
United Kingdom
www.plasticlogic.com
Products: flexible LCD displays

Portable Welders
2 Wedgwood Road
Bicester
Oxon OX26 4UL
United Kingdom
www.portablewelders.com
*Products: portable resistance welding
equipment*

Potatopak
Theocrest House
Binley Industrial Estate
Coventry CV3 2SF
United Kingdom
www.potatopak.org
Products: biodegradable packaging

Potatopak NZ
34 Inkerman Street
Renwick
Marlborough
New Zealand
www.potatoplates.com
Products: as above

Pro-Pac Packaging
6 Rich Street
Marrickville, NSW 2204
Australia
www.pro-pac.com.au
*Products: biodegradable loose-fill
packaging*

Rexite
7 Via Edison
1-20090 Cusago (MI)
Italy
www.rexite.it
*Products: furniture, storage and desktop
accessories*

Remarkable Pencils
The Remarkable Factory
Midland Road
Worcester WR5 1DS
United Kingdom
www.remarkable.co.uk
Products: stationery

Rolls-Royce International
65 Buckingham Gate
London SW1E 6AT
United Kingdom
www.rolls-royce.com
Products: power systems and services

Saluc
Rue de Tournai 2
7604 Callenelle
Belgium
www.saluc.com
*Products: billiard balls; bowling balls;
trackballs and other precision balls*

Sei Global
450 West 15th Street
New York, NY 10011
USA
www.seiwater.com
Products: drinking water

Skimeter
82 rue des Tattes
74500 Publier
France
www.skimeter.com
*Products: skiing equipment and
accessories*

Spyker Cars
Edisonweg 2
3899 AZ Zeewolde
The Netherlands
www.spykercars.com
Products: cars

Swarovski Crystal Palace
2nd Floor
14–15 Conduit Street
London W1S 2XJ
United Kingdom
www.swarovskisparkles.com
*Products: fashion; gemstones; lighting;
miniatures*

Thonet
Michael-Thonet-Straße 1
PO Box 1520
35059 Frankenberg
Germany
www.thonet.de
Products: furniture

Vertu
Beacon Hill Road
Church Crookham
Hampshire GU52 8DY
United Kingdom
www.vertu.com
Products: mobile phones

William Levene
Bridge House, Eelmoor Road
Farnborough
Hampshire GU14 7UE
United Kingdom
www.williamlevene.com
Products: cookware; gadgets; kitchen products

ARCHITECTS, DESIGNERS AND RESEARCH INSTITUTES

Ansel Thompson
Unit 1C Enterprise House
Tudor Grove
London E9 7QL
United Kingdom
www.ansel.co.uk

AnthonyQuinn London
16A Wyndham Road
London SE5 0UH
United Kingdom
www.anthonyquinndesign.com

Atelier Bellini
Piazza Arcole 4
20143 Milan
Italy
www.bellini.it

BarberOsgerby
35–42 Charlotte Road
London EC2A 3PG
United Kingdom
www.barberosgerby.com

Beckman Institute
Dept of Aerospace Engineering
University of Illinois at Urbana-Champaign, 205C Talbot Lab
104 South Wright Street
Urbana, IL 61801
USA
www.autonomic.uiuc.edu

Bertjan Pot
Korte Haven 135
3111 BH Schiedam
The Netherlands
www.bertjanpot.nl

Black+Blum
2.07 Oxo Tower Wharf
Bargehouse Street
London SE1 9PH
United Kingdom
www.black-blum.com

Charlie Davidson
117A Whitecross Street
London EC1Y 8JH
United Kingdom
www.charlie-davidson.com

Curiosity
2–13–16 Tomigaya
Shibuya-Ku
Tokyo 151-0063
Japan
www.curiosity.jp

D-Bros
Japan
www.d-bros.jp

Droog Design
Staalstraat 7a–7b
1011 JJ Amsterdam
The Netherlands
www.droogdesign.nl

Ellis Williams Architects
Exmouth House
Pine Street
London EC1R 0JH
United Kingdom
www.ewa.co.uk

Front Design
Tegelviksgatan 20
S-112 55 Stockholm
Sweden
www.frontdesign.se

Fusion Design
2.08 Oxo Tower Wharf
Bargehouse Street
London SE1 9PH
United Kingdom
www.studiofusion.co.uk

Future Factories
52 Rauceby Drive
South Rauceby NG34 8QB
United Kingdom
www.futurefactories.com

Georg Baldele
The Dove Centre, Unit 15
109 Bartholomew Road
London NW5 2BJ
United Kingdom
www.georgbaldele.com

Heatherwick Studio
16 Acton Street
London WC1X 9NG
United Kingdom
www.heatherwick.com

Hopkins Architects
27 Broadley Terrace
London NW1 6LG
United Kingdom
www.hopkins.co.uk

Hulger
8 Elder Street
London E1 6BT
United Kingdom
www.hulger.com

Jackie Choi
13D Lordship Park
London N16 5UN
United Kingdom
www.jackiechoi.com

Jet Propulsion Laboratory
Jet Propulsion Laboratory
4800 Oak Grove Drive
Pasadena, California 91109
USA
www.jpl.nasa.gov

JPRA Architects
31000 Northwestern Highway, Suite 100
Farmington Hills, MI 48334
USA
www.jpra.com

Moldflow
492 Old Connecticut Path, Suite 401
Framingham, MA 01701
USA
www.moldflow.com

Nasa
Suite 1M32
Washington, DC 20546-0001
USA
www.nasa.gov

PD Design Studio
42-7 Fukakusa Nakanoshima-cho
Fushimi-ku
Kyoto 612-0049
Japan
www.pd-design-st.com

Product Partners Design
The Old Warehouse
Church Street, Biggleswade
Bedfordshire SG18 0JS
United Kingdom
www.productpartners.co.uk

Quigley Design
Westgate House
Hills Lane
Shrewsbury SY1 1QU
United Kingdom
www.kqd.co.uk

Rachel Galley
Studio E15 Cockpit Yard
Northington Street
London WC1N 2NP
United Kingdom
www.rachelgalley.com

Raul Barbieri Design
Milan, Italy
www.raulbarbieri.com

Retrouvius
2A Ravensworth Road
London NW10 5NR
United Kingdom
www.retrouvius.com

Shunsuke Ishikawa
Meguro Sunny Hive room 303
4–13–14 Meguro Meguro-ku
Tokyo 153-0063
Japan
www.light-d.jp

Studio Dillon
28 Canning Cross
London SE5 8BH
United Kingdom
www.studiodillon.com

Studio Job
Stijfselstraat 10
2000 Antwerp
Belgium
www.studiojob.be

TWI
Granta Park
Great Abington
Cambridge CB21 6AL
United Kingdom
www.twi.co.uk

Vexed Generation
Unit 26 Ada Street Workshop 8
Andrews Road
London E8 0QN
United Kingdom
www.vexed.co.uk

Virgile and Stone Associates
25 Store Street
London WC1E 7BL
United Kingdom
www.virgileandstone.com

Vujj
Mäster Nilsgatan 1
21-126 Malmö
Sweden
www.vujj.com

Will Alsop Architects
41 Parkgate Road
London SW11 4NP
United Kingdom
www.alsoparchitects.com

Yoyo Ceramics
Studio E16 Cockpit Yard
Northington Street
London WC1N 2NP
United Kingdom
www.yoyoceramics.co.uk

Organizations and Other Sources of Information

GENERAL

Azom
www.azom.com
Information is provided about a range of materials and processes, and there is a directory of manufacturers and suppliers worldwide.

Designboom
www.designboom.com
A vast and growing online resource, which provides designers with information about other designers, projects, competitions and trade shows.

Design inSite
www.designinsite.dk
Design inSite provides designers with information about a wide range of manufacturing processes, materials and examples of products where they are used.

Engineers Edge
www.engineersedge.com
This is useful for engineers and has information, charts and tables for making calculations in product development.

Institute of Materials, Minerals and Mining (IOM³)
www.iom3.org
This organization represents the international materials, minerals and mining community. Among other things, they produce a monthly publication called *Materials World*, hold annual competitions and provide technical advice to students and companies.

Institute of Packaging (IoP)
www.pi2.org.uk
This is a division of IOM³ and provides a range of information about materials in packaging, current issues, events and journals.

Materials Research Society
www.mrs.org
An organization made up of researchers from universities, government and industry, which publishes monthly web bulletins and journals with information about material news.

MatWeb Material Property Data
www.matweb.com
This site provides a comprehensive guide to more than 60,000 materials, including thermoplastics, thermosetting plastics, composites, metals and ceramics. Searches can also be carried out using a material trade name or manufacturer.

Modern Plastics Worldwide
www.modplas.com
A US-based magazine dedicated to plastic news, markets, technology and trends.

Rematerialise: Eco Smart Materials
www.rematerialise.org
This is an online database of materials and processes for eco design. It is operated by Kingston University, UK, and is the culmination of many years of research.

Society for the Advancement of Material and Process Engineering
www.sampe.org
SAMPE is dedicated to new materials and technologies, and it provides information useful to designers, engineers, scientists and academics.

Society of Manufacturing Engineers
www.sme.org
SME provides useful information about materials and processes across a range of industries. It produces a range of journals.

Technology Review
www.techreview.com
A magazine published by the Massachusetts Institute of Technology (MIT).

Transstudio
www.transstudio.com

This website, accompanied by the book *Transmaterial: A Catalogue of Materials, Products and Processes that Redefine Our Physical Environment* edited by Blaine Brownell, features exciting new materials and developments from companies, universities and research facilities around the world.

Waste and Resources Action Programme
www.wrap.org.uk

WRAP is based in the UK and is a promotion campaign working with companies to help create better awareness of the economic advantages of recycling and finding new markets for recycled products.

Wikipedia
www.wikipedia.org

Wiki is a free encyclopedia that is submitted to and edited by volunteers. At the time of writing it had more than 1.5 million articles in English covering a range of topics, including materials and processes.

PLASTICS AND RUBBER

American Plastics Council
www.plastics.org

This is the plastics division of the American Chemistry Council (ACC) and it represents many leading plastic manufacturers.

Association of Rotation Molders International
www.rotomolding.org

This US-based organization provides information about rotation molding and manufacturers from all over the world.

British Plastics Federation
www.bpf.co.uk

The BPF is the leading trade organization for plastic producers and converters in the UK. The website provides an introduction to a range of plastic materials and processes, which are sponsored by key manufactures.

Center for the Polyurethanes Industry
www.polyurethane.org

A US-based promotion campaign for polyurethane in a wide range of applications. It provides useful information about the environmental impacts of polyurethane, safety and standards.

Distrupol
www.distrupol.com

This European plastic distribution company has most of its materials online with useful information about their properties and applications.

Injection Molding Magazine
www.immnet.com

US-based monthly injection molding magazine.

JEC Composites
www.jeccomposites.com

A promotion campaign for composite materials, including a range of featured materials and manufacturers, trade shows and competitions.

Plastics Foodservice Packaging Group
www.polystyrene.org

US-based promotion campaign for polystyrene in food packaging. It provides useful information and educational resources about polystyrene and its environmental impacts.

Plastics Technology
www.plasticstechnology.com

This monthly magazine provides information about a range of plastics, processes and design innovations.

Rubber Manufacturers Association
www.rma.org

A US-based trade organization that represents manufacturers of rubbers and elastomers.

Society of Plastics Engineers
www.4spe.org

This international organization promotes the use of plastic through the use of trade shows, books, publications and seminars.

Vinyl
www.vinyl.org

This website is dedicated to the promotion of vinyl with links to international organizations, design and applications.

METAL

Aluminium Federation
www.alfed.org.uk
A UK-based trade association, which has an online technical library, educational material, a database of suppliers and manufacturers and useful information about aluminium and related applications.

Aluplanet
www.aluplanet.com
This online periodical provides information about aluminium, manufacturing and suppliers.

Cast Metals Federation
www.castmetalsfederation.com
This association represents the majority of UK metal foundries and provides a single point of contact for casting enquiries, buying guidance and other useful information.

European Aluminium Association
www.alueurope.eu
This organization provides a range of useful information about aluminium and suppliers across Europe. Member companies are involved in mining, rolling, extruding, recycling and foil.

GalvInfo Center
www.galvinfo.com
This website has extensive information about galvanized materials and the associated benefits of galvanizing.

International Metalworkers' Federation
www.imfmetal.org
This organization is based in Switzerland and represents many of the world's metalworkers in more than 100 countries. It publishes a range of interesting journals.

International Stainless Steel Forum
www.worldstainless.org
Useful information about the benefits of stainless steel, applications, statistics and news are supplied by this not-for-profit research organization.

International Zinc Association
www.zincworld.org
The IZA is paired with the American Zinc Association. It provides information about the benefits of zinc in a wide range of applications.

Key to Metals
www.key-to-metals.com
This is a comprehensive database that with subscription provides information about ferrous and non-ferrous metals, suppliers and manufacturers globally.

Magnesium
www.magnesium.com
This website has useful information about magnesium, applications, manufacturers working with magnesium and suppliers.

Titanium Information Group
www.titaniuminfogroup.co.uk
This UK-based organization provides useful information about titanium, datasheets and companies who work with it.

World-Aluminium
www.world-aluminium.org
This website provides useful information about mining bauxite, manufacturing aluminium products, applications, statistics and news.

WOOD

Aktrin Wood Information Centre
www.wood-info.com
Reports, books and downloads relating to global wood markets and trends.

American Hardwood Information Centre
www.hardwoodinfo.com
This is an US-based promotion campaign that provides useful information about a range of American hardwood species, their properties and a range of possible applications.

APA The Engineering Wood Association
www.apawood.org
This not-for-profit organization was founded in 1933 as the Douglas Fir Plywood Association, and later changed its name to The American Plywood Association. It is now known as APA The Engineering Wood Association and is involved in researching and developing engineering timbers with its members. It is a leading resource for information about engineering timbers.

Forest Certification Resource Centre
www.certifiedwood.org
Information is provided about certified forests and products all over the world, including Forest Stewardship Council Sustainable Forestry Initiative and Canadian Standards Association programs.

Forest Stewardship Council
www.fsc.org
FSC is an international, not-for-profit organization. Its aim is to promote forest management systems that are sustainable, economic and beneficial to the environment and local population.

Sustainable Forestry Initiative
www.sfiprogram.org
This is a North American-based, independent, forest-certification organization. SFI labels identify the percentage of certified wood content, percentage of recycled fibre content, source of wood and chain of custody.

Timber Trade Federation
www.ttf.co.uk
The TTF promotes the use of wood as a sustainable building material. Its website provides useful information about buying wood and wood products from certified sources as well as a directory of suppliers.

Timber Trades Journal
www.ttjonline.com
A UK-based magazine with international distribution that has a lot of useful information about wood products, manufacturing and a buyer's guide.

Wood for Good
www.woodforgood.com
A UK promotion campaign designed to maximize the awareness of the benefits and sustainability of wood-based materials in architecture.

Wood Magazine
www.woodmagazine.com
Magazines and books about wood and furniture making.

Wood Works
www.wood-works.org
North American wood promotion campaign that is operated by the Canadian Wood Council.

Woodweb
www.woodweb.com
Online directory of wood-related books, suppliers, forums and other useful resources.

CERAMICS AND GLASS

American Ceramic Society
www.ceramics.org
This organization is based in the US and has useful information ceramic materials and publishes monthly journals.

British Glass
www.britglass.org.uk
This organization represents the UK glass industry and has a wealth of useful information about all types of glass and mass production techniques.

Ceramics
www.ceramics.com
This is a US-based website with links to manufacturers working with ceramic materials.

Corning Museum of Glass
www.cmog.org
The Corning Museum of Glass, which is based in New York, has a lot of useful information about studio glass and the history of glassblowing on its website.

Glass Magazine
www.glassmagazine.net
A US-based monthly publication that serves the architectural glass market.

Glass on Web
www.glassonweb.com
This is an online directory of glass suppliers and manufactures from all over the world and covers a range of glass industries and product categories.

Glass Pac
www.glasspac.com
This organization, operated by British Glass, promotes the use of glass in packaging applications.

National Glass Association
www.glass.org
This organization's website has information about its members, who are made up of architectural and automotive glass suppliers in the US.

Performance Materials
www.performance-materials.net
This online directory has information about new material developments, mostly in the field of high performance ceramics and composites.

Society of Glass Technology
www.societyofglasstechnology.org.uk
The SGT website has useful information about books, publications and events related to glass technology.

US Glass
www.usglassmag.com
A US-based industry magazine with information about architectural glass products and applications.

Further Reading

DESIGN AND ENGINEERING

Adams, Vince and Abraham Askenazi, *Building Better Products with Finite Element Analysis* (Santa Fe: High Mountain Press, 1998)

Alessi, Alberto, *The Dream Factory: Alessi Since 1921* (Milan: Electa/Alessi, 1998)

Antonelli, Paola, *Objects of Design from The Museum of Modern Art* (New York: The Museum of Modern Art, 2003)

Ashby, Mike and Kara Johnson, *Materials and Design: The Art and Science of Material Selection in Product Design* (Oxford: Butterworth-Heinemann, 2002)

Baxter, M. R., *Product Design: Practical Methods for the Systematic Development of New Products* (Boca Raton: CRC Press, 1995)

Betsky, Aaron, *Landscapes: Building With the Land* (London: Thames & Hudson, 2002)

Beukers, Adriaan and Ed van Hinte, *Flying Lighness: Promises for Structural Elegance* (Rotterdam: 010 Publishers, 2005)

Beukers, Adriaan and Ed van Hinte, *Lightness: The Inevitable Renaissance of Minimum Energy Structures* (Rotterdam: 010 Publishers, 2001)

Brower, Cara, Rachel Mallory and Zachary Ohlman, *Experimental Eco Design* (Mies: RotoVision, 2005)

Brown, David J., *Bridges: Three Thousand Years of Defying Nature* (London: Mitchell Beazley, 1999)

Byars, Mel, *50 Chairs: Innovations in Design and Materials* (Crans-Près-Céligny: RotoVision, 1996)

Byars, Mel, *50 Lights: Innovations in Design and Materials* (Crans-Près-Céligny: RotoVision, 1997)

Byars, Mel, *50 Products: Innovations in Design and Materials* (Crans-Près-Céligny: RotoVision, 1998)

Byars, Mel, *50 Tables: Innovations in Design and Materials* (Crans-Près-Céligny: RotoVision, 1998)

Croft, Tony and Robert Davison, *Mathematics for Engineers: A Modern Interactive Approach*, 2nd edn (Boston: Prentice Hall, 2003)

Cummings, Neil and Marysia Lewandowska, *The Value of Things* (Basel: Birkhäuser, 2000; London: August Media, 2000)

Datschefski, Edwin, *The Total Beauty of Sustainable Products* (Crans-Près-Céligny: RotoVision, 2001)

Denison, Edward and Richard Cawthray, *Packaging Prototypes* (Crans-Près-Céligny: RotoVision, 1999)

Denison, Edward and Guang Yu Ren, *Thinking Green: Packaging Prototypes 3* (Hove: RotoVision, 2001)

Edgerton, David, *The Shock of the Old: Technology and Global History Since 1900* (London: Profile Books, 2006)

Fishel, Catharine, *Design Secrets: Packaging, 50 Real-Life Projects Uncovered* (Massachusetts: Rockport, 2003)

Fuad-Luke, Alastair, *The Eco-Design Handbook: A Complete Sourcebook for the Home and Office* (London: Thames & Hudson, 2002)

Gershenfeld, Neil, *When Things Start to Think* (London: Hodder & Stoughton, 1999)

Hinte, Ed van, *Eternally Yours: Time in Design* (Rotterdam: 010 Publishers, 2004)

Hinte, Ed van, *Eternally Yours: Vision on Product Endurance* (Rotterdam: 010 Publishers, 1996)

Hinte, Ed van and Conny Bakker, *Trespassers: Inspirations for Eco-Efficient Design* (Rotterdam: 010 Publishers, 1999)

IDSA (Industrial Designers Society of America), *Design Secrets: Products, 50 Real-Life Projects Uncovered* (Massachusetts: Rockport, 2003)

Julier, Guy, *The Thames & Hudson Dictionary of 20th-Century Design and Designers* (London: Thames & Hudson, 1993)

Kwint, Marius, Christopher Breward and Jeremy Aynsley (eds), *Material Memories; Design and Evocation* (New York: Berg, 1999)

Lupton, Ellen, *Skin: Surface, Substance + Design* (London: Laurence King, 2002)

Manzini, Ezio, *The Material of Invention: Materials and Design* (Milan: Arcadia, 1986)

Mason, Daniel, *Experimental Packaging* (Crans-Près-Céligny: RotoVision, 2001)

McDonough, William and Michael Braungart, *Cradle to Cradle: Remaking the Way We Make Things* (New York: North Point Press, 2002)

Mollerup, Per, *Collapsibles: A Design Album of Space-Saving Objects* (London: Thames & Hudson, 2001)

Papanek, Victor, *Design for the Real World: Human Ecology and Social Change*, 2nd edn (London: Thames & Hudson, 2000)

Papanek, Victor, *The Green Imperative: Ecology and Ethics in Design and Architecture* (London: Thames & Hudson, 1995)

Ramakers, Renny, *Less+More: Droog Design in context* (Rotterdam: 010 Publishers, 2002)

Ramakers, Renny and Gijs Bakker (eds), *Droog Design: Spirit of the Nineties* (Rotterdam: 010 Publishers, 1998)

Schönberger, Angela (ed.), *Raymond Loewy: Pioneer of American Industrial Design* (Munich: Prestel-Verlag, 1990)

Whiteley, Nigel, *Design for Society* (London: Reaktion Books, 1994)

MATERIALS AND PROCESSES

Addington, Michelle and Daniel L. Schodek, *Smart Materials and Technologies for the Architecture and Design Professions* (Burlington: Architectural Press, 2004)

Ball, Philip, *Made to Measure: New Materials for the 21st Century* (Princeton: Princeton University Press, 1997)

Ballard Bell, Victoria and Patrick Rand, *Materials for Architectural Design* (London: Laurence King, 2006)

Beylerian, George M. and Andrew Dent, *Material Connexion: The Global Resource of New and Innovative Materials for Architects, Artists and Designers* edited by Anita Moryadas (London: Thames & Hudson, 2005)

Braddock, Sarah E. and Marie O'Mahony, *Techno Textiles: Revolutionary Fabrics for Fashion and Design* (London: Thames & Hudson, 1998)

Brownell, Blaine (ed.), *Transmaterial: A Catalogue of Materials, Products and Processes that Redefine Our Physical Environment* (Princeton: Princeton University Press, 2006)

Fournier, Ron and Sue Fournier, *Sheet Metal Handbook* (New York: HP Books, 1989)

Guidot, Raymond (ed.), *Industrial Design Techniques and Materials* (Paris: Flammarion, 2006)

Hara, Kenya et al., *Haptic: Tokyo Paper Show 2004* (Tokyo: Masakazu Hanai, 2004)

Harper, Charles A., *Handbook of Materials for Product Design*, 3rd edn (Columbus: McGraw-Hill, 2001)

IDTC (International Design Trend Centre), *How Things Are Made: Manufacturing Guide for Designer* (Seoul: Agbook, 2003)

Joyce, Ernest, *The Technique of Furniture Making*, 4th edn, revised by Alan Peters (London: Batsford, 2002)

Lesko, Jim, *Industrial Design: Materials and Manufacturing Guide* (New York: John Wiley & Sons, 1999)

Marzano, Stefano, Josephine Green, Clive van Heerden, Jack Mama and David Eves, *New Nomads: An Exploration of Wearable Electronics by Philips* (Rotterdam: 010 Publishers, 2000)

McQuaid, Matilda, *Extreme Textiles: Designing for High Performance* (London: Thames & Hudson, 2005)

Mori, Toshiko (ed.), *Immaterial Ultramaterial: Architecture, Design and Materials* (New York: Harvard Design School/George Braziller, 2002)

Mostafavi, Mohsen and David Leatherbarrow, *On Weathering: The Life of Buildings in Time*, 2nd edn (Massachusetts: The MIT Press, 1997)

O'Mahony, Marie and Sarah E. Braddock, *Techno Textiles: Revolutionary Fabrics for Fashion and Design* (London: Thames & Hudson, 2002)

Onna, Edwin van, *Material World: Innovative Structures and Finishes for Interiors* (Amsterdam: Frame Publishers, 2003; Basel: Birkhäuser, 2003)

Rossbach, Ed, *Baskets as Textile Art* (Toronto: Studio Vista, 1973)

Stattmann, Nicola, *Ultra Light Super Strong: A New Generation of Design Materials* (Basel: Birkhäuser, 2003)

Wilkinson, Gerald, *Epitaph for the Elm* (London: Arrow, 1979)

Illustration Credits

Rob Thompson photographed the processes, materials and products in this book. The author would like to acknowledge the following for permission to reproduce photographs and CAD visuals:

INTRODUCTION

Page 11 (Bellini chair): Atelier Bellini
Page 11 (in-mold decoration): Yoshida Technoworks
Page 12 (Entropia): Future Factories
Page 13 (Roses on the Vine): Swarovski Crystal Palace
Page 14 (laser-cut T-shirt): Vexed Generation
Page 14 (Biomega MN01, Extravaganza): Biomega
Page 15 (Rage paint): Cambridgeshire Coatings/US Chemical and Plastics
Page 17 (Finite Element Analysis): W. H. Tildesley

PROCESSES

Vacuum Casting
Page 43 (image 10): CMA Moldform

Injection Molding
Pages 53–5 (all images): Product Partners Design
Page 55 (image 10): Product Partners Design
Pages 56–7 (all images): Moldflow
Page 58 (image 1): Magis
Page 60 (image 1): Crowcon Detection Instruments
Page 62 (image 1): Luceplan

Reaction Injection Molding
Page 66 (image, bottom left): Boss Design
Page 67 (image 7): Boss Design

Panel Beating
Page 73 (image, above right): Coventry Prototype Panels
Page 75 (image 1): Spyker Cars
Page 77 (images 14 and 15): Spyker Cars

Metal Spinning
Page 81 (image 12): Mathmos

Metal Stamping
Page 84 (blank preparation images 2 and 4): Alessi
Page 84 (metal stamping image 1): Alessi
Pages 86–7 (images 1–4): Alessi

Deep Drawing
Page 91 (image 12): Raul Barbieri Design and Rexite

Superforming
Pages 96–7 (all images): Superform Aluminium

Tube and Section Bending
Page 100 (image 1): Thonet

Die Casting
Page 125 (image, below right): Magis
Page 129 (image 9): Magis

Investment Casting
Page 132 (images, top): Bernard Morrissey and Deangroup International

Metal Injection Molding
Page 137 (images, below right): Metal Injection Mouldings
Page 138–9 (all images): Metal Injection Mouldings

Glassblowing
Page 159 (images 2, 8 and 9): Beatson Clark

Wood Laminating
Page 194 (image 4): Isokon Plus
Page 195 (image 4): Isokon Plus
Page 196 (image 4): Barber Osgerby
Page 197 (image 4): Isokon Plus

Steam Bending
Page 200 (image 1): Thonet

Paper Pulp Molding

Page 205 (image, bottom right): Cullen
   Packaging

Composite Laminating

Page 206 (title image): Lola Cars
   International
Page 210 (image 1): Ansel Thompson
Page 214 (all images): Lola Cars
   International
Page 215 (image 1): Lola Cars International
Page 217 (images 15, 17 and 18): Lola Cars
   International

Filament Winding

Page 223 (image, middle right): Crompton
   Technology Group
Page 226 (images 1 and 2): Crompton
   Technology Group

3D Thermal Laminating

Pages 230–1 (all images): North Sails
   Nevada

Rapid Prototyping

Pages 236–41 (all images): CRDM

Punching and Blanking

Page 261 (image, top right): Alessi
Page 263 (image 1): Alessi

Die Cutting

Page 269 (image 1): Black+Blum

Arc Welding

Pages 282–6 (all images): TWI

Power Beam Welding

Pages 288–93 (all images): TWI

Friction Welding

Pages 294–7 (all images except Bang &
   Olufsen BeoLab): TWI
Page 297 (images, above left): Morten
   Larsen and Bang & Olufsen

Ultrasonic Welding

Page 303 (image, above right): Product
   Partners

Resistance Welding

Page 310 (images, middle, right): Portable
   Welders

Soldering and Brazing

Page 314 (image 1): Alessi

Joinery

Page 326 (image 1): Isokon Plus
Page 330 (housing joints in the donkey,
   image 1): Isokon Plus

Weaving

Page 333 (image, middle right): Lloyd
   Loom of Spalding
Page 334 (image 1): Lloyd Loom of
   Spalding
Page 336 (image 1): Thonet

Upholstery

Page 341 (image 1): Boss Design
Page 342 (image 1): Boss Design

Timber Frame Structures

Page 347 (images 11 and 12): Trus Joist

Spray Painting

Page 354 (image 1): Duncan Cubitt
Page 354 (image 2): Pioneer Aviation
Page 355 (images 9–13): Hydrographics

Anodizing

Page 361 (image, below right): Jesper
   Jørgen and Bang & Olufsen

Galvanizing

Page 369 (image, below right): Medway
   Galvanising

Grinding, Sanding and Polishing

Page 381 (manual polishing images 1–4):
   Alessi

Electropolishing

Page 386 (image 2): Fusion Glass
   Designs

MATERIALS

Introduction

Page 418 (Cosmos): Swarovski Crystal
   Palace
Page 419 (Crochet table): supplied by
   Moooi, photography Maarten van
   Houten
Page 420 (Remarkable): Remarkable
   Pencils
Page 420 (Self-healing plastic
   microcapsule): supplied by
   Beckman Institute, photography
   Michael Kessler, University of
   Illinois
Page 421 (Self-healing plastic sample):
   supplied by Beckman Institute,
   photography Chris Brown
   Photography
Page 421 (table by Insects): supplied by
   Droog Design, photography Anna
   Lönnerstam

Acknowledgments

The technical detail and accuracy of the manufacturing case studies is the result of the extraordinary generosity of many individuals and organizations. Their knowledge of materials and processes, and in most cases their years of hands on experience were invaluable for understanding the opportunities of the various technologies. I would like to give personal thanks to the following in order of their contribution: Graham Shaddock at RS Bookbinders; David Taylor and Vikki Shaw at Polimoon; David Whitehead, Marc Ommeslagh and Andrew Carver at Kaysersberg Plastics; Orietta Rosso and Birgit Augsburg at Magis; Kevin Buttress and David Buttress at CMA Moldform; Nigel Hill at Rubbertech2000; Ray McLaughlin and Edith Cornfield at Cromwell Plastics; Richard Gamble at ENL and Paul Neal at Product Partners Design; Jessica Castelli and Caroline Martin at Moldflow; Steen Gunderson at Hymid Multi-Shot; Rosi Guadagno at LucePlan; Nick Reid at Interfoam; Gordon Day, Dave Clarke and Phill Gower at Cove Industries; Brendan O'Toole and Matt Rose at Coventry Prototype Panels; Cressida Granger at Mathmos; Gloria Barcellin and Danilo Alliata at Alessi; Rino Pirovano and Roberto Castiglioni at Rexite; Stuart Taylor at Superform Aluminium and Kevin Quigley at Quigley Design; Susanne Korn and Stefan Wocadlo at Thonet; Nick Crossley at Pipecraft; Bryan Elliott at Elmill Group; Gordon Wright at Blagg & Johnson; John Tildesley and Bruce Burden at W. H. Tildesley; Alan Baldwin at Chiltern Casting; Christopher Dean at Deangroup International; Brian Mills at Metal Injection Mouldings; Richard Lewis at BJS Royal Silversmiths; Jill Ellinsworth and Stephanie Moore at The National Glass Centre; Peter Layton and Layne Row at London Glassblowing; Charlotte Muscroft and Tim Sweatman at Beatson Clark; Reece Bramley at Dixon Glass; Jack Evans at S&B Evans & Sons; Frances Chambers and Cynthia Whitehurst at Hartley Greens & Co. (Leeds Pottery); Edward Tadros, Vicky Tadros and Floris van den Broecke at Ercol Furniture; Chris McCourt at Isokon Plus; Ken Blake at Cullen Packaging; Darren at Radcor; Sam Smith, Paul Rennie and Ian Handscombe at Lola Cars International; Leon Houseman at Crompton Technology Group; Jim Allsopp and Bill Pearson at North Sails Nevada; Andrew Mitchell at CRDM; Tom Hutton at Mercury Engraving; Jamie Hale at Zone Creations; Chris Sears at PFS Design & Packaging; Gregg Botterman at Instrument Glasses; Dave McKeown, Penny Edmundson and Roy Smith at TWI; Peter Wells and Roderich Knoche at Branson Ultrasonics; Chris McCourt at Windmill Furniture; Henry Harris at Lloyd Loom of Spalding; Mark Barrell at Boss Design; Roger Smith at Marlows Timber Engineering; Jon Sykes at Hydrographics; Phil Roberts at Medway Galvanising; Andy Robinson at Heywood Metal Finishers; Paul Taylor at VMC; Kirsty Davies at Professional Polishing Services; David Nicol and Iain M Barker at Firma-Chrome; Ian Carey at Impressions Foil Blocking; and Chi Lam at Distrupol.

The book's content would not have been so rich and colourful if were not for the extraordinary generosity of individuals, organizations and professional photographers that have supplied images of products and materials. I would like to give personal thanks to the following: Martin Thompson; Ansel Thompson; Alexander Åhnebrink at Atelier Bellini; Haruki Yoshida of Yoshida Technoworks; Lionel Dean of Future Factories; Pip Kyriacou at Swarovski Crystal Palace; Vexed Generation; Biomega; Cambridgeshire Coatings/US Chemical & Plastics; W. H. Tildesley; CMA Moldform; Product Partners Design; Moldflow; Magis; Crowcon Detection Instruments; LucePlan; Boss Design; Coventry Prototype Panels; Spyker Cars; Mathmos; Alessi; Raul Barbieri Design and Rexite; Superform Aluminium; Thonet; Bernard Morrissey at Deangroup International; Metal Injection Mouldings; Beatson Clark; Isokon Plus; Barber Osgerby; Cullen Packaging; Lola Cars International; Crompton Technology Group; North Sails Nevada; CRDM; Black+Blum; TWI; Bang & Olufsen; Portable Welders; Lloyd Loom of Spalding; Trus Joist; Duncan Cubitt and Pioneer Aviation; Hydrographics; Medway Galvanising; Fusion Glass Designs; Moooi; Remarkable Pencils; Beckman Institute; Droog Design; Plastic Logic; Andrew Wilkins and DuPont™ Engineering Polymers; Flos; Nicolas Roope at Hulger; Duncan Riches and Vujj; Kei Tominaga and PD Design Studio; Toby Summerskill and Charlie Davidson; Vertu; Bianchi; Rolls-Royce International; KME; Rachel Galley; Candidus Prugger; Vicky Tadros at Ercol Furniture; Retrovius; Georg Baldele; Fusion Design; Áron Losonczi at Litracon; Jet Propulsion Laboratory; Mark Pinder; Helen Johannessen/Yoyo Ceramics; John Baldwin at Küppersbusch; and Julie Woodward at Pilkington Group.

Without the support, encouragement and input from colleagues, family and friends this book would not have happened. I would like to thank the book's designer Chris Perkins for his dedication to the project and excellent design skills, the editors Joanna Chisholm and Candida Frith-Macdonald for working through the text with incredible patience and their valuable insights, and Thames & Hudson for believing in such an ambitious project and then supporting it. My Dad, Martin Thompson, was on hand with invaluable help and advice regarding the photography and image content. His photographic technical ability and attention to detail is unrivalled. Also, thanks to Selwyn Taylor, for his thoughts on the layouts and energetic encouragement throughout this project. Shunsuke Ishikawa and Kei Tominaga were very helpful with information about Japanese manufacturers, designers and new technologies. I am privileged to have Simon Bolton, the co-founder of London-based design group Creative Resource Lab and Course Director of Product Design at Central Saint Martins, London, as a mentor and friend. He has challenged and inspired me throughout. Lastly, I would like to thank Molly Taylor and my family, Lynda, Martin, Ansel and Murray, for their strength and inspiration.

Index